# Nematode Diseases of Crops and their Management

Parvatha P. Reddy

# Nematode Diseases of Crops and their Management

Springer

Parvatha P. Reddy
Indian Institute of Horticultural Research
Bangalore, Karnataka, India

ISBN 978-981-16-3241-9    ISBN 978-981-16-3242-6 (eBook)
https://doi.org/10.1007/978-981-16-3242-6



This Springer imprint is published by the registered company Springer Nature Singapore Pte Ltd.
The registered company address is: 152 Beach Road, #21-01/04 Gateway East, Singapore 189721, Singapore

# Preface

The destructive plant-parasitic nematodes are one of the major limiting factors in the production of crop plants throughout the world. For centuries, man has been plagued by these microscopic organisms feeding on the roots of crop plants essential to his survival and well-being. Roots damaged by the nematodes are not efficient in the utilization of available moisture and nutrients in the soil resulting in reduced functional metabolism. Visible symptoms of nematode attack often include reduced growth of individual plants, varying degrees of chlorosis, wilting of the foliage, and sometimes death of plants. Furthermore, roots weakened and damaged by nematodes are easy prey to many types of fungi and bacteria which invade the roots and accelerate root decay. These deleterious effects on plant growth result in reduced yields and poor quality of crops. Nematode management is, therefore, important for high yields and quality that are required by the high cost of modern crop production.

The information on various aspects of nematode diseases of crop plants and their management is very much scattered and there is no book at present which comprehensively and exclusively deals with the above aspects. The present book on *Nematode Diseases of Crops and their Management* gives a detailed description of the causal organism, distribution, nature of damage and symptoms, crop losses, host range, biology and life cycle, interaction with other organisms, spread and survival, and management of nematode diseases of cereal, pulse, sugar, fiber, oil seed, vegetable, fruit, plantation, spice, tuber, ornamental, medicinal, and aromatic crops.

The book is divided into nine parts. Part I describes the importance of nematode diseases in agriculture and presents a historical review, economic importance, emerging nematode problems, interaction with other pathogens (fungi, bacteria, and viruses), and nematode management methods (regulatory, physical, cultural, chemical, biological, host resistance, and integrated methods).

The nematode diseases of large grained cereal crops such as rice, wheat, barley, and maize and small grained cereal crop like sorghum are discussed in detail in Part II.

In Part III, the nematode diseases of pulse crops (chickpea, pigeon pea, green gram, and black gram) and oil seed crops (groundnut, castor, soybean, and sunflower) are dealt with in a very systematic manner.

The nematode diseases of fiber crops (cotton and jute) and sugar crops (beetroot and sugarcane) are discussed in detail in Part IV.

In Part V, the nematode diseases of fruit crops such as tropical (banana, citrus, papaya, pineapple, and mulberry), subtropical (guava and grapevine), temperate (peach, strawberry, and apple), and semiarid (pomegranate and fig) are discussed in detail.

Nematode diseases of vegetable crops such as solanaceous (tomato, brinjal, and chili), malvaceous (okra), leguminous (French bean, cowpea, and pea), cucurbitaceous (cucumber, pointed gourd, bottle gourd, and bitter gourd), root (carrot), bulbous (onion and garlic), cruciferous (cabbage and cauliflower), leafy (lettuce and celery) mushrooms, and vegetables grown under protected cultivation (tomato, bell pepper, cucumber, and lettuce) are envisaged in Part VI.

In Part VII, ornamental crops grown under open (tuberose, gladiolus) and protected (carnation, gerbera, chrysanthemum, lilies, orchids, and anthuriums) conditions; medicinal crops (ashwagandha, coleus, brahmi, soda apple, and henbane); and aromatic crops (mints, basil, jasmine, patchouli, davana, scented geranium, and chamomile) are discussed in detail.

Nematode diseases of plantation (coconut, areca nut, coffee, and tea), spice (black pepper, cardamom, ginger, and turmeric), and tuber (taro, sweet potato, yam, Chinese potato, yam bean, winged bean, and elephant foot yam) crops are dealt with in Part VIII.

Part IX deals with future thrusts and conclusion.

This book is mainly intended for postgraduate students specializing in Plant Nematology, Plant Pathology, and Agricultural Entomology. It will be of immense value to the scientific community involved in teaching, research, and extension activities related to crop protection. The book can also serve as a very useful reference to policy makers and practicing farmers. Suggestions to improve the contents of the book are most welcome (E-mail: reddypp42@gmail.com). The publisher, Springer Nature Singapore Pte Ltd., Singapore, deserves commendation for their professional contribution.

Bangalore, India
March 30, 2021

Parvatha P. Reddy

# About the Book

The destructive plant-parasitic nematodes are one of the major limiting factors in the production of crop plants throughout the world. Annual estimated crop losses due to nematodes in India have been worked out to be about ₹102 billion. For centuries, man has been plagued by these microscopic organisms feeding on the roots of crop plants essential to his survival and well-being. Roots damaged by the nematodes are not efficient in the utilization of available moisture and nutrients in the soil resulting in reduced functional metabolism. Visible symptoms of nematode attack often include reduced growth of individual plants, varying degrees of chlorosis, wilting of the foliage, and sometimes death of plants. Furthermore, roots weakened and damaged by nematodes are easy prey to many types of fungi and bacteria which invade the roots and accelerate root decay. These deleterious effects on plant growth result in reduced yields and poor quality of crops. Nematode management is, therefore, important for high yields and quality that are required by the high cost of modern crop production.

To impart basic knowledge about the nematode diseases of cereal, pulse, oilseed, sugar, fiber, fruit, vegetable, ornamental, medicinal, aromatic, plantation, spice, and tuber crops, a detailed description of the causal organism, distribution, nature of damage and symptoms, crop losses, host range, biology and life cycle, interaction with other organisms, spread and survival, and management is provided.

This book is mainly intended for postgraduate students specializing in Plant Nematology, Plant Pathology, and Agricultural Entomology. It will be of immense value to the scientific community involved in teaching, research, and extension activities related to crop protection. The book can also serve as a very useful reference to policy makers and practicing farmers.

## Contents

# About the Author

**Parvatha P. Reddy** Obtained his M.Sc. (Agri.) degree from Karnatak University, Dharwad, and Ph. D. degree jointly from the University of Florida, Gainesville, USA, and the University of Agricultural Sciences, Bangalore.

Dr. Reddy served as the Director of the prestigious Indian Institute of Horticultural Research (IIHR) at Bangalore from 1999 to 2002 during which period the Institute was honored with "ICAR Best Institution Award." He also served as the Head of the Division of Entomology and Nematology at IIHR and gave tremendous impetus and direction to research, extension, and education in developing bio-intensive integrated pest management strategies in horticultural crops. Dr. Reddy has about 34 years of experience working with horticultural crops and involved in developing an F1 tomato hybrid "Arka Varadan" resistant to root-knot nematodes.

Dr. Reddy has over 250 scientific publications to his credit, which also include about 35 books. He has also guided two Ph.D. students at the University of Agricultural Sciences, Bangalore.

Dr. Reddy is serving as Senior Scientific Advisor to Dr. Prem Nath Agricultural Science Foundation, Bangalore. He had served as Chairman, Research Advisory Committee of Indian Institute of Vegetable Research, Varanasi; Member of the Research Advisory Committees of the National Centre for Integrated Pest Management, New Delhi; National Research Centre for Citrus, Nagpur; and the Project Directorate of Biological Control, Bangalore. Dr. Reddy served as a Member of QRT to review the progress of the Central Tuber Crops Research Institute, Trivandrum; AICRP on Tuber Crops; AICRP on Nematodes, and AINRP on Betel vine. He also served as a Member of the Expert Panel for monitoring the research program of National Initiative on Climate Resilient Agriculture (NICRA) on the theme of Horticulture including Pest Dynamics and Pollinators. He is the *Honorary Fellow* of the Society for Plant Protection Sciences, New Delhi; and *Founder President* of the Association for Advancement of Pest Management in Horticultural Ecosystems (AAPMHE), Bangalore.

Dr. Reddy has been awarded with the prestigious "Association for Advancement of Pest Management in Horticultural Ecosystems Award," "Dr. G.I. D'souza Memorial Award," "Prof. H.M. Shah Memorial Award," and "Hexamar Agricultural Research and Development Foundation Award" for his unstinted efforts in developing sustainable, bio-intensive, and eco-friendly integrated pest management strategies in horticultural crops.

Dr. Reddy has organized the "Fourth International Workshop on Biological Control and Management of *Chromolaena odorata*," "National Seminar on Hitech Horticulture," "First National Symposium on Pest Management in Horticultural Crops: Environmental Implications and Thrusts," and "Second National Symposium on Pest Management in Horticultural Crops: New Molecules and Biopesticides."

# Part I

# Introduction

# Nematode Diseases of Crop Plants: An Overview

# 1

**Abstract**

Plant parasitic nematodes (PPNs) have emerged as a serious biotic stress and significantly impacted the yield potentials of agricultural and horticultural crops. Overall, plant parasitic nematodes cause 21.3% crop losses amounting to ₹102 billion (US$ 1577 million) annually in India. Nematodes induce mechanical injuries and physiological alterations in the plant systems also facilitate the infection of other pathogens. The changing cropping patterns, introduction of new crops, crop diversification, agronomic practices, etc. also change the spectrum of pests and pathogens, including plant parasitic nematodes. In the near future, the management strategies like use of regulatory (seed certification), physical (soil solarization and hot water treatment of planting material), cultural (crop rotation, organic amendments), chemical (naturally occurring nematicides), and biological (natural enemies) methods and host resistance (induced resistance, interruption in recognition of host, and genetically modified crops) will form components of integrated nematode management.

**Keywords**

Economic importance · Emerging nematodes · Interaction with other pathogens · Nematode management

## 1.1 Introduction

Nematodes constitute one of the most important groups of organisms which inhabit the soil around the roots of plants and which frequently play a vital role in their growth and production. Rarely any crop is free from their attacks, yet we usually are unaware of their presence because of their microscopic size and protected position within the soil. Plant parasitic nematodes (PPNs) have emerged as a serious biotic


P. P. Reddy, *Nematode Diseases of Crops and their Management*,
https://doi.org/10.1007/978-981-16-3242-6_1

stress and significantly impacted the yield potentials of agricultural and horticultural crops. Nematode-induced mechanical injuries and physiological alterations in the plant systems also facilitate the infection from other pathogens.

These slender, active, worm-like creatures are so numerous that Cobb (1914) aptly remarked "...if all the matter in the universe except the nematodes were swept away, our world would still be dimly recognizable.... We would find its mountains, hills, valleys, rivers, lakes, and oceans represented by a film of nematodes." According to Thorne (1961), "Each year these minute organisms exact an ever increasing toll from almost every cultivated acre in the world: a bag of rice in Burma, a pound of tea in Ceylon, a ton of sugar beets in Germany, a bag of potatoes in England, a bale of cotton in Georgia, a bushel of corn in Iowa, a box of apples in New York, a sack of wheat in Kansas, or a crate of oranges in California."

The destructive plant parasitic nematodes are one of the major limiting factors in the production of agricultural and horticultural crops throughout the world. Some plant parasitic nematodes (PPNs) are capable of causing disease on many economically important crops grown throughout the world and attained the status of pests for substantial reduction of crop yield. For centuries, humans have been plagued by these microscopic organisms feeding on the roots of crop plants essential to their survival and well-being. Roots damaged by the nematodes are not efficient in the utilization of available moisture and nutrients in the soil resulting in reduced functional metabolism. Stunting of individual plants (reduced growth), yellowing of leaves (varying degrees of chlorosis), wilting of the foliage, and sometimes death of plants are some of the visible symptoms of nematode attack. Further, nematodes interact with soil-borne pathogens in inducing disease complexes. Reduced yields and poor quality of crops are some of the deleterious effects on plant growth. In view of the high cost of modern crop production, nematode management is therefore important for high yields and quality of the produce.

## 1.2 Historical Importance

### 1.2.1 International Scenario

The first plant parasitic nematode that seems to have been to come to the attention of the early investigators was the seed gall nematode of wheat, *Anguina tritici*, discovered by Needham (1743). It was not until 1855 that Berkeley from England found a root-knot nematode *Meloidogyne* sp. causing galls on the root system of greenhouse grown cucumbers. Kuhn (1857) noticed the stem and bulb nematode *Ditylenchus dipsaci* infesting the heads of teasel. From Germany, the sugar beet cyst nematode *Heterodera schachtii* was reported by Schacht (1859). Other historical highlights on plant parasitic nematodes are listed in Table 1.1.

**Table 1.1** A list of historical highlights on plant parasitic nematodes worldwide, in chronological order

| Year | The early records of plant parasitic nematodes |
|---|---|
| 1873 | The morphology of free-living nematodes was first described by Butschii |
| 1881 | Kuhn—First soil fumigation experiments using $CS_2$ for the control of the sugar beet cyst nematode *Heterodera schachtii* |
| 1884 | In the Netherlands, soil and fresh water nematodes taxonomic monograph was published by DeMan |
| 1888 | Strubell—Detailed morphology of *H. schachtii* |
| 1889 | Atkinson and Neal—Independently published on root-knot nematodes in the USA |
| 1892 | Atkinson—Root-knot nematode and *Fusarium* wilt disease complex in vascular wilt of cotton |
| 1907 | Cobb—Joined the USDA, considered to be the father of American nematology |
| | Titus—Reported *H. schachtii* in the USA |
| 1914 | The book "Contributions to a Science of Nematology" was published by Cobb |
| 1918 | Cobb—Developed methods and apparatus used in nematology |
| 1933 | The book "Plant Parasitic Nematodes and the Diseases they Cause" was published by T. Goodey |
| 1934 | S. Stekhoven (1941) translated a book on "Nematodes that are of Importance for Agriculture" published by Filipjev (1934) from Russian to English under the title "A Manual of Agricultural Helminthology" |
| 1943 | Carter—Nematicidal value of D-D which initiated the era of soil fumigation |
| 1945 | Christie—Nematicidal value of EDB |
| 1948 | World's first formal university course in nematology was taught by Allen at the University of California, Berkeley |
| 1950 | The book "The Potato Nematode, A Dangerous parasite to Potato Monoculture" was published by Oostenbrink |
| 1951 | Christie and Perry—Role of ectoparasitic nematodes as plant pathogens |
| | The book "Soil and Fresh Water Nematodes" was published by T. Goodey |
| | FAO—First International Nematology Course and Symposium held at Rothamsted Experiment Station, England |
| 1954 | Holdeman and Graham—*Fusarium* wilt of cotton augmented by *Belonolaimus longicaudatus* |
| 1955 | European Society of Nematologists was founded |
| 1956 | *Nematologica*—First journal devoted entirely to nematology papers published |
| 1958 | Hewitt, Raski, and Goheen—Transmission of a soil-borne plant virus (grapevine fan leaf) by a nematode (*Xiphinema index*) |
| 1961 | G. Thorne—Book on "*principles of nematology*" |
| | Society of Nematologists founded in the USA |
| 1967 | Organization of tropical American Nematologists was founded |
| 1969 | The Society of Nematologists, USA, first published the *Journal of Nematology* |
| 1973 | *Nematologia Mediterranea* published from Italy |
| 1978 | *Revue de Nematologie* published from France |

**Table 1.2** A list of historical highlights on plant parasitic nematodes in India, in chronological order

| Year | The early records of plant parasitic nematodes |
|---|---|
| 1901 | Barber—Root-knot nematode on tea from South India |
| 1913 | Butler—'Ufra" disease on rise from Bengal caused by *Ditylenchus angustus* |
| 1919 | Milne—Ear cockle disease of wheat from Punjab |
| 1926 | Ayyar—Root knot of vegetables and other crops from South India |
| 1936 | Dastur—White tip of rice caused by *Aphelenchoides besseyi* from Central India |
| 1956 | Thirumala Rao—Root-knot nematodes on citrus from India |
| 1959 | Prasad, Mathur, and Sehgal—Cereal cyst nematode from India |
| 1961 | Jones (from Rothamsted Experimental Station, UK) recorded potato cyst nematodes from Ootacamund (Nilgiri Hills) in Tamil Nadu which boosted the development of nematology in India |
| | M.R. Siddiqi—Citrus nematode from India |
| 1966 | Nair, Das, and Menon—Reported the burrowing nematode on banana from Kerala, India |
| 1968 | First South East Asia Post-graduate Nematology course held in India |
| 1969 | The Nematological Society of India founded. First All India Nematology Symposium held at New Delhi |
| 1971 | *Indian Journal of Nematology*—first published |
| 1972 | New Delhi hosted the First All India Nematology Workshop |
| 1976 | Summer Institute in Phytonematology held at Allahabad (India) |
| 1983 | Parvatha Reddy—Publication of book on "Plant Nematology" comprehensively covering the subject for the first time from India |
| 1987 | Parvatha Reddy—Publication of book entitled "A Treatise on Phytonematology" |

### 1.2.2 Indian Scenario

Although the first plant parasitic nematode from India was reported in 1901, their economic importance to agriculture was realized only during 1960–70s with the interception of "molya" disease of wheat and barley in Rajasthan, golden nematode of potato in Nilgiri Hills (Tamil Nadu), and the burrowing nematode of banana in Kerala. Since then, there has been a spurt in the research efforts on applied aspects of nematode problems in agricultural and horticultural crops. Some of the early records of plant parasitic nematodes are listed below in chronological order (Table 1.2):

## 1.3 Economic Importance

### 1.3.1 International Scenario

On a worldwide basis, the 10 most important genera of plant parasitic nematodes were reported, as listed in Table 1.3.

The estimated overall average annual yield loss of the world's major crops due to damage by plant parasitic nematodes is 12.3% (Table 1.4). For the 20 crops (left-hand column) that stand between man and starvation (life-sustaining crops), an

**Table 1.3** The 10 most important genera of plant parasitic nematodes on a worldwide basis (Sasser and Freckman 1987)

| Sl. no. | Important genera | Sl. no. | Important genera |
|---|---|---|---|
| 1 | *Meloidogyne* | 6 | *Tylenchulus* |
| 2 | *Pratylenchus* | 7 | *Xiphinema* |
| 3 | *Heterodera* | 8 | *Radopholus* |
| 4 | *Ditylenchus* | 9 | *Rotylenchulus* |
| 5 | *Globodera* | 10 | *Helicotylenchus* |

**Table 1.4** Estimated annual yield losses due to damage by plant parasitic nematodes on a worldwide basis (Sasser and Freckman 1987)

| Life-sustaining crops | Loss (%) | Economically important crops | Loss (%) |
|---|---|---|---|
| Banana | 19.7 | Cocoa | 10.5 |
| Barley | 6.3 | Citrus | 14.2 |
| Cassava | 8.4 | Coffee | 15.0 |
| Chickpea | 13.7 | Cotton | 10.7 |
| Coconut | 17.1 | Cowpea | 15.1 |
| Corn | 10.2 | Eggplant | 16.9 |
| Field bean | 10.9 | Forages | 8.2 |
| Millet | 11.8 | Grapes | 12.5 |
| Oat | 4.2 | Guava | 10.8 |
| Peanut | 12.0 | Melons | 13.8 |
| Pigeon pea | 13.2 | Misc. other | 17.3 |
| Potato | 12.2 | Okra | 20.4 |
| Rice | 10.0 | Ornamentals | 11.1 |
| Rye | 3.3 | Papaya | 15.1 |
| Sorghum | 6.9 | Pepper | 12.2 |
| Soybean | 10.6 | Pineapple | 14.9 |
| Sugar beet | 10.9 | Tea | 8.2 |
| Sugarcane | 15.3 | Tobacco | 14.7 |
| Sweet potato | 10.2 | Tomato | 20.6 |
| Wheat | 7.0 | Yam | 17.7 |
| Average | 10.7% | Average | 14.0% |
| Overall average—12.3% | | | |

estimated annual yield loss of 10.7% is reported. For the 20 crops (right-hand column) that represent a miscellaneous group important for food or export value, an estimated annual yield loss of 14% is reported (Sasser and Freckman 1987).

Based on 1984 production figures and prices, the nematodes were responsible for monetary crop losses to the extent of US$ 77 billion annually on 21 crops, 15 of which are life sustaining. These figures are staggering, and the real figure, when all crops are considered, probably exceeds US$ 100 billion annually. The losses are 5.8% greater in developing countries than in developed countries (Sasser and Freckman 1987).

Abad et al. (2008) reported that the crop losses caused by phytonematodes in economic terms were estimated to be US$ 157 billion annually to the world agriculture.

### 1.3.2 Indian Scenario

The avoidable yield losses due to plant parasitic nematodes in horticultural crops are presented in Table 1.5.

In India, the crop losses caused by phytonematodes were estimated at about ₹2100 million annually (Jain et al. 2007).

A critical analysis of crop losses caused by major nematodes to various crop plants in India was made by the different centers (located throughout India) of All India Coordinated Project on Nematodes (Walia and Chakraborty 2018). Overall, plant parasitic nematodes cause 21.3% crop losses amounting to ₹102039.79 million (US$ 1577 million) annually (Table 1.6). The losses in 19 horticultural crops were assessed at ₹50224.98 million, while for 11 field crops, it was estimated at ₹51814.81 million. Rice root-knot nematode *Meloidogyne graminicola* was economically most important causing yield loss of ₹23272.32 million in rice. Citrus (₹9828.22 million) and banana (₹9710.46 million) among fruit crops and tomato (₹6035.2 million), brinjal (₹3499.12 million), and okra (₹2480.86 million) among the vegetable crops suffered comparatively more losses. The details of crop losses incited by major nematodes in different crops are provided in Table 1.6.

## 1.4 Emerging Nematode Problems

The changing cropping patterns, introduction of new crops, crop diversification, agronomic practices, etc. also change the spectrum of pests and pathogens, including plant parasitic nematodes. For example, the adoption of water-saving techniques like System of Rice Intensification (SRI) in rice and drip irrigation in horticultural crops; diversification toward horticultural crops, particularly protected cultivation systems; and widespread and unchecked movement of planting materials from horticultural nurseries have led to the emergence of new nematode problems in newer areas and intensification of existing nematode problems. Adoption of the intensive cropping systems led to the emergence of a number of new nematode problems.

Some of the emerging nematode problems due to introduction of new nematode pests under globalization of agricultural produce, impact of climate change, threat to lack of management strategy in standing crops, etc. include root-knot nematode problem on rice, pomegranate, guava, and vegetables and flowers grown under protected cultivation, cyst nematodes problem on potato, and many others.

**Table 1.5** Avoidable yield losses in horticultural crops due to plant parasitic nematodes in India

| Crop | Nematode(s) | Yield loss (%) | Reference(s) |
|---|---|---|---|
| Banana | *Radopholus similis* | 38.00 | Rajagopalan and Naganathan (1977b) |
| | | 32.00 | Parvatha Reddy et al. (1996) |
| | | 41.00 | Nair (1979) |
| | *Meloidogyne incognita* | 30.90 | Jonathan and Rajendran (2000) |
| Sweet orange | *Tylenchulus semipenetrens* | 69.00 | Baghel and Bhatti (1983a) |
| Lemon | *T. semipenetrans* | 29.00 | Mukhopadhyaya and Suryanarayana (1969) |
| Sweet lime | *T. semipenetrans* | 19.00 | Mukhopadhyaya and Dalal (1971) |
| Grapevine | *M. incognita* | 55.00 | Rajagopalan and Naganathan (1977a) |
| | *M. javanica* | 53.00 | Baghel and Bhatti (1983b) |
| Papaya | *Rotylenchulus reniformis* | 28.00 | Rajendran and Naganathan (1981) |
| Pomegranate | *Meloidogyne* sp. | 24.64–27.45 | Singh et al. (2003) |
| Peach | *Mesocriconema xenoplax* | 33.00 | Anon (1990a) |
| Plum | *M. xenoplax* | 10.00 | Anon (1990a) |
| Potato | *M. incognita* | 42.50 | Prasad (1989) |
| | *Globodera rostochiensis* | 99.50 | Prasad (1989) |
| Tomato | *M. incognita* | 30.57–46.92 | Bhatti and Jain (1977)<br>Reddy (1981)<br>Darekar and Mahse (1988) |
| | *M. javanica* | 77.50 | Anon (1993a, b) |
| | *R. reniformis* | 42.25–49.02 | Subramanyam et al. (1990) |
| Brinjal | *M. incognita* | 27.30–48.55 | Bhatti and Jain (1977)<br>Parvatha Reddy and Singh (1981)<br>Darekar and Mahse (1988) |
| Chili | *Meloidogyne* sp. | 24.54–28.00 | Singh et al. (2003) |
| Okra | *M. incognita* | 90.90<br>28.08 | Bhatti and Jain (1977)<br>Parvatha Reddy and Singh (1981) |
| | *M. javanica* | 20.20–41.20 | Jain et al. (1986) |
| French bean | *M. incognita* | 19.38–43.48 | Das (1994)<br>Parvatha Reddy and Singh (1981) |

(continued)

**Table 1.5** (continued)

| Crop | Nematode(s) | Yield loss (%) | Reference(s) |
|---|---|---|---|
| | *M. javanica* | 30–40 | Sharma et al. (2002) |
| Cowpea | *M. incognita* | 28.60 | Parvatha Reddy and Singh (1981) |
| | *R. reniformis* | 13.20–32.00 | Palanisamy and Sivakumar (1981)<br>Hasan and Jain (1998) |
| Peas | *M. incognita* | 20.0–50.61 | Parvatha Reddy (1985)<br>Upadhyay and Dwivedi (1987)<br>Sharma (1989) |
| | *R. reniformis* | 15.8 | Dalal and Vats (1998) |
| Carrot | *M. incognita* | 56.64 | Devi (1993) |
| Bitter gourd | *M. incognita* | 36.72 | Darekar and Mahse (1988) |
| Pointed gourd | *M. incognita* | 30–40 | Verma (2001) |
| Water melon | *M. incognita* | 18–33 | Hasan and Jain (1998) |
| Mushroom | *Ahelenchoides sacchari* | 40.6–100.0 | Singh et al. (2003) |
| | *A. composticola* | 35–60 | Laqman Khan (2001) |
| | *A. avenae* | 25.8–53.5 | Bajaj and Jain (2001) |
| Crossandra | *M. incognita* | 21.64 | Khan and Parvatha Reddy (1994) |
| Tuberose | *M. incognita* | 13.78 | Khan and Parvatha Reddy (1994) |
| Patchouli | *M. incognita* | 47.00 | Prasad and Reddy (1984) |
| Davana | *M. incognita* | 50 | Haseeb and Pandey (1989) |
| Lemon grass | *M. incognita* | 20 | Pandey (1994) |
| Carnation | *M. incognita* | 27 | Nagesh and Parvatha Reddy (2000) |
| Gerbera | *M. incognita* | 31 | Nagesh and Parvatha Reddy (2000) |
| *Coleus forskohlii* | *M. incognita* | 70.2 | Senthamarai et al. (2006) |
| Chethikoduveli (*Plumbago rosea*) | *M. incognita* | 29.00–43.96 | Santhosh Kumar and Sheela (2004) |
| Kacholam (*Kaempferia galanga*) | *M. incognita* | 18–64 | Sheela and Rajani (1998) |
| Colocasia | *M. incognita* | 24 | Anon (1990b) |
| Menthol mint | *M. incognita* | 30 | Pandey (2003) |
| Cardamom | *M. incognita* | 32–47 | Ali (1986) |
| Betel vine | *M. incognita* | 21.1–38 | Saikia (1992)<br>Jonathan et al. (1990) |
| Ginger | *M. incognita* | 29.60–33.35<br>74.10<br>46.40 | Ramana et al. (1998)<br>Koshy (2002)<br>Charles and Kurian (1979) |
| | *R. similis* | 39–73 | Sundararaju et al. (1979) |

(continued)

**Table 1.5** (continued)

| Crop | Nematode(s) | Yield loss (%) | Reference(s) |
|---|---|---|---|
| Turmeric | *M. incognita* | 18.6–25.0 | Poornima and Vadivelu (1993) |
| | *R. similis* | 46–75 | Sosamma et al. (1979) |
| Coriander | *M. incognita* | 51 | Midha and Trivedi (1991) |
| Cumin | *M. incognita* | 34 | Midha and Trivedi (1991) |
| Fennel | *M. incognita* | 39 | Midha and Trivedi (1991) |
| Black pepper | *R. similis* and *M. incognita* | 38.5–64.6 | Mohandas and Ramana (1991) |
| | *R. similis* | 59 | Mohandas and Ramana (1991) |
| | *M. incognita* | 46 | Mohandas and Ramana (1991) |
| Coconut | *R. similis* | 30 | Koshy and Geetha (1992) |

### 1.4.1 Root-Knot and Foliar Nematodes on Rice

Due to rice cropping intensification and increasing scarcity of water, the root-knot nematode *Meloidogyne graminicola* emerged as a serious threat for the successful rice production in nurseries, uplands, deep water, and irrigated fields in eastern, north-eastern, and southern states of India. Prasad and Somasekhar (2009) reported that the extent of losses due to *M. graminicola* have been estimated to be 16–32% and yield loss due to poorly filled kernels to be 17–30%. The widespread detection of rice white tip nematode *Aphelenchoides besseyi* in the southern and eastern states of India was reported. The rice white tip nematode is emerging as a serious pest causing 17–54% yield reduction in rice.

### 1.4.2 Root-Knot Nematode on Groundnut

The root-knot nematodes *Meloidogyne arenaria* and *M. javanica* are emerging as important constraints to groundnut production in Gujarat which are responsible for 13–50% and 10–23% of yield losses, respectively. It is evident that root-knot nematodes are important constraints to groundnut production in Gujarat (Patel et al. 1996).

### 1.4.3 Root-Knot Nematode on Acid Lime

The root-knot nematode *Meloidogyne indica* is devastating the acid lime in Gujarat. Severe nematode infestation of the root-knot nematode M. indica was observed especially in Banaskantha district in North Gujarat acid lime orchards. Severe infection of M. indica was also observed in Banaskantha, Mahesana, and Anand districts of Gujarat (Patel et al. 1999).

**Table 1.6** Estimated losses due to economically important plant parasitic nematodes to various major crops in India (2014–2015) (Walia and Chakraborty 2018)

| Crop | Nematode | Production (million tons) | Yield loss (%) | Price per metric ton (₹) | Monetary loss (₹ in million) |
|---|---|---|---|---|---|
| *Fruit crops* | | | | | |
| Banana | *Meloidogyne incognita* | 02.92 (29.22) | 15 | 22,170 | 9710.46 |
| Citrus | *Tylenchulus semipenetrans* | 01.16 (11.65) | 27 | 31,380 | 9828.22 |
| Grapes | *M. incognita* | 00.28 (02.82) | 30 | 46,910 | 3940.44 |
| Guava | *Meloidogyne* spp. | 00.40 (03.99) | 28 | 20,990 | 2350.88 |
| Papaya | *M. incognita* + *Rotylenchulus reniformis* | 00.49 (04.91) | 30 | 17,120 | 2516.64 |
| Pomegranate | *Meloidogyne* spp. | 00.18 (01.78) | 23 | 73,030 | 3023.44 |
| Mean yield loss in fruit crops—25.5% | | Total monetary loss in fruit crop—₹31370.08 million | | | |
| *Vegetable crops* | | | | | |
| Bitter gourd | *M. incognita* | 00.08 (00.77) | 13.5 | 23,410 | 252.82 |
| Bottle gourd | *M. incognita* | 00.19 (01.82) | 22 | 9660 | 403.78 |
| Brinjal | *Meloidogyne* spp. | 01.25 (12.58) | 21 | 13,330 | 3499.12 |
| Capsicum | *Meloidogyne* spp. | 00.02 (00.18) | 10 | 26,460 | 52.92 |
| Carrot | *Meloidogyne* spp. | 00.10 (00.96) | 34 | 22,180 | 754.12 |
| Chili | *Meloidogyne* spp. | 00.20 (01.99) | 15 | 24,830 | 744.90 |
| Cucumber | *Meloidogyne* spp. | 00.07 (00.67) | 12 | 13,150 | 110.46 |
| Okra | *Meloidogyne* spp. | 00.57 (05.70) | 19.5 | 22,320 | 2480.86 |
| Potato[a] | *Globodera* spp. | 0.032 (0.032) | 26 | 15,270 | 127.04 |
| Tomato | *Meloidogyne* spp. | 01.64 (16.38) | 23 | 16,000 | 6035.20 |
| Mean yield loss in vegetable crops—19.6% | | Total monetary loss in vegetable crops—₹14461.22 million | | | |
| *Spice crops* | | | | | |
| Ginger | *M. incognita* | 00.08 (00.76) | 29–33 | 75,170 | 1894.28 |

(continued)

**Table 1.6** (continued)

| Crop | Nematode | Production (million tons) | Yield loss (%) | Price per metric ton (₹) | Monetary loss (₹ in million) |
|---|---|---|---|---|---|
| Black pepper | *Radopholus similis* | 0.006 (00.06) | 24 | 605,450 | 871.84 |
| Turmeric | *M. incognita* | 00.08 (00.83) | 33 | 61,650 | 1627.56 |
| Mean yield loss in spice crops—29.5% | | Total monetary loss in spice crops—₹4393.68 million | | | |
| *Cereal crops* | | | | | |
| Maize | *Heterodera zeae* | 02.41 (24.17) | 12 | 13,100 | 3788.52 |
| Rice | *M. graminicola* | 10.54 (105.48) | 16 | 13,800 | 23272.32 |
| Wheat[b] | *H. avenae* | 02.17 (21.71) | 23.5 | 14,500 | 8967.52 |
| Mean yield loss in cereal crops—18.80% | | Total monetary loss in cereal crops—₹36028.36 million | | | |
| *Pulse crops* | | | | | |
| Black gram | *M. incognita* | 00.19 (01.96) | 19 | 43,500 | 1570.35 |
| Chickpea | *M. incognita* | 00.73 (07.33) | 21 | 31,750 | 4867.27 |
| Green gram | *M. incognita* | 00.15 (01.50) | 29 | 46,000 | 2001.00 |
| Mean yield loss in pulse crops—23.00% | | Total monetary loss in pulse crops—₹8438.62 million | | | |
| *Oilseed crops* | | | | | |
| Castor | *R. reniformis* | 00.18 (01.87) | 15 | 40,103 | 1082.78 |
| Groundnut[c] | *M. arenaria* | 00.22 (02.20) | 4.5 | 40,000 | 396.00 |
| Sunflower | *M. incognita* | 00.04 (00.43) | 16 | 37,500 | 240.00 |
| Mean yield loss in oilseed crops—11.80% | | Total monetary loss in oilseed crops—₹1718.78 million | | | |
| *Fiber crops* | | | | | |
| Cotton | *M. incognita/ M. javanica* | 00.59 (05.91) | 20.5 | 39,000 | 4717.05 |
| Jute | *Meloidogyne* spp. | 00.20 (02.00) | 19 | 24,000 | 912.00 |

(continued)

**Table 1.6** (continued)

| Crop | Nematode | Production (million tons) | Yield loss (%) | Price per metric ton (₹) | Monetary loss (₹ in million) |
|---|---|---|---|---|---|
| Mean yield loss in fiber crops—19.75% | | Total monetary loss in fiber crops—₹5629.05 million | | | |
| Summary | | | | | |
| Mean yield loss in field crops—18.23 | | Total monetary loss in field crops—₹51814.81 million | | | |
| Mean yield loss in horticultural crops—23.03% | | Total monetary loss in horticultural crops—₹50224.98 million | | | |
| Overall mean yield loss—21.30% | | Grand total monetary loss —₹102039.79 million | | | |

Loss estimations are based mainly on work done at various centers of AICRP on nematodes
Only 10% area of total production infested with phytonematodes has been considered for estimation of national losses
Figures in parentheses are total production of the crops in country
[a]*Globodera* spp. are widespread only in Nilgiri Hills (Tamil Nadu), hence the area, production of Nilgiri Hills only has been considered
[b]Production in Rajasthan and Haryana only taken into consideration
[c]Production in Gujarat only taken into consideration

### 1.4.4 Root-Knot Nematode on Pomegranate

Khan et al. (2005) observed that the pomegranate trees that were severely infected with root-knot nematode *Meloidogyne incognita* (Race 2) died within a period of 3 months after the onset of symptoms. This nematode was introduced mainly through infected planting stock from commercial nurseries. Besides causing direct damage, the root-knot nematode species are responsible for causing wilt disease complexes in association with soil-borne fungal plant pathogens. The root-knot nematode complex has similarly created serious situation with *Ceratocystis* fimbriata on pomegranate in Maharashtra, Karnataka, and North Gujarat states.

### 1.4.5 Root-Knot Nematode on Guava

Until recently, *Meloidogyne incognita* and *M. javanica* have been reported to infect guava in various parts of the country (Ansari and Khan 2012); however, these were not considered highly pathogenic to guava. From Ayyakudi area (Dindigul district) of Tamil Nadu, Poornima et al. (2016) reported an exotic species of root-knot nematode *Meloidogyne enterolobii* (synonym *M. mayaguensis*) on guava. Young guava trees witnessed heavy mortality (30–50%) within 3 months of first appearance of the symptoms (Poornima et al. 2016; Ashokkumar and Poornima 2019). Subsequent surveys revealed its occurrence in many districts of Tamil Nadu (Ashokkumar et al. 2019) besides 10 other states (Andhra Pradesh, Telangana, Karnataka, Gujarat, Kerala, Haryana, Uttar Pradesh, Rajasthan, Uttarakhand, and West Bengal) of India (AICRP-Nematodes Centers Reports).

### 1.4.6 Root-Knot Nematode on Mulberry

Reduction of life span of mulberry plants and herbage yield and quality of leaves is caused by the root-knot nematode *Meloidogyne incognita*, which is a major constraint in its cultivation. It is responsible for 10–12% herbage yield loss and adversely affects the silk industry in Karnataka (Govindaiah et al. 1991).

### 1.4.7 Cyst Nematodes on Potato

Potato cyst nematodes (*Globodera rostochiensis* and *G. pallida*), which were hitherto restricted to south Indian states such as Tamil Nadu (Nilgiri Hills) and Kerala (Kodai Hills), have now been detected in north Indian state Himachal Pradesh (Shimla, Mandi, Kullu, Chamba, and Sirmaur districts), which is a serious concern with ramification on export and quarantine issues (Ganguly et al. 2010). Recently, the potato cyst nematodes have also been recorded from Jammu and Kashmir and Uttarakhand states.

### 1.4.8 Floral Malady on Tuberose

The major limiting factor in successful production of tuberose *Polianthes tuberosa* includes the foliar nematode *Aphelenchoides besseyi*. It induces a typical "floral malady" symptom, which is a prime threat for growing quality flowers particularly in major tuberose-growing areas of West Bengal and Odisha in India (Das and Khan 2007). The foliar nematode is responsible for 38% loss in spike yield and 59% loss in loose flower yield.

### 1.4.9 Nematode Problems on Polyhouse Crops

Vegetable crops (tomato, bell pepper, cucumber, and okra) and flower crops (carnations and gerbera) are being grown throughout India under protected cultivation (in polyhouses/greenhouses/shade nets) which are seriously infested with nematodes such as *Meloidogyne incognita*, *M. javanica* (root-knot nematodes), and *Rotylenchulus reniformis* (reniform nematode). Nematode problems on all these crops under protected conditions have assumed alarming proportions leading to huge losses (up to 80%) in select crops. The nematode infestations exacerbate severity of fungal diseases leading to complete crop losses. *M. incognita* infection makes the plants highly susceptible to *Fusarium oxysporum* f. sp. *dianthi* attack. The complex disease induced by the root-knot nematode *M. incognita* in association with root-rot fungal pathogen *Phytophthora parasitica* was responsible for 40–60% reduction in yield.

## 1.5 Interaction with Other Pathogens

Many plant diseases are influenced by associated organisms, since nature does not work with pure cultures. The nematodes interact with soil-borne disease pathogens in causing disease complexes. The bacterium-nematode and the fungus-nematode interactions, and weakly parasitic fungal and bacterial parasites, can cause considerable damage once they gain entry into plant roots in the presence of feeding nematodes. Some species of nematodes belonging to five genera *Xiphinema, Longidorus, Paralongidorus, Trichodorus*, and *Paratrichodorus* have a unique and important role as pathogens of plants, because they also transmit certain viruses to their host plants.

### 1.5.1 Fungi

Much experimental evidence indicates a biological interaction between nematodes and certain soil-borne fungal pathogens. The nematodes predispose the plants to the soil-borne fungal pathogens by modifying the host substrate suitable for fungal multiplication and enhance the pathogenicity mechanism in the host plants.

Chahal and Chhabra (1984) reported that the incidence and severity of root rot of okra, tomato, and brinjal caused by the soil-borne fungi such as *Phomopsis vexans*, *Rhizoctonia solani*, and *R. bataticola* was increased in the presence of *M. incognita*. Likewise, Mai and Abawi (1987) found that endoparasitic nematodes such as root-knot and cyst nematodes have long been known as primary pathogens for their ability to predispose plants to infection by secondary pathogens such as several species of *Rhizoctonia*, *Fusarium,* and *Phytophthora*.

The kinds of effects observed in the nematode-fungus etiological relationships are:

- Fungus disease aggravated.
- Host growth suffered.
- Resistance to fungus reduced.
- Fungus suppressed by nematode.
- Nematode suppressed by fungus.
- Susceptibility to fungus increased (Norton 1978).

### 1.5.2 Bacteria

Nematodes are likely to predispose plants to bacterial disease complexes. The role of nematodes in relation to bacterial pathogens has usually been regarded as providing wounds in the roots through which bacteria may enter the plant. This appeared to be the case in the association of the root-knot nematodes *Meloidogyne incognita* and *M. hapla* with *Ralstonia solanacearum* and *Agrobacterium tumefaciens*, respectively. There is evidence that the bacterium *Rhodococcus fascians* attaches specifically to the cuticle of the nematode *Aphelenchoides ritzemabosi* and is thus transported to the strawberry host plant. Various grades of leaf deformation are induced by both the pathogens together (Hawn 1971).

*M. incognita* is responsible for breakdown of bacterial wilt (*Ralstonia solanacearum*) resistance in resistant cultivars of tomato and eggplant. The survival rate of wilt-resistant tomato plants was reduced to 33–36% in the presence of both the pathogens. Naik (2004) reported that both bell pepper and eggplant are prone to many soil-borne diseases among which the bacterial wilt (*R. solanacearum*) in combination with root-knot nematode (*M. incognita*) cause severe crop losses.

### 1.5.3 Viruses

Hewitt et al. (1958) provided first experimental evidence that *Xiphinema index* acted as vector of grapevine fan leaf virus. Nepoviruses, which are polyhedral, isodiametric particles about 28 nm in diameter, are transmitted by three genera *Xiphinema, Longidorus*, and *Paralongidorus* belonging to the Longidoridae. Grapevine fan leaf and strawberry and raspberry ring spot viruses are transmitted by the Longidoridae. Throughout the world, grapevine is closely associated as a host to the dagger nematode *X. index*.

Tobraviruses, which are rod-shaped, short and long (45–115 nm and 180–210 nm) particles, are transmitted by two genera *Trichodorus* and *Paratrichodorus* belonging to the Trichodoridae. Tobacco rattle and pea early browning viruses, transmitted by the Trichodoridae, infect a wide range of wild and cultivated plants. Tobacco rattle induces diseases of economically important bulbous flower crops in Europe and potato crops in Europe and USA. Browning disease has spread in Europe and induces severe stunting and premature death in large areas of pea crops (Taylor 1980).

## 1.6 Nematode Management

The overall goal of nematode management is to keep the population density as low as possible, since nematodes cannot be eradicated, and that we must live with them. In view of the high cost of modern crop production, nematode management is important for high yields and quality. Increased quantity and quality of the produce; improved health of plants, thereby reducing their susceptibility to plant pathogens and increasing the ability to withstand adverse growing conditions; and a better utilization of nutrients and moisture are the direct and indirect benefits of nematode management.

Regulatory, physical, cultural, chemical, biological, and host resistance are some of the nematode management methods in the field that can be effectively employed to keep nematode population to a minimum level. Efficient management requires the carefully integrated combinations of several practices, since it is not advisable to depend on a single method to control nematodes.

## 1.6.1 Regulatory Methods

### 1.6.1.1 Plant Quarantine

Legal quarantine is the most effective method for prevention, if a harmful species of nematode is not present in an area. By means of plant quarantine, numerous attempts have been made to prevent the introduction of nematodes into countries or provinces. In order to prevent the introduction of new pests, legislative or other government authorities usually give quarantine authorities power to establish and enforce plant quarantine for regulations purpose. Such regulations prevent bringing infected planting material into protected areas where similar crops are vulnerable to infection.

Restrictions for the import or movement of plants and plant material in order to prevent the introduction of insect, fungus, or other pests which may be destructive to crops are enacted by The Destructive Insects and Pests Act, 1914. The rules permit the Plant Protection Advisor to the Government of India or any officer authorized by them to undertake inspection and treatments. Specific regulations for nematodes have been made against the coconut red ring nematode *Bursaphelenchus cocophilus* and potato cyst nematodes *Globodera rostochiensis* and *G. pallida*. Other nematodes of quarantine significance to India include foliar nematode *Aphelenchoides fragariae* and *A. ritzemabosi* on ornamental plants belonging to Family Compositae; stem and bulb nematode *Ditylenchus dipsaci* in bulbs of ornamental plants, seeds of alfalfa, beans, rye, oats, onion, garlic, etc.; pine wood nematode *Bursaphelenchus xylophilus* on *Pinus* spp. and *Cedrus* spp.; sugar beet cyst nematode *Heterodera schachtii* on plants belonging to families of Chenopodiaceae and Cruciferae; and soybean cyst nematode *Heterodera glycines* on Soybean. Domestic quarantine regulations have also been imposed to restrict the movement of potato both for seed and table purposes in order to prevent the spread of potato cyst nematode from Tamil Nadu to other states and Union Territories in India. In spite of these restrictions, the potato cyst nematode has already spread to Himachal Pradesh, Jammu and Kashmir, and Uttarakhand.

### 1.6.1.2 Seed Certification

Seed certification to produce nematode-free planting material through hot water treatment or meristem tissue culture is another legal procedure to manage pests. Vegetative seed of banana free of the burrowing nematode (*Radopholus similis*), lesion nematode (*Pratylenchus coffeae*), and root-knot nematode (*Meloidogyne* spp.); strawberry of bud and leaf nematode (*Aphelenchoides fragariae*); potato of cyst nematodes (*Globodera rostochiensis, G. pallida*); and garlic of stem and bulb nematode (*Ditylenchus dipsaci*) can be produced commercially by seed certification.

## 1.6.2 Physical Methods

Heat treatment is one of the oldest methods of nematode control. Plant parasitic nematodes can be killed by heat, desiccation, irradiation, high osmotic pressure, etc., but it is more difficult to employ physical methods for killing nematodes in soil and planting materials. Field-scale treatment of soil is also difficult.

#### 1.6.2.1 Hot Water Treatment of Planting Material

It is widely used for the killing of nematodes within plant tissues before planting and has proved useful with nematode-infested bulbs, corns, tubers, seeds, and roots of plants such as chrysanthemums, strawberries, bananas, and citrus. First, the time temperature mortality curves of the nematode have to be determined. This entails immersing the nematode-infested material into thermostatically controlled water baths at different temperatures for different lengths of time. Information must also be obtained on the plants' reaction to temperature, because if the treatment is to be successful, the plants must be able to tolerate temperatures lethal to nematodes (Table 1.7).

Nematodes can be killed at 44–48 °C. The nematode enzymes are inactivated at short time exposure to about 50 °C in vitro. Bare root dip in hot water has to be specifically determined for different nematode species.

#### 1.6.2.2 Solarization

Another method of killing nematodes with heat is by soil solarization. This technique consists of covering moist soil with a plastic film during periods of intense sunshine and heat (Fig. 1.1), thereby capturing radiant heat energy from the sun, causing physical, chemical, and biological changes in the soil. The soil temperature is increased to levels (by 5–15 °C) lethal to many soil-borne fungi, bacteria, nematodes, and weed seeds. The short-wave incident solar radiation penetrates the polyethylene sheet, but the long-wave radiation is prevented from soil, thus trapping the heat resulting in thermal inactivation and production of heat shock proteins and irreversible heat injury. Other advantages of soil solarization include the enhancement of the available N and other elements to improve the plant nutrition.

Soil solarization is one of the effective ways of suppressing root-knot nematode populations (*Meloidogyne javanica* and *M. indica*) on citrus and can be employed mostly during hot weather days.

A 37–100% reduction in population of *Macroposthonia xenoplax* and other nematodes on peach has been achieved by soil solarization in South Africa (Barbercheck and von Broembsen 1986).

White polyethylene mulching for 15 days on beds prepared for planting betel vine revealed high reduction in nematode populations due to increased soil temperature to 44.1 °C (Sivakumar and Marimuthu 1987).

Soil solarization using thin transparent polyethylene mulches for 6 weeks caused as much as 72.6 and 88.5% decrease in the population densities of *Tylenchorhynchus vulgaris* and *Hoplolaimus indicus*, respectively, in nursery beds (Table 1.8). Kamra and Gaur (1998) reported that the solarization effect to reduce the nematode population was increased by soil application of mahua cake prior to solarization.

**Table 1.7** Time and temperature requirements of hot water treatment for producing nematode-free planting material

| Nematode | Planting material | Time min. | Temp. °C |
|---|---|---|---|
| *Tylenchulus semipenetrans* | Citrus rooted cuttings | 10<br>25 | 46.7<br>45.0 |
| *Radopholus similis* | Banana suckers | 10 | 50.0 |
| *R. citrophilus* | Citrus rooted cuttings | 10 | 50.0 |
| *Aphelenchoides besseyi* | Rice seeds | 10 | 52.0 |
| *A. ritzemabosi* | Chrysanthemum stools | 15<br>30 | 47.8<br>43.0 |
| *A. fragariae* | Easter lily bulbs | 60 | 44.0 |
| | Strawberry runners | 10 | 46.0 |
| *Pratylenchus penetrans* | Dormant strawberry plants | 7.5<br>17.5 | 51.0<br>49.4 |
| *Ditylechus dipsaci* | Narcissus bulbs | 240 | 43.0 |
| | Onion | 120 | 43.5 |
| *D. destructor* | Irish bulbs | 180 | 43.0 |
| *Anguina tritici* | Wheat seeds | 10<br>30 | 54.0<br>50.0 |
| *Meloidogyne* spp. | Cherry rootstocks | 5–10 | 50.0–51.1 |
| | Sweet potatoes | 65 | 46.7 |
| | Peach rootstocks | 5–10 | 50.0–51.1 |
| | Tuberose tubers | 60 | 49.0 |
| | Grapes rooted cuttings | 10<br>30 | 50.0<br>47.8 |
| | Begonia tubers | 30<br>60 | 48.0<br>45.0 |
| | Caladium tubers | 30 | 50.0 |
| | Yam tubers | 30 | 51.0 |
| | Ginger rhizomes | 10 | 55.0 |
| | Strawberry roots | 5 | 52.8 |
| | Rose roots | 60 | 45.5 |
| | Potato tubers | 120 | 46.0–47.5 |

### 1.6.3 Cultural Methods

Cultural practices are usually the operations that will be normal carried out at little or no extra expense. Basically, cultural management involves depriving the nematode of a suitable host and thereby reducing the nematode population by starvation. Nematode management can be achieved by adopting several cultural methods. Crop rotation is one of the major cultural practices employed to manage nematodes. Other cultural methods used for nematode management include fallowing, flooding, mulching, trap cropping, time of planting, use of organic amendments to the soil, intercropping with antagonistic plants, and influence of fertilizers.

**Fig. 1.1** Laying of plastic sheets in strips by machines for soil solarization

**Table 1.8** Management of economically important plant nematodes in major crops using soil solarization

| Crop | Nematode | Duration | Results |
|---|---|---|---|
| Citrus | Root-knot nematodes *Meloidogyne javanica* and *M. indica* | 7–10 days | Suppressed nematode population |
| Tomato nursery | Root-knot nematodes *Meloidogyne* spp. | 15 days | Decreased root-knot disease and weeds by 66% and 93% |
| French bean | Root-knot nematodes *Meloidogyne* spp. | 4–8 weeks | Nematodes effectively controlled |
| Cardamom nursery | Root-knot nematodes *Meloidogyne* spp. | 40–45 days | Nematodes suppressed |
| Betel vine | Root-knot nematode *Meloidogyne incognita* | 15 days | High reduction in nematode population |
| Turmeric | Root-knot nematodes *Meloidogyne* spp. | 40 days | Reduced nematode population |

Botanicals and their products play an important role in the reduction of nematode population in soil. Significant reduction of several plant parasitic nematodes in infested soils can be obtained by addition of organic materials (green manures, oil cakes, crop residues, cellulosic soil amendments, etc.) or growing of antagonistic plants (marigold, mustard, sesame, asparagus, etc.).

#### 1.6.3.1 Crop Rotation

Crop rotation attempts to keep nematode populations to a level at which crop damage is reduced to a minimum. By growing nonhost crops, the numbers of plant nematodes are reduced and the number of years for this to occur depending on the initial population and the rate of population decrease. Rotations must also provide economically useful crops which are not susceptible to other nematodes affecting the main crop.

More than 3-year rotations with wheat, strawberry, cabbage, cauliflower, peas, maize, and beans reduce the potato cyst nematode population to a safe level; while 3-year rotations with gram, fenugreek, and carrot control the cereal cyst nematode *Heterodera avenae.*

Inclusion of nonhosts/resistant varieties (viz., garlic, onion, Hisar Lalit variety of tomato, Shree Bhadra of sweet potato, etc.) in vegetable-based cropping systems has led to reduction in population of root-knot nematodes and enhancement of crop yields.

#### 1.6.3.2 Trap Cropping

Trap crops are the more preferred host when grown before the main crop. Trap crops can increase the efficiency of control by concentrating the nematode pests in one location and by destroying the trap crops and associated nematode pests through uprooting and burning.

Cowpea causes the root-knot nematode eggs to hatch; the larvae enter the roots and develop to immobile stage. Before the nematodes attain the adult stage, the crop in field is destroyed by burning. *Crotalaria* is highly susceptible to invasion by the root-knot nematode but is resistant to the development of larvae into adults.

*Solanum sisymbriifolium* acts as a trap crop and can be used as sustainable and effective alternative to chemical control of potato cyst nematode (PCN). The crop acts as a completely risk-free trap crop for PCN by triggering hatch in the soil but preventing completion of the pest's life cycle, however long the crop is left in the ground.

#### 1.6.3.3 Cover Crops

Cover/green manure crops are planted between main crops which are valuable tools for managing crop nematode pests, which positively impact yields over time. The warm season legume cover crops are effective in reducing populations of certain plant parasitic nematodes by breaking their life cycles (Vargas-Ayala et al. 2000). In greenhouse and field tests, the cover crop velvet bean reduced the population levels of several root-knot nematode species present simultaneously (Rodriguez-Kabana et al. 1992).

### 1.6.4 Chemical Methods

The primary advantage of chemical control over other methods is that the nematode population is reduced to a very low density within a matter of days after the chemical is applied, enabling the grower to plant a crop soon after treatment or in some cases at the time of treatment. Most crops are especially vulnerable to nematode attack during the seedling stage. In the case of annuals, the crop develops good root systems in treated soil and matures before residual population of nematodes has increased to a level which would cause damage.

The use of chemicals on a field scale for control of plant parasitic nematodes was not possible until early 1940s when effective and economical soil fumigants like D-D and EDB were discovered which made it possible to provide growers with spectacular differences in growth and yield through the effective control of nematodes and other soil pests. Several effective nematicides belonging to organo-phosphate and carbamate groups were developed and improvements in methods of application provided more economic control. Nematode management through soil treatment is an established farm practice in many parts of the world.

#### 1.6.4.1 Halogenated Hydrocarbons

Nematodes are sensitive to the halogens, and most of our successful nematicides are halogenated hydrocarbons which include chloropicrin, methyl bromide, D-D, EDB, and DBCP.

#### 1.6.4.2 Organophosphates

The need for nonphytotoxic nematicides led to the investigation of this group of compounds (Phorate, Fensulfothion, Fenamiphos, and Ethoprophos) in the hope of finding one that would control nematodes at rates not injurious to plants.

#### 1.6.4.3 Dithiocarbamates

Certain chemicals of this group (Metham sodium, Aldicarb, Carbofuran, Methomyl, and Oxamyl) are effective in killing nematodes.

The different modes of action of nematicides have been summarized in Table 1.9.

### 1.6.5 Host Resistance

One of the most economically feasible practices for the management of nematodes in low acre value crops is the use of resistant and tolerant varieties which contain one or

**Table 1.9** Probable modes of action of different groups of nematicides

| Nematicide | Mode of action |
|---|---|
| *Fumigants* (water required to activate toxicity in soil) [MIT (methyl isothiocyanate), Metham (Vapam), Dazomet (Mylone)] | MIT may react with amino acids, oxidases, and nucleophilic sites on proteins; Metham and Dazomet form MIT |
| *Fumigants* (air required to activate toxicity in soil) [1, 3-D (1, 3-dichloropropene); EDB (ethylene dibromide); DBCP (Nemagon); MB (methyl bromide); Chloropicrin] | Alkyl halides may react with nucleophilic sites on proteins and oxidize iron porphyrins and hemeproteins |
| *Organophosphates* [Ethoprop (Mocap), Fensulfothion (Dasanit), Fenamiphos (Nemacur)] | Irreversible binding of acetyl cholinesterase, esterase inhibition, and various pharmacologic actions |
| *Carbamates* [Carbofuran (Furadan), Aldicarb (Temik), Oxamyl (Vydate)] | Reversible binding of acetyl cholinesterase esterase inhibition, and various pharmacologic actions |

more genes conferring resistance. Practical results have been obtained in breeding for resistance to a number of major nematodes (Walia and Chakraborty 2018) (Table 1.10).

Biotechnology has a role to play in incorporation of resistance against nematodes and biological control of plant nematodes. Embryo rescue technique, protoplast fusion or somatic hybridization, and recombinant DNA technology have been used in nematology to overcome interspecific plant breeding problems with a view to incorporate nematode resistance.

### 1.6.6 Biological Methods

Effective nematode management can be achieved by the use of biological control agents [natural enemies like predators, parasites, and pathogens (fungi and bacteria)]. Biological control can be enhanced by application of organic amendments like FYM/compost, oil cakes, etc. to the soil.

#### 1.6.6.1 Bacteria

Biological suppression of plant parasitic nematodes with antagonistic bacteria (*Pseudomonas fluorescens, Pastueria penetrans, Bacillus* spp.) and Actinomycetes

**Table 1.10** Nematode-resistant/−tolerant varieties of major crops developed/identified (Walia and Chakraborty 2018)

| Crop | Nematode | Resistant/tolerant varieties |
|---|---|---|
| Cotton | *Meloidogyne incognita* | Bikaneri Nerma, Sharda, paymaster |
| Chili | *M. incognita/ M. javanica* | NP-46A, Pusa Jwala, Mohini |
| Cowpea | *M. incognita/ M. javanica* | GAU-1 |
| Grapevine | *M. incognita/ M. javanica* | Khalili, Kishmish Beli, Banquabad, Cardinal, Early Muscat, Loose Perlette |
| Green gram | *M. incognita/ M. javanica* | ML-30, ML-62 |
| Potato | *Globodera rostochiensis* | Kufri Swarna |
| Rice | *Ditylenchus angustus* | Basudeb |
| Sweet potato | *M. incognita* | Shree Bhadra |
| Tomato | *M. incognita/ M. javanica* | PNR-7, NT-3, NT-12, Hisar Lalit |
| Tomato (Polyhouse) | *M. incognita* | Regi, Moscotel 74–101 |
| Tuberose | *Aphelenchoides besseyi* | Prajwal, Phule Rajani, Shringar |

(*Streptomyces avermitilis*) is gaining increasing importance as a result of realization that many environmental and health hazards are linked with the use of chemicals.

#### 1.6.6.2 Fungi

Use of antagonistic fungi (*Purpureocillium lilacinum, Pochonia chlamydosporia, Trichoderma harzianum,* and *T. viride*), mycorrhizae (*Glomus mossece, G. fasciculatum*) for the biological suppression of plant parasitic nematodes is quite promising.

Standardization of methods for effective utilization of biocontrol agents is very important for evolving ecologically sound integrated nematode management strategies.

Some of the successful biocontrol agents which are demonstrated at farmers' fields and included in the package of practices of SAUs are as follows (Walia and Chakraborty 2018):

- Soil application with *Purpureocillium lilacinum* at 2.5 kg/ha at sowing significantly reduced the population of reniform nematode infecting soybean.
- Use of local strain of *Pseudomonas fluorescens* has given promising results in the management of *Meloidogyne incognita* as well as *Sclerotium rolfsii* on soybean.
- Effective management of *Pratylenchus thornei* and *Fusarium* wilt disease complex in chickpea can be managed by soil application of *Trichoderma* harzianum at 25 kg/ha along with *Pochonia chlamydosporia* at 10 kg/ ha 1 week prior to sowing.
- Effective management of *Rotylenchulus reniformis* in castor was obtained with soil incorporation of *Pseudomonas fluorescens* at 2.5 kg/ha.

### 1.6.7 Integrated Nematode Management

Integrated nematode management (INM) includes research, development, technology transfer, and implementation needed to integrate two or more control measures to manage one or more nematode species. The aim of INM is to reduce and stabilize the nematode pest population below damaging levels, resulting in favorable long-term socioeconomic and environmental consequences. Biological monitoring; environmental monitoring; agricultural production systems; nematode crop ecosystem models; system design; and implementation are some of the components of INM systems. The proceedings available for INM system design and implementation are based on the principles of exclusion, population reductions, and tolerance (Fig. 1.2).

### 1.6.8 Biointensive Integrated Nematode Management

Biointensive integrated nematode management (BINM) is a systems approach to pest management based on an understanding of nematode ecology. In order to keep the nematode populations within threshold levels, BINM relies on correct diagnosis

**Fig. 1.2** Different components of integrated nematode management

and then a range of preventive tactics and biological controls. Reduced-risk nematicides are used if other tactics have not been adequately effective, as a last resort, and with care to minimize risks. In BINM approach, the emphasis is on proactive measures to redesign the agricultural ecosystem to the advantage of its parasite and predator complex and the disadvantage to nematodes.

Proactive or reactive measures for nematode management are the two options for BINM.

#### 1.6.8.1 Proactive Options

Proactive options, such as crop rotations and creation of habitat for beneficial organisms, permanently lower the carrying capacity of the farm for the nematode. The carrying capacity is determined by the factors like food, shelter, natural enemy's complex, and weather, which affect the reproduction and survival of a nematode species. Proactive options include different cultural methods used under field conditions. Proactive practices include crop rotation, resistant crop cultivars including transgenic plants, disease-free seeds and plants, crop sanitation, spacing of plants, altering planting dates, mulches, etc.

The proactive strategies (cultural controls) include:

- Healthy, biologically active soils (increasing below-ground diversity).
- Habitat for beneficial organisms (increasing above-ground diversity).
- Appropriate plant cultivars.

#### 1.6.8.2 Reactive Options

The reactive options mean that the grower responds to a situation, such as an economically damaging population of nematodes, with some type of short-term suppressive action. Physical and mechanical methods, organic amendments, chemical nematicides, and applied biological control methods are some of the reactive options.

Low-cost, eco-friendly, and practically feasible INM technologies developed, demonstrated at farmers' fields, and included in the package of practices of SAUs are as follows (Walia and Chakraborty 2018):

- Effective management against rice root-knot nematode *Meloidogyne graminicola* can be achieved by combination of solarization of the nursery beds with 25μm polythene sheets for 15 days during May/June (transplanted rice) along with soil application of Carbofuran at 1 kg a.i./ha at 45 DAT.
- Soil fumigation with Dazomet at 40 g and Metham sodium at 40 ml/$m^2$ applied 25 and 15 days prior to transplanting, respectively, gave highest control of root-knot nematode in tomato, cucumber, capsicum, and carnations under protected conditions.
- Combined application of neem cake at100 g/$m^2$ + *Purpureocillium lilacinum* at 50 g/$m^2$ has been found most effective in reducing the *Meloidogyne incognita* population in soil and also increasing the yield of cucumber grown in polyhouses.
- Enhancement of yield and reduction of the root-knot nematode population in banana cv. Grand Naine was obtained with integration of paring, hot water treatment at 55 ° C, soil application of neem cake at 1 kg/plant + *Pseudomonas fluorescens* at 20 g/$m^2$ + carbofuran at 0.5 g a.i./plant.
- Use of *Purpureocillium lilacinum* at 20 kg/tree + castor cake at 2 ton/ha in root zone of pomegranate at regular interval of 6 months reduced the root-knot nematode population below economic threshold level.

## 1.7 Conclusion

Throughout the world, plant parasitic nematodes have been recognized as one of the important limiting factors in production of crops. Estimated crop losses due to plant parasitic nematodes in developed countries include 8.8% as compared to 14.6% in developing countries. In the present shift in cropping systems, development of practical integrated pest and disease management (including nematode management) is quite essential. With this in mind, it is essential that the full spectrums of crop protection limitations are considered appropriately, including the often-overlooked nematode constraints. There is a need for development of sustainable nematode management strategies in view of changes in crop production systems.

Development of economic damage/action thresholds is the major need of the hour in order to quantify crop losses. Control measures based on chemicals, genetic resistance, and cultural practices require a greater knowledge of nematode biology to achieve satisfactory results. The nematode management methods presently available include physical methods (hot water treatment of planting stock and soil solarization), cultural methods (crop rotation, intercropping), biomanagement (biocontrol agents and biopesticides), and host plant resistance (nematode-resistant cultivars). Ensuring that the imported seed and planting stock is free from nematode infection through regulatory methods is emphasized. Significant reduction in crop losses due to infected seeds and planting materials can be obtained by following the above methods in different countries. Nematode management approaches which need further refinement include crop rotation and soil solarization. The development and deployment of host resistance for nematode control is given a very high priority in many countries. Research in developed countries is targeted at the deployment of integrated management strategies that minimize the use of chemicals. Interdisciplinary collaboration between various researchers including social scientists is an important aspect that needs to be considered very seriously for effective management of nematode pests in different crops. Future developments of sustainable management systems for preventing major economical agricultural losses due to nematodes should be focused on strategies that limit production costs, enhance crop yields, and protect the environment.

## References

Abad P, Gouzy J, Aury JM et al (2008) Genome sequence of the metazoan plant-parasitic nematode *Meloidogyne incognita*. Nat Biotechnol 26:909–915

Ali SS (1986) Root-knot nematode problem in cardamom and its management. In: Proceedings of the second group discussion on the nematological problems of plantation crops, Central Coffee Research Institute, Balehonnur, pp 10–12

Anon (1990a) Final report of research scheme on nematode problems in some selected temperate fruit crops and their management. Y.S. Parmar University of Horticulture & Forestry, Solan, 54 pp

Anon (1990b) Annual report of the central tuber crops research institute. Trivandrum

Anon (1993a) QRT report of AICRP on nematodes (1989–1993), Department of Nematology, Haryana Agricultural University, Hisar, 149 pp

Anon (1993b) Biennial report (1991–93) of AICRP on plant parasitic nematodes with integrated approach for their control, Department of Nematology, Haryana Agricultural University, Hisar, 80 pp

Ansari RA, Khan TA (2012) Parasitic association of root-knot nematode, Meloidogyne incognita on guava. e-J Sci Technol (e-JST) 5:65–67

Ashokkumar N, Poornima K (2019) Occurrence and distribution of root knot nematode, *Meloidogyne enterolobii* in guava (*Psidium guajava* L.) in Tamil Nadu. J Pharmacogn Phytochem 8:1922–1924

Ashokkumar N, Poornima K, Kalaiarasan P, Kavino M (2019) Screening and histological characterization of guava (*Psidium guajava* L.) cultivars against root knot nematode, *Meloidogyne enterolobii*. Pest Mangmt Hort Ecosyst 25:84–92

Ayyar PNK (1926) A preliminary note on the root nematode, *Heterodera radicicola* Mitter and its economic importance in South India. Sci Cult 22:391–393

Baghel PPS, Bhatti DS (1983a). Evaluation of pesticides for the control of phytonematodes on citrus. In: Third all india nematology symposium, Yashwant Singh Parmar University of Horticulture and Forestry , Solan, pp 38–39

Baghel PPS, Bhatti DS (1983b) Relative efficacy of nematicides for control of phytonematodes on grapevine varieties. In: Third all india nematology symposium, Yashwant Singh Parmar University of Horticulture and Forestry, Solan, p 39

Bajaj HK, Jain RK (2001) CCS Haryana Agriculture University, Hisar, Haryana. In: Dhawan SC, Gaur HS, Pankaj, Kaushal KK, Ganguly S, Chawla G, Singh RV, Sirohi A (eds) Indian nematology–problems and perspectives. Division of Nematology Indian Agricultural Research Institute, New Delhi, pp 78–95

Barber CA (1901) Dept Land Records and Agriculture. Madras Agricultural Branch 2, Bulletin No 45, pp 227–234

Barbercheck ME, von Broembsen SL (1986) Effects of soil solarization on plant-parasitic nematodes and *Phytophthora cinnamomi* in South Africa. Plant Dis 70:945–950

Bhatti DS, Jain RK (1977) Estimation of loss in okra, tomato and brinjal yield due to *Meloidogyne incognita*. Indian J Nematol 7:37–41

Butler EJ (1913) An Eelworm disease of rice. Bulletin 34, Agricultural Research Institute, Pusa, India

Chahal PPK, Chhabra HK (1984) Interaction of *Meloidogyne incognita* with *Rhizoctonia solani* on tomato. Indian J Nematol 14:56–57

Charles JSK, Kurian KJ (1979) Studies on nematode incidence in ginger. In: Proceedings of the second symposium on plantation crops, pp 50–57

Cobb NA (1914) Nematodes and their relationships. In: USDA year book, pp 457–490

Dalal MR, Vats R (1998) Estimation of loss in pea (*Pisum sativum* L.) due to *Rotylenchulus reniformis*. In: Mehta UK (ed) Nematology – challenges and opportunities in 21st century. Sugarcane Breeding Institute, Coimbatore, pp 8–9

Darekar KS, Mahse NL (1988) Assessment of yield losses due to root-knot nematode, *Meloidogyne incognita* race 3 in tomato, brinjal and bitter gourd. Int Nematol Netw Newsl 5(4):7–9

Das AK (1994) Assessment of yield loss due to *Meloidogyne incognita* on French bean and its management. In: M. Sc. (Agriculture) thesis, Assam Agriculture University, Jorhat

Das TK, Khan MR (2007) Occurrence and distribution of white tip nematode, *Aphelenchoides besseyi* in West Bengal, India. Indian J Nematol 27:94–97

Dastur JP (1936) A nematode disease of rice in the central provinces. Proc Indian Acad Sci 4:108–121

Devi G (1993) Pathogenicity, crop loss assessment and management of *Meloidogyne incognita* (Kofoid and White, 1919) chitwood, 1949 on carrot (Dacus carota L.). M. Sc. (Agriculture) thesis, Assam Agriculture University, Jorhat

Ganguly S, Singh M, Ganguly AK (2010) Record of potato cyst nematode, *Globodera rostochiensis* and *G. pallida* in Shimla, Himachal Pradesh, India. Indian J Nematol 40:96–102

Govindaiah, Dandin SB, Sharma DD (1991) Pathogenicity and avoidable leaf yield loss due to *Meloidogyne incognita* in mulberry (*Morus alba* L.). Indian J Nematol 21:52–57

Hasan N, Jain RK (1998) Nematological research in Uttar Pradesh: an overview. In: Trivedi PC (ed) Phytonematology in India. C.B.S. Publishers and Distributors, New Delhi, pp 150–169

Haseeb A, Pandey R (1989) Root-knot disease of henbane, *Hyscyamus niger* - a new disease record. Trop Pest Mangmt 35:212–213

Hawn EJ (1971) Mode of transmission of *Corynebacterium insidiosum* by *Ditylenchus dipsaci*. J Nematol 3:420–421

Hewitt WB, Raski DJ, Goheen AC (1958) Nematode vector of soil-borne fanleaf virus of grapevines. Phytopathology 48:586–595

Jain RK, Paruthi IJ, Gupta DC, Dhankar BS (1986) Appraisal of losses due to root-knot nematode *Meloidogyne javanica* on okra under field conditions. Trop Pest Mangmt 32:341–342

Jain RK, Mathur KN, Singh RV (2007) Estimation of losses due to plant parasitic nematodes on different crops in India. Indian J Nematol 37:219–221

Jonathan EI, Rajendran G (2000) Assessment of avoidable yield loss in banana due to root-knot nematode, *Meloidogyne incognita*. Indian J Nematol 30:162–164

Jonathan EI, Nagalakshmi S, Padmanabhan D (1990) Estimation of yield losses in betel vine due to *Meloidogyne incognita*. Int Nematol Network Newsl 7(4):26
Jones FGW (1961) The potato root eelworm *Heterodera rostochiensis* Woll. In India. Curr Sci 30:187
Kamra A, Gaur HS (1998) Control of nematodes, fungi and weeds in nursery beds by soil solarization. Indian J Nematol 28:46–52
Khan RM, Parvatha Reddy P (1994) Nematode problems of ornamental crops and their management. In: Prakash J, Bhandary KR (eds) Floriculture-technology, trades and trends. Oxford & IBH Publishing Co. Pvt. Ltd., New Delhi, pp 468, 667 pp–472
Khan A, Shaukat SS, Siddiqui IA (2005) A survey of nematodes of pomegranate orchards in Baluchistan province, Pakistan. Nematol Medit 33:25–28
Koshy PK (2002) Nematode pests of spices and condiments and their control. In: Alam MM, Sharma N (eds) Nematode control in crops. International Book Distributing Company, Lucknow, pp 187–200
Koshy PK, Geetha SM (1992) Nematode pests of palms and cocoa. In: Bhatti DS, Walia RK (eds) Nematode pests of crops. CBS Publishers & Distributors, Delhi, pp 214–227
Kuhn J (1857) Uber des Vorkommen von Anguillule an erkrankten Bluhten koppen von *Dipsacus fullonum L.* Zeitscher Wiss Zool 9:129–137
Laqman Khan M (2001) Dr. Y.S. Paramar University of Horticulture and Forestry, Solan, Himachal Pradesh. In: Dhawan SC et al (eds) Indian nematology-progress and prospectives. Division of Nematology, Indian Agricultural Research Institute, New Delhi, pp 153–158
Mai WF, Abawi GS (1987) Interaction among root-knot nematodes and Fusarium wilt fungi on host plants. Annu Rev Phytopathol 25:317–338
Midha RL, Trivedi PC (1991) Estimation of losses caused by *Meloidogyne incognita* on coriander, cumin and fennel. Curr Nematol 2:159–162
Milne D (1919) Ear cockle in wheat. Dept Agric Punjab Bull (Bot) 1:1–5
Mohandas C, Ramana KV (1991) Pathogenicity of *Meloidogyne incognita* and Radopholus similis on black pepper (Piper nigrum L.). J Plantn Crops 19:41–43
Mukhopadhyaya MC, Dalal MR (1971) Effect of two nematicides on *Tylenchulus semipenetrans* and on sweet lime yield. Indian J Nematol 1:95–97
Mukhopadhyaya MC, Suryanarayana D (1969) Citrus decline in Haryana: role of *Tylenchulus semipenetrans* and its control. Indian Phytopath 22:495–497
Nagesh M, Parvatha Reddy P (2000) Crop loss estimation in carnation and gerbera due to the root-knot nematode, *Meloidogyne incognita* (Kofoid & White) Chitwood. Pest Mangmt Hort Ecosyst 6:158–159
Naik D (2004) Biotechnological approaches for the management of wilt disease complex in capsicum (Capsicum annuum L.) and egg plant (Solanum melongena L.) with special emphasis on biological control. PhD thesis, Kuvempu University, Shimoga, 201 pp
Nair KKR (1979) Studies on the chemical control of banana nematodes. Agri Res J Kerala 17:232–235
Nair MRGK, Das NM, Menon MR (1966) On the occurrence of the burrowing nematode, *Radopholus similis* (Cobb, 1893) Thorne, 1949 on banana in Kerala. Indian J Ent 28:553–554
Needham T (1743) A letter concerning chalky tubulous concretions called malm: with some microscopical observations on the fauna of the red lily and of worms discovered in smutty corn. Philos Trans Roy Soc London 42(173–174):634–641
Norton DC (1978) Ecology of plant parasitic nematodes. Wiley, New York
Palanisamy S, Sivakumar CV (1981) Assessment of avoidable yield loss in cowpea, black gram, maize and finger millet. In: Second all India nematology symposium, Tamil Nadu Agricultural University, Coimbatore, p 58
Pandey R (1994) Bionomics of phytonematodes in relation to medicinal and aromatic plants. Indian J Pl Path 12:16–23
Pandey R (2003) Mint nematology – current status and future needs. In: Trivedi PC (ed) Advances in nematology. Scientific Publishers, Jodhpur, pp 155–166

Parvatha Reddy P (1983) Plant nematology. Agricole Publishing Academy, New Delhi, 287 pp
Parvatha Reddy P (1985) Estimation of crop losses in peas due to *Meloidogyne incognita*. Indian J Nematol 15:226
Parvatha Reddy P (1987) A treatise on phytonematology. Agricole Publishing Academy, New Delhi, 381 pp
Parvatha Reddy P, Singh DB (1981) Assessment of avoidable yield loss in okra, brinjal, French bean and cowpea due to root-knot nematodes. In: International symposium on plant pathology, New Delhi, pp 93–94
Parvatha Reddy P, Rao MS, Nagesh M (1996) Crop loss estimation in banana due to the burrowing nematode, *Radopholus similis*. Pest Mangmt Hort Ecosyst 2:85–86
Patel BA, Patel DJ, Sharma SB, Patel HV (1996) Nematode problems of groundnut and their management in Gujarat, India. Int Arachis Newsl 16:38–39
Patel D, Patel BA, Patel SK et al (1999) Root-knot nematode, *Meloidogyne indica* on kagzi lime in North Gujarat. Indian J Nematol 29(2):185–205
Poornima K, Vadivelu S (1998) Pathogenicity of *Meloidogyne incognita* to turmeric (Curcuma longa L.). In: Mehta UK (ed) Nematology-challenges and opportunities in 21st century. Sugarcane Breeding Institute, Coimbatore, pp 29–33
Poornima K, Suresh P, Kalaiarsan P et al (2016) Root knot nematode, *Meloidogyne enterolobii* in guava (*Psidium guajava* L.) - a new record from India. Madras Agric J 103:359–365
Prasad KSK (1989) Nematological problems and progress of research on potato in India. In: Fourth group meeting on nematological problems of plantation crops, University of Agricultural Sciences, Bangalore
Prasad PRK, Reddy DDR (1984) Pathogenicity and analysis of crop losses on patchouli due to *Meloidogyne incognita*. Indian J Nematol 14:36–38
Prasad JS, Somasekhar N (2009) Nematode pests of rice: diagnosis and management. Technical Bulletin No 38, Directorate of Rice Research (ICAR), Rajendranagar, Hyderabad, 29 pp
Prasad N, Mathur RL, Sehgal SP (1959) Molya disease of wheat and barley in Rajasthan. Curr Sci 28:453
Rajagopalan P, Naganathan TG (1977a) Studies on nematode parasites of grapevine. Tamil Nadu Agric Univ Ann Rept 6:129
Rajagopalan P, Naganathan TG (1977b) Studies on nematode parasites of banana. Tamil Nadu Agric Univ Ann Rept 6:131
Rajendran G, Naganathan TG (1981) Evaluation of systemic chemicals for the control of *Rotylenchulus reniformis* on papaya. Indian J Nematol 11:130
Ramana KV, Eapen SJ, Sarma YR (1998) Effect of *Meloidogyne incognita, Pythium aphanidermatum* and *Ralstonia solanacearum* alone and in combinations in ginger. In: Mehta UK (ed) Nematology-challenges and opportunities in 21st century. Sugarcane Breeding Institute, Coimbatore, pp 87–93
Reddy DDR (1981) Analysis of crop losses in tomato due to Meloidogyne incognita (Abstract). In: Second all India nematology symposium, Coimbatore, p 60
Rodriguez-Kabana R, Pinochet J, Robertson DG et al (1992) Horse bean (*Canavalia ensiformis*) and crotalaria (*Crotalaria spectabilis*) for the management of *Meloidogyne* spp. Nematropica 22:29–35
Saikia B (1992) Pathogenicity, crop loss assessment and biological control of *Meloidogyne incognita* (Kofoid and White, 1919) Chitwood, 1949 on Betel Vine (*Piper betle* L.). M. Sc. (Agriculture) thesis, Assam Agriculture University, Jorhat
Santhosh Kumar T, Sheela MS (2004) Status of nematode pests of Chethikoduveli, *Plumbago rosea* L. In: National symposium on paradigms in nematological research for biodynamic farming, University of Agricultural Sciences, Bangalore, pp 28–29
Sasser JN, Freckman DW (1987) A world perspective on nematology: the role of the society. In: Veech JA, Dickson DW (eds) Vistas on nematology. Society of Nematologists, Hyattsville, pp 7–14

Schacht H (1859) Ueber einige Feinde and Krankheithen der Zuckerribe. Zeitschr Ver Ruben ercker Ind Zoolver 9:390

Senthamarai M, Poornima K, Subramanian S (2006) Assessment of avoidable yield loss on *Coleus forskohlii* due to *Meloidogyne incognita*. Indian J Nematol 36:296–297

Sharma GL (1989) Estimated losses due to root-knot nematodes, *Meloidogyne incognita* and *M. javanica* in pea crop. Int Nematol Netw Newsl 6(1):28–29

Sharma SB, Sharma HK, Pankaj (2002) Nematode problems in India. In: Prasad D, Puri SN (eds) Crop pest and disease management: challenges for the millennium. Jyoti Publishers, New Delhi, pp 267–275

Sheela MS, Rajani TS (1998) Status of phytonematodes as a part of medicinal plants in Kerala. In: Mehta UK (ed) Nematology – challenges and opportunities in 21$^{st}$ century. Sugarcane Breeding Institute, Coimbatore, pp 2–3

Siddiqi MR (1961) Occurrence of the citrus nematode, *Tylenchulus semipenetrans* Cobb, 1913, and the reniform nematode, *Rotylenchulus reniformis* in India (Abstract). In: Proceedings of the 48th Indian Science Congress Part III, p 504

Singh RV, Midha SK, Kumar V (2003) Major nematode pests of economically important crops in India and their management. In: Trivedi PC (ed) Advances in nematology. Scientific Publishers, Jodhpur, pp 233–256

Sivakumar M, Marimuthu T (1987) Preliminary studies on the effect of soil solarization on phytonematodes of betel vine. Indian J Nematol 17:58–59

Sosamma VK, Sundararaju P, Koshy PK (1979) Effect of *Radopholus similis* on turmeric. Indian J Nematol 9:27–31

Subramanyam S, Rajendran G, Vadivelu S (1990) Estimation of loss in tomato due to *Meloidogyne incognita* and *Rotylenchulus reniformis*. Indian J Nematol 20:239–240

Sundararaju P, Sosamma VK, Koshy PK (1979) Pathogenicity of *Radopholus similis* on ginger. Indian J Nematol 9:91–94

Taylor CE (1980) Nematodes. In: Harris KF, Marmorosch K (eds) Vectors of plant pathogens. Academic Press, New York, pp 375–416

Thirumala Rao V (1956) Stray notes on some crop pest outbreaks of South India. Indian J Ent 18:123–126

Thorne G (1961) Principles of nematodes. McGraw Hill, New York and London, 553 pp

Upadhyay KD, Dwivedi K (1987) Analysis of crop losses in pea and gram due to *Meloidogyne incognita*. Int Nematol Netw Newsl 4(4):6–7

Vargas-Ayala R, Rodriguez-Kabana R, Morgan-Jones G et al (2000) Shifts in soil micro-flora induced by velvet bean (*Mucuna deeringiana*) in cropping systems to control root-knot nematodes. Biol Control 17:11–22

Verma AC (2001) Narendra Deva University of Agriculture and Technology, Kumarganj, Faizabad, Uttar Pradesh. In: Dhawan SC et al (eds) Indian nematology-progress and perspectives. Division of Nematology, Indian Agricultural Research Institute, New Delhi, pp 121–125

Walia RK, Chakraborty PK (eds) (2018) Nematodes problems of crops in India. In: ICAR-all india coordinated research project on nematodes in agriculture, New Delhi, 402 pp

# Part II

# Cereal Crops

# 2 Cereal Crops

**Abstract**

The most important genera of plant parasitic nematodes on cereal crops such as rice, wheat, barley, maize, and sorghum include species of *Meloidogyne* (root-knot nematodes), *Heterodera* (cyst nematodes), and *Pratylenchus* (root-lesion nematodes) that account for most of the global crop damage. Root-invading nematodes enter into disease complexes with plant-pathogenic fungi that cause root diseases. Nematode management strategies that are effective include combinations of seed treatment chemicals or biological agents, planting nonhost crops, or placing land into prolonged periods of fallow between plantings of susceptible crops. Use of genetic resistance is especially important because it is the control strategy that is environmentally and socially most acceptable for minimizing yield losses caused by plant parasitic nematodes.

**Keywords**

Cereal crops · Crop losses · Symptoms · Survival · Spread · Management

## 2.1 Rice, *Oryza sativa*

Investigations on the nematode diseases of rice have been concentrated on the five genera, *Aphelenchoides, Ditylenchus, Hirschmanniella, Meloidogyne,* and *Heterodera.*

### 2.1.1 Root-Knot Nematode, *Meloidogyne graminicola*

In view of rice cropping intensification and increasing scarcity of water, the root-knot nematode *Meloidogyne graminicola*, which was prevalent in eastern,


P. P. Reddy, *Nematode Diseases of Crops and their Management*, https://doi.org/10.1007/978-981-16-3242-6_2

north-eastern, and southern states of India, has emerged as a serious threat for the successful rice production in nurseries, uplands, deep water, and irrigated fields.

#### 2.1.1.1 Crop Losses

In India, the losses due to *M. graminicola* have been estimated to be 16–32% and yield loss due to poorly filled kernels to be 17–30%, and in severe cases up to 64%. Walia and Chakraborty (2018) reported 10.54% loss in rice yield due to *M. graminicola* amounting to ₹*23272.32 million annually.*

#### 2.1.1.2 Distribution

The nematode is distributed throughout India.

#### 2.1.1.3 Symptoms

Symptoms include patches of stunted and chlorotic rice plants with less vigor, yellowing, and curling of leaves in infested rice fields (Fig. 2.1). There is reduction in chlorophyll content of leaves (Swain and Prasad 1988). The galls incited by *Meloidogyne* spp. on rice are terminal, spiral, or horse shoe shaped (Fig. 2.1). The crop has poor stand, less tillering, few number of effective tillers, lesser and smaller spikes, sparsely filled grains, and poor yields.

#### 2.1.1.4 Biology and Life Cycle

Soil-borne second-stage juveniles infect root tips and lay egg masses after maturation. The second-stage juveniles emerge from egg masses and infect fresh feeder roots. The life cycle is completed within 20–25 days.

#### 2.1.1.5 Survival and Spread

The nematode can survive in egg stage inside root gall or reproduce on various weeds in crop field and disseminated through infected seedlings and weed hosts.

**Fig. 2.1** *Left*—Symptoms of root-knot nematode injury in nursery showing yellowing of foliage. *Middle and Right*—Rice roots showing terminal and horse-shoe galls caused by *Meloidogyne graminicola*

### 2.1.1.6 Management

#### 2.1.1.6.1 Physical Methods

The nematode population in nursery bed may effectively be reduced by soil polarization. The nursery bed is covered for 4–6 weeks with a plastic polyethylene tarp (25 am) prior to sowing. Soil solarization raises temperature by 4 °C as compared to non-solarized soil to reach 40–50 °C at 10 cm depth. In the upper 15 cm soil layer, the nematodes get killed and become undetectable. This results in 80% reduction of *Meloidogyne* spp. population in rice beds and improves seedling growth (Ganguly et al. 1996).

#### 2.1.1.6.2 Cultural Methods

For raising nursery, root-knot nematode-free area should be selected. The seedlings which are free from nematode galls must be used for transplanting.

##### Crop Rotation

In order to reduce nematode populations, crop rotation with resistant or poor hosts of *M. graminicola* could be used. Crop rotation with resistant varieties (TKM-6, Patna 6, Dumai, Ch 47, and Hamsa) and nonhost crops like jute, mustard, chickpea, castor, cowpea, sweet potatoes, soybeans, sunflower, sesame, onion, cauliflower, turnip, beans, and okra reduces *M. graminicola* infestations (Rao 1985; Rao et al. 1984, 1986). Long rotations, more than 12 months, will be needed to reduce *M. graminicola* soil population to low levels. Introducing a bare fallow free of weed hosts into the crop rotation will also give effective control of the nematodes.

Rice following with two crops of cowpea had increased the yield by 34% when treated with Carbofuran and by 26% in untreated plants. An yield increase of 85% was observed when single crop of cowpea followed rice crop. Cowpea cv. Iron Clay which has high degree of resistance was shown to be an effective rotation crop to control some root-knot species. The reduction in nematode population and increase in rice yield by 30–80% was noticed by following one or two consecutive crops of cowpea or a season of fallow before a rice crop. It is recommended that low density of *M. graminicola* population should be maintained by nonhost crop rotations or fallows, ideally for two seasons before planting rice to ensure higher rice yields (Soriano and Reversat 2003).

An alternate rotation of rice with a nonhost crop or a season of fallow seems to be the only solution at the moment for managing *M. graminicola* population. The effective crop rotation sequences to reduce nematode development include rice-mustard-rice, followed by rice-maize-rice and rice-fallow-rice (Kalita and Phukan 1996). The reduction in the population and development of rice root-knot nematode was obtained with rotation of rice with marigold (*Tagetes* spp.) and sun hemp (*Crotalaria incana* and *C. mucronata*). For overall management of *M. graminicola*, planting of *Tagetes* or *Crotalaria* species in nematode-infested soil is feasible and can be used.

#### Fallowing

Soriano and Reversat (2003) reported that the rice yield increased by 41% with one season of fallow (120 days) and by only 31% with two consecutive fallows (240 days). In the third season, the most efficient strategy to increase rice yields includes two seasons of fallow along with Carbofuran treatment before rice.

#### Flooding

Bridge and Page (1982) found that damage to the crop can be avoided by raising rice seedlings in flooded soils thus preventing root invasion by *M. graminicola*. In Vietnam, high effectiveness in control of *M. graminicola* was obtained with continuous flooding (Kinh et al. 1982). When the rice crop is flooded early and kept flooded until a late stage of development, the yield losses may be minimized (Garg et al. 1995; Soriano et al. 2000).

By growing healthy and disinfested nursery and following summer plowing and puddling of main fields before transplanting, *M. graminicola* problem in rice can be managed. Gaur (2003) and Pankaj et al. (2006) noticed several advantages of good puddling before transplanting that include water retention in soil which reduces aeration and also reduces nematode movement and invasion of fresh roots in soil by infective juveniles.

#### Organic Amendments

Roy (1976) reported that the root-knot nematode infestation can be reduced and plant growth increased by soil amendment with decaffeinated tea waste or water hyacinth compost (300 or 600 g/4.5 kg soil).

Raising of marigold and incorporating into soil before planting rice suppressed nematode root galling and increased rice grain yield by 46%.

Amarasinghe (2012) found that the nematode infestation level in rice plants can be reduced and soil fertility improved by occasional soil application of poultry manure as a nematicide cum fertilizer.

*M. graminicola* population and damage can be reduced by seed treatment with neem-based pesticides (Das and Deka 2002).

### 2.1.1.6.3 Chemical Methods

The effectiveness of Carbofuran to control *M. graminicola* in rice has been confirmed by several studies. Soriano and Reversat (2003) reported that the application of Carbofuran increased rice yield by 10% for the first rice crop. During the second season, 85% yield increase was observed when rice was planted after cowpea; 75% after cowpea with Carbofuran application; 41% after fallow; and 48% after fallow with Carbofuran. In the third season, when all plots were planted to rice, 49 and 31% yield increases were obtained, with and without Carbofuran, respectively, from plots that had been left fallow for two consecutive seasons.

The possible treatments in the present scenario with the nonavailability of other nematicides could be:

- Seed soaking in 500 ppm Carbosulfan for 12 h.; root-dip in 200 ppm of Carbofuran for 12 h.
- Two split dose application at 15 and 45 days after planting of Carbofuran at 1 kg a.i./ha.

#### 2.1.1.6.4 Biological Methods

Pathak and Kumar (2003) reported that significant improvement in plant growth and reduction in number of galls and larval penetration per root system was obtained with soil treatment of biocontrol agents like *Trichoderma harzianum* and *T. virens* at 4 and 8 g/kg soil, respectively. *T. virens* was comparatively more suppressive to nematode population in host tissue than *T. harzianum* (Pathak and Kumar 2003).

In aerobic rice, Shanmuga Priya (2015) opined that *T. viride* can be used as an alternative biomanagement strategy for root-knot nematode on rice. She reported that *T. viride* significantly reduced extent of galling (12.80), gall index (1), yield (3260 kg/ha), and final soil population (209.28).

The bacterial biocontrol agent *Pseudomonas fluorescens* was also found significantly effective in reducing *M. graminicola* population in rice root tissues. Anita and Samiyappan (2012) reported that 7, 14, and 21 days after *P. fluorescens* inoculation, there was 40.7%, 68.4%, and 75.1% reduction in the nematode population in the treated plants, respectively. They concluded from the present study that induced systemic resistance and decrease in nematode infection collectively contribute to earlier and higher accumulation of phenols and defense enzymes, viz., peroxidase, polyphenol oxidase, phenylalanine ammonia lyase, and chitinase, in rice root tissue treated with *P. fluorescens* in response to invasion by *M. graminicola* (Anita and Samiyappan 2012).

Renitha Govind and Ushakumari (2010) found that rice root-knot nematode can be effectively controlled during early stages by using seedlings raised from seeds treated with *Pseudomonas fluorescens, Trichoderma viride, P. fluorescens – T viride,* and *Bacillus subtilis*. However, *B. subtilis* showed superior effect in enhancing the shoot and root characters at the time of harvest. For the management of rice root-knot nematode population in root and soil, *B. subtilis* was found to be equally superior as compared to Carbofuran.

#### 2.1.1.6.5 Host Resistance

Prasad et al. (1986) and Rao et al. (1986) reported that several rice cvs. have been found resistant to the rice root-knot nematode. Three rice accessions of *O. glaberrima* (TOG7235, TOG5674, and TOG5675) and one accession of *Oryza longistaminata* represented by two individuals (WL02-2 and WL02-15) were found resistant to *M. graminicola* (Soriano et al. 1998).

#### 2.1.1.6.6 Integrated Methods

In Indo-Gangetic Plains, *M. graminicola* infects both rice and wheat which are grown in sequence. In order to manage root-knot nematode in rice-wheat cropping system, the following methods could be used in combination along with the judicious use of chemicals.

Nursery Bed Treatment

- Selection of nursery area free from the root-knot nematode.
- Soil solarization of nursery beds, for 3–4 week in April–May (severe summer months).
- Before sowing, soil application of Carbofuran at 3.3 g/m$^2$ of nursery bed.
- Gaur (2003) reported effective reduction in root-knot nematode infestation in rice nursery by soil solarization of nursery beds for 3–4 weeks in summer and nursery bed treatment with Carbofuran or Phorate at 2 kg a.i./ha.

Main Field Treatment

- Two summer plowing of the main field at 14 days interval.
- Before transplanting, proper puddling of the main field.
- Avoid direct sowing of rice in the main field.
- Crop rotation with mustard or other nonhost crops.
- Late sowing of wheat up to late November till mid-December using suitable variety.
- If time permits, growing of green manure crops like *Crotalaria* spp. or *Sesbania* spp. and their incorporation in soil.

### 2.1.2 White Tip Nematode, *Aphelenchoides besseyi*

Dastur (1936) published an account of a widespread rice disease in the central provinces of India, known as "white tip," where it had reached epidemic proportions in 1934.

#### 2.1.2.1 Crop Losses

Atkins and Todd (1959) reported that the annual yield loss caused by *Aphelenchoides besseyi* amounts to as much as 17.4–54.1%.

#### 2.1.2.2 Symptoms

Seed germination is delayed in root-knot infected seed beds and produce small seedlings. In the tillering stage, the upper 2–5 cm leaf tips turn white or pale yellow, then brown, necrotic, and frayed (Fig. 2.2). Characteristic shortening and twisting of flag leaves are noticed at their apical portions. Other symptoms noticed include general stunting of the plant, leaf injury, shorter panicles, reduction in number of spikelets and terminal tissues of the panicle, production of small deformed kernels, delayed maturity of panicles, and emergence of secondary panicles from lower nodes if the panicle is sterile.

#### 2.1.2.3 Biology and Life Cycle

The nematode is seed-borne. In between husk and kernel, about 5–6 quiescent nematodes are embedded which become active with availability of moisture in soil. Nematodes climb to the growing points of young seedlings and finally reach

**Fig. 2.2** Rice leaves infected with *Aphelenchoides besseyi* showing "white tip" symptoms

the inflorescence and panicle where they become endoparasites. Life cycle duration is short and only 10–12 days.

### 2.1.2.4 Host Range

Besides rice, *Boehmeria nivea,* chrysanthemum, *Cyperus iria, Ficus elastica, Hibiscus bracheridgii,* hydrangea, *Panicum sanguinale, Pennisetum, Setaria italica, S. viridis, Sporobolus poirettii,* strawberry, tuberose, and *Vanda orchid* have been reported as hosts of the white tip nematode.

### 2.1.2.5 Survival and Spread

The nematode can survive for 3 years as second-stage larvae in quiescent stage outside or inside the husk. They do not survive in soil and die in 4 months on grain left in the field. However, the nematodes may survive from one season to the next on rice seedlings growing from shed grain. *A. besseyi* is carried beneath the hull of the rice kernel in a quiescent, immature, and usually preadult in dormant stage and can survive as long as 23 months in this state.

Long distance spread of the nematode is largely by planting infected seeds, while the short distance spread is through irrigation water.

### 2.1.2.6 Management

#### 2.1.2.6.1 Regulatory Methods

The white tip nematode can be eliminated by the use of certified seed.

#### 2.1.2.6.2 Physical Methods

Good control of the nematode can be obtained by seed treatment in hot water maintained at 55° to 61 °C for 15 min.

#### 2.1.2.6.3 Cultural Methods

The nematode population can be significantly reduced by seeding of rice in water as compared to drilled and flooded (Cralley 1957). Probably the quiescent nematodes revive when immersed, begin moving about, leave the seed, and are lost in the water before the seeds sprout.

Nematode control can be achieved by using healthy/certified seed and by destroying infected stubbles/weeds/straw/debris, followed by keeping the fields dry by plowing.

#### 2.1.2.6.4 Chemical Methods

Seed soaking in 0.1% Carbosulfan (25 EC) for 6 h followed by foliar spray with 0.02% Carbosulfan (25EC) at 40 days after transplanting was found effective.

Spray of Monocrotophos 36 SL at 1000 ml/ha at boot leaf stage is also beneficial.

#### 2.1.2.6.5 Host Resistance

Atkins and Todd (1959) reported that varieties like Blue Bonnet 50, Fortune, and Texas Patna were very resistant to the white tip nematode. Goto and Fukatsu (1956) found that Toson 38 showed no symptoms of white tip during 3 years of field testing.

## 2.1.3 Rice Stem Nematode, *Ditylenchus angustus*

Butler (1913) first reported that the rice stem nematode was the causal agent of "ufra" disease, from rice producing region north of the Bay of Bengal and east of the Ganges River (Bangladesh). The nematode was responsible for as much as 50% reduction in yield from India and 20–90% in Thailand.

### 2.1.3.1 Symptoms

Rice plants that are more than 2 months old exhibited the earliest sign of chlorosis or streaks on the upper leaves in the field. Two distinct types of symptoms include (1) the "swollen ufra," where the panicle remains enclosed within the leaf sheath and there is a strong tendency toward branching of the stem in the infected portions; and (2) the "ripe ufra," where the panicle emerges and produces normal grain only near the tip (Fig. 2.3). Infected plants show symptoms like stunted plant growth, wilting of leaves, dark brown peduncle, and unfertilized lower parts of flowers.

### 2.1.3.2 Life Cycle

As the plants mature and dry, cottony masses of preadult nematodes (fourth-stage larvae) are formed. They become active, climb up on the stems, and invade the growing point, during humid periods. The nematodes feed ectoparasitically on the epidermal cells of the young seed head, the peduncle, the part of the stems just above the upper nodes, and the young leaves rolled about the bud above the growing point. Enormous numbers of nematodes in all stages from eggs to adults are found in developing heads. They are not found living endoparasitically in plant tissues at any time (Butler 1913; Voung 1969).

**Fig. 2.3** Symptoms of different types of *ufra* disease

#### 2.1.3.3 Host Range

Besides *Oryza* spp., *Leersia hexandra* has been recorded as a host of *D. angustus* in Madagascar.

#### 2.1.3.4 Survival and Spread

As the plants mature and dry at the end of the growing season, cottony masses of nematode in a stage of anabiosis are formed. When the crop is harvested, some of these nematodes remain on the stubble. They revive and infect the new crop when the rainy season starts and the humidity is high. The nematodes can also be spread in irrigation water. They can survive desiccation for more than 15 months.

#### 2.1.3.5 Management

##### 2.1.3.5.1 Cultural Methods

Some of the cultural control practices used for the management of *D. angustus* include using nematode-free seeds, improving drainage as the disease increase in water-logged areas, rotation of rice with jute (Dastur 1936), destruction of stubbles, and elimination of volunteer plants.

##### 2.1.3.5.2 Chemical Methods

The nematodes are completely controlled within 72 h by spraying Diazinon 100 ppm on the soil in the rice crop.

#### 2.1.3.5.3 Host Resistance

Rice cv. "Khao Tah Oo" variety besides being resistant to the leaf blast disease was also found highly resistant to the rice stem nematode in Thailand. The cv. Rayada 116-06 was also found resistant.

#### 2.1.3.5.4 Integrated Methods

Integration of seed treatment with Carbosulfan (25 EC) at 3% a.i. (w/w) and foliar spray with Carbosulfan at 0.2% a.i. at 40 and 120 days after transplanting was found effective against *D. angustus*.

### 2.1.4 Rice Root Nematodes, *Hirschmanniella* spp.

Rice root rot nematodes *Hirschmanniella mucronata, H. oryzae,* and *H. gracilis* are important nematode pests in transplanted rice, which can survive in standing water.

#### 2.1.4.1 Distribution

The nematode is distributed throughout the country.

#### 2.1.4.2 Crop Losses

The average yield loss due to this nematode is estimated around 7–20.2%, which corresponds to ₹26.85–52 million annually.

#### 2.1.4.3 Symptoms

The symptoms include patchy appearance in crop stand in the field (Fig. 2.4), retarded plant growth and reduced tillering, occasional yellowing of plants, delayed flowering and maturity, yellow to brown necrosis in roots (Fig. 2.4), and root decay under heavy infestation. Symptom of zinc deficiency is more pronounced in heavily

**Fig. 2.4** *Left*—Patchy appearance of rice field infected with root nematode showing reduced plant vigor. *Right*—Drastic reduction in root system (3 plants on the right) compared to healthy root (1 plant on the left)

infested soil. Diseased symptoms are observed throughout the year in medium and lowland.

#### 2.1.4.4 Biology and Life Cycle

The nematode is a migratory endoparasite. Infection starts from the soil and population builds up inside the roots during vegetative growth. All stages are infective and oviposition takes place inside the root cortex. Nematodes tend to leave the roots and move into soil near crop maturity.

#### 2.1.4.5 Dissemination

Spread occurs through infected seedlings, irrigation water, soil adhering to farm implements, and field workers.

#### 2.1.4.6 Management

##### 2.1.4.6.1 Physical Methods

Solarization of nursery beds of rice for 15 days during mid-summer season.

##### 2.1.4.6.2 Cultural Methods

Two to three deep summer plowings after premonsoon shower are beneficial. Crop rotation with nonhosts/antagonistic crops like groundnut, black gram, mustard, potato, and groundnut in rice-based cropping sequence is effective against *Hirschmanniella* spp. (Saikia 1991). The nematode population can also be reduced by growing of nonhost crops such as wheat, linseed, potato, mustard, gram, and berseem in *rabi* season.

In order to maintain nematode population below economic threshold level, fallowing and weed control are good practices.

Growing and incorporating two green manure legume crops like *Sesbania rostrata* and *Sphenoclea zeylanica* can also give good nematode control with the additional benefit of adding soil nitrogen.

##### 2.1.4.6.3 Chemical Methods

Nursery treatment with Carbofuran at 1 kg a.i. /ha and seedling root-dip in 0.2% Dimethoate for 6 h before transplanting is beneficial for managing rice root nematode.

Soil application of Carbofuran at 1–2 kg a.i. /ha in the main field decreased *Hirschmanniella* spp. population by 29% with corresponding increase in yield by 10%.

##### 2.1.4.6.4 Host Resistance

Growing rice cvs. resistant to *H. oryzae* like TKM 9, CR 142-3-2, CR 52, N 136, and W 136 seem to be beneficial.

#### 2.1.4.6.5 Integrated Methods

Solarization of nursery beds of rice for 15 days during peak summer and nursery bed treatment with Carbofuran 3G at 30 g/$m^2$ and field application of Carbofuran at 1 kg a.i. /ha after 40 days of transplanting gives effective control of *Hirschmanniella* spp.

Application of Carbofuran at 1 kg a.i. /ha in nursery bed 7 days before uprooting the seedlings followed by application of Carbofuran 3G at 1 kg a.i. /ha in the main field 45 days after transplanting is found beneficial (Choudhury et al. 2011).

## 2.1.5 Cyst Nematode, *Heterodera oryzicola*

*H. oryzicola* causes the failure of successive cropping of upland rice in Japan.

### 2.1.5.1 Distribution

The rice cyst nematode is present in India, Japan, and Ivory Coast.

### 2.1.5.2 Yield Losses

Kumari and Kuriyan (1981) estimated that yield losses due to *H. oryzicola* infestations varied from 21% to 42%. In order to cause 10% loss threshold level, the number of infective juveniles required per 30-day-old plant was up to 85–100 (Rao 1985).

### 2.1.5.3 Symptoms

Two weeks after germination, the nematode retards plant growth and causes leaf yellowing and a reduction in the number of tillers of upland rice seedlings. The characteristic symptoms include fewer root hairs and a brownish discoloration on roots (Fig. 2.5). The inflorescence emergence is delayed in severe cases. The nematode is responsible for reduction in the root and grain weight.

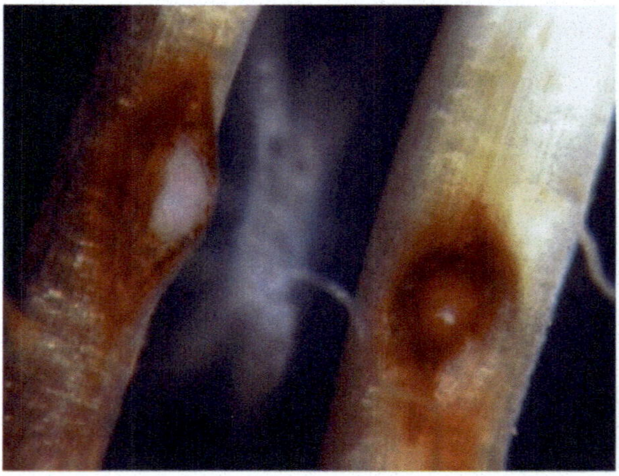

**Fig. 2.5** Cyst nematode *Heterodera oryzicola* emerging from rice root

#### 2.1.5.4 Life Cycle

The nematode life cycle is short. During the 26th and 30th day after inoculation, the second-stage larvae emerge from the cysts. Tylose formation is there around the female body. The absence of true giant cell formation may be peculiar to some monocotyledonous roots. Males are essential for nematode reproduction. In the rice growing season (4 months), two to three nematode generations may take place.

#### 2.1.5.5 Host Range

*H. oryzicola* has a limited host range which include *Panicum crusgalli* and *P. crusgalli* var. *frumentaceum, Cynodon dactylon,* and *Brachiaria* sp. besides rice and banana being major hosts. Charles and Venkitesan (1990) have reported that the weed *Kyllinga monocephala* is a poor host of this nematode.

#### 2.1.5.6 Interaction with Other Organisms

The synergistic interaction exists between *H. oryzicola* and the root rot fungus *Sclerotium rolfsii.* The fungal infection was enhanced in the roots during the penetration of cyst nematode in roots. Jayaprakash and Rao (1984) reported that the cyst formation was inhibited in the seedlings inoculated with the fungus. Antagonistic nematode-nematode interaction occurred between *M. graminicola* and *H. oryzicola* (Rao et al. 1984). When inoculated together, *M. graminicola* establishes faster and suppresses the multiplication of *H. oryzicola.*

#### 2.1.5.7 Management

##### 2.1.5.7.1 Cultural Methods

The moderate nematode population in the upper layers of the soil can be reduced by 1- or 2-year crop rotation with nonhost plants or fallowing.

The cultural method of suppression of the nematodes with tractor wheels followed by an application of calcium cyanamide has been developed in Japan.

##### 2.1.5.7.2 Chemical Methods

Rao (1985) reported that soaking rice seeds in 0.2% solution of Oxamyl or Carbosulfan at 250 ppm reduces cyst development in *H oryzicola*. The incidence of the nematode is reduced by 70% and increase grain yield by 28% with soil application of Carbofuran or Phorate at 1 kg a.i. /ha, at 7 and 50 days after planting (Kuriyan 1985).

##### 2.1.5.7.3 Host Resistance

Rice cvs. Lalnakanda-41, CR 143-2-2, Ratna, Hamsa, Mtu-17, and Mtu-4 were found resistant to *H. oryzicola* (Jayaprakash and Rao 1983).

## 2.2 Wheat, *Triticum* spp. and Barley, *Hordeum vulgare*

At present, two species of nematodes—the cereal cyst nematode causing the "molya" disease and the seed gall nematode causing the "ear cockle" and "tundu" diseases—are of major importance on wheat.

### 2.2.1 Cereal Cyst Nematode, *Heterodera avenae*

Kuhn (1874) first observed in Germany that the cereal cyst nematode was as a parasite of cereals. Prasad et al. (1959) reported this nematode on wheat for the first time in India. The nematode is responsible for 6.3–95.3% loss in wheat in India (Seshadri and Sethi 1978), 60% loss in the Netherlands, and 30–40% loss in England (Gair 1965). The cereal cyst nematode is responsible for causing "molya" disease of wheat and barley.

#### 2.2.1.1 Distribution

The nematode is distributed in Rajasthan, Haryana, and some parts of Himachal Pradesh, Punjab, and western UP.

#### 2.2.1.2 Symptoms

The affected fields usually give a patchy appearance. Such patches gradually spread and may cover the whole field in 3 to 4 years with continuous cropping of wheat or barley. Above-ground symptoms of infected plants include stunted growth, general chlorosis, stiffer, thinner and narrower leaf blades, less tillering, thinner and weaker culms, premature flowering, and ears with very few grains. Damage is most severe during dry seasons.

The below-ground symptoms of infected plants include short roots with multiple branches, bunchy appearance, and often bearing small gall like formations. Large numbers of cysts show on the roots in February under Rajasthan conditions (Fig. 2.6).

**Fig. 2.6** *Left*—Infected (Bushy roots) and uninfected wheat plants with *Heterodera avenae*. Right—*H. avenae* infected roots showing pearly white females

The infected plants can be easily pulled out of the ground, since the fibrous root system is almost negligible.

#### 2.2.1.3 Life Cycle

The eggs hatch and second-stage juveniles emerge in soil. This coincides with germination of wheat seedlings with the advent of wheat sowing season. By the end of January/February, the juveniles invade the roots, develop, and reach female stage. The females lay eggs inside, turn into brown cysts, detach from roots, and fall in soil, as the crop matures.

#### 2.2.1.4 Host Range

Besides wheat and barley, the cereal cyst nematode infects maize, oats, rye, spelt, and other graminaceous plants. The only non-graminaceous host recorded is *Senebiera pinnatifida,* belonging to Cruciferae.

#### 2.2.1.5 Ecology

The critical threshold population was determined to be 2 larvae per 1 g of light soil. Gair (1965) showed that an initial inoculum level of 40 cysts per 100 g soil (32 eggs per g of soil) caused 30–40% reduction in wheat yield in England.

#### 2.2.1.6 Survival and Spread

The cysts can remain viable in the soil without a host for several years. Nematode survives during off season in the form of cysts enclosing eggs inside. The cysts remain dormant in soil till next *rabi* season.

Cultural operations, irrigation water, implements, and labors' feet can spread the cereal cyst nematode.

#### 2.2.1.7 Biotypes

Mathur et al. (1974) reported the presence of at least five biotypes in Rajasthan, based on the following host differentials (Table 2.1):

**Table 2.1** Biotypes of *Heterodera avenae* present in Rajasthan, India

| | Cyst formation on host differentials | | | | | |
|---|---|---|---|---|---|---|
| | | Barley | | | Oat | |
| Biotypes | *Lolium perenne* | Barley 191 | Herta | Drost | Sun II | *Dactylis glomerata* |
| Biotype 1 | + | – | – | – | – | – |
| Biotype 2 | – | + | + | – | – | – |
| Biotype 3 | – | – | + | – | – | – |
| Biotype 4 | – | – | + | – | – | + |
| Biotype 5 | – | – | + | + | – | – |

+ Cyst formed. – Cysts not formed

#### 2.2.1.8 Management

##### 2.2.1.8.1 Cultural Methods

Early Sowing

Early sown crop suffers less nematode damage because the crop attains better growth before sufficient numbers of juveniles are available to infect after nematode hatching from cysts. Hence, sowing of wheat in November first fortnight is recommended which yields better and can tolerate the nematode attack.

Crop Rotation

There is much scope for adopting crop rotation as a method of control, as the cereal cyst nematode is highly host specific. Crop rotation with nonhost crops like onion, mustard, carrot, fenugreek, and gram has shown that the nematode population gets reduced by 47–55% at the end of first year and by 75% by the end of second year. Wheat yield increased by 83% after 1 year and by 135% after 2 years of crop rotation. Crop rotation with mustard, chickpea, and seed spices for 2–3 years was also found beneficial.

Opting for crop rotation with rice reduced the number of cysts substantially. The population of this nematode has considerably been reduced by the introduction of rice in rice-wheat cropping system in Indo-Gangetic Plains. In Punjab, when an infested field with 36 cysts of *H. avenae*/250 cc soil was brought under paddy-wheat rotation, the cyst population decreased to 5 cysts/250 cc soil within 5 years.

Intercropping

The nematode population can be reduced by intercropping wheat with *Brassica campestris* (crucifer).

##### 2.2.1.8.2 Chemical Methods

Application of Carbofuran (3G) at 1 kg a.i. /ha at sowing was found beneficial.

##### 2.2.1.8.3 Biological Methods

Seed treatment with *Azotobacter chroococcum* (strain HT 54) was found effective.

##### 2.2.1.8.4 Host Resistance

Host resistance and PV 18, and barley variety JB 226 tolerant to the nematode, could be raised on infested soil since they give comparatively higher yields (Mukhopadhyaya et al. 1972). Wheat variety Raj MR-1 and barley cvs. C-164 and Raj Kiran were also found resistant.

##### 2.2.1.8.5 Integrated Methods

Ahamad and Ahamad (2014) found that the best control of cereal cyst nematode was obtained by integration of Fenamiphos (4.0 kg a.i. /ha) and urea (600 kg/ha in three split doses). This treatment resulted in the most significant effect in decreasing the number of cysts/root system and increasing the growth of nematode-infected wheat plants.

### 2.2.2 Seed Gall Nematode, *Anguina tritici*

As early as in 1743, Needham reported the first plant parasitic nematode (seed gall nematode). In India, this nematode is still one of the most important plant parasites of wheat. Gupta and Swarup (1972) found that the seed gall nematode, besides being responsible for the ear cockle disease, is often associated with a bacterium *Rathayibacter tritici* in causing a yellow ear rot commonly referred to as the "tundu" disease.

#### 2.2.2.1 Crop Losses

The sowing of seeds contaminated with galls at 2.5%, 6.5%, and 8.5% of total weight resulted in yield decreases of 30%, 54%, and 69%, respectively. *Tundu* disease is responsible for causing 52% yield losses in wheat.

#### 2.2.2.2 Symptoms

Basal part of the stem is slightly enlarged in infested seedlings. The leaves emerging from such seedlings are twisted and crinkled, often folded frequently, with their tips held near the growing point (Fig. 2.7). Stunted growth and premature death occur in severely infested seedlings. The infested plants generally show profuse tillering and may produce ears almost 30–40 days in advance compared to the healthy plants. Other usual symptoms of infected plants include shorter ears, glumes spread farther apart by the galls that replace the kernels, and smaller dark seeds (Fig. 2.7).

The symptoms of "tundu" disease include the production of a light-yellow slime or gum on the leaf surfaces of young plants, and also on the abortive ears (Fig. 2.8). In humid weather, this yellow slime can be seen trickling down the tissues, which

**Fig. 2.7** Symptoms of *Anguina tritici* infestation on wheat seedling (twisted and crinkled seedling leaves) (*Left*), ear heads (*Middle*), and seeds (*Right*) (*Top*—infected seeds; *Bottom*—healthy seeds)

**Fig. 2.8** Symptoms of "tundu" disease of wheat (twisting and distortion of ear head and rotting of spikelets with profuse oozing of yellow liquid from the infected tissue)

becomes hard, brittle, and brown on drying. Infested culms either die at the young stages or grow until heading. In the latter case, either it fails to emerge out of the boot-leaf or the emerging spike is narrower and shorter, with the grains partially or completely replaced by the bacterial mass. When the ear shows the bacterial symptoms, the stalk is always distorted.

#### 2.2.2.3 Life Cycle

During harvest, the seed galls fall to the ground and breakdown in the soil to release second-stage larvae. The number of larvae in a gall may vary from 3000 to 12,000 (with an average of 6000) depending upon the gall size. The larvae invade seedlings of the host plant, making their way up the shoot to live ectoparasitically round the growing point and its leaf sheath. The nematodes invade the embryonic flower tissues and change from ectoparasitic to an endoparasitic existence. In the developing seed, larvae develop into adults which become a gall. The females lay thousands of eggs after the mating occurs. The eggs hatch after the adults die. The larvae molt once in the gall, so that when they leave it to infect new plants, they are in the second stage. The second-stage larvae in seed galls can live for many years.

#### 2.2.2.4 Host Range

Hosts susceptible to injury by the seed gall nematode include wheat, rye, emmer (*Triticum dicoccum*), and spelt (*T. spelta*). Oats and barely are immune for all practical purposes.

#### 2.2.2.5 Survival and Spread

The most important method of spread and the only one whereby this nematode is likely to be carried for any considerable distance is by galls in seed grains.

In dried seed galls, the second-stage larvae can live for many years. The larvae from stored galls were revived after 25 (Needham 1743) and 28 years (Fielding 1951).

### 2.2.2.6 Management

#### 2.2.2.6.1 Regulatory Methods

Dry Cleaning of Seeds or Brine Flotation

The use of clean seed free of nematode galls is the surest way to control the diseases, since the source of inoculum for both ear cockle and "tundu" diseases being the seed material contaminated with nematode galls. These diseases can be managed by adopting dry-cleaning methods like sieving and winnowing, or by seed floating in plain water, or, preferably, in 20% brine solution.

Seed Certification

Use of certified seed free from nematode contamination helps to control these diseases. In order to prevent the recurrence of the disease, seed supplying agencies, both in the private and public sectors, have to ensure the distribution of certified seeds.

#### 2.2.2.6.2 Physical Methods

Hot water treatment of seed presoaked for 4–6 h in cold water at 54 °C for 10 min gives effective control.

#### 2.2.2.6.3 Cultural Methods

Crop Rotation

Crop rotation with nonhost crops for 1 or 2 years effectively eliminates nematodes remaining in the field after harvest.

Rogueing

Infected plants can be easily detected in early stages and should be removed and destroyed from the field.

#### 2.2.2.6.4 Host Resistance

Winter wheat cv. "Kenred-hard" is reported to be resistant and useful to control the seed gall nematode.

## 2.3 Maize, *Zea mays*

### 2.3.1 Cyst Nematode, *Heterodera zeae*

The maize cyst nematode (*Heterodera zeae*) is considered to be one of the most important nematode pests of maize in India (Mehta et al. 2016).

#### 2.3.1.1 Distribution

The nematode is distributed in Rajasthan, Haryana, Himachal Pradesh, Punjab, Bihar, and UP.

#### 2.3.1.2 Crop Losses

The nematode is responsible for yield losses of maize to the extent of 17–29% in India at varied inoculum levels and soil conditions (Srivastava and Chawla 2005).

#### 2.3.1.3 Symptoms

- Stunted growth in patches (Fig. 2.9).
- Yellowing of leaves (Fig. 2.9).
- Root appears bushy and poorly developed.
- Tassel earlier, bear small cobs with fewer grains.
- Female cyst nematodes (Fig. 2.10) are attached to the roots.

#### 2.3.1.4 Host Range

Maize and sorghum are susceptible to *Heterodera zeae*.

#### 2.3.1.5 Biology and Life Cycle

The second-stage larvae (the infective stage) of maize cyst nematode *H. zeae* penetrated sweet corn roots at 48 h of inoculation. After penetration, the third- and fourth-stage larvae were observed on sixth and 13th day of inoculation, respectively. Adult females and mature cyst of *H. zeae* were observed on 19th and 25th day after

**Fig. 2.9** Rows of stunted plants in maize cyst nematode-infested field

**Fig. 2.10** Cysts of *Heterodera zeae*

inoculation, respectively. The maize cyst nematode completes its life cycle on sweet corn in 25 days after inoculation of second-stage larvae.

Second-stage juveniles hatch out in soil and invade maize roots during *kharif* season. During *rabi* season, the cysts remain in soil protecting eggs inside. In one maize cropping season, the nematode completes 5–6 generations, being multivoltine.

#### 2.3.1.6 Survival and Spread

At crop maturity, the females with eggs inside turn into cysts and fall into soil and remain dormant till next *kharif* season. The nematode spreads through irrigation water and agronomic practices.

#### 2.3.1.7 Management

- Deep summer plowing 2–3 times at 10–15 days' interval.
- Crop rotation for 2 years with non-cereal crops like vegetables, oilseeds, and pulses would bring down the nematode population below economic threshold levels.
- Maize cvs. Ageti 71 and Karnal-1 are reported to be moderately resistant to the cyst nematode.
- Seed treatment with Carbosulfan (25DS) at 3% a.i. w/w or neem-based formulation (Achook or neem seed kernel powder) at 5%.
- Soil application of Carbofuran (3G) at 1–2 kg a.i. /ha at the time of sowing.
- The practice adopted by farmers in the eastern parts of the country that combines soil application of mustard cake along with tobacco dust at 250 kg/ha was as efficient as soil treatment with Carbofuran at 2 kg a.i. /ha in increasing crop yield and suppressing the nematode populations of *H. zeae*.
- Baheti et al. (2017) found that integration of *Purpureocillium lilacinum* at 2% w/w along with neem leaf powder at 2 g/plant was found most effective for the management of maize cyst nematode followed by *Pochonia*

*chlamydosporia* + neem leaf powder, and *Purpureocillium lilacinum* + karanj leaf powder.

### 2.3.2 Root-Knot Nematodes, *Meloidogyne incognita* and *M. javanica*

#### 2.3.2.1 Symptoms

Stunting, yellowing, and wilting are the common symptoms of severe root-knot nematode injury in above-ground corn plant parts. The nematode may not cause any perceptible symptoms but can cause significant damage to the crop and reduction in yield. Under-ground symptoms observed on the roots include swollen areas or galls (Fig. 2.11), lack of fine roots, and reduced root branching.

#### 2.3.2.2 Management

One of the best methods for managing root-knot nematodes includes rotation of corn with peanut. Similarly, resistant varieties of soybean and cotton could also be successfully rotated with corn to reduce nematode populations. Likewise, root-knot nematode populations can be reduced by more diverse rotations including alfalfa or oats. Riekert (1996) reported that soil application of Carbofuran nematicide was found effective where *M. incognita* and *M. javanica* dominated as either single or mixed populations and increased the yield, ranging from 50 to 500 kg $ha^{-1}$.

Seed treatment with biocontrol agents like Avicta 500FS (that contains secondary metabolites of the soil-inhabiting bacterium *Streptomyces avermitilis* as a.i.) and Poncho (VOTiVo) (with *Bacillus firmus* as the a.i.) were found effective against root-knot nematodes (Anon 2016).

**Fig. 2.11** *Left*—Galled and stunted maize roots. *Right*—Root-knot nematode females (indicated by white circle) visible at the junction of the tap and secondary roots

**Fig. 2.12** *Left*—Injured patches of maize resulting from lesion nematode feeding. *Right*—Lesions on maize roots due to lesion nematode infection

### 2.3.3 Lesion Nematode, *Pratylenchus zeae*

#### 2.3.3.1 Symptoms

Symptoms induced by the lesion nematode include above-ground plant stunting, leaf yellowing and discoloration, reduction in plant height and chlorophyll content in the leaves, lesions on roots, and severe root pruning (Fig. 2.12).

The root damage by lesion nematodes can often be diagnosed by the presence of small lesions on the root surface. The poor uptake of nutrients and water by the plant is mainly due to severe necrosis and damage to root hairs.

#### 2.3.3.2 Management

Crop rotation with nonhost crops has been shown to be effective at reducing lesion nematode numbers. Soil application of Carbofuran resulted in yield increases ranging from 28% to 42% where *P. zeae* dominated (Walters 1979).

## 2.4 Sorghum, *Sorghum bicolor*

### 2.4.1 Root-Knot Nematode, *Meloidogyne incognita*

#### 2.4.1.1 Crop Losses

The root-knot nematode is responsible for yield losses ranging from 15% to 30%.

#### 2.4.1.2 Symptoms

The root-knot nematode symptoms on grain sorghum include production of very small galls that may be difficult to identify. The galls are usually produced on the small lateral roots and not on the brace roots (Fig. 2.13).

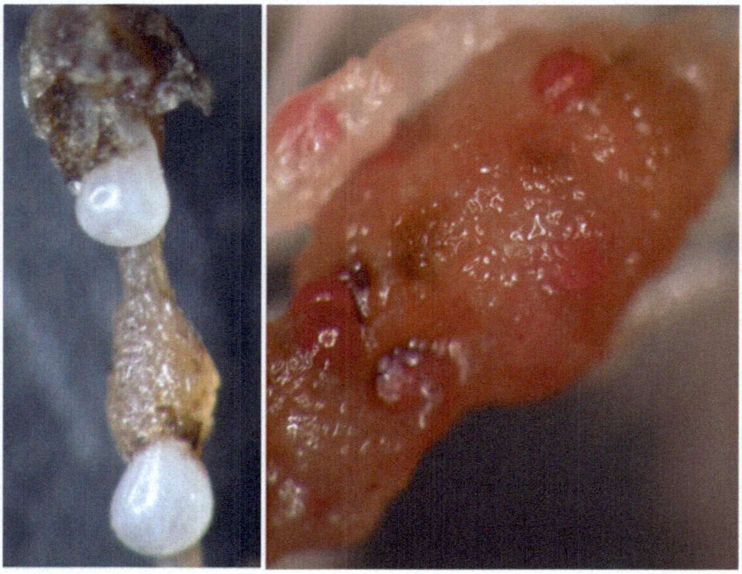

**Fig. 2.13** *Left*—Small galls on sorghum root with protruding developing females. *Right*—Larger galls with multiple root-knot females (stained red) inside

#### 2.4.1.3 Management

Seed treatment with nematicides such as Counter 15G or 20G and Poncho/Votivo was found effective against root-knot nematodes. Use of moderately resistant sorghum cv. REV RV9782 is beneficial against *M. incognita*.

### 2.4.2 Lesion Nematode, *Pratylenchus neglectus*

#### 2.4.2.1 Symptoms

The above-ground symptoms include stunting of plants, yellowing of lower leaves, poor tillering and wilting in the plant shoots, and nutrient deficiency. This may be due to the inefficiency of the damaged roots in taking up water and nutrients—particularly nitrogen (N), phosphorus (P), and zinc (Zn). The below-ground symptoms include dark discolored lesions on the root system, poor root structure (Fig. 2.14), and yield loss.

#### 2.4.2.2 Management

Crop rotation of grain sorghum with other nonhost crops like cotton, corn, green gram, sunflower, or soybeans helps to reduce nematode pressure from *Pratylenchus neglectus*. The nematode populations can also be reduced by crop rotation with tolerant and partially resistant wheat cultivars, together with crops such as sunflower, millets, and canary seed.

**Fig. 2.14** Lesions on sorghum roots caused by the lesion nematode

Other management practices that were found to limit crop damage include fallowing, deep plowing 2–3 times, and seed treatment with Carbofuran. Soil amendment with organic matter helps in improving soil health and biological suppression of root-lesion nematodes.

### 2.4.3 Cyst Nematode, *Heterodera sorghi*

#### 2.4.3.1 Symptoms

Affected crop gives unhealthy appearance with pale greenish yellow stunted plants. On uprooting the plants, elongated main root with excessive branching at the tip gives a twiggy appearance. The sorghum cyst nematode females develop into cysts with the onset of plant senescence (Fig. 2.15).

#### 2.4.3.2 Life Cycle

The larvae penetrate the root after emergence from the cysts at a wide range of temperatures (28–36 °C). The sorghum cyst nematode *H. sorghum* completes its life cycle in 20–24 days at 28–36 °C (Srivastava 1985). The nematode produces up to four generations per season. The males are vermiform and are inhabitants of soil, while adult female nematode is sedentary and remains in or attached to the roots. Mature female nematodes exhibit a unique characteristic of retaining eggs and larvae within their body. After its death, nematode cuticle gets hardened, protecting the

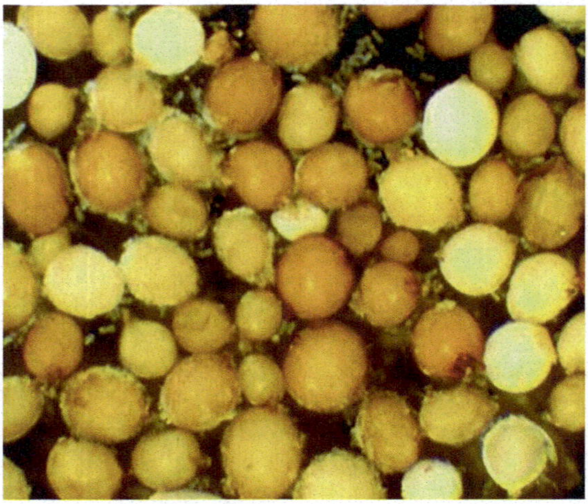

**Fig. 2.15** Cysts of *Heterodera sorghi*

infective units from adverse environmental stress. Each cyst contains about 200 eggs.

#### 2.4.3.3 Survival

The eggs in brown cysts can survive in the soil for few years.

#### 2.4.3.4 Management

The strategies that can be used to manage nematode include crop rotation with nonhost crops, deep plowing, and summer fallow.

## 2.5 Conclusion

In order to keep plant parasitic nematode population below damage threshold levels, the most important challenge to nematologists is to find suitable, effective, and sustainable measures for cereal crops. The uncertainty of how widely economically important plant parasitic nematodes are distributed and what the effects of different species have on each crop is an important nematological aspect relating to all cereal crops produced. Concrete proof and sound economic bases of damage could change general perceptions and bring greater benefits to farmers and related concerns in the respective crop industries. The aspect which should receive high priority in terms of research includes finding differences in crop-genotype susceptibility to the main nematode pests.

The effect that different forms of soil tillage might have on total nematode community structures and compositions under various cropping systems and abiotic conditions is another prominent knowledge gap. The need for investigations on interactions between root pathogens (specifically fungi) and nematode pests is another closely related aspect. Another important variable is to investigate that involve all possible crops and rotation-system variations under different environmental conditions. The development and adoption of approaches like precision agriculture (site specific treatment), population dynamics of nematodes, and several other aspects would provide invaluable basic information about plant nematology that would previously have been very difficult to justify investigating.

## References


Ahamad SAH, Ahamad AMD (2014) Effect of urea and certain NPK fertilizers on the cereal cyst nematode (Heterodera avenae) on wheat. Saudi J Bio Sci 21:191–196

Amarasinghe LD (2012) An integrated approach to the management of rice root knot nematode, *Meloidogyne graminicola* in Sri Lanka. J Sci Univ Kelaniya Sri Lanka 6:55–63

Anita B, Samiyappan R (2012) Induction of systemic resistance in rice by *Pseudomonas fluorescens* against rice root knot nematode *Meloidogyne graminicola*. J Biopest 5:53–59

Anon (2016) Biennial report (2014–2015). AICRP on Nematode pests of crops and their control, Assam Agricultural University, Jorhat

Atkins JG, Todd EH (1959) White tip disease of rice III. Yield tests and varietal resistance. Phytopathology 49:189–191

Baheti BL, Dodwadiya M, Bhati SS (2017) Eco-friendly management of maize cyst nematode, *Heterodera zeae* on sweet corn (*Zea mays* L. *saccharata*). J Entomol Zool Stud 5(6):989–993

Bridge J, Page SLJ (1982) The rice root-knot nematode, *Meloidogyne graminicola*, on deep water rice (*Oryza sativa* sub sp. *indica*). Revue de Nematologie 5:225–232

Butler EJ (1913) *Ufra* disease of rice. Agric J India 8:205–220

Charles JSK, Venkitesan TS (1990) Host records of the rice cyst nematode *Heterodera oryzicola* Indian J Nematol 20(2):222–224

Choudhury BN, Bhagawati B, Bora BC (2011). Three decades of nematology in Assam (1977–2011). AICRP on plant parasitic nematodes with integrated approach for their control. Department of Nematology, Assam Agricultural University, Jorhat, 49 pp

Cralley EM (1957) The effect of seeding methods on the severity of white tip of rice (Abstr.). Phytopathology 47:7

Das P, Deka BC (2002) Efficacy of neem-based pesticides against *Meloidogyne graminicola* on rice as seed treatment. Indian J Nematol 32:203–205

Dastur JP (1936) A nematode disease of rice in the central provinces. Proc Indian Acad Sci 4:108–121

Fielding MJ (1951) Observations on the length of dormancy in certain plant infecting nematodes. Proc Helminth Soc Wash 18(2):110–112

Gair R (1965) Cereal root eelworm. In Southey JF (ed) Plant nematology. Technical Bulletin No. 7, Ministry of Agriculture, Fisheries Foundation, London, pp 199–211

Ganguly AK, Pankaj S, Sirohi A (1996) Effect of soil solarization of rice nursery beds to suppress plant parasitic nematodes. Int Rice Res Notes 21:80–81

Garg RN, Gaur HS, Singh GC (1995) Effect of tillage practices and water regimes on the soil physical properties and plant parasitic nematodes in rice. Ann Plant Prot Sci 3:121–126

Gaur HS (2003) Root-knot disease of rice and wheat: problem and management. Technical Bulletin (TB – ICN: 1/2003), Indian Agricultural Research Institute, New Delhi, 23 pp

Goto K, Fukatsu R (1956) Studies on the white tip of rice plant. Inst Agri Sci (Japan) Ser C, No 6

Gupta P, Swarup G (1972) Ear-cockle and ear root diseases of wheat. II nematode - bacterial association. Nematologica 18:320–324

Jayaprakash A, Rao YS (1983) Reaction of rice cultivars against the cyst nematode, *Heterodera oryzicola*. Indian J Nematol 13:117–118

Jayaprakash A, Rao YS (1984) Cyst nematode, *Heterodera oryzicola* and seedling blight fungus, *Sclerotium rolfsii* disease complex in rice. Indian J Nematol 14(1):58–59

Kalita M, Phukan PN (1996) Effect of crop rotation on the development of *Meloidogyne graminicola* on rice. Indian J Nematol 25:206–207
Kinh DN, Houng NM, Ut NU (1982) Root-knot disease of rice in the Mekong Delta, Vietnam. Int Rice Res Newsl 7(4):15
Kuhn J (1874) Uber des Vorkommen von Ruben - Nematoden an den wurzeln der Halmfruchte. Landw Jbr 3:47–50
Kumari U, Kuriyan KJ (1981) Cyst nematode, *Heterodera oryzicola*, on rice in Kerala. I. Estimation of loss in rice due to *H. oryzicola* infestation, in field conditions. Indian J Nematol 11:106
Kuriyan KJ (1985) Annual research progress report of AICRP on nematode pests of crops and their control. Kerala Agricultural University, Vellayani, Kerala
Mathur VK, Arya HC, Mathur RL, Handa DK (1974) The occurrence of biotypes of cereal cyst nematode (*Heterodera avenae*) in the light soils of Rajasthan and Haryana, India. Nematologica 20:19–26
Mehta SK, Baheti BL, Nama CP, Rathore BS (2016) Farmer friendly integrated management strategy for maize cyst nematode, *Heterodera zeae* on maize (*Zea mays* L.). Prog Res 11 (Special-I):407–410
Mukhopadhyaya MC, Dalal MR, Saran S, Kharub SS (1972) Studies on the 'molya' disease of wheat and barley. Indian J Nematol 2:11–20
Needham T (1743) A letter concerning chalky tubulous concretions called malm: with some microscopical observations on the fauna of the red lily and of worms discovered in smutty corn. Philos Trans of Roy Soc Lond 42(173–174):634–641
Pankaj, Ahlawat JS, Saha M (2006) Predominant nematode pests of rice nursery in North-Western India. In: Second International Rice Congress, Indian Agricultural Research Institute, New Delhi
Pathak KN, Kumar B (2003) Effect of culture filtrates of *Gliocladium virens* and *Trichoderma harzianum* on the penetration of rice roots by *Meloidogyne graminicola*. Indian J Nematol 33:149–151
Prasad N, Mathur RL, Sehgal SP (1959) Molya disease of wheat and barley in Rajasthan. Curr Sci 28:453
Prasad JS, Panwar MS, Rao YS (1986) Screening of some rice cultivars against the root-knot nematode, *Meloidogyne graminicola*. Indian J Nematol 16:112–113
Rao YS (1985) Research on rice nematodes. In: Padmanabhan SY (ed) Rice in India. ICAR Monograph, New Delhi, pp 591–615
Rao YS, Prasad JS, Yadava CP, Padalia CR (1984) Influence of rotation crops in rice soils on dynamics of plant parasitic nematode populations. Biol Agri Hort 2:69–78
Rao YS, Prasad JS, Panwar MS (1986) Nematode problems in rice: crop losses, symptomatology and management. In: Swarup G, Dasgupta DR (eds) Plant parasitic nematodes of India – problems and progress. Indian Agricultural Research Institute, New Delhi, pp 279–299
Renitha Govind, Ushakumari R (2010) Efficacy of different bioagents in controlling the root-knot nematode (*Meloidogyne graminicola* Golden and Birchfield) of rice under dry condition. National conference on Innovations in nematological research for agriculture sustainability – challenges and A Roadmap Ahead, Tamil Nadu Agricultural University, Coimbatore
Riekert HF (1996) Economic feasibility of nematode control in dry land maize in South Africa. Afr Crop Sci J 4:477–481
Roy AK (1976) Effect of decaffeinated tea waste and water hyacinth compost on the control of *Meloidogyne graminicola* on rice. Indian J Nematol 6(1):73–77
Saikia H (1991) Management of Rice Root Nematode *Hirschmanniella oryzae*. M.Sc. thesis, Assam Agricultural University, Jorhat, 88 pp
Seshadri AR, Sethi CI (1978) Nematode problems. In: Wheat research in India. ICAR, New Delhi, pp 168–187
Shanmuga Priya M (2015) Biomanagement of rice root knot nematode, *Meloidogyne graminicola* Golden and Brichfield in aerobic rice. Int J Mangmt Soc Sci 3(4):591–598

Soriano IR, Reversat G (2003) Management of *Meloidogyne graminicola* and yield of upland rice in South-Luzon, Philippines. Nematology 5(6):879–884

Soriano IR, Espiritu MJ, Schmit V et al (1998) Resistance to rice root-knot nematode *Meloidogyne graminicola* in *Oryza longistaminata* and *Oryza glaberrima*. Philippine J Crop Sci 23:89

Soriano IR, Prot JC, Matias DM (2000) Expression of tolerance for *Meloidogyne graminicola* in rice cultivars as affected by soil type and flooding. J Nematol 32:309–317

Srivastava AN (1985) On the biology of *Helerodera sorghi* (Abstr.). Indian J Nematol 15:292

Srivastava AN, Chawla G (2005) Maize cyst nematode, *Heterodera zeae*: a key nematode pest of maize and its management. Indian Agricultural Research Institute, New Delhi. 18 pp

Swain BN, Prasad JS (1988) Chlorophyll content in rice as influenced by the root-knot nematode, *Meloidogyne graminicola* infection. Curr Sci 57:895–896

Voung HH (1969) The occurrence in Madagascar of the rice nematodes, *Aphelenchoides besseyi* and *Ditylenchus angustus*. Tech Comm Commonw Bur Helminthol 40:274–288

Walia RK, Chakraborty PK (eds) (2018) Nematodes problems of crops in India. ICAR-All India Coordinated Research Project on Nematodes in Agriculture, New Delhi, 402 pp

Walters MC (1979) Present status of knowledge of nematode damage and control in South Africa. Technical Communication No 152. Department of Agricultural Technical Services, Pretoria, pp 62–68

# Part III

# Pulse and Oil Seed Crops

# Pulse Crops

# 3

**Abstract**

Nematode diseases are one of the main biotic constraints in reducing the quantity and quality of pulse crops. Root-knot nematodes (*Meloidogyne* spp.), cyst nematodes (*Heterodera* spp.), reniform (*Rotylenchulus reniformis*), and lesion nematodes (*Pratylenchus* spp.) are the major nematode pests that attack pulse crops (chickpea, pigeon pea, green gram, and black gram). Nematodes have also been found to interact with pathogenic soil-borne fungi and increase the severity of the disease pathogens. Seed treatment can be the option to minimize the chemical use for nematode management. For effective management of nematodes under field conditions, growing of marigold, *Tagetes* spp., as an intercrop was found effective. Several methods have been used to manage the pulse nematodes by integrating different farming practices like using cultural and physical control methods, encouraging naturally occurring biological control agents, and host plant resistance.

**Keywords**

*Meloidogyne* spp. · *Heterodera* spp. · *Pratylenchus* spp. · Symptoms · Management

## 3.1 Chickpea, *Cicer arietinum*

The most important nematode pests affecting chickpea crop include root-knot nematodes (*Meloidogyne incognita* and *M. javanica*), the lesion nematode (*Pratylenchus thornei*), and the reniform nematode (*Rotylenchulus reniformis*)



P. P. Reddy, *Nematode Diseases of Crops and their Management*, https://doi.org/10.1007/978-981-16-3242-6_3

## 3.1.1 Root-Knot Nematodes, *Meloidogyne incognita* and *M. javanica*

### 3.1.1.1 Crop Losses

The root-knot nematode is responsible for annual yield loss of 21% in chickpea amounting to ₹4867.27 million in India (Walia and Chakraborty 2018).

Ali (2009) reported 25.6% chickpea yield loss in a field heavily infested with *M. javanica*. Infestation of *M. incognita* causes yield losses from 12% to 23% in sandy soil than sandy loam or loamy sand soils (Ali 1992).

### 3.1.1.2 Symptoms

The above-ground common symptoms which can be observed under field conditions are poor and uneven growth of chickpea plants in patches (Fig. 3.1), stunted growth of plants with lesser branches, and pale green leaves. The presence of knots or galls (uneven swellings) on the root system is the common below-ground symptom. Galls are produced mostly on root tips and on their vicinity, but all along the roots the beaded appearance of multiple galls (the result of coalescing of adjacent galls) is also common (Fig. 3.2). Plants which are heavily galled show wilting symptoms under field conditions. A severe infestation by root-knot nematode finally results in poor pod formation with deformed seeds that are less developed, and smaller in size and weight (Ali 1995). The roots of infested plants bear fewer rhizobium nodules in comparison to healthy ones. Interestingly, bacterial nodules of heavily galled root were also found infested with root-knot nematode.

**Fig. 3.1** Sparse, uneven stunted growth and yellowing of leaves of chickpea infested with root-knot nematode

**Fig. 3.2** Chickpea roots showing root-knot galls

#### 3.1.1.3 Life Cycle

*M. incognita* and *M. javanica* complete their life cycle in 27–32 days at an optimum temperature of 25–30 °C. During the cropping season, more than one life cycle is completed by the nematode. Fertilized females lay on average 200–500 eggs in a gelatinous matrix.

#### 3.1.1.4 Interaction with Other Pathogens

The root-knot nematode interacts and aggravates the severity of many diseases such as *Fusarium* wilt, black root rot, collar rot, and dry root rot in chickpea. *M. incognita* or *M. javanica* can break down resistance of the wilt fungus *F. oxysporum* f. sp. *ciceri* in wilt-resistant genotypes of chickpea (Mani and Sethi 1987; Uma Maheswari et al. 1995, 1997; Krishna Rao and Krishnappa 1996). *M. javanica* modifies the resistance of chickpea cv. Avrodhi to *Fusaruim oxysporum*, and probably the nematode predisposes the plant to fungal infection in northern India (Upadhyay and Dwivedi 1987; Ali and Gurha 1995).

#### 3.1.1.5 Management

##### 3.1.1.5.1 Cultural Methods

The root-knot nematode population can be reduced by late sowing of chickpea and intercropping of chickpea with mustard (2:1).

The inclusion of nonhost crop like sesame, mustard, groundnut, and winter cereals for 2–3 years in the cropping system may be useful in reducing the population of root-knot nematodes *M. incognita* and *M. javanica* in chickpea (Sharma et al. 1992a; Sikora et al. 2005).

#### 3.1.1.5.2 Chemical Methods

Seed treatment with Carbosulfan 25DS at 3% a.i. w/w was found effective for the control of root-knot nematode.

Seed treatment of Carbofuran 25ST at 3% a.i. w/w + Carbendazim at 0.2% gave effective control of root-knot nematode and wilt disease complex.

Kaushik and Bajaj (1981) reported that seed treatment with Carbofuran and Fenamiphos at 1, 2, and 4% w/w reduced the number of root galls on chickpea 42 days after sowing.

Soil application of Carbofuran at 1–2 kg a.i. /ha in root-knot nematode-infested soil and increased the yield of chickpea.

#### 3.1.1.5.3 Biological Methods

The root-knot nematode can be commercially and very successfully managed by using fungal biocontrol agents like *Aspergillus niger, Trichoderma harzianum, Purpureocillium lilacinum,* and *Pochonia chlamydosporia* (Askary 2015).

Pant et al. (2004) found that *Glomus fasciculatum* (AMF) was very effective for the management of *M. incognita.*

#### 3.1.1.5.4 Host Resistance

Three chickpea cultivars (N 31, N 59, and ICCC 42) and a promising chickpea breeding line (ICCV 90043) were found tolerant to the root-knot nematode.

Chickpea accessions such as Phule G-00108, Phule G-00109, Phule G-94259, Phule G-96006, and PDG-84-16 were found resistant, while Phule G-00110, Phule G-94091, H-82-2, IPCK-256, IPC-2001-02, and HR-00-299 were moderately resistant to both *M. incognita* and *F. oxysporum* (Haseeb et al. 2006).

#### 3.1.1.5.5 Integrated Methods

Integration of soil application of neem cake at 100 kg along with Trichoderma *viride* ($2 \times 10^6$ cfu/g) at 2.5 kg + 200 kg FYM/ha at sowing gave effective management of root-knot nematodes.

Pandey et al. (2005) found that integration of *P. lilacinum* along with neem cake, besides causing a reduction in root galling, improved the plant growth characters and chlorophyll content.

Integration of *P. lilacinum* KIA with *Rhizobium* was found most effective in reducing the root galls and nematode multiplication (Siddiqui and Akhtar 2009).

Chakrabarti and Mishra (2001) reported that integration of seed treatment with neem seed powder at 10% w/w and soil application of systemic nematicides (Carbofuran at 1 kg. a.i. /ha) was found most effective against *M. incognita* infecting chickpea.

Integration of farmyard manure (FYM) (5 t/ha), neem cake (150 kg/ha), and fungal biocontrol agents *P. lilacinum* and *Aspergillus niger* proved most effective in suppressing the soil population of root-knot nematode *M. incognita* in a chickpea field experiment (Singh et al. 2011).

## 3.1.2 Lesion Nematode, *Pratylenchus thornei*

### 3.1.2.1 Distribution

The nematode is restricted to Bundelkhand (parts of Uttar Pradesh and Madhya Pradesh).

### 3.1.2.2 Crop Losses

The lesion nematode was responsible for 13% yield loss in chickpea (Walia and Chakraborty 2018). Di Vito et al. (1992) reported that the nematode recorded an estimated yield loss of 58% in chickpea at 2 nematodes/$cm^3$ soil.

### 3.1.2.3 Symptoms

In young seedlings, the most visible above-ground symptoms observed in the field are stunted growth and pale green foliage, which becomes more prominent as the plant ages. Root-lesion nematodes are migratory endoparasites. The below-ground symptoms include dark brown to black necrotic lesions on the epidermal, cortical, and endodermal cells of chickpea roots (Fig. 3.3). With the passage of time, these necrotic spots coalesce, leading to the development of necrosis in the entire root.

### 3.1.2.4 Biology and Life Cycle

The root-lesion nematodes *Pratylenchus* spp. complete their entire life cycle inside the roots, since they are migratory endoparasites. The life cycle is completed in about 1 month under suitable environmental conditions. Askary et al. (2012) recorded that four molts take place: the first within the egg and three outside. All the life stages except the J1 are parasitic.

### 3.1.2.5 Host Range

The root-lesion nematode infects chickpea, soybean, and wheat.

### 3.1.2.6 Interaction with Other Pathogens

The nematode plays the role of a predisposer to fungal infection. *P. thornei* interacts with *Fusarium oxysporum* f. sp. *ciceri* leading to accentuation of the problem.

**Fig. 3.3** Necrotic lesions on chickpea roots due to lesion nematode infection

#### 3.1.2.7 Survival and Spread

The nematode undergoes anhydrobiotic stage and survives under stress conditions, especially a dry fallow season of soil (Talavera and Valor 2000).

The nematodes spread through cultural practices and irrigation water.

#### 3.1.2.8 Management

##### 3.1.2.8.1 Cultural Methods

*Pratylenchus thornei* in chickpea can be effectively managed by seed treatment with NSKP at 10% w/w.

The reduction in root populations of *P. thornei* as well as significant increase in the shoot and root mass of chickpea was achieved by soil application of mustard cake (Sebastian and Gupta 1996).

##### 3.1.2.8.2 Chemical Methods

Seed treatment of chickpea with Carbosulfan at 0.1% and 3% w/w significantly reduced the population of *Pratylenchus thornei* by 40–60% and increased the yield by 35–45%.

Sebastian and Gupta (1997) found that application of Carbofuran and Phorate each at 2 kg a.i. /ha also reduced the population of *P. thornei* in chickpea to the extent of 62% and increased the yield by 21%.

##### 3.1.2.8.3 Biological Methods

Seed treatment of with *Trichoderma viride* at 10 g/kg seed is effective against *Pratylenchus thornei*, while soil treatment with *T. harzianum* at 2.5 kg/ha gave 62% reduction in population of lesion nematode and increased the yield of chickpea by 15%.

##### 3.1.2.8.4 Host Resistance

Tiwari et al. (1992) recorded that the chickpea lines GNG 543, GF 88428, and PKG-24 were found highly resistant to *P. thornei*.

##### 3.1.2.8.5 Integrated Methods

Integration of seed treatment with neem seed kernel powder at 5 g along with *T. viride* at 5 g/kg seed effectively reduced the population of lesion nematode by and 60–71%, thereby increasing the yield of chickpea by 20–53%.

Soil application of neem cake at 10 g/m$^2$ along with *T. viride* at 2.5 kg effectively reduced the population of lesion nematode by 64–68% and increased the yield of chickpea by 24%.

Tiyagi and Ajaz (2004) reported that integration of *Purpureocillium lilacinum* with neem cake was found to be promising in reducing the population of root-lesion nematode in chickpea under field conditions.

Significant reduction of nematode population and subsequent increase in the yield by 15% was obtained by summer plowing of chickpea field infested with *P. thornei* during peak summer coupled with soaking of chickpea seed with Carbosulfan at 0.1% for 4 h.

Soil application of *Trichoderma harzianum* at 2.5 kg/ha along with *Pochonia chlamydosporia* at 10 kg/ha and FYM 1 week before sowing of seed was found effective against root rot of chickpea caused by *Pratylenchus thornei* in combination with *Fusarium* species. In farmers' practice, the net profit was about ₹15,200/ha, and through adoption of this demonstrated technology, it was about ₹26,700/ha.

Dwivedi et al. (2008) recorded that the decrease of soil nematode population by 34.8% and 22.3% and increase in yield from 25.0% to 29.5% over untreated control were obtained by integration of soil application of *Pseudomonas fluorescens* and *Trichoderma viride* at 2.5 kg/ha soil along with seed treatment at 5 g/kg seed.

### 3.1.3 Reniform Nematode, *Rotylenchulus reniformis*

Ali (1993) opined that the reniform nematode is one of the most economically important species which has been reported on chickpea from India.

#### 3.1.3.1 Crop Losses

The nematode was responsible for the loss in yield in chickpea to the extent of 80% in the presence of 10 nematodes/g soil (Mahapatra and Padhi 1986).

#### 3.1.3.2 Symptoms

The above-ground symptoms on infected plants include patchy field appearance, stunted growth, early senescence, pale green leaves, and uneven plant growth at the younger stage. The remarkable symptom of infection caused by reniform nematode on chickpea plants is the presence of premature females on rootlets. A reduction in the number of rhizobium nodules was recorded due to infection by *R. reniformis* on chickpea root as compared with the uninfected ones (Tiyagi and Parveen 1992).

#### 3.1.3.3 Biology and Life Cycle

From egg to egg, the reniform nematode completes its life cycle in about 25–30 days. Ali (1995) reported that the egg laying starts at the 9th to 14th day after infection, while the gelatinous matrix is secreted on the 12th day. Sivakumar and Seshadri (1971) found that a single female lays 60–200 eggs into the gelatinous matrix, which flow out from the vulva.

#### 3.1.3.4 Interaction with Other Pathogens

Ali (1995) reported from Kanpur, India that *R. reniformis* interacts with *Fusarium oxysporum* f. sp. *ciceri* (the causal agent of wilt disease) and *Rhizoctonia solani* (the causal agent of root-rot disease). Combined inoculation of *R. reniformis* with *F. oxysporum* f. sp. *ciceri* resulted in more deleterious effects on the plant (Siddiqui and Mahmood 1994).

#### 3.1.3.5 Management

##### 3.1.3.5.1 Physical Methods

A significant reduction in population densities of *R. reniformis*, parasitic to chickpea, was observed by soil solarization (by covering the soil with transparent polythene sheets) during summer months (April to June) (Sharma and Nene 1990a).

##### 3.1.3.5.2 Cultural Methods

Mojumder (1999) found that chickpea seed treatment with powdered Neem formulation (20% w/w) resulted in a significant reduction in the soil populations of *R. reniformis* and resulted in increasing the grain yield of chickpea under field conditions.

##### 3.1.3.5.3 Biological Methods

Anver and Alam (1999) reported that soil application of *Purpureocilium lilacinum* reduced the rate of multiplication of *R. reniformis* and caused less damage to chickpea.

##### 3.1.3.5.4 Chemical Methods

The reduction in the population density of *R. reniformis* by up to 87.3% and a significant increase in the yield of chickpea was recorded by application of Carbofuran at 1 kg a.i./ha at two split dosages (one at the time of sowing of chickpea and the second 40 days after seed germination) (Ali 1988).

##### 3.1.3.5.5 Integrated Methods

Ashraf and Khan (2008) found fruit wastes of papaya when applied in combination with the fungal biocontrol agent *Purpureocillium lilacinum* at 2 g (mycelium +spores)/ plant resulted in an increase in plant growth of chickpea and population reduction of *R. reniformis*.

## 3.2 Pigeon Pea, *Cajanus cajan*

The important nematode pests of pigeon pea include the cyst nematode (*Heterodera cajani*), root-knot nematodes (*Meloidogyne incognita* and *M. javanica*), and the reniform nematode (*Rotylenchulus reniformis*) (Sharma et al. 1992b).

### 3.2.1 Cyst Nematode, *Heterodera cajani*

Koshy (1967) reported for the first time the pigeon pea cyst nematode *Heterodera cajani* from India. It is a major pest of pigeon pea in India.

#### 3.2.1.1 Distribution

The pigeon pea cyst nematode is mostly restricted to India and Pakistan. *H. cajani* is the most widely distributed cyst nematode of pigeon pea in India (Koshy and Swarup 1971a; Varaprasad et al. 1997).

#### 3.2.1.2 Crop Losses

On a global basis, the annual yield loss to pigeon pea caused by plant parasitic nematodes has been estimated at 13.2% (Abd-Elgawad and Askary 2015). *Heterodera cajani* was found to cause 16–34% yield losses (Walia and Chakraborty 2018).

The reduction in plant growth and grain yield were about 25–27% (Saxena and Reddy 1987). *H. cajani* infection suppresses plant growth of pigeon pea by 28% and reduces grain yield by 24%.

In a greenhouse experiment, 14–22% yield loss was recorded due to infestation on pigeon pea plants caused by *H. cajani* (Sharma et al. 1993b).

#### 3.2.1.3 Symptoms

The above-ground symptoms of nematode injury appear in the form of stunted plant growth, yellowing of leaves, reduction in plant height and vigor, reduced leaf lamina size, yellowing of cotyledonary leaves, patchy appearance, unthrifty growth of plants in field, and reduced size of pods (Sharma 1993). The roots are not well developed, and there is drastic reduction in root system. The number and size of flowers and pods are also reduced.

Below-ground symptoms are a pearly white or lemon-shaped female attached with roots at seedling stage (30–40 days) of plant (Fig. 3.4). Such infested plants show a reduction in rhizobial nodulation, which is responsible for nitrogen fixation.

**Fig. 3.4** Cyst nematode females and egg masses on pigeon pea roots

#### 3.2.1.4 Host Range

Pigeon pea, hyacinth bean (*Dolichos lab-lab*), moth bean (*Phaseolus aconitifolius*), green gram (*Phaseolus lathyroides*), black gram (*Vigna radiata*), cowpea (*V. unguiculata*), and sesame (*Sesamum indicum*) were the most favored hosts and showed extensive damage from *H. cajani* attack (Koshy and Swarup 1973). Bhatti and Gupta (1973) recorded guar (*Cyamopsis tetragonoloba*) as an additional host of *H. cajani*.

#### 3.2.1.5 Life Cycle

Koshy and Swarup (1971a) studied the life cycle of *H. cajani* on pigeon pea plants at an average soil temperature of 29 °C (range 27–36 °C). Within 48 h of inoculation, second-stage juveniles ($J_2$) penetrate both tap and lateral roots, not necessarily near the root tips. On the third day, molting began from the anterior part and completing on the fourth day. On the tenth day, preadult juveniles were found with well-developed reflexed ovaries. On the 12th day, lemon-shaped adult females were noticed, while the males were observed on the 12th or 13th day. Many eggs were seen in the egg sacs as well as inside the white females on the 14th day. On the 16th day, hatched second-stage juveniles were collected from soil as well as from cysts. Emergence of some males from the cysts, apparently embedded in the egg mass, was also observed. From the 20th day onward, the change in color of females from white to bright yellow took 8 days. The color of egg sac turned opaque and yellowish, sometimes purple. On the 38th day after inoculation, brown cysts were noticed. Sometimes females reproduced without males, although males are thought to be necessary for reproduction.

The cyst nematode completed its life cycle in 16 days at 29 °C, while the life cycle took 45–80 days for completion during cooler conditions (10–25 °C). The nematode completed about 8–9 generations during one crop season (Koshy and Swarup 1971a, b).

#### 3.2.1.6 Spread and Survival

Transportation results mainly from flooding, drainage or transfer of infested seeds and plants, from soil washings, and from soil attached to farm machinery, livestock, tools, or people (Mathur 1986; Sharma 1998).

The cysts stored in air-dried soil under laboratory conditions remained viable for at least 2 years (Koshy and Swarup 1971c).

#### 3.2.1.7 Biotypes

Three races of *H. cajani* have been reported by Siddiqui and Mahmood (1993) from 14 populations from 9 hosts collected from different localities in seven districts of Uttar Pradesh (India) (Table 3.1).

#### 3.2.1.8 Interaction with Other Microorganisms

The pigeon pea cyst nematode interacts with soil-borne fungus *Fusarium udum* (*Gibberella indica*) in causing wilt disease complex (Hasan 1984).

**Table 3.1** Races of *H. cajani* present in Uttar Pradesh (Siddiqui and Mahmood 1993)

| Host differentials | Races of *H. cajani* | | |
|---|---|---|---|
| | Race 1 | Race 2 | Race 3 |
| *Cajanus cajan* | + | + | + |
| *Cyamopsis tetragonoloba* | + | – | – |
| *Crotalaria juncea* | + | + | – |

### 3.2.1.9 Management

#### 3.2.1.9.1 Cultural Methods

The following cultural methods have provided some control:

- Deep summer plowing of fields.
- Intercropping of pigeon pea with sorghum (2:1) for management of root-knot and cyst nematodes.
- Growing of nonhost crops like wheat, maize, rice, etc. reduces the population of pigeon pea cyst nematodes.
- A rotation of 2–3 years with maize, rice, pearl millet, sorghum, groundnut, castor, and cotton may suppress the multiplication of nematode and thereby the harmful effect on the pigeon pea crop (Sharma et al. 1992b). Rotation of pigeon pea plots with sorghum, safflower, and chickpea reduced the population of eggs and juveniles of *H. cajani* than did plots previously planted to pigeon pea, cowpea, or *Vigna radiata*. Growing sorghum continuously in the rainy season and safflower in the post-rainy season has been found to reduce the population density of *H. cajani*.
- Intercropping sorghum with a tolerant pigeon pea cultivar could be effective in increasing the productivity of traditional production systems in *H. cajani*-infested regions (Sharma et al. 1996).
- Essential oils of *Mentha piperita, Ocimum sanctum, O. basilicum, Cymbopogon martini, C. nardus, C. winterianus,* and *C. flexuosus* were found effective (Gokte et al. 1993).
- Root extract of *Xanthium strumarium* (Malik et al. 1987) and powdered leaf extracts of *Terminelia arjuna* on pigeon pea (Singh and Singh 1992) were found effective.
- Green manures of *Vigna radiata*, cowpea (*Vigna unguiculata*), sun hemp (*Crotalaria juncea*), *Sesbania bispinosa* (Devi and Gupta 1995), fenugreek (*Trigonella foenum-graecum*), wild methi (*Senji* species), berseem (*Trifolium alexandrinum*), and drum stick (*Moringo pterygosperma*) (Devi 1997) gave effective control.
- Powdered leaf extracts of *Argemone mexicana, Cannabis sativa,* and *Datura metel, Nerium indicum* (Mojumdar et al. 1989) and plant extract of *Solanum xanthocarpum* were found effective (Bhatti et al. 1997).
- Oil cakes of neem, mustard, and mahua gave effective control (Rai and Singh 1995, 1996; Devi and Gupta 1996).

#### 3.2.1.9.2 Chemical Methods

Seed dressing or seed soaking with Carbosulfan 25 DS at 3.0% a.i. (w/w) significantly reduced *H. cajani* by 13.5–32.7% and increased the grain yield.

Seed soaking in Monocrotophos 0.1% significantly increased pigeon pea seed yield by 165.5%. Another effective chemical treatment includes Carbofuran 3G at 1 kg a.i. applied to soil.

Zaki and Bhatti (1986) reported that pigeon pea seed soaking for long time in high concentration of Fensulfothion was found highly effective against *H. cajani*.

#### 3.2.1.9.3 Biological Methods

Seed treatment with *Pseudomonas fluorescens* and *Trichoderma viride* formulations each at 5 g/kg seeds decreased pigeon pea cyst nematode and wilt complex and increased yield significantly.

Application of *P. fluorescens* at 2.5 kg/ha at sowing significantly decreased *H. cajani* population in roots at 45 DAS by 18.32% and final nematode population by 29.55% and thereby gave 35.93% and 32.46% higher pod and grain yield, respectively.

Soil application of *T. harzianum* at 2.5 kg/ha along with *Pochonia chlamydosporia* at 10 kg/ha at the time of sowing gave effective control.

#### 3.2.1.9.4 Host Resistance

Pigeon pea varieties P-1397, P-1071, P-1-1072, P-4555, and P-9-143-3 were found less susceptible to *H. cajani*.

#### 3.2.1.9.5 Integrated Methods

Summer plowing of field and seed dressing with Carbosulfan 3% a.i. w/w gave effective control.

Combination treatment of *P. fluorescens* + *T. viride* at 5 + 5 g/kg seed significantly reduced *H. cajani* population by 32.5% and increased the yield.

Application of NSKP + *T. viride* at 5 + 5 g/kg seed effectively checked *H. cajani* on pigeon pea.

Integration of *T. harzianum* at 5 kg/ha along with *P. chlamydosporia* at 2 kg/ha significantly increased the plant height and pigeon pea yield (0.56 kg/plant as compared to 0.40 kg/plant in control) and reduced the seedling mortality and egg and cyst population by parasitizing them.

Combined treatment of 10% neem seed kernel powder + *Trichoderma viride* at 10 g/kg seeds was found to reduce *H. cajani* in pigeon pea by 58% and increased yield by 32% (Patel et al. 2010).

Patel et al. (2010) reported 62% increase in yield by soil incorporation of neem cake at 100 kg/ha + *Trichoderma viride* at 2.5 kg/ha and reduced the pigeon pea cyst nematode.

Seed treatment with neem seed kernel powder at 10% w/w + *Pseudomonas fluorescens* at 10 g/kg seed was highly effective which recorded 30.11% reduction in nematode population and 29.10% increase in yield over control. Soil application of neem cake at 10 g/m$^2$ + *Pseudomonas fluorescens* at 2.5 kg/ha was also highly effective which recorded 31.63% reduction in nematode population and 17.88% increase in yield over control (Subramanian and Sivakumar 2010).

An integrated control approach involving summer plowing, crop rotation, seed treatment, limited applications of chemicals, and the use of resistant varieties should be adopted for the control of cyst nematodes (Yadav 1986).

Two different management schedules—(a) soil application of neem seed powder at 50 kg/ha + soil solarization (transparent polythene sheet of 400 gauge thickness for a period of 4 weeks) + AMF at 100 kg/ha and (b) seed treatment with neem seed powder at 10% w/w + soil solarization + AMF at 100 kg/ha—were found equally effective against *H. cajani* infesting pigeon pea under field conditions. The cost–benefit ratio was 1:2.54 and 1:2.60 for treatments (a) and (b), respectively, which indicates that both the treatment schedules were economically viable (Nageswari and Mishra 2005).

#### 3.2.1.10 Success Story

Management of the pigeon pea cyst nematode was demonstrated in farmer's field at village Attwa, block Chaubepur of Kanpur district in Central UP using the technology of deep summer plowing and seed dressing with Carbosulfan (25 DS) 3% a.i. (w/w). Through this practice, yield of pigeon pea was increased up to 14% and the monetary gain was up to ₹3000/ha compared to farmers' practice.

## 3.2.2 Interaction of Cyst Nematode with *Fusarium* Wilt

### 3.2.2.1 Symptoms

The wilt disease complex caused by *Fusarium udum* in association with *Heterodera cajani* has been reported as the most severe constraint in the cultivation of pigeon pea. Inoculation with *F. udum* and *H. cajani* together significantly increased wilt severity in pigeon pea seedlings compared with inoculation of the fungus alone.

### 3.2.2.2 Management

#### 3.2.2.2.1 Chemical Methods

Carbofuran and Phorate reduced disease severity both in plants inoculated with *H. cajani* and *F. udum* and in plants inoculated with fungus alone. Greatest reduction occurred at the highest dose (20 mg a.i./kg soil) (Hasan 1989).

#### 3.2.2.2.2 Biological Methods

In a farmers' field trial conducted at Arania, Bulandshahr district of Uttar Pradesh during 2001, wilting of plants was not seen in plots treated with Carbofuran, neem seed powder, neem jivan, *Trichoderma harzianum*, and *Aspergillus niger* and 1–5% wilting was observed in plots treated with *Pupureocillium lilacinum* (Haseeb and Shukla 2005).

#### 3.2.2.2.3 Integrated Methods

Integration of neem seed powder and *T. harzianum*; and neem seed powder + Dimethoate/*T. harzianum*/*P. lilacinum*/latex prevented wilting of plants due to disease complex under field conditions. For enhancing the yield and decreasing the nematode and fungal pathogens, the most effective treatment was neem seed powder + *T. harzianum*, followed by Carbofuran, neem seed powder + *P. lilacinum*, neem seed powder + Dimethoate and neem seed powder + latex. Data regarding *F. udum* infection in roots gathered at the preflowering stage (90 DAS) indicated that all the treatments maintained significant protection of the roots as compared to control (Haseeb and Shukla 2005).

The disease complex on pigeon pea involving the combination of sedentary endoparasite *H. cajani* and the fungus *F. udum* were reduced by application of the biocontrol agents [*Bacillus subtilis, Bradyrhizobium japonicum* and *Glomus fasciculatum* (Siddiqui and Mahmood 1995a); *P. lilacinum, P. chlamydosporia* and *Gigaspora margarita* (Siddiqui and Mahmood 1995b); and *Trichoderma harzianum, P. chlamydosporia* and *Glomus mosseae* (Siddiqui and Mahmood 1996)].

### 3.2.3 Root-Knot Nematodes, *Meloidogyne incognita* and *M. javanica*

Root-knot nematodes *Meloidogyne* spp. are one of the major limiting factors to pulse production.

#### 3.2.3.1 Crop Losses

Root-knot nematodes are responsible for 14.70% yield loss worth ₹1,57,458,168 in pigeon pea. Ali (1997) and Sharma et al. (1993a) reported 14–29% loss in yield due to root-knot nematodes in pigeon pea.

In India, Patel and Patel (1993) recorded 14.2% loss in yield due to combined infection of *M. incognita* and *M. javanica* in Gujarat in pigeon pea cv. Pusa Ageti.

#### 3.2.3.2 Symptoms

Patchy appearance in field, stunting of plants, yellowing of leaves, and reduction in size of leaves and pods are some of symptoms observed in root-knot nematode infected fields. Pods may ripen and dry prematurely and remain partially filled and undersized (Reddy et al. 1990). Poor emergence and death of young seedlings may occur in heavily infested soil, but the death of full-grown plants is rare unless there is an association of fungus or bacteria to form a disease complex.

The discernible and most characteristic symptom of root-knot nematode infestation is formation of gall on the plant roots (Fig. 3.5). A reduction in the number and size of rhizobial nodules were observed on plant roots.

**Fig. 3.5** *Left*—Sparse, uneven stunted growth of pigeon pea infested with root-knot nematode. *Right*—Pigeon pea roots infested with root-knot nematode

#### 3.2.3.3 Biology and Life Cycle

Reproduction is parthenogenetic and when the conditions are favorable, life cycle from egg to egg is completed in about 3 weeks. Generally, the life cycle of *M. incognita* or *M. javanica* is completed in 21–25 days at an optimum temperature of 26–27 °C, but it may take as long as 80 days when the temperature is low (14–16 °C).

#### 3.2.3.4 Interaction with Other Pathogens

The root-knot nematode *Meloidogyne* spp., when in combination with wilt fungus *Fusarium udum*, can cause severe damage to pigeon pea crop. The wilting of pigeon pea plants caused by *F. udum* is aggravated in the presence of *M. javanica* (Singh et al. 2004; Askary and Ali 2012).

The presence of *M. javanica* with *F. udum* applied to the roots of pigeon pea seedlings caused a susceptible reaction to *Fusarium* wilt in all the five resistant accessions (DPPA 85-5, DPPA 85-11, DPPA 85-14, Banda Palera, and Sujata) used in the study. Wilt severity was highest in DPPA 85-14 (46.67%) followed by DPPA 85-11 (40.0%) (Askary and Ali 2012).

#### 3.2.3.5 Management

##### 3.2.3.5.1 Cultural Methods

Intercropping of pigeon pea with sorghum (2:1 ratio) was found effective.

In a field experiment, pigeon pea cv. Narendra Arhar 1 seed treatment with various neem products, neem oil (2%), Neemarin (2%), neem seed powder (5%), and crude neem leaf extract (5%), the gall index, and soil population of *M. javanica* was minimal and significantly lower than the control (Singh 2009).

#### 3.2.3.5.2 Chemical Methods

Seed treatment of Carbosulfan 25DS at 3% a.i. w/w proved effective for the control of root-knot nematode.

Mishra et al. (2003) recorded enhanced plant growth characters and yield, and reduction in root-knot nematode population by seed treatment with Dimethoate at 8 ml/kg seed, Chlorpyriphos at10 ml/kg seed, and Triazophos at 1%.

Seed treatment with Chlorpyriphos and Dimethoate at 1 ml/kg seed resulted in causing a significant reduction in soil population of *M. javanica*, and increasing the plant growth characters and yield of pigeon pea cultivar UPAS 120 as compared to untreated control under field conditions (Askary 2012).

#### 3.2.3.5.3 Biological Methods

Use of bio-agents *Pseudomonas fluorescens* or *Trichoderma viride* at 10 g/kg seed, is effective against root-knot nematode.

Askary et al. (2012) reported that presowing seed coating with *P. lilacinum* and *Aspergillus niger* was found significantly effective in reducing nematode infection on plants and increasing the plant growth characters and yield as compared to the untreated control.

#### 3.2.3.5.4 Host Resistance

Pigeon pea varieties found resistant to *M. incognita* include CG-28 of cowpea, AL-15, Prabhat, 4-64, 4-83, BDM-2, T-21, Sebore-197, and UG-300.

In a field with mixed populations of *M. javanica* and *M. incognita*, pigeon pea lines 77-1 and 18-1 were found highly resistant (Patel et al. 1987).

#### 3.2.3.5.5 Integrated Methods

Treatment constituting farmyard manure (FYM), oilseed cake of *Pongamia pinnata*, and arbuscular mycorrhizal fungus (AMF) has been found to reduce the disease incidence caused by root-knot nematode–*Fusarium* complex in pigeon pea to a significant level and improve the plant growth parameters (Goswami et al. 2007).

## 3.2.4 Interaction of Root-Knot Nematode with *Fusarium* Wilt

### 3.2.4.1 Symptoms

The root-knot nematode *M. incognita* is a major limiting factor among plant parasitic nematodes, while the wilt fungus *F. udum* is one of the most severe diseases of pigeon pea. The most serious problem in pigeon pea is the disease complex incited by *M. incognita* along with *F. udum*.

Simultaneous or sequential inoculation of *M. incognita* and *F. udum* increased the severity of the disease. Simultaneous inoculation of *M. incognita* and *F. udum* was responsible for maximum reduction in fresh/dry weight and plant height, followed by *M. incognita* prior and *F. udum* 7 days later and *F. udum* prior and *M. incognita* 7 days later, respectively (Perveen et al. 1998). The varieties developed as resistant to *Fusarium* wilt when grown under farmers' field conditions become susceptible to wilt due to the presence of root-knot nematodes.

Increase in wilting was observed in five pigeon pea accessions (identified as resistant to *Fusarium* wilt), namely ICP 8859 (50%), AWR 74/15 (60%), KPL 44, ICPL 89049 (50%), and ICPL 12745 (50%), against combined infection of *M. javanica* and *F. udum* (Singh et al. 2004).

#### 3.2.4.2 Management

##### 3.2.4.2.1 Biological Methods

Pigeon pea seed treatment with *Trichoderma harzianum* and *Pochonia chlamydosporia* gave maximum control of the disease complex in which wilt severity decreased by 79–81%, root galling by 44–49%, and egg mass production by 33–35% (Khan et al. 2010).

##### 3.2.4.2.2 Host Resistance

Five accessions of pigeon pea which were resistant to *F. udum* (KPL 43, PI 397430, BWR 370, GPS 33, and ICPL 89048) were also found resistant to combined inoculation of *M. javanica* and *F. udum*, which indicated that wilt resistance in these accessions was not influenced by the presence of *M. javanica* (Singh et al. 2004).

##### 3.2.4.2.3 Integrated Methods

The treatment constituting FYM, Karanj oilseed cake, and arbuscular mycorrhizal fungus *Glomus fasciculatum* reduced the disease incidence caused by root-knot nematode *M. incognita* and root wilt fungus *F. udum* on pigeon pea to a great extent with the most promising improvement in plant growth parameters (Goswami et al. 2007).

### 3.2.5 Reniform Nematode, *Rotylenchulus reniformis*

#### 3.2.5.1 Crop Losses

The reniform nematode is responsible for 32.84% crop loss in pigeon pea (Walia and Chakraborty 2018).

Ali (1996) reported yield loss of 19% in pigeon pea due to infestation by *R. reniformis*. However, the loss may vary from 14% to 29%, depending on the initial population level of nematode and duration of the crop (Ali and Singh 2005).

#### 3.2.5.2 Symptoms

Pigeon pea plants infested by *R. reniformis* show the symptoms such as yellowing of new leaves, dieback of twigs and main stem, and premature death of plants (Hutton and Hammerton 1975). The patches of stunted plant growth can be observed in the field, and the number of such patches increases under water-stress conditions (Ali and Singh 2005). The dirty appearance of roots due to the soil particles clinging to

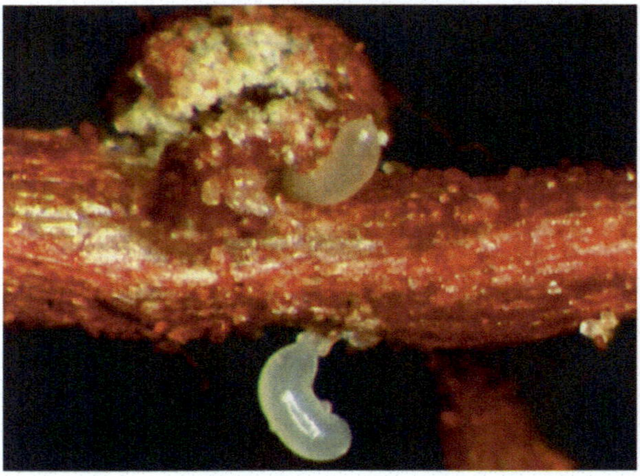

**Fig. 3.6** Pigeon pea root infected with reniform nematode

the gelatinous matrix of egg masses that do not easily dislodge by shaking or gently washing the root with water is the characteristic symptom of reniform nematode infection (Fig. 3.6) (Sharma et al. 1992a).

#### 3.2.5.3 Survival

Sharma and Nene (1992) reported that the reniform nematode can survive as long as 300 days in soil without loss in infectivity in the absence of host.

#### 3.2.5.4 Interaction with Other Pathogens

Sharma and Nene (1990b) reported that the presence of *R. reniformis* along with *F. udum* accelerated the death of wilt-susceptible pigeon pea cultivars. Jain and Sharma (1996) observed that wilt-resistant pigeon pea cultivar ICPL 270 loses its resistance in presence of *R. reniformis.*

#### 3.2.5.5 Management

##### 3.2.5.5.1 Cultural Methods

Nonhost crops such as groundnut and sesame, when intercropped with pigeon pea, has been found successful in bringing down the soil population of *R. Reniformis*, reducing the number of galls and ultimately increasing the yield of pigeon pea (Upadhyay et al. 1997).

##### 3.2.5.5.2 Biological Methods

Soil application of *P. lilacinum* on pigeon pea plants resulted in reducing the multiplication rate of *R. reniformis* and plant damage (Anver and Alam 1997).

##### 3.2.5.5.3 Host Resistance

The reniform nematode-tolerant pigeon pea lines include ICP 16329, ICP 16330, ICP 16331, ICP 16332, and ICP 16333 (Sharma et al. 2000).

The pigeon pea accession KM-137 was found resistant to *R. reniformis* (Anver and Alam 2001).

#### 3.2.5.5.4 Integrated Methods

A combined application of *P. lilacinum* with *Zea mays* and *Sesbania aculeata* as green manuring proved highly effective for the management of *R. reniformis* on pigeon pea (Mahmood and Siddiqui 1993).

Application of Carbofuran at 1 kg/ha to deep summer plowed field gave good control of reniform nematode in pigeon pea crop.

## 3.3 Green Gram, *Vigna radiata*

The root-knot nematode *Meloidogyne incognita* and the reniform nematode *Rotylenchulus reniformis* are the major problems on green gram. Castillo et al. (1977) recorded a yield loss of 28% in mung bean grown in a field infested with mixed population of *M. incognita* and *R. reniformis*.

### 3.3.1 Root-Knot Nematode, *Meloidogyne incognita*

#### 3.3.1.1 Crop Losses

Root-knot nematode is responsible for 29% yield loss in green gram worth ₹2001 million in India (Walia and Chakraborty 2018). Yield losses of 31% in *kharif* green gram were estimated due to root-knot nematode.

#### 3.3.1.2 Symptoms

The above-ground symptoms appear as nutritional deficiency in plants with stunting, yellowing, and loss of yield. The roots of infected plants show characteristic symptoms of gall formation (Fig. 3.7). Sometimes rotting is also found due to secondary fungal infection.

#### 3.3.1.3 Management

##### 3.3.1.3.1 Cultural Methods

The farmers of middle Gujarat (AES III) growing green gram during *kharif* season in root-knot nematode-infested soil are advised to adopt crop rotation with cabbage in *rabi* and cluster bean (vegetable purpose) in summer for 2 years to manage root-knot nematodes.

Cropping sequence green gram/black gram-*toria*-paddy is effective against root-knot nematode.

Intercropping of green gram with sesame (2:1) or application of neem cake/neem-based organic amendment at 1 t/ha were found effective.

**Fig. 3.7** Root-knot nematode-infested green gram plant (*left*) and roots (*right*)

Wani and Bhat (2012) reported that soil amended with Nimin-coated urea (neem-based product with neem-triterpenes) was found effective against *M. incognita* and improved plant growth characteristics and chlorophyll content of plant leaves.

Soil application of neem cake at 2 t/ha gave effective control of *M. incognita* infesting green gram (Sumita and Das 2014).

#### 3.3.1.3.2 Chemical Methods

Seed treatment with Carbosulfan 25DS at 3% a.i. w/w before sowing is recommended for the control of root-knot nematode.

Seed soaking in Carbosulfan 25EC at 0.01% for 6 h is effective (Choudhury et al. 2011).

Soil application of Carbofuran 3G at 1–2 kg a.i. /ha in root-knot nematode-infested soil increased the yield of green gram.

#### 3.3.1.3.3 Biological Methods

Seed treatment with *Trichoderma viride* at 5 g/kg seed.

Soil application of Biofor-pf (a combination of *T. harzianum* and *Pseudomonas fluorescens*) at 100 kg/ha gave effective control of *M. incognita* infesting green gram (Sumita and Das 2014).

#### 3.3.1.3.4 Host Resistance

Grow green gram varieties (T-44, L-62, K-851, PDM-14, ML 1265, and COGG-912) resistant to root-knot nematode (*Meloidogyne incognita*).

#### 3.3.1.3.5 Integrated Methods

Seed treatment with NSKP at 5 g/kg seed + *Trichoderma viride* ($2 \times 10^6$ cfu/g) at 5 g/kg seed is effective in reducing the infestation of *M. incognita* in green gram (Choudhury et al. 2011).

To manage root-knot nematode in green gram, sow Carbosulfan 25DS at 3% treated seed in deep summer plowed field.

Soil application of neem cake at 100 kg/ha along with *Trichoderma viride* ($2 \times 10^8$ spores/g) at 2.5 kg/ha at the time of sowing was found effective to manage root-knot nematodes and to increase in green gram yield.

## 3.3.2 Interaction of Root-Knot Nematode with *Fusarium* Wilt

### 3.3.2.1 Interaction

Nematode-fungus disease complex particularly of *Meloidogyne incognita* and *Fusarium oxysporum* poses a great problem to the cultivation of green gram (*Vigna radiata*) by inflicting severe yield losses (Mahapatra and Swain 2001).

### 3.3.2.2 Management

Maximum increase in all the plant growth and yield parameters and suppression of nematode reproduction, i.e., reproduction factor (Rf) and root-knot index (RKI) were found in Carbofuran-treated plants (Rf—0.46 as compared to 1.74 in control, RKI—0.25 as compared to 3.00 in control) followed by neem seed powder (Rf—0.99 as compared to 1.74 in control, RKI—0.50 as compared to 3.00 in control) treated plants as compared to untreated inoculated plants (Haseeb et al. 2005).

## 3.3.3 Reniform Nematode, *Rotylenchulus reniformis*

### 3.3.3.1 Crop Losses

Reniform nematode is responsible for 21.19% crop loss in green gram.

### 3.3.3.2 Management

#### 3.3.3.2.1 Host Resistance

Grow green gram varieties (Pusa-103, ML-62, ML-80, and PDM-14) resistant to reniform nematode.

#### 3.3.3.2.2 Integrated Methods

For integrated control of reniform nematode in green gram, seed treatment with 5 g/kg each of *Trichoderma viride* and *Pseudomonas fluorensens* or 10 g neem seed kernel powder was found effective.

## 3.4 Black Gram, *Vigna mungo*

The root-knot nematode *Meloidogyne incognita* and the reniform nematode *Rotylenchulus reniformis* are the major problems on black gram.

### 3.4.1 Root-Knot Nematode, *Meloidogyne incognita*

#### 3.4.1.1 Crop Losses

The root-knot nematode is responsible for annual yield loss of 19% in black gram amounting to ₹1570.35 million in India (Walia and Chakraborty 2018). *Meloidogyne incognita* is responsible for 36.94% crop loss in black gram.

#### 3.4.1.2 Symptoms

- Infected plants in patches in the field.
- Formation of galls on host root system is the primary symptom (Fig. 3.8).
- Roots branch profusely starting from the gall tissue causing a "beard root" symptom.
- Infected roots become knobby and knotty.
- Significant reduction or absence of feeder roots under severe infestation.
- The functions of uptake and transport of water and nutrients by roots are seriously hampered.
- During the afternoon when temperatures are high, the plants wilt even though enough soil moisture is present in soil.
- Nematode infection predisposes plants to fungal and bacterial root pathogens.

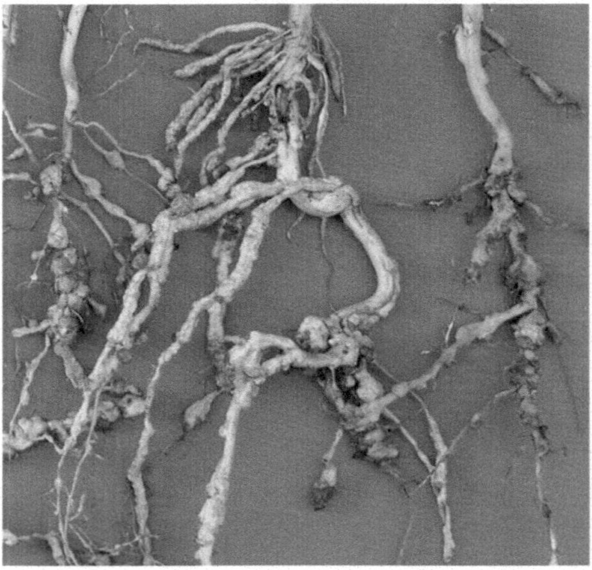

**Fig. 3.8** Root-knot nematode-infested black gram roots

### 3.4.1.3 Management

#### 3.4.1.3.1 Cultural Methods

Intercropping of black gram with sesame (2:1) or application of neem cake/neem-based organic amendment at 1 t/ha was found effective.

Cropping sequence green gram/black gram-*toria*-paddy is effective against root-knot nematode.

#### 3.4.1.3.2 Chemical Methods

Seed treatment with Carbosulfan 25ST at 3% (w/w) before sowing in black gram is effective against *M. incognita*.

Seed soaking in Carbosulfan 25EC at 0.1% for 6 h in black gram is effective (Choudhury et al. 2011).

#### 3.4.1.3.3 Biological Methods

In black gram, *Trichoderma harzianum* and *Pochonia chlamydosporia* at 20 g/m$^2$ ($1 \times 10^6$ spores/g) recorded lowest population of *M. incognita* with minimum galls and also recorded higher yield.

Akhtar et al. (2012) reported that plants inoculated with bacterial biocontrol agents *Pseudomonas fluorescens* or *Bacillus subtilis* resulted in improving plant growth parameters and decreasing the number of root galls.

Treatment of *Bradyrhizobium* and *P. lilacinum* significantly reduced the *M. incognita* damage to plant growth of black gram (Bhat et al. 2012).

#### 3.4.1.3.4 Host Resistance

Varieties of black gram found resistant/tolerant against *M. incognita* include OBG-31 and TU-26-1.

Moderate resistance to *M. javanica* was observed in black gram cvs. KU-300 and KU-303. Black gram genotypes TPU-94-2, KU-315, and K-98 showed moderate resistance to *M. incognita* (Goel 2004).

#### 3.4.1.3.5 Integrated Methods

To control root-knot nematode in black gram, seed treatment with Carbosulfan 25DS at 3% and soil treatment with *Trichoderma harzianum* at 10 kg per ha is effective.

Seed treatment with NSKP at 5 g/kg seed + *Trichoderma viride* at 5 g/kg seed is effective in reducing infestation of *M. incognita* in black gram (Choudhury et al. 2011).

Integration of deep summer plowing (3 plowings) and seed treatment with Carbosulfan 25ST t 3% (w/w) is effective against root-knot nematode in black gram.

### 3.4.2 Reniform Nematode, *Rotylenchulus reniformis*

#### 3.4.2.1 Life Cycle

Within 24 h after inoculation, the preadult female penetrated host roots and showed subsequent swelling of roots and females on 3rd day after inoculation. Seven days after inoculation, fully developed adult females were observed which secreted gelatinous matrix on 8th day and produced eggs up to 15th day after inoculation. The nematode completed its life cycle in 10–15 days after inoculation (Midha and Trivedi 1990).

#### 3.4.2.2 Management

##### 3.4.2.2.1 Chemical Methods

Seed treatment with Carbosulfan at 3.0% w/w gave 37% increase in yield and 78% reduction in reniform nematode population.

##### 3.4.2.2.2 Host Resistance

Grow reniform nematode-resistant varieties of black gram like UG-135, UG-201, and Ratanpur-1.

## 3.5 Conclusion

For an effective and precise management of plant parasitic nematodes attacking pulse crops, a prerequisite step should be the correct diagnoses of the nematode up to species level and also their pathogenic variants. There is a need to use newly developed technologies, based on biochemical and DNA analyses, for correct diagnoses of the nematode up to species level and also their pathogenic variants. Major nematode pests on pulse crops include species of *Meloidogyne*, *Heterodera*, *Pratylenchus*, and *Rotylenchulus*. Mapping of major pulse nematode pests in pulse-growing regions through extensive surveys on a global basis is essential for real assessment of yield losses due to nematodes. There is a need for estimation of crop losses due to nematodes in different major pulse crops on a global basis. Hot spots of major nematode pests have to be identified in pulse-growing areas on a regional basis through extensive surveys. While using a breeding program to evolve a variety resistant to fungi or bacteria, breeders should also keep in mind that it should also be resistant to nematodes. Currently, very few reports are available of pulse crops resistant to both nematode and wilt fungi. Genotypes of pulse crops having high-level resistance against nematode pests need to be sought out. There is a need to integrate various management techniques in order to prepare an integrated nematode management (INM) module to disseminate them to the pulse growers by conducting multilocational trials at farm level (Askary 2015). The management of nematodes must be based on strategies that integrate the use of available host-resistant or host-tolerant cultivars with crop rotation or inclusion of a nonhost crop in cropping systems.

## References


Abd-Elgawad MMM, Askary TH (2015) Impact of phytonematodes on agriculture economy. In: Askary TH, Martinelli PRP (eds) Biocontrol agents of phytonematodes. CAB International, Wallingford, pp 3–49

Akhtar A, Hisamuddin, Abbasi (2012) Interaction between *Meloidogyne incognita*, *Pseudomonas fluorescens* and *Bacillus subtilis* and its effect on plant growth of black gram (*Vigna mungo* L.). Int J Plant Path 3(2):66–73

Ali SS (1988) Investigations on plant parasitic nematodes associated with pulse crops. Annual report 1987–1988. Directorate of Pulses Research, Indian Council of Agricultural Research (ICAR), Kanpur, pp 77–79

Ali SS (1992) Investigation on plant parasitic nematodes associated with pulse crops. Annual report 1991–1992. Directorate of Pulses Research, ICAR, Kanpur, pp 55–73

Ali SS (1993) Prevalence of plant parasitic nematodes associated with chickpea in Gwalior district of Madhya Pradesh. Int. Chickpea Newsl. 28:11

Ali SS (1995) Nematode problems in chickpea. Pawel Graphics Private Limited, Kanpur, 184 pp

Ali SS (1996) Estimation of yield losses in pigeonpea due to reniform nematodes. Indian J Pulses Res 9(2):209–210

Ali SS (1997) Status of nematode problems and research in India. In: Diagnosis of key nematode pests of chickpea and pigeonpea and their management. Proceedinhgs of Regional Training Course, 25–30 November, 1996. ICRISAT, Patancheru, Hyderabad, pp. 74–82

Ali SS (2009) Estimation of unavoidable yield losses in certain *rabi* pulse crops due to the root-knot nematode, *Meloidogyne javanica*. Trends Biosci 2(2):48–49

Ali SS, Gurha SN (1995) Role of *Meloidogyne javanica* in Fusarium wilt of chickpea. Indian J. Pulses Res 8:201–203

Ali SS, Singh B (2005) Nematodes of pigeonpea and their management. In: Ali M, Kumar S (eds) Advances in pigeonpea research. DK Agencies Pvt Ltd, New Delhi, pp 284–314

Anver S, Alam MM (1997) Control of *Meloidogyne incognita* and *Rotylenchulus reniformis* singly and concomitantly on pigeonpea with Paecilomyces lilacinus. Indian J Nematol 27(2):209–213

Anver A, Alam MM (1999) Control of *Meloidogyne incognita* and *Rotylenchulus reniformis* singly and concomitantly on chickpea and pigeonpea. Arch Phytopathol Plant Prot 32(2):161–172

Anver S, Alam MM (2001) Reaction of pigeonpea accessions to root-knot nematode *Meloidogyne incognita* and reniform nematode *Rotylenchulus reniformis*. In: Chickpea Pigeonpea Newsl 8:41–42

Ashraf MS, Khan TA (2008) Biomanagement of reniform nematode, *Rotylenchulus reniformis* by fruit wastes and *Paecilomyces lilacinus* on chickpea. World J Agri Sci 4(4):492–494

Askary TH (2012) Management of root-knot nematode *M. javanica* in pigeonpea through seed treatment. Indian J Ecol 39(1):151–152

Askary TH (2015) Nematophagous fungi as biocontrol agents of phytonematodes. In: Askary TH, Martinelli PRP (eds) Biocontrol agents of phytonematodes. CAB Int, Wallingford, Oxfordshire, pp 81–125

Askary TH, Ali SS (2012) Effect of *Meloidogyne javanica* and *Fusarium udum* singly and concomitantly on wilt resistant accessions of pigeonpea. Indian J Plant Prot 40(3):167–170

Askary TH, Banday SA, Iqbal U et al (2012) Plant parasitic nematode diversity in pome, stone and nut fruits. In: Lichtfouse E (ed) Agroecology and strategies for climate change. Springer, Heidelberg, pp 237–268

Bhat MY, Wani AH, Fazal M (2012) Effect of *Paecilomyces lilacinus* and plant growth promoting rhizobacteria on *Meloidogyne incognita* inoculated black gram, *Vigna mungo* plants. J Biopest 5 (1):36–43

Bhatti DS, Gupta DC (1973) Guar an additional host of *Heterodera cajani*. Indian J Nematol 3:160

Bhatti DS, Dutt R, Verma KK (1997) Larval emergence from cysts of *Heterodera avenae* and *H. cajani* as affected by plant leaf extracts. Indian J Nematol 27(1):63–69

Castillo MB, Alejar MS, Litsinger JA (1977) Pathological reactions and yield loss of mung bean to known populations of *Rotylenchulus reniformis* and *Meloidogyne acrita*. Philipp Agric 61:12–24

Chakrabarti U, Mishra SD (2001) Seed treatment with neem products for integrated management of *Meloidogyne incognita* infecting chickpea. Curr Nematol 12(1/2):15–19

Choudhury BN, Bhagawati B, Bora BC (2011) Three decades of nematology in Assam (1977–2011) AICRP on plant parasitic nematodes with integrated approach for their control. Department of Nematology, Assam Agricultural University, Jorhat, 49 pp

Devi LS (1997) Evaluation of green manures against the pigeon pea cyst nematode *Heterodera cajani*. Nat Acad Sci Lett 20(1/2):1–2

Devi S, Gupta P (1995) Effect of four green manures against Heterodera cajani on pigeon pea sown with or without Rhizobium seed treatment. Indian J Mycol Pl Path 25(3):254–256

Devi S, Gupta P (1996) Larval emergence from egg sacs of *Heterodera cajani* in extracts of cakes in various media and their effect on cowpea. Indian J Nematol 25:190–193

Di Vito M, Greco N, Saxena MC (1992) Pathogenicity of *Pratylenchus thornei* on chickpea in Syria. Nematol Medit 20:71–73

Dwivedi K, Upadhyay KD, Verma RA, Ahmad F (2008) Role of bioagents in management of *Pratylenchus thornei* infecting chickpea. Indian J Nematol 38(2):138–140

Goel SR (2004) Reaction of certain urdbean (*Vigna mungo* L.) genotypes to root-knot nematode, *Meloidogyne javanica* and *Meloidogyne incognita*. Ann Agric Res 25(4):626–627

Gokte N, Maheswari ML, Mathur VK (1993) Nematicidal activity of new essential oils against root-knot and cyst nematode species. Indian J Nematol 21(2):123–127

Goswami BK, Pandey RK, Goswami J, Tewari DD (2007) Management of disease complex caused by root knot nematode and root wilt fungus on pigeonpea through soil organically enriched with Vesicular Arbuscular Mycorrhiza, karanj (*Pongamia pinnata*) oilseed cake and farmyard manure. J Envt Sci Health 42(8):899–904

Hasan A (1984) Synergism between *Heterodera cajani* and Fusarium udum attacking Cajanus cajan. Nematol Medit 12(1):159–162

Hasan A (1989) Efficacy of certain non-fumigant nematicides on the control of pigeon pea wilts involving *Heterodera cajani* and *Fusarium udum*. Phytopathol Z 126:335–342

Haseeb A, Shukla PK (2005) Wilt disease complex of pigeon pea and its management. In: Nehra S (ed) Plant diseases: biocontrol management. Avishkar Publishers, Distributors, Jaipur, pp 84–96

Haseeb A, Sharma A, Shukla PK (2005) Studies on the management of root-knot nematode, Meloidogyne incognita-wilt fungus, Fusarium oxysporum disease complex of green gram, Vigna radiata cv. ML-1108. J Zhejiang Univ Sci 6B:736–742

Haseeb A, Sharma A, Abuzar S, Kumar V (2006) Evaluation of resistance in different cultivars of chickpea against *Meloidogyne incognita* and *Fusarium oxysporum* f. sp. *ciceri* under field conditions. Indian Phytopathol 59(2):234–236

Hutton DG, Hammerton JL (1975) Investigating the role of *Rotylenchulus reniformis* in a decline of pigeonpea. Nematropica 5(2):24

Jain KC, Sharma SB (1996) Loss of *Fusarium* wilt resistance in a pigeonpea line ICPL 270 in reniform nematode infested soil at ICRISAT Asia Centre. Int Chickpea Pigeonpea Newsl 3:90

Kaushik HD, Bajaj HK (1981) Control of root-knot nematodes *Meloidogyne javanica* and crop yield. Haryana Agri Univ J Res 22(1):40–45

Khan MM, Khan MR, Reshu (2010) Evaluation of effectiveness of seed treatment with pesticides and biocontrol agents against the disease complex of pigeon pea in pots. In: National conference on innovations in nematological research for agricultural sustainability – challenges and a roadmap ahead, Tamil Nadu Agricultural University, Coimbatore, p 101

Koshy PK (1967) A new species of *Heterodera* from India. Indian Phytopathol 20:272–274

Koshy PK, Swarup G (1971a) Investigations on the life history of the pigeon-pea cyst nematode, *Heterodera cajani*. Indian J Nematol 1:44–51

Koshy PK, Swarup G (1971b) On the number of generations of *Heterodera cajani*, the pigeon-pea cyst nematode in a year. Indian J Nematol 1:88–90

Koshy PK, Swarup G (1971c) Factors affecting emergence of larvae from cysts of *Heterodera cajani* Koshy, 1967. Indian J Nematol 1:209–219

Koshy PK, Swarup G (1973) Susceptibility of plants to pigeon-pea cyst nematode, *Heterodera cajani*. Indian J Nematol 2:1–6

Krishna Rao V, Krishnappa K (1996) Interaction of *Fusarium oxysporum* f. sp. ciceri with Meloidogyne incognita on chickpea in two soil types. Indian Phytopath 49:142–147

Mahapatra BC, Padhi NN (1986) Inoculum potential of the reniform nematode in relation to growth of chickpea. Int. Chickpea Newsl 15:15

Mahapatra SN, Swain PK (2001) Interaction between *Meloidogyne incognita* and *Fusarum oxysporum* on blackgram. Ann Plant Prot Sci 9:92–94

Mahmood I, Siddiqui ZA (1993) Integrated management of *Rotylenchulus reniformis* by green manuring and *Paecilomyces lilacinus*. Nematol Medit 21(2):285–287

Malik MS, Sangwan NK, Dhindsa KS, Bhatti DS (1987) Nematicidal activity of extracts of *Xanthium strumarium*. Pesticides 21(10):19–20

Mani A, Sethi CL (1987) Interaction of root-knot nematode *Meloidogyne incognita* with *Fusarum oxysporum* f. sp. ciceris and F. solani on chickpea. Indian J Nematol 17:1–6

Mathur VK (1986) Quarantine: an effective and possible method to control plant parasitic nematodes. In: Swarup G, Dasgupta DR (eds) Plant-parasitic nematodes of India, problems and progress. Indian Agricultural Research Institute, New Delhi, pp 490–497

Midha RL, Trivedi PC (1990) Post infection life cycle of the reniform nematode (*Rotylenchulus reniformis*) on black-gram (*Vigna mungo* L.). Pakistan J Nematol 8(1):23–24

Mishra SD, Dhawan SC, Tripathi MN, Nayak S (2003) Field evaluation of biopesticides, chemicals and bioagents on plant parasitic nematodes infesting chickpea. Curr Nematol 14(1, 2):82–89

Mojumdar V, Mishra SD, Haque MM, Goswami BK (1989) Nematicidal efficacy of some wild plants against pigeon pea cyst nematode, *Heterodera cajani*. Int Nematol Network Newsl 5 (2):21–24

Mojumder V (1999) Effect of seed treatment of chickpea with crude neem products and neem-based pesticides on nematode multiplication in soil and the grain yield. Int J Nematol 9(1):76–79

Nageswari S, Mishra SD (2005) Integrated nematode management schedule incorporating neem products, VAM and soil solarisation against *Heterodera cajani* infesting pigeonpea. Indian J Nematol 35(1):68–71

Pandey RK, Pant H, Yadav S et al (2005) Application of biocontrol agents and neem cake for the management of *Meloidogyne incognita* on chickpea (*Cicer arietinum* L.). Pakistan J Nematol 23 (1):57–60

Pant H, Pandey G, Shukla DN (2004) Effect of different concentrations of bio-control agents on root-knot disease of chick pea and its rhizosphere microflora. Pakistan J. Nematol. 22 (1):103–109

Patel GA, Patel DJ (1993) Avoidable yield losses in pigeonpea cv. Pusa Ageti due to *Meloidogyne javanica*. Int. Pigeonpea Newsl. 17:26–27

Patel BA, Chavda JC, Patel ST, Patel DJ (1987) Susceptibility of some pigeonpea lines to root-knot nematodes (*Meloidogyne incognita* and *M. javanica*). Int Pigeonpea Newsl 6:55–57

Patel DJ, Patel BA, Patel SK, Patel AD (2010) Innovations in extension nematology. In: National conference on innovations in nematological research for agricultural sustainability – challenges and a roadmap ahead, Tamil Nadu Agricultural University, Coimbatore, pp 63–64

Perveen S, Ehteshamul-Haque S, Ghaffar A (1998) Efficacy of *Pseudomonas aeruginosa* and *Paecilomyces lilacinus* in the control of root rot-root knot disease complex on some vegetables. Nematol Medit 26:209–212

Rai PK, Singh KP (1995) Efficacy of certain oilcake amendments on *Heterodera cajani*, *Fusarium udum* and associated wilt of pigeonpea. Int J Trop Pl Dis 13(2):213–219

Rai PK, Singh KP (1996) Efficacy of certain oilcake amendments on *Heterodera cajani*, *Fusarium udum* and associated wilt of pigeonpea. Int J Trop Pl Dis 14(1):51–58

Reddy MV, Sharma SB, Nene YL (1990) Pigeonpea disease management. In: Nene YL, Susan DH, Sheila YK (eds) The pigeonpea. CAB International, Wallingford, Oxfordshire, pp 303–347

Saxena R, Reddy DDR (1987) Crop losses in pigeonpea and mungbean by pigeonpea cyst nematode, *Heterodera cajani*. Indian J Nematol 17(1):91–94

Sebastian S, Gupta P (1996) Evaluation of some oil-cakes against *Pratylenchus thornei* on chickpea. Int Chickpea Pigeonpea Newsl 3:40–41

Sebastian S, Gupta P (1997) Crop loss trial of chickpea infested with *Pratylenchus thornei*. Indian JNematol 27(1):142–143

Sharma SB (1993) Pearly root of pigeonpeas caused by *Heterodera cajani*. Indian J Nematol 21:169

Sharma SB (ed) (1998) The cyst nematodes. Kluwer Academic Publishers, Dordrecht

Sharma SB, Nene YL (1990a) Effects of soil solarisation on nematodes parasitic to chickpea and pigeonpea. J Nematol 22(4S):658–664

Sharma SB, Nene YL (1990b) Effect of *Fusarium udum* alone and in combination with Rotylenchulus reniformis or Meloidogyne spp. on wilt incidence, growth of pigeonpea and multiplication of nematodes. Int J Trop Pl Dis 8:95–101

Sharma SB, Nene YL (1992) Spatial and temporal distribution of plant parasitic nematodes on pigeonpea in alfisols and vertisols. Nematropica 22:13–20

Sharma SB, Patel HV, Patel BA et al (1992a) First report of *Heterodera cajani* on pigeonpea in Gujarat State in India. Int Pigeonpea Newsl 16:17

Sharma SB, Smith DH, McDonald DI (1992b) Nematode constraints of chickpea and pigeonpea production in the semi-arid tropics. Plant Dis 76(9):868–874

Sharma SB, Ali SS, Patel DJ et al (1993a) Distribution and importance of plant parasitic nematodes associated with pigeonpea in Gujarat state, India. Afro-Asian J Nematol 3:55–59

Sharma SB, Nene YL, Reddy MV, McDonald D (1993b) Effect of *Heterodera cajani* on biomass and grain yield of pigeonpea on vertisol in pot and field experiments. Plant Pathol 42 (2):163–167

Sharma SB, Ali SS, Upadhyay KD, Ahmad F (1996) Potential nematode constraints of pigeonpea in U.P. in northern India. Afro-Asian J Nematol 6:151–155

Sharma SB, Jain KC, Lingaraju S (2000) Tolerance to reniform nematode (*Rotylenchulus reniformis*) race A in pigeonpea (*Cajanus cajan*) genotypes. Ann Appl Biol 136(3):247–252

Siddiqui ZA, Akhtar MS (2009) Effect of plant growth promoting rhizobacteria, nematode parasitic fungi and root-nodule bacterium on root-knot nematodes *Meloidogyne javanica* and growth of chickpea. Biocontrol Sci Tech 19(5):511–521

Siddiqui ZA, Mahmood I (1993) Occurrence of races of *Heterodera cajani* in Uttar Pradesh, India. Nematol Medit 21(2):185–186

Siddiqui ZA, Mahmood I (1994) Interactions of *Meloidogyne javanica*, *Rotylenchulus reniformis*, *Fusarium oxysporum* f. sp. ciceri and Bradyrhizobium japonicum on the wilt disease complex of chickpea. Nematol Medit 22(2):135–140

Siddiqui ZA, Mahmood I (1995a) Biological control of *Heterodera cajani* and *Fusarium udum* by *Bacillus subtilis*, *Bradyrhizobium japonicum* and *Glomus fasciculatum* on pigeonpea. Fund Appl Nematol 18(6):559–566

Siddiqui ZA, Mahmood I (1995b) Some observations on the management of the wilt disease complex of pigeonpea by treatment with a vesicular arbuscular fungus and biocontrol agents for nematodes. Bioresour Technol 54(3):227–230

Siddiqui ZA, Mahmood I (1996) Biological control of *Heterodera cajani* and *Fusarium udum* on pigeonpea by *Glomus mosseae*, *Trichoderma harzianum*, and *Verticillium chlamydosporium*. Israel J Plant Sci 44(1):49–56

Sikora RA, Greco N, Veloso Silva JF (2005) Nematode parasites of food legumes. In: Luc M, Sikora RA, Bridge J (eds) Plant parasitic nematodes in subtropical and tropical agriculture. CABI Publishing, Wallingford, pp 259–318

Singh B (2009) Management of root-knot nematode *Meloidogyne javanica* by seed treatment with neem based products in pigeonpea. J Food Legumes 22(1):71–72

Singh KP, Singh VK (1992) *Terminelia arjuna* leaf powder reduces population density of *Heterodera cajani*. Int Pigeonpea Newsl 16:17–18

Singh B, Ali SS, Naimuddin. and Askary, T.H. (2004) Combined effect of *Fusarium udum* and *Meloidogyne javanica* on wilt resistant accessions of pigeonpea. Ann Plant Prot Sci 12 (1):130–133

Singh S, Bhagawati B, Goswami BK (2011) Biomanagement of root-knot disease of chickpea caused by *Meloidogyne incognita*. Ann Plant Prot Sci 19(1):159–163

Sivakumar CV, Seshadri AR (1971) Life history of the reniform nematode, *Rotylenchulus reniformis* Linford and Oliveira, 1940. Indian J Nematol 1:7–20

Subramanian S and Sivakumar M. 2010 Management of Heterodera cajani in pigeon pea by non-chemical approaches. National Conf. on Innovations in Nematological Research for Agricultural Sustainability – Challenges and A Roadmap Ahead. Tamil Nadu Agri. Univ., Coimbatore, p. 93

Sumita K, Das D (2014) Management of root-knot nematode, *Meloidogyne incognita* on green gram through bioagents. Int J Plant, AnimEnvt Sci 4(4):287–289

Talavera M, Valor H (2000) Influence of the previ-ous crop on the anhydrobiotic ability of *Pratylenchus thornei* and *Merlinius brevidens*. Nematol Medit 28:77–81

Tiwari SP, Vadhera I, Shukla BN, Bhatt J (1992) Studies on the pathogenicity and relative reactions of chickpea lines to *Pratylenchus thornei* (Filipjev, 1936) Sher & Allen, 1953. Indian J Mycol Pl Pathol 22(3):255–259

Tiyagi SA, Ajaz S (2004) Biological control of plant parasitic nematodes associated with chickpea using oil cakes and *Paecilomyces lilacinus*. Indian J. Nematol 34(1):44–48

Tiyagi SA, Parveen M (1992) Pathogenic effect of root-lesion nematode *Pratylenchus thornei* on plant growth, water absorption capability, and chlorophyll content of chickpea. Int Chickpea Newsl 26:18–20

Uma Maheswari T, Sharma SB, Reddy DDR, Haware MP (1995) Co-infection of wilt resistant chickpeas by *Fusarium oxysporum* f. sp. *ciceri* and *Meloidogyne javanica*. J Nematol 27:649–653

Uma Maheswari T, Sharma SB, Reddy DDR, Haware MP (1997) Interaction of *Fusarium oxysporum* f. sp. *ciceris* and *Meloidogyne javanica* on Cicer arietinum. J Nematol 29:117–126

Upadhyay KD, Dwivedi K (1987) Effect of interaction between *Meloidogyne javanica* and Fusarium oxysporum f. sp. ciceri on chickpea. Indian J Nematol 17:145–146

Upadhyay KD, Dwivedi K, Srivastava SK (1997) Effect of intercropping on pigeonpea infested with *Meloidogyne incognita*. Indian J Nematol 27(2):270

Varaprasad KS, Sharma SB, Loknathan TR (1997) Nematode constraints to pigeonpea and chickpea in Vidarbha region of Maharashtra in India. Int. J. Nematol 7(2):152–157

Walia RK, Chakraborty PK (eds) (2018) Nematodes problems of crops in India. ICAR-All India Coordinated Research Project on Nematodes in Agriculture, New Delhi, 402 pp

Wani AH, Bhat MY (2012) Control of root-knot nematode. *Meloidogyne incognita* by urea coated with Nimin or other natural oils on mung, *Vigna radiata* (L.) R. Wilczek. J Biopest 5:255–258

Yadav BS (1986) Nematode problems of pulse crops. In: Swarup G, Dasgupta DR (eds) Plant-parasitic nematodes of India, problems and progress. Indian Agricultural Research Institute, New Delhi, pp 328–335

Zaki FA, Bhatti DS (1986) Control of pigeon-pea cyst nematode, *Heterodera cajani* Koshy, 1967 by chemical seed treatment. Indian J Nematol 16(1):106–108

# Oilseed Crops

# 4


**Abstracts**

Economically important nematodes associated with oilseed crops like groundnut, castor, soybean, and sunflower in India include root-knot nematodes (*Meloidogyne* spp.), reniform nematode (*Rotylenchulus reniformis*), and cyst nematode (*Heterodera* spp.). In this chapter, symptoms of damage inflicted by particular plant parasitic nematodes on oilseed crops, crop losses, survival, and spread of the nematodes are described, as are the management strategies available. These typically include the use of regulatory, physical, cultural, chemical, biological, host plant resistance, and integrated methods.



**Keywords**

*Meloidogyne* spp. · *Rotylenchulus reniformis* · *Heterodera* spp. · Symptoms · Management


## 4.1 Groundnut, *Arachis hypogea*

The root-knot nematode *Meloidogyne arenaria* and the stunt (Kalahasthy malady) nematode *Tylenchorhynchus brevilineatus* are the serious problems associated with groundnut.

### 4.1.1 Root-Knot Nematodes, *Meloidogyne arenaria* and *M. javanica*

The peanut root-knot nematode *Meloidogyne arenaria* is the most common and severe species infecting peanut (Dickson 1998). Patel et al. (1988)) and Sakhuja and Sethi (1985) reported from different parts of the world including India the presence




P. P. Reddy, *Nematode Diseases of Crops and their Management*, https://doi.org/10.1007/978-981-16-3242-6_4

of *M. javanica* infecting groundnut. *Meloidogyne arenaria* and *M. javanica* are the major constraints to groundnut damage in Gujarat, India (Patel et al. 1996).

#### 4.1.1.1 Distribution

Root-knot nematodes are distributed in Gujarat, Rajasthan, and Uttar Pradesh.

The root-knot nematode *M. arenaria* is widely distributed in the groundnut growing areas of Gujarat such as Supedi, Sanala, Kathrota, Upleta, and Patanvav regions in Rajkot district; and Khadia and Dhoraji in Junagadh district of Saurashtra region. In Betawada, Khalvada, Lalpur, Muvada, Narsipur, Pariana, and Telnar regions of Napadvanj mandal of Kaira district of Gujarat, the presence of *M. javanica* has been reported. A pathotype of *M. javanica* which reproduces on groundnut and produces severe galls is present in Kapadvanj area (Patel et al. 1993). It is evident that root-knot nematodes are important constraints to groundnut production in Gujarat.

#### 4.1.1.2 Crop Losses

*Meloidogyne arenaria* and *M. javanica* are responsible for 13–50% and 10–23% of yield losses, respectively. Losses caused by the nematode (*M. arenaria*) are about 8.55–14.15% depending on the nematode population.

*M. arenaria* is responsible for 4.5% annual yield loss in groundnut amounting to ₹396 million in Gujarat, India (Walia and Chakraborty 2018).

#### 4.1.1.3 Symptoms

Areas of root-knot nematode-infected peanuts are usually round to oblong in shape. Root-knot nematode-affected plants show stunting, reduced number of branches and internodes, burning of leaf margins, reduced leaf size, yellowing of foliage, and will wilt more readily during afternoon when the sun's heat is maximum in infested field. Rows of infected plants may never, or not so quickly, meet as those of healthy plants (Fig. 4.1).

**Fig. 4.1** *Left*—Peanut fields exhibiting symptoms of extensive damage due to a severe infestation of the root-knot nematode. *Middle*—Galling of roots. *Right*—Galling of peanut pods and pegs due to *Meloidogyne arenaria* infection

Symptoms of peanut root-knot nematodes include the presence of galls (knots) on both roots and pods (single or multiple wart-like growths) (Fig. 4.1). Further damage and ultimate death of plants is caused as root-knot nematode infection progresses, and the secondary root and pod rotting occurs.

Abdel-Momen and Starr (1997) reported that at the initial nematode (*M. arenaria*) population density of 1–10 eggs and second-stage juveniles per 500 $cm^3$ soil (economic damage threshold), the yields begin to decline on susceptible peanut cultivars.

#### 4.1.1.4 Biology and Life Cycle

Root-knot nematode is soil-borne; second-stage juveniles are present in soil, invade the roots, and become sedentary endoparasites. The nematode feeds on vascular tissues of roots and tender pods and develops to sac-like adult female inside the root galls. The nematode lays its eggs commonly on the root surface. However, the egg masses may be embedded inside the compound galls also. The nematode takes about a month to complete its life cycle. It completes several generations within a crop season, since the nematode is multivoltine. The juveniles hatch and emerge from galls and reinfect feeder roots.

#### 4.1.1.5 Interaction with Other Microorganisms

The presence of *M. arenaria* along with Fusarium solani advances wilting of groundnut plants (Patel et al. 1985) and reduces the number of *Rhizobium* nodules.

#### 4.1.1.6 Dissemination

The nematodes spread through irrigation water and agronomic practices.

#### 4.1.1.7 Management

##### 4.1.1.7.1 Cultural Methods

Crop Rotation

Rotation of a peanut crop for 1 or 2 years with the most effective crops such as grasses such as Bahia grass, Bermuda grass, millets, and sorghum was effective for the management of *M. arenaria* on groundnut. Though some root-knot nematode reproduction could be expected on field corn, this crop is considerably less susceptible than peanut and is generally effective in the reduction of root-knot nematode soil populations. Rotation of peanut with cotton is beneficial, since peanut is susceptible to *M. incognita*. Similarly, cotton is not susceptible to *M. arenaria*. Hence, rotation of peanut with cotton is effective for the management of root-knot nematodes infecting both crops.

A 2-year Bahia grass rotation is sufficient to manage plant parasitic nematodes in peanut crop provided weeds in the Bahia grass planting are controlled early and regularly during the first year and continued throughout the cycle of rotation.

A 2-year rotation with cotton, Bahia grass, or velvet bean (Rodríguez-Kábana et al. 1991a, b) was also found effective.

#### Cover Crops

The use of winter cover crops is helpful to provide competition against volunteer peanuts and spring weeds, and also the cover crop planting process helps destroy those plants growing in the fall. Only nonhost or poor host cover crops should be used for nematode management. Winter cereals are most suitable for managing root-knot nematode in this regard.

#### Organic Amendments

Organic cakes of castor, mustard, and neem (1 t $ha^{-1}$ and above) 1 week before sowing under crop row significantly reduce the root-knot disease and increase pod and haulm yields. Neem cake is more effective than other cakes.

#### Early Maturing Varieties

Late maturing varieties have more potential for damage than short-season Spanish market type's cultivation of varieties.

### 4.1.1.7.2 Chemical Methods

Management of root-knot nematodes on peanut has traditionally relied primarily on treatment of infested fields with nematicides such as granular Carbofuran. A split application of Carbofuran 3G with a band or in-furrow application at planting followed by a band application at peg initiation is an acceptable usage.

Seed treatment with 6% Carbofuran reduces *M. arenaria* damage (Patel et al. 1986). Groundnut seeds soaked for 12 h in Fenamiphos, Monocrotophos, Phosalane, or Oxamyl (each at 125, 250, and 500 ppm) inhibit the penetration of *M. arenaria* larvae in roots and increase root biomass. Fenamiphos at 250 and 500 ppm is highly effective.

Application of Carbofuran at 2 kg $ha^{-1}$ in *M. arenaria* or *M. javanica*-infested soils reduces the root-knot disease by 16–96% and increases the pod yields by 15–33% (Patel 1983).

### 4.1.1.7.3 Biological Methods

Seed treatment with *Purpureocillium lilacinum* at 10 g/kg seed before sowing was found effective.

The biocontrol agent, NemOut, is a formulation of spores of the fungus, *Purpureocillum lilacinum*, and this product used as both an in-furrow and as an at-pegging treatment has shown a moderate level of nematode suppression in University of Florida trials. It should only be used in conjunction with other recommended nematode management practices including good crop rotation and weed control.

#### 4.1.1.7.4 Host Resistance

The sources of resistance against root-knot nematodes on groundnut in Gujarat have been identified (Table 4.1) which can be used in breeding programs.

Varieties G-201, Japtin-220-15, Local-256, Local Ambali, and 4018 are highly resistant against *M. javanica*, and N-C-4x revealed resistance against *M. arenaria*. Five cultivars of groundnut, viz. C-212, C-371, C-396, BP-1, and US-74, were found to be moderately resistant to *M. arenaria*.

Under Florida, USA growing conditions, the peanut variety "Tifguard" showed resistance to both root-knot nematode and tomato spotted wilt virus. The variety has a good yield potential and governed by single dominant gene which provides excellent resistance to the peanut root-knot nematode.

The cultivar COAN with resistance to *M. arenaria* and *M. javanica* was released by the Texas Agricultural Experiment Station (Simpson and Starr 2001). In 2002, the cultivar NemaTam, which has greater yield potential than COAN with the same level of resistance to *M. arenaria* and *M. javanica*, was released by the Texas Agricultural Experiment Station. The availability of these nematode-resistant cultivars gives growers an additional option for management of *M. arenaria* and *M. javanica* and will reduce the growers' reliance on nematicides.

Peanut-resistant cultivars COAN and NemaTAM are recommended only for fields known to be infested with *M arenaria* or *M. javanica*, since the yield of both varieties are not comparable to susceptible varieties commercially grown.

An additional benefit of the resistance of COAN and NemaTAM is that reproduction of the root-knot nematode is inhibited relative to reproduction on a susceptible cultivar (Koenning and Barker 1992). Therefore, these resistant cultivars not only protect the peanut crop from yield loss due to nematode parasitism, but they also suppress the nematode's population density at crop harvest. Any crop planted after the resistant peanut will be subjected to less disease pressure than it would experience following a susceptible peanut crop.

#### 4.1.1.7.5 Integrated Methods

Use of organic amendments, neem or castor cakes, at 1 t/ha 7 days prior to sowing has been found to reduce root-knot nematode damage in groundnut. Further

**Table 4.1** Groundnut lines identified as resistant to *Meloidogyne arenaria* or *M. javanica* pathotype 2 in Gujarat, India

| Root-knot nematode species | Highly resistant | Resistant |
|---|---|---|
| *Meloidogyne arenaria* | None | A3, Abuarbaa, Ah 25, Ah 3328, Ah 4515, Ah 6902, Ah 6719, Ah 7188, Ah 7299, Ak 10-2, 55-437, C 83, C 149, EC 24118, Kigung, Khargaon 3, NC Ac 50 |
| *M. javanica* | ICG 5341, ICG 6330 | Apexy, B 1, C 162, C 166, Dangi, EC 85994, ICGs 411, 852, 859, 1268, 2248, 2496, 3053, 3104, 6323, 6826, 10,047, JH 223, KG 61-22, PI 263594, PI 270787, S 7-2-1, S 7-24-3, U 4-7-3, No 923, No 523 |

reduction in nematode damage and increase in yield can be obtained by combinations with Carbosulfan seed treatment.

Application of Carbofuran (3G) at 1 kg a.i./ha and neem oil at 5 l/ha (at 2000 l of water/ha with 0.1% detergent powder) 1 day before sowing.

For effective management of root-knot nematodes infecting groundnut, farmers of South Saurashtra region of Gujarat have been advised to sow groundnut with castor as an intercrop (row ratio 2:1) along with soil application of Carbofuran at 1.0 kg a.i./ha.

Groundnut plants treated with *P. fluorescens* as both seed treatment + soil application recorded least number of galls/plant, egg masses/plant, eggs/egg mass, and soil population/200 ml soil with corresponding reduction of 70.9, 34.4, 20.3, and 68.1%, respectively over untreated control (Ramakrishnan 2003).

Combined application of *P. fluorescens* with Carbofuran 3G recorded maximum reductions in number of galls/plant and soil population/200 ml soil to the tune of 66.1 and 45.2%, respectively, from control compared to only 61.1 and 40.5% reduction in plants treated with *P. fluorescens* alone. The root colonization of both *P. fluorescens* and *T. viride* was not affected by combined application with Carbofuran 3G. It is very clear from the present results that Carbofuran does not affect biocontrol potential of *P. flourescens* and *T. viride* against *M. arenaria* in groundnut (Ramakrishnan 2003).

The following practices have been recommended for the integrated management of root-knot nematodes in groundnut:

- In order to control root-knot nematodes on groundnut, farmers of South Saurashtra region of Gujarat have been advised to sow groundnut with castor as an intercrop (row ratio 2:1) along with soil application of Carbofuran at 1.0 kg a. i./ha.
- Late maturing varieties have more potential for damage than short-season Spanish market types cultivation of varieties.

### 4.1.2 Stunt Nematode, *Tylenchorhynchus brevilineatus*

During 1975–1976, a severe disease of groundnut characterized by reduction in pod size and brownish discoloration of pod surface was noticed near Kalahasti village of Andhra Pradesh, and since then, it is popularly known as *Kalahasti* malady. It has also been observed in Nellore district of Andhra Pradesh. This disease was reportedly caused by the nematode *Tylenchorynchus brevilineatus* (Reddy et al. 1984).

#### 4.1.2.1 Distribution

The nematode has been reported only from Andhra Pradesh in India.

#### 4.1.2.2 Crop Losses

Estimation of crop losses due to *Kalahasti* malady was limited to visual estimation and was accounted for 40–60%.

#### 4.1.2.3 Symptoms

Infected plants appear in patches in the field, stunted in growth with greener than normal foliage, presence of small and necrotic lesions on roots, pegs, and developing young pods. Host cells around lesions are proliferated and the margins of the lesions appear to be slightly elevated. Peg length is reduced, and in advanced stages of the disease, the entire pod surface becomes blackened (Fig. 4.2). Discoloration can also be observed on roots, but this is less conspicuous than pod discoloration. The kernels inside the affected pods are slightly smaller and healthy.

#### 4.1.2.4 Host Range

*Tylenchorhynchus brevilineatus* has been reported on several crops, viz. *Cicer arietinum, Citrus sinensis, Cuminum cyminum, Dahlia* spp., and *Sorghum* spp.

#### 4.1.2.5 Management

##### 4.1.2.5.1 Cultural Methods

Crop rotation with rice was found effective.

Rotation of groundnut with marigold *Tagetes erecta* (poor host for *T. brevilineatus*) was responsible for the enhancement of groundnut pod yield and reduction in pod disease severity and the nematode population in soil followed by mustard crop in rotation (Naidu et al. 2000a).

Poultry manure (5 t $ha^{-1}$), farmyard manure (10 t $ha^{-1}$), and saw dust (2.5 t $ha^{-1}$) reduced population levels of *T. brevilineatus* and disease severity, and increased groundnut yield over nonamended control plots. Greatest nematode control (33.5%) and highest increase in pod yields (50.3%) were obtained in poultry manure

**Fig. 4.2** Groundnut plant infected with *Tylenchorhynchus brevilineatus* showing small severely discolored pods

amendment followed by neem cake (1 t $ha^{-1}$). Naidu et al. (2000b) reported effective management of the stunt nematode with soil incorporation of poultry manure (benefit: cost ratio of 8.7), followed by farmyard manure (benefit: cost ratio of 3.6). Mustard cake significantly increased pod yield (Naidu et al. 2000b).

Application of gypsum at 200 kg/ha at the time of earthing-up was found effective.

#### 4.1.2.5.2 Chemical Methods

Application of Carbofuran 3G at 1.0 kg a.i./ha 25–30 days after sowing along with irrigation water decreased both soil populations of *T. brevilineatus* and of the percentage of diseased pods and increased plant height and yield of pods, pod, and kernel weights.

#### 4.1.2.5.3 Host Resistance

Grow resistant varieties. Mehan et al. (1993) reported that a high-yielding breeding line TCG 1518 resistant to *T. brevilineatus* which is in the process of release in nematode-infested areas in Andhra Pradesh.

## 4.2 Castor, *Ricinus communis*

The reniform nematode *Rotylenchulus reniformis* is a major pest of castor in India.

### 4.2.1 Reniform Nematode, *Rotylenchulus reniformis*

The reniform nematode was first reported on castor by Seshadri and Sivakumar (1963) from Tamil Nadu, India. The nematode-infested plants produce seeds of inferior quality containing lesser amounts and inferior quality oil.

#### 4.2.1.1 Distribution

The nematode is distributed in Andhra Pradesh, Gujarat, Rajasthan, and Tamil Nadu.

#### 4.2.1.2 Crop Losses

The reniform nematode is responsible for annual yield loss of 15% in castor amounting to ₹1082.78 million in India (Walia and Chakraborty 2018).

#### 4.2.1.3 Symptoms

The reniform nematode causes growth reduction, shedding of leaves, early flowering, malformation, and discoloration of seeds. Seshadri and Sivakumar (1963) reported dieback symptoms and stunting in heavily infested castor fields (Fig. 4.3).

**Fig. 4.3** Reniform nematode symptoms on castor plants (yellowing and stunting of plants in patches) (*left*) and roots (females attached to roots) (*right*)

### 4.2.1.4 Life Cycle

The life cycle of *R. reniformis* on castor is studied in detail by Sivakumar and Seshadri (1971). From egg to egg, the nematode completes its life cycle in 24–29 days. The nematode secretes a gelatinous matrix through the vulval opening in which about 83 eggs are laid.

After penetration, the young female nematode (only infective stage) establishes feeding site, feeds on the cells, and matures into kidney-shaped adult female projecting most of the body outside the root surface. Adult female lays about 30–200 eggs embedded in a gelatinous sac. Depending on the host suitability and temperature, the nematode completes its life cycle in 3–4 weeks.

### 4.2.1.5 Ecology

Soil moisture from 25% to 30% and 30 °C soil temperature were optimum for its reproduction on castor (Khan and Khan 1973). Sivakumar and Seshadri (1972) found that while there is good multiplication of the nematode in sandy loam, brown loam, and black clay loam, the symptom expression was more severe in black clay loam. Mukhopadhyay and Haque (1980) reported that laterite (sandy clay loam) and alluvial (clay loam) soils supported maximum number of reniform nematodes while sandy soil supported the least. Vigorous plant growth in these soil types provided the nematodes with larger feeding areas which might account for higher rate of nematode multiplication.

### 4.2.1.6 Host Range

The nematode is polyphagous, but it is a major pest of castor, cotton, and cowpea.

#### 4.2.1.7 Survival and Spread

Sivakumar and Seshadri (1976) reported that the reniform nematode (juvenile female) lives for about 6 months in desiccated soil and 7 months in moist soil. There is an indication that the fourth-stage larvae are better adapted for survival under adverse conditions, being protected by the previous larval cuticles which are not shed.

The nematodes are disseminated through irrigation water and agronomic practices.

#### 4.2.1.8 Management

##### 4.2.1.8.1 Physical Methods

Soil solarization with LLDPE (25μm) for 3 weeks reduces incidence of reniform nematode fungus wilt complex.

##### 4.2.1.8.2 Cultural Methods

Crop rotation with cereals (sorghum, maize, wheat, rice, millets, and wheat) and mustard.

Khan and Khan (1973) reported onion, garlic, turnip, bell pepper, and carrot as nonhosts for *R. reniformis* in India. Growing of these nonhost crops for 2 years between castors gives fair to good control of the reniform nematode.

Application of FYM at 10 t/ha or neem cake at 500 kg/ha was also found effective against reniform nematode on castor.

##### 4.2.1.8.3 Chemical Methods

Soil application of Carbofuran 3G at 1 kg a.i./ha at the time of sowing was found effective.

##### 4.2.1.8.4 Biological Methods

Soil application of Pseudomonas fluorescens *Pf1* at 2.5 kg/ha proved beneficial.

##### 4.2.1.8.5 Host Resistance

Nandwana et al. (1981) tested 70 varieties of castor against the reniform nematode and found only one variety GH-3 to be highly resistant and 21 cultivars to be tolerant.

In severely infected fields, use of nematode-resistant castor cultivars/lines like RG 5, RG 450 (VI 2-1), RG 460 (VI 18), SHB 110, SHB 118, SKI 50; SPS 43-2, JI 5, JI 59; RC 879/1, RC 13676, RC 1203, RC 55212, and AS 7 is beneficial.

##### 4.2.1.8.6 Integrated Methods

Seed treatment with *Pseudomonas fluorescens* at 20 g/kg and its soil application at 2.5 kg/ha was found effective for managing reniform nematode.

Seed treatment of castor cv. TMV 5 with Carbosulfan 2% a.i. (w/w) combined with application of Rugby at 1 kg a.i./ha at sowing reduced the reniform nematode population significantly by 22.5%.

Application of *Purpureocillium lilacinum* at 2.5 kg/ha at the time of sowing along with farm yard manure (FYM) was also found effective.

## 4.3 Soybean, *Glycine max*

### 4.3.1 Cyst Nematode, *Heterodera glycines*

#### 4.3.1.1 Crop Losses

In heavily infested fields, soybean cyst nematode (SCN) can cause soybean yield losses of more than 30%, and in some sandy soils, complete yield loss can occur, especially in a drought year.

#### 4.3.1.2 Symptoms

Poor patches in field, stunting of plant growth, and leaf yellowing are some of the symptoms exhibited by soybean plants infected with the cyst nematode in sandy soils (Fig. 4.4). The indirect above-ground symptoms include early senescence in fields. Infected plants have poorly developed fibrous roots and with fewer *Rhizobium* nodules. Small, white to yellow spheres (bodies of female nematodes about the size of coarse sugar grains) attached to the root surface can be observed by close root examination during early in the growing season (Fig. 4.5). The nematode is responsible for severe yield losses without showing symptoms on the above-ground plant parts.

#### 4.3.1.3 Life Cycle

SCN overwinters as eggs encased in the female cyst, which helps protect them from the environment and predators. After hatching and a first molt, second-stage juveniles (J2) emerge and penetrate soybean roots using a spear-like mouth structure called a stylet and then enter the root tissue. After penetration, J2 establishes a

**Fig. 4.4** Symptoms of cyst nematode injury in soybean field (patches of yellow and stunted plants)

**Fig. 4.5** Cyst nematode females attached to soybean roots

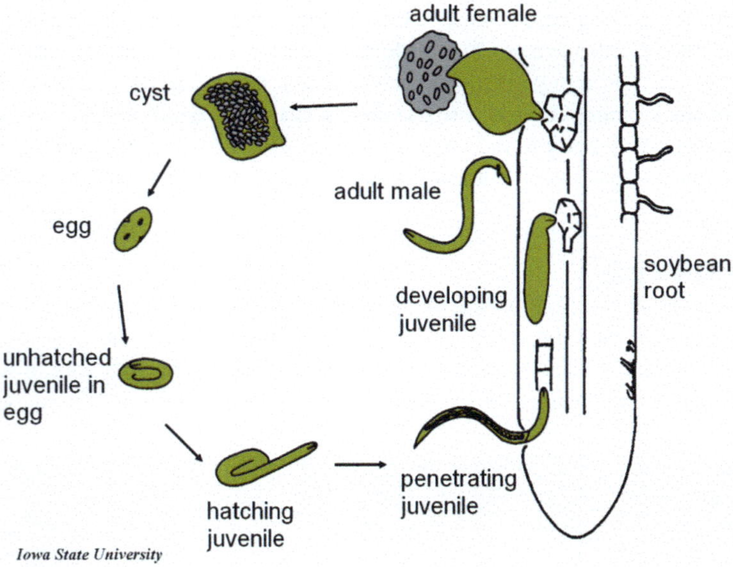


**Fig. 4.6** Life cycle of soybean cyst nematode

syncytium (a feeding site) and becomes sedentary. After feeding, J2 undergoes three molts and become adult. During its entire life cycle, the adult female remains attached to the feeding site. The body enlarges to become lemon shaped containing about 500 eggs, after the female is fertilized. The enlarging nematode body breaks through the epidermis of the root. When the plant matures, the female body becomes a brown cyst after its death. The eggs are protected for several years inside the brown cyst with a strong structure. Under favorable conditions, a life cycle can be completed about once a month (Fig. 4.6).

#### 4.3.1.4 Management

##### 4.3.1.4.1 Cultural Methods

Rotating soybeans with a nonhost crop is the simplest and least expensive method to reduce SCN populations. Nonhost crops include corn, peanut (Fig. 4.7), cotton, small grains, and alfalfa. Host plants, on which the SCN population can maintain, include all types of beans, lespedeza, and hairy vetch. SCN is capable of reproducing well on several weeds, including henbit, purple dead nettle, and common mullein. One year in a weed-free nonhost crop can reduce an SCN population as much as 55%. Starving all the nematodes is not possible, as some of the eggs may remain unhatched in the cyst for years.

##### 4.3.1.4.2 Host Resistance

Almost every seed company now offers SCN-resistant varieties. The source of resistance for almost 97% of the current resistant cultivars is the same (PI 88788 and Peking) (Fig. 4.8).

##### 4.3.1.4.3 Integrated Methods

Long-term effective management of SCN will rely on an integrated program that includes resistant soybean varieties, crop rotation, and possibly alternative strategies such as soil fertility management and biological control.

### 4.3.2 Root-Knot Nematodes, *Meloidogyne incognita* and *M. javanica*

#### 4.3.2.1 Crop Losses

Yield losses ranging from 25% to 70% have been reported for soybean as a result of root-knot nematode parasitism (Fourie et al. 2010).

**Fig. 4.7** Effect of crop rotation on soybean cyst nematode

**Fig. 4.8** Effect of resistant (back) and susceptible (front) soybean cultivars on cyst nematode

**Fig. 4.9** Field infestation of root-knot nematodes showing patches of poorly growing soybean plants

#### 4.3.2.2 Symptoms

The above-ground symptoms include patches of poorly growing soybean plants in a field (where high population densities of root-knot nematodes occurred) and stunted plants with yellow leaves (Fig. 4.9). Galling as a result of root-knot nematode parasitism is characterized by elongated to roundish, knot-like protuberances that form on soybean roots (Fig. 4.10).

#### 4.3.2.3 Management

##### 4.3.2.3.1 Chemical Methods

Seed treatment with Oxamyl SL, Terbufos GR, and Abamectin resulted in a significant ($P \leq 0.05$) reduction of root-knot nematode (*M. incognita*) population levels. These nematicides also generally resulted in a higher income per hectare compared to the untreated control.

**Fig. 4.10** Galls on soybean roots due to root-knot nematode infection

#### 4.3.2.3.2 Host Resistance

Soybean cv. Gazelle is resistant to *M. javanica* (monogenic) and cv. LS5995 is resistant to *M. incognita* (polygenic).

## 4.4 Sunflower, *Helianthus annuus*

### 4.4.1 Root-Knot Nematode, *Meloidogyne incognita*

#### 4.4.1.1 Crop Losses

*M. incognita* is responsible for 16% annual yield loss in sunflower amounting to ₹240 million in India (Walia and Chakraborty 2018).

#### 4.4.1.2 Symptoms

The most common above-ground symptoms caused by nematodes include chlorosis (yellowing), stunting, and wilting during the day (with the plant recovering during evening hours) (Fig. 4.11). Root galls are the below-ground symptoms (Fig. 4.12).

#### 4.4.1.3 Management

The economical control of root-knot nematode requires a weed-free rotation to monocot crops such as corn, sorghum, or wheat.

Rugby 10 G proved to be the best in reducing root-knot disease and increasing plant vigor, followed by Carbofuran-3 G (Rehman et al. 2006b).

The sunflower varieties Beimisal-205, Hyson-33, and Super-25 exhibited tolerance to *M. incognita* (Rehman et al. 2006a).

**Fig. 4.11** Symptoms of root-knot nematode infection on sunflower (note stunting and wilting of plants in a patch)

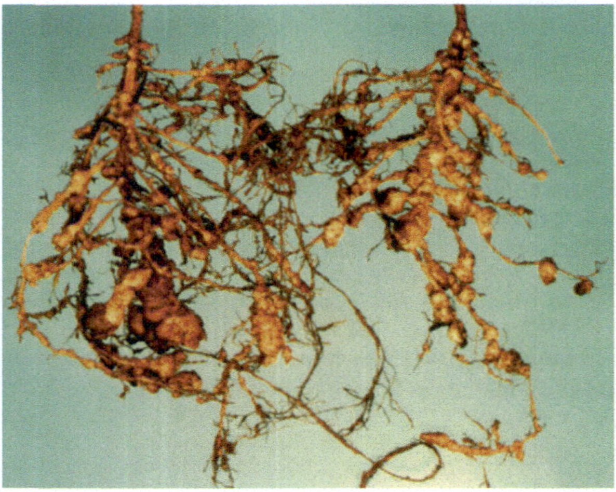

**Fig. 4.12** Sunflower roots infected with root-knot nematodes

## 4.5 Conclusion

The nematode pests like root-knot (*Meloidogyne* spp.), cyst (*Heterodera* spp.), and reniform (*Rotylenchulus reniformis*) nematodes are important limiting factors for successful production and expansion of area under oilseed crops. Research on the impact of plant parasitic nematodes on oilseed crops and the development of management strategies to limit damage inflicted by such pests has to be pursued. There is a need for research work on aspects like distribution and impact of nematode diseases on different oil seed crops. Initiatives to reduce population levels of plant parasitic nematodes in oilseed crops with practical and cost-effective strategies are the need of the hour to ensure sustainable crop production. The current approach toward environmental-friendly strategies to combat diseases and pests, particularly plant parasitic nematodes, increases the pressure on scientists and related industries to coordinate related research initiatives.

## References

Abdel-Momen SM, Starr JL (1997) Damage functions for three *Meloidogyne* species on *Arachis hypogaea* in Texas. J Nematol 29:478–483

Dickson DW (1998) Peanut. In: Barker KR, Peterson GA, Windham GL (eds) Plant and nematode interaction. American Society of Agronomy, Madison, WI, pp 523–566

Fourie H, Mc Donald AH, De Waele D (2010) Relationships between initial population densities of *Meloidogyne incognita* race 2 and nematode population development in terms of variable soybean resistance. J Nematol 42:55–61

Khan FA, Khan AM (1973) Studies on the reniform nematode, *Rotylenchulus reniformis* I. Host range and population changes. Indian J Nematol 3:24–30

Koenning SR, Barker KR (1992) Relative damage functions and reproductive potentials of *Meloidogyne arenaria* and *M. hapla* on peanut. J Nematol 24:187–192

Mehan VK, Reddy DDR, McDonald D (1993) Resistance in groundnut genotypes to Kalahasti malady caused by the stunt nematode, *Tylenchorhynchus brevilineatus*. Int J Pest Mangmt 39:201–203

Mukhopadhyay MC, Haque MS (1980) Influence of soil types on *Rotylenchulus reniformis* on castor. Indian J Nematol 10:89–91

Naidu PH, Mosas GJ, Reddy DDR (2000a) Influence of intercropping on Kalahasti malady (*Tylenchorhynchus brevilineatus*) in groundnut. J Mycol Plant Pathol 30:207–209

Naidu PH, Mosas GJ, Sitaramaiah K (2000b) Control of groundnut Kalahasti malady (*Tylenchorhynchus brevilineatus*) through organic and inorganic soil amendments. J. Mycol Plant Pathol. 30(2):180–183

Nandwana RP, Pathak AK, Yadav BS (1981) Susceptibility of varieties of certain oil seed crops to the reniform nematode *Rotylenchulus reniformis*. In: Second all India nematology symposium. Coimbatore, p 49

Patel HR (1983) Studies on root-knot nematodes of groundnut. Ph. D. thesis, Gujarat Agricultural University, Sardarkrushinagar, Gujarat, 125 pp

Patel HR, Vaishnav MU, Dhruj IU (1985) Interaction of *Meloidogyne arenaria* and Fusarium solani on groundnut. Indian J Nematol 15:98–99

Patel HR, Vaishnav MU, Dhruj IU (1986) Efficacy of aldicarb sulfone and carbofuran flowable seed treatment on plant growth and against *Meloidogyne arenaria* on groundnut. Pesticides 20:29–31

Patel DJ, Patel BA, Chavda JC, Patel HV (1988) Record of *Meloidogyne javanica* on groundnut in Gujarat, India. Int Arachis Newsl 3:16–17

Patel HR, Patel BA, Patel HV (1993) Pathotypes of *Meloidogyne javanica* in India. Nematol Medit 21:207–208

Patel BA, Patel DJ, Sharma SB, Patel HV (1996) Nematode problems of groundnut and their management in Gujarat, India. Int Arachis Newsl 16:38–39

Ramakrishnan S (2003) Studies on the management of root-knot nematode, *Meloidogyne arenaria* (Neal, 1989) in groundnut (*Arachis hypogaea* L.). M. Sc. (Agriculture) thesis, Department of Nematology, Tamil Nadu Agricultural University, Coimbatore

Reddy DDR, Subrahmanyam P, Sankara Reddy GH et al (1984) A nematode disease of peanut caused by *Tylenchorhynchus brevilineatus*. Plant Dis 68:526–529

Rehman A, Bibi R, Ullah MH (2006a) Screening of different sunflower cultivars against root-knot nematode (*Meloidogyne incognita*). J Agric Soc Sci 2(3):182–184

Rehman A, Bibi R, Ullah MH (2006b) Evaluation of different chemicals against root knot nematode (*Meloidogyne incognita*) on sunflower. J Agric Soc Sci 2(3):185–186
Rodríguez-Kábana R, Robertson DG, Weaver CF, Wells L (1991a) Rotations of bahiagrass and castorbean with peanut for the management of *Meloidogyne arenaria. Suppl.* J Nematol 23:658–661
Rodríguez-Kábana R, Robertson DG, Wells L et al (1991b) Cotton as a rotation crop for management of *Meloidogyne arenaria* and *Sclerotium rolfsii* in peanut. Suppl J Nematol 23:652–657
Sakhuja PK, Sethi CL (1985) Frequency of occurrence of various plant-parasitic nematodes and root-rot fungi on groundnut in Punjab. Indian J Nematol 15:191–194
Seshadri AR, Sivakumar C (1963) A preliminary note on the reniform nematode (*Rotylenchulus reniformis* Linford and Oliveira, 1940) on a number of cultivated crops in South India. Madras Agric J 50:134–137
Simpson CE, Starr JL (2001) Registration of 'COAN' peanut. Crop Sci 41:918
Sivakumar CV, Seshadri AR (1971) The life history of the reniform nematode, *Rotylenchulus* reniformis Linford and Oliveira, 1940. Indian J Nematol 1:7–20
Sivakumar CV, Seshadri AR (1972) Effect of soil texture on the reniform nematode, *Rotylenchulus* reniformis. Indian J Nematol 2:83–86
Sivakumar CV, Seshadri AR (1976) Longevity of the reniform nematode, *Rotylenchulus reniformis* in host free soil. Indian J Nematol 6:138–144
Walia RK, Chakraborty PK (eds) (2018) Nematodes problems of crops in India. ICAR-All India Coordinated Research Project on Nematodes in Agriculture, New Delhi. 402 pp

# Part IV

# Fiber and Sugar Crops

# Fiber Crops

# 5

**Abstract**

In India, the major nematode problems in different cotton- and jute-growing areas which affect the production and quality include root-knot (*Meloidogyne incognita* and *M. javanica*) and reniform (*Rotylenchulus reniformis*) nematodes. In the northern cotton-growing areas, the root-knot nematode (*M. incognita*) is a major problem, while the reniform nematode *R. reniformis* is more common in the southern and central India. In jute, the root-knot nematode (*M. incognita*) is a major problem. Their geographical distribution, economic importance, symptoms, biology, interaction with other soil-borne pathogens, survival and spread, and management methods are discussed.

**Keywords**

*Meloidogyne* spp. · *Rotylenchulus reniformis* · Symptoms · Interactions · Management

## 5.1 Cotton, *Gossypium* spp.

Root-knot nematodes *Meloidogyne* spp. and the reniform nematode *Rotylenchulus reniformis* are the major cotton pests in loamy and clay soils of India.

### 5.1.1 Root-Knot Nematodes, *Meloidogyne incognita* and *M. javanica*

Varying degrees of damage is caused by the root-knot nematodes on cotton. They are damaging to cotton as a single pest problem and as part of the *Fusarium* wilt race 1 and race 4 disease complexes.


P. P. Reddy, *Nematode Diseases of Crops and their Management*, https://doi.org/10.1007/978-981-16-3242-6_5

#### 5.1.1.1 Distribution

Root-knot nematodes are distributed in all cotton-growing areas of the country.

#### 5.1.1.2 Crop Losses

Root-knot nematodes are responsible for annual yield loss of 20.5% in cotton amounting to ₹4717.05 million in India (Walia and Chakraborty 2018).

#### 5.1.1.3 Symptoms

The most common symptoms in cotton include stunting, yellowing, and wilting of plants (Fig. 5.1). The plant height within certain areas or across a field may be very irregular. Within the field, the cotton matures during different times of the year due to premature death of cotton plants. The root-knot nematode produces galls that are clearly visible on the roots making this nematode easily identifiable (Fig. 5.2). Root

**Fig. 5.1** Field symptoms of root-knot-infested cotton showing a patch of stunted plants

**Fig. 5.2** Cotton plant (*left*) and roots (*right*) infected with root-knot nematodes

galling can be detected at any time during the growing season, but more commonly 30–40 days after planting.

#### 5.1.1.4 Biology and Life Cycle

Root-knot nematode is soil-borne; second-stage juveniles are present in soil, invade the roots, and become sedentary endoparasites. After penetration, the nematode establishes a feeding site in vascular system, and the adult female develops into a sac-like structure. The mature female commonly lays egg mass on the surface of root. However, in compound galls, the nematode may also lay eggs embedded within the root. It takes about a month's time to complete the nematode's life cycle. The nematode completes several generations within a crop growing cycle, since it is multivoltine. The emergence of second-stage juveniles and infection of feeder roots continue.

Generally, life cycle of *M. incognita* in cotton lasts for 33–38 days and 32 days on *Gossypium barbadense* and *G. hirsutum*, respectively. Each female lays about 300–400 eggs.

#### 5.1.1.5 Spread

Root-knot nematodes are disseminated through irrigation water and agronomic practices.

#### 5.1.1.6 Management

Good field sanitation helps prevent infestations from spreading, and weed control is important in eliminating weed hosts supporting root-knot nematodes. Under extensive severe nematode infestation conditions, there is a need for soil fumigation, or growing of cotton cvs. Resistant to root-knot nematodes. The nematode can also be managed by a resistant variety of another crop such as tomato or alfalfa in rotation with cotton.

##### 5.1.1.6.1 Cultural Methods

Clean fallowing (i.e., weed-free) during years when land is to be left unplanted is effective in controlling root-knot nematodes on cotton, but clean fallow for a period of 1 year may reduce the beneficial endomycorrhizae population.

For the management of cotton root-knot nematode, rotation of cotton with root-knot nematode-resistant soybean variety may be adopted.

Crop rotation is effective in managing cotton root-knot nematode populations. Certain grass crops or pastures, grain sorghum, and peanut can effectively reduce cotton root-knot after a 2–3-year period. Management of susceptible weeds is a prerequisite for the crop rotation to be successful. Rotation crops that help to reduce cotton root-knot nematodes include alfalfa, cowpeas (cvs. Resistant to root-knot nematodes like CB 5, CB 27, CB 46, CB 50, and California Blackeye), peanut, sorghum, Sudan grass, tomatoes (processing types with root-knot nematode resistance), and winter small grains.

Management of both root-knot and reniform nematodes can be achieved by following the cropping sequence of mustard, kulfa (*Portulaca oleracea*), methi, zinnia (*Zinnia elegans*), turnip, green gram, wheat, and barley. Crops such as mustard (*Brassica* spp.), sesame (*Sesamum indicum*), sun hemp (*Crotolaria spectabilis*), asparagus, and African marigold have antagonistic effect which suppresses root-knot nematodes.

Peanut (*Arachis hypogaea*) is a nonhost for *M. incognita* and *R reniformis* and provides an attractive rotational crop for managing these nematodes (Koenning et al. 2004). *M. incognita* and *R. reniformis* have little or no reproduction on grain crops such as corn or grain sorghum (Robinson et al. 1997).

Trap crop such as sun hemp (*Crotolaria spectabilis*) which traps root-knot larvae can be grown and used as a green manure.

In California, the disease complex involving the root-knot nematode and *Fusarium* wilt can be managed by late planting of upland cotton in spring season (Jeffers and Roberts 1993).

#### 5.1.1.6.2 Chemical Methods

In zone-III of middle Gujarat, the effective root-knot nematode (*M. javanica*) management was achieved by seed dressing with Carbosulfan (25DS) at 0.75% a.i.

Seed dressing with Carbosulfan (25DS) at 3% a.i. w/w along with soil application with Carbofuran (3G) at 1 kg a.i. /ha proved effective.

Carbofuran at 1 kg a.i. /ha was most effective in reducing root-knot nematode attack and increasing the yield.

#### 5.1.1.6.3 Biological Methods

Seed treatment at 20 g/kg of seed + soil application at 2.5 kg/ha of *Pseudomonas fluorescens* reduced nematode population.

#### 5.1.1.6.4 Host Resistance

Germplasm procured from Tamil Nadu, Gujarat, and Maharashtra revealed high degree of resistance against *M. incognita* in EL-405 and resistance in ISC-67E-3, ISC-33, ISC-77, BC-68-G-3, G-cot-13, EL-192, EL-958, PK-558, and B-61-1862.

Root-knot nematode-resistant varieties of cotton under field conditions include Deltapine 174RF, Deltapine 1454RN B2RF, Fibermax 2011GT, Phytogen 367WRF, and Stoneville 4946GLB2. In severely infested fields, cotton cv. Phytogen 417WRF greatly reduces nematode reproduction.

Resistance in cotton cvs. Acala NemX, LA 887, and H1560 suppress root-knot nematode (*M. incognita*) reproduction resulting in reduced nematode population densities in North Carolina (Fig. 5.3) (Koenning et al. 2004). Ogallo et al. (1999) reported that cotton cultivars resistant to root-knot nematode can be grown in rotation with susceptible crops to protect them against root-knot nematodes.

In order to reduce crop losses and initial root-knot nematode population of cotton plants in infested field in the San Joaquin Valley of California, a cotton cv. NemX HY has been developed. Because reproduction of the southern root-knot nematode on roots of NemX HY is limited, the population of this nematode in the soil in fall is much lower than the level in soil following susceptible cotton varieties.

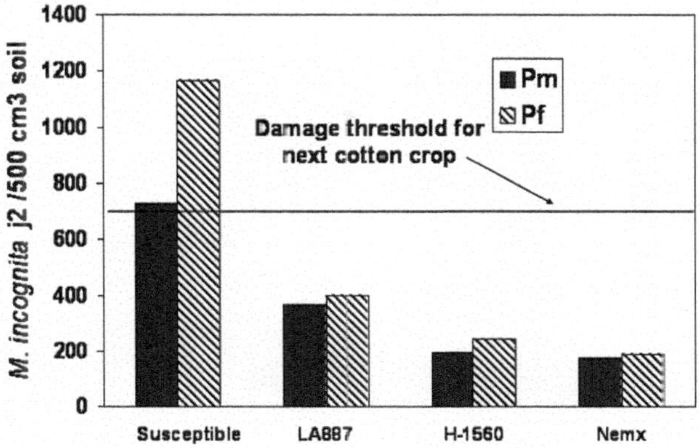


**Fig. 5.3** Effect of resistant cotton cultivars LA 887, H1560, and Acala NemX on midseason (Pm) and final (Pf) population of *M. incognita* compared to susceptible cotton cultivar

The root-knot nematode-resistant cotton cv. NemX HY can be grown in rotation with other nematode susceptible crops to reduce their damage potential by preventing the buildup of root-knot nematode population. In addition to resisting the southern root-knot species *Meloidogyne incognita,* it also acts as a nonhost for the other common species of root-knot nematodes, including *M. arenaria, M. hapla* and *M. javanica.*

### 5.1.2 Reniform Nematode, *Rotylenchulus reniformis*

Seshadri and Sivakumar (1963) have reported this nematode on cotton for the first time from India. In addition to reducing cotton yield, this nematode also caused a delay in maturity, a reduction in size of boll, and in some years a reduction in lint percentage.

#### 5.1.2.1 Crop Losses

In India, the reniform nematode (*R. reniformis*) on cotton is responsible for 14.7% on crop loss The reniform nematode causes 10–15% loss in cotton yield (Palanisamy and Balasubramanian 1981). Field trials on avoidable yield losses conducted at CICR regional Station, Coimbatore showed yield increase by 9.5–17.4% when the nematicide Metham sodium (Vapam, Sistan) was applied.

#### 5.1.2.2 Symptoms

- Symptoms of reniform nematode infection include dwarfing, chlorosis, purpling of leaves (Fig. 5.4), premature decay and loss of secondary roots and plant mortality.
- Causes severe pruning of seedling roots.
- When infested roots were seen under microscope, they show several semi-endoparasitic kidney-shaped female nematodes with egg masses (Fig. 5.5).

**Fig. 5.4** Chlorosis of cotton leaves due to reniform nematode infection

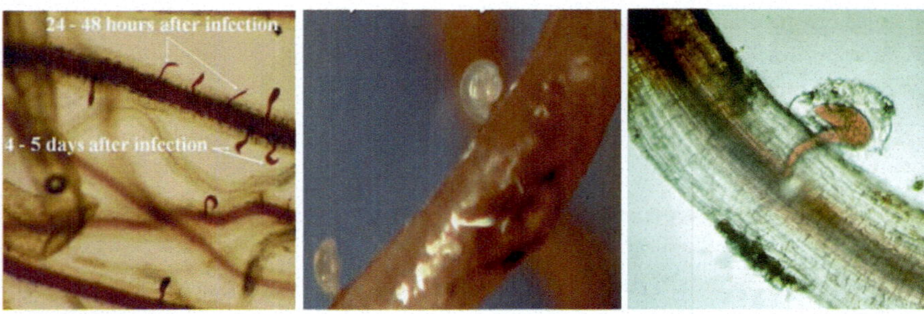


**Fig. 5.5** Cotton roots infected with reniform nematode females

- *R. reniformis*-infested roots show brownish discoloration especially at the point of infection. These roots were thin and dried with brown lesions at the point of infection.
- Affected plants show smaller and lesser bolls.
- An easy method of identifying *R. reniformis* infection in cotton roots is to dip the roots in water to which few drops of fountain pen ink were added. The egg masses and nematodes are stained blue in color.
- Fewer bolls, reduced yields, and delayed maturity are common symptoms of damage caused by reniform nematodes.

#### 5.1.2.3 Life Cycle

The adult female of *R reniformis* is an obligate, sedentary, semi-endoparasite of roots, while the male is nonparasitic. The species is bisexual and reproduces by amphimixis. The complete life cycle takes about 17–23 days on cotton. After infecting the roots, the young female feeds, grows, becomes obese, and starts laying eggs in 7–9 days; an average of 66 eggs per egg mass is recorded on cotton.

#### 5.1.2.4 Ecology

Soil moisture from 25% to 30% and 30 °C soil temperature were optimum for its reproduction on cotton (Khan and Khan 1973). Sivakumar and Seshadri (1972) found that while there is good multiplication of the nematode in sandy loam, brown loam, and black clay loam, the symptom expression was more severe in black clay loam. Mukhopadhyay and Haque (1980) reported that laterite (sandy clay loam) and alluvial (clay loam) soils supported maximum number of reniform nematodes while sandy soil supported the least. Vigorous plant growth in these soil types provided the nematodes with larger feeding areas which might account for higher rate of nematode multiplication.

#### 5.1.2.5 Host Range

Besides infecting cotton, the reniform nematode also infects bananas, citrus, papaya, pineapple, mango, jack fruit, coffee, tea, clove, tobacco, potato, tomato, okra, cowpea, bean, cabbage, cauliflower, carrot, pea, lettuce, radish, cucumber, brinjal, groundnut, sugarcane, rice, and maize.

#### 5.1.2.6 Interaction with Other Pathogens

In the region of Baton Rouge, Louisiana, *R. reniformis* predisposed the cotton plants to *Fusarium* wilt (Neal 1954). In a glasshouse experiment, 81.4% of the plants of a susceptible variety (Half and Half) developed wilt when both the nematode and the fungus were present in the soil, as against 10% when only the fungus was present. However, the presence of the nematode did not appreciably increase the incidence of wilt in a wilt-resistant variety (Delfos 425–920). Besides *Fusarium* wilt, the reniform nematode interacts with *Verticillium* wilt in causing disease complex in cotton (Birchfield 1962).

#### 5.1.2.7 Spread and Survival

They can be spread by anything that can move contaminated soil—farm equipment, birds, flooding, or even dust.

*R. reniformis* survived in air-dried soil (3.3% moisture) stored for 7 months at 20–25 °C.

#### 5.1.2.8 Physiological Races

At least two races, A and B, have been detected in India, the former multiplying on cowpea, castor, and cotton, while the latter reproducing only on cowpea (Dasgupta and Seshadri 1971).

#### 5.1.2.9 Management

##### 5.1.2.9.1 Physical Methods

Nematodes in soil may be killed by soil solarization. Tarping of moist soil during peak summer with two layers of polythene sheet raises soil temperature to kill nematodes. This is very effective for top soil in hot tropical summer months. Eggs and juveniles of reniform nematode get killed by exposure for 1–24 h at 41–47 °C, and repeated exposure to lethal temperature for sublethal period has been observed to have cumulative lethal effect.

#### 5.1.2.9.2 Cultural Methods

Deep plowing up to a depth of 20 cm followed by fallowing for 15 days after breaking the clods reduced the population of *R. reniformis* in soil.

Several nonhosts such as mustard, oats, onion, chili, rice, sorghum, sugarcane, turnip, peanuts, wheat, corn, rice, peas, cabbage, and cauliflower have been reported for *R. reniformis*. Growing of nonhost crops for 2 years between susceptible crops may give fair to good control of the reniform nematode infecting cotton. A 1- or 2-year rotation with grain sorghum, corn, peanuts, and other nonhost crops is as effective as using nematicides.

Population of reniform nematode gets reduced by 80%, when chili and other nonhost crops are grown. Crop rotation with corn, sugarcane, marigold, and zinnia reduced the reniform nematode population. Safflower (*Carthamus tinctorius*) was found to decrease reniform population by 96–100% 45 days after the sowing.

Robinson et al. (1997) reported that nonhost plants like maize or grain sorghum are the poor hosts of the reniform nematode. In order to enhance the cotton crop yield and to reduce reniform nematode population, rotation of cotton with winter grain crops or resistant soybean cvs. is recommended (Fig. 5.6) (Davis et al. 2003).

Rotating with nonhost crops such as wheat or sorghum also reduce the reniform nematode base populations.

Mulato grass (*Brachiaria ruziziensis* x *B. brizantha*) and forage sorghum (*Sorghum bicolor*) in particular, used as autumn or winter cover crops in fields infested by *R. reniformis*, proved to reduce the nematode population and thereafter increase the yield of seed and cotton fiber (Asmus et al. 2008).

#### 5.1.2.9.3 Chemical Methods

Spot application of Carbofuran at 1.0 kg a.i. ha$^{-1}$ 15 days after sowing reduced nematode population and increased the yield by about 11–32%.

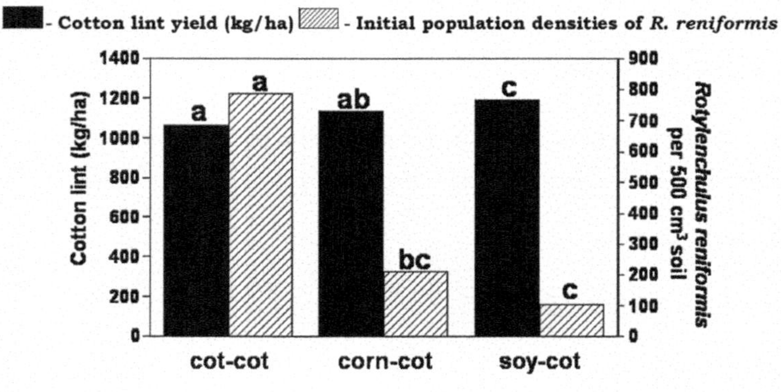


**Fig. 5.6** Effect of rotation of cotton with a resistant soybean cultivar (soy-cot), nonhost corn (corn-cot), or continuous cotton (cot-cot) on lint yield and population of *Rotylenchulus reniformis*

Seed-dressing treatment with Carbosulfan (25 DS) at 3% a.i. w/w + soil application with Carbofuran at 1.0 kg a.i. $ha^{-1}$ is effective against *Rotylenchulus reniformis* and *Meloidogyne incognita.*

#### 5.1.2.9.4 Biological Methods

Seed dressing with plant growth-promoting rhizobacterium, *Gluconacetobacter diazotrophicus* strain 35–47 ($2 \times 10^8$ cfu/g) prepared in charcoal powder at 20 g/kg cotton seed was found to be effective against both root-knot nematodes and the reniform nematode.

Seed treatment with *Pseudomonas fluorescens* at 20 g/kg combined with soil application of the same bioagent at 2.5 kg/ha recorded the lowest incidence of reniform nematodes, highest plant growth parameters, and highest *kapas* yield of cotton. The benefit–cost ratio is 5.40.

Seed treatment with *P. flouresens* and split soil application of *P. flouresens* strain showed maximum decrease in soil and root nematode population to the level of 74.2 and 53.9%, respectively, over control (Jayakumar et al. 2004).

Soil application of *P. fluorescens* or *Purpureocillium lilacinum* (cfu $2 \times 10^6$/g) at 2.5 kg/ha through 100 kg FYM/ha at sowing was found effective for the management of reniform nematode.

*Pasteuria* spp. have demonstrated potential as seed treatments for the control of reniform nematodes. Seed treatment trial with $1.0 \times 10^3$ *Pasteuria* Pr3 endospores/seed or thiodicarb/imidacloprid showed 58.6% and 64.3% decrease in total reniform nematodes, respectively, as compared to the untreated control (Fig. 5.7). The bioagent reduced reniform nematode females by 82.3% as compared to the untreated control. Nematode population control was comparable to a seed-applied nematicide

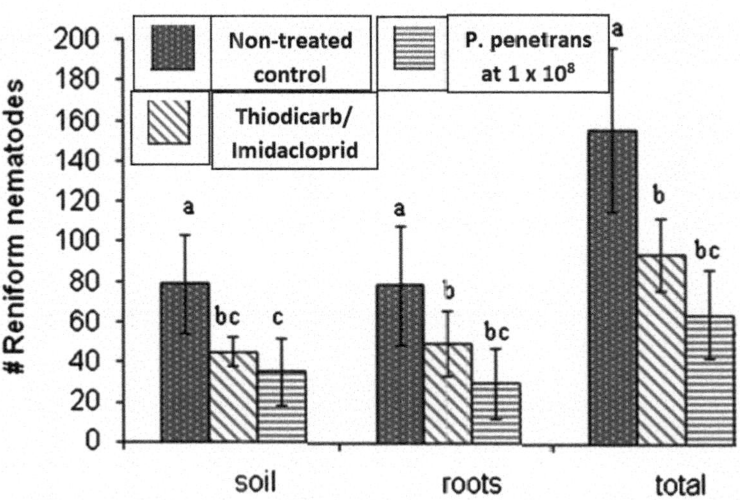


**Fig. 5.7** Effect of seed bio-priming (*Pasteuria* Pr3 endospores) or nematicide/insecticide (thiodicarb/imidacloprid) on number of *Rotylenchulus reniformis* in cotton 4 weeks after treatment

(thiodicarb) at a seed coating application rate of $1.0 \times 10^8$ spores/seed (Schmidt et al. 2010).

#### 5.1.2.9.5 Host Resistance

Three Bt cotton hybrids [Chiranjeevi and Tulsi-4 (BG-I) and RCH-2 (BG-II)] showed a disease index of 1.0. Twenty of the cultivars showed a disease index of 2.0 (resistant) and two cultivars, for example, MRC-7918 and Tulsi-117 of BG-II, were moderately resistant (Sonavane 2010).

#### 5.1.2.9.6 Integrated Methods

Seed treatment with Carbosulfan (ST 3%) followed by soil application of Carbosulfan (1 kg a.i. /ha) in cotton MCU 5 increased the yield of cotton kapas by two folds and decreased the population of *R. reniformis* by 71.43%.

The demonstration trials in cotton showed that seed dressing with Carbosulfan 25ST at 3% w/w followed by soil application of Carbofuran 3G at 1.0 kg a.i. /ha significantly reduced the population of *R. reniformis*, increased the kapas yield, and the benefit–cost ratio was 2.86 and 3.02 in two locations.

## 5.2 Jute, *Corchorus* spp.

Root-knot nematodes *Meloidogyne incognita* and *M. javanica* are the two most important nematode species of jute crop.

### 5.2.1 Root-Knot Nematodes, *Meloidogyne incognita* and *M. javanica*

#### 5.2.1.1 Crop Losses

Root-knot nematodes are responsible for the annual yield loss of 19% in jute amounting to ₹912 million in India (Walia and Chakraborty 2018). Phukan and Roy (1983) reported that yield loss of jute fiber weight was up to 50% due to attack of root-knot nematode.

Recent yield losses have been estimated to be 13–53%, while 21–54% in jute (cultivar JRO 524) has been reported earlier. *M. incognita* alone can cause an average yield loss of 15% in jute.

#### 5.2.1.2 Symptoms

Infected plants exhibit symptoms of general mineral deficiency, yellowing, stunting, and wilting during hotter part of the day, chlorosis, and premature shedding of leaves resulting in low fiber yield (Fig. 5.8). The infection of root-knot nematode produces characteristic "root gall" or "knotted roots" (Fig. 5.9).

**Fig. 5.8** Symptoms of field infestation of root-knot nematode on jute (*left*), and nematode-infected plant showing root galling (*right*)

**Fig. 5.9** Severe galling on jute roots due to root-knot nematode infection

#### 5.2.1.3 Biology and Life Cycle

Root-knot nematode is soil-borne; second-stage juveniles are present in soil, invade the roots, and become sedentary endoparasites. After penetration, the nematodes feed on vascular tissues and develop to become sac-like adult female inside the roots. The mature female commonly lays egg mass on the surface of root. However, in compound galls, the nematode may also lay eggs embedded within the root. It takes about a month's time for the root-knot nematode to complete its life cycle. The nematode completes several generations within a crop growing cycle, since it is multivoltine. The emergence of second-stage juveniles and infection of feeder roots continue.

#### 5.2.1.4 Interaction with Other Microorganisms

The quick rotting of root system, wilting, and premature death of plants (locally known as Hooghly wilt in West Bengal) resulted in the presence of fungus

(*Macrophomina phaseolina* and *Fusarium solani*) and bacterium (*Ralstonia solanacearum*).

#### 5.2.1.5 Spread

Root-knot nematodes are disseminated through infested soil, irrigation water, and other agronomic practices.

#### 5.2.1.6 Management

##### 5.2.1.6.1 Cultural Methods

Crop rotation with wheat, maize, mustard, and marigold was found effective.

Growing of sun hemp as a trap crop before jute was found effective in reducing root galling (91.43%), increasing plant height (14.76%), fiber yield (51.43%), and stick yield (35.53%) (Haque et al. 2008)

Amendment of soil with jute seed powder, potash, and sulfur reduced infestation of root-knot nematode (*Meloidogyne* spp.) on jute (Variety D-154) and also increased growth of plants (Begum et al. 1994).

##### 5.2.1.6.2 Chemical Methods

Seed treatment with Carbosulfan 25 ST (Marshal 25ST) at 3% w/w is a most effective and economical approach. The root-knot nematode was also effectively controlled by soil incorporation of Carbofuran 3G at 2 kg a.i. /ha.

Seed treatment with Carbosulfan (25 DS) at 3.0% a.i. (w/w) followed by soil application of Fenamiphos 10G, Phorate l0G, or Sebuphos I0G at 2 kg a.i. /ha was found effective. A reduction of root galling was observed in plots treated with Sebuphos at 2 kg a.i. ha (Khan 2004).

##### 5.2.1.6.3 Integrated Methods

Seed dressing with Carbosulfan 25ST (Marshal 25DS) at 3% w/w followed by soil application of Carbofuran 3G at 1 kg a.i./ha 25 DAS proved highly effective against root-knot nematode *Meloidogyne incognita* and increased the yield up to 24.6% with an additional return of about ₹8000 per ha (Anon 2012).

Application of Carbofuran (3G) at 1 kg a.i. /ha + *Pseudomonas fluorescens* (Biofor Pf) at 20 g/m$^2$ with vermicompost at 1:10 ratio was found effective.

### 5.2.2 Interaction of Root-Knot Nematode with Bacterial Wilt

The quick rotting of root system, wilting, and premature death of plants (locally known as Hooghly wilt in West Bengal) resulted in the presence of root-knot nematode *M. incognita* and wilt bacterium *Ralstonia solanacearum*. Infestation of *M. incognita* is severe at early stage of the crop, while nematode-bacterium complex is severe at maturity stage of the crop.

#### 5.2.2.1 Crop Losses

*M. incognita* alone can cause an average yield loss of 15% in jute; however, in association with the bacterium *Ralstonia solanacearum*, it can cause up to 30% loss in fiber yield. The monetary loss is estimated to the tune of ₹30 million.

*M. incognita* alone, *R. solanacearum* alone, and both pathogens together caused 18.32%, 26.11%, and 35.6% losses in fiber yield of jute, respectively, at a preplant population densities of 328 $J_2$/250 ml soil for the nematode and $7.2 \times 10^8$ cfu/g soil for the bacteria. Similarly, at preplant population densities of 264 $J_2$/250 ml soil and $7.6 \times 10^8$ cfu/g soil, the losses were estimated to be 16.49%, 25.27%, and 31.25% due to nematode, bacteria, and for the both pathogens, respectively (Hazarika 2002).

#### 5.2.2.2 Symptoms

Sole infestation of *M. incognita* in jute produces typical symptoms of stunted growth, yellowing of leaves, patchiness, and formation of galls in the roots. When both nematode and bacterium are associated, in addition to heavy root galling, drying of twigs are observed. First the tips of the plants are dried and become black in color. The blackish color is extended to the base of the stem and ultimately the whole plant dries.

#### 5.2.2.3 Management

##### 5.2.2.3.1 Cultural Methods

Crop rotation schedules of jute-fallow-mustard-jute and jute-fallow-wheat-jute were found equally effective in reducing the root-knot and wilt disease complex and increasing the fiber yield of jute (Hazarika 2002; Hazarika et al. 2006).

Sun hemp (*Crotalaria juncea*) was found suitable trap crop for increasing plant height, fiber and stick yield, and reducing gall formation (Haque et al. 2008).

Soil application of neem cake at 2 t/ha in combination with bleaching powder at 12 kg/ha was found most effective in reducing the nematode population in soil (Hazarika and Bora 2007).

##### 5.2.2.3.2 Chemical Methods

Application of Carbofuran at 2 kg a.i. /ha at sowing was found beneficial.

##### 5.2.2.3.3 Integrated Methods

Combined application either with Carbofuran at 1 kg a.i. /ha plus Streptocycline sulfate 2 kg/10 liter water or neem cake plus Streptocycline sulfate 2 g/m$^2$ is effective against the disease complex (Anon 2005).

*Pseudomonus fluorescens* (Biofor Pf) at 20 g/m$^2$ with Vermicompost and Streptocycline sulfate at 4 g/10 L of water is also effective against this disease complex (Anon 2005).

Application of Carbofuran 3G at 1 kg a.i. /ha + Biofor-Pf at 20 g/m$^2$ with Vermicompost at 1:10 ratio increase the yield up to 22.3% with an additional return of ₹7300 per ha (Choudhury and Bhagawati 2015).

### 5.2.3 Interaction of Root-Knot Nematode with Root Rot

#### 5.2.3.1 Interaction

The interaction of root-knot nematodes (*M. incognita/M. javanica*) with fungus *Macrophomina phaseolina* (= *Rhizoctonia bataticola*) in jute is popularly known as "Hooghly wilt" as it was first reported from Hooghly district of West Bengal. Plants inoculated with *M. incognita* and *Macrophomina phaseoli* were more severely damaged (Fig. 5.10) (Haque and Mukhopadhyaya 1979).

#### 5.2.3.2 Management

##### 5.2.3.2.1 Cultural Methods

Removal and safe destruction of old stubbles after harvest of jute eliminates residual nematode population from the infested crop field. Early (at least 10 days) harvesting of crop is suggested when the jute crop suffers from combined infestation of *Macrophomina phaseolina* and root-knot nematode.

##### 5.2.3.2.2 Integrated Methods

Application of neem cake at 1 t/ha + Mancozeb 75WP at 0.25% soil drench at sowing time was considered best to manage *Meloidogyne incognita-Macrophomina phaseolina* complex in jute (Mukhopadhyay and Roy 2005).

**Fig. 5.10** *Meloidogyne—Macrophomina* disease complex resulting in root rot of jute

## 5.3 Conclusion

The root-knot (*M. incognita*) and reniform (*R. reniformis*) nematodes are major problems limiting the production of fiber crops like cotton and jute. In order to reduce the impact of above economically important nematode problems on crop yield, development and use of good quality resistant varieties shall find widespread acceptance among farmers. For the development of efficient, high-throughput marker-assisted selection protocols, there is a need for private sector breeders. In future nematode management systems in cotton, the important role of host resistance shall be determined by the grower acceptance based on the fiber quality and yield potential.

For more efficient adoption, there is a need for more efficacious and environmentally safe nematicides. Efficient placement of nematicides poses a serious challenge in view of clustered distribution of most nematodes within a field. Nematode management using rotation with nonhost crops (corn and soybean) is likely to be limited. The use of crop rotation as a management tactic depends on the potential rotation crop's enhanced economic value. However, the use of soil organic amendments and crop rotation will form an important profitable component in future nematode management strategy.

## References


Anon (2005) Biennial report 2003–2005. All India coordinated research project on plant parasitic nematodes with integrated approach for their control, Assam Agricultural University, Jorhat

Anon (2012) Biennial report 2010–2011. All India coordinated research project on plant parasitic nematodes with integrated approach for their control, Assam Agricultural University, Jorhat

Asmus GL, Inomoto MM, Cargnin RA (2008) Cover crops for reniform nematode suppression in cotton: greenhouse and field evaluations. Trop Plant Pathol 33(2):85–89

Begum HA, Sultana K, Alauddin S, Khardker S (1994) Control of root knot nematode disease of jute through soil amendment (in Bangladesh) (1991). Agri 16(12):9–13

Birchfield W (1962) Host parasitic relations of *Rotylenchulus reniformis* on *Gossypium hirsutum*. Phytopathology 52:862–865

Choudhury BN, Bhagawati B (2015) Integrated management of *Meloidogyne incognita* and *Ralstonia solanacearum* complex of jute (Abstract). In: Proceedings of national seminar on "harnessing science for social development. Assam Science Society, Assam Agricultural University Branch, p 79

Dasgupta DR, Seshadri AR (1971) Races of the reniform nematode. *Rotylenchulus reniformis* Linford and Oliveira, 1940. Indian J Nematol 1:21–24

Davis RF, Koenning SR, Kemerait RC et al (2003) *Rotylenchulus reniformis* management in cotton with crop rotation. J Nematol 35:58–64

Haque MS, Mukhopadhyaya MC (1979) Pathogenicity of *Macrophomina phaseoli* on jute in the presence of *Meloidogyne incognita* and *Hoplolaimus indicus*. J Nematol 11(4):318–321

Haque SMA, Mosaddeque HQM, Sultana K et al (2008) Effect of different trap crops against root knot nematode disease of jute. J Innov Dev Strategy 2(3):42–47

Hazarika K (2002) Interrelationship of *Meloidogyne incognita* and *Ralstonia solanacearum* on jute and management of the disease complex caused by them. Ph. D. thesis, Department of Nematology, Assam Agricultural University, Jorhat, Assam, 143 pp

Hazarika K, Bora LC (2007) Management of disease complex of jute caused by *Meloidogyne incognita* (Kofoid and White 1919) Chitwood 1949 and *Ralstonia solanacearum* Yabuuchi. J Plant Prot Envt 4(1):111–115

Hazarika K, Rahman MF, Bora LC (2006) Effect of crop sequences on incidence of root-knot and wilt disease complex of jute caused by *Meloidogyne incognita* and *Ralstonia solanacearum.* Indian J Nematol 36:81–84

Jayakumar J, Ramakrishnan S, Rajendran G (2004) Biological control of cotton reniform nematode, *Rotylenchulus reniformis* with *Pseudomonas fluorescens*. Indian J Nematol 34(2):230–231

Jeffers DP, Roberts PA (1993) Effect of planting date and host genotype on the root-knot nematode-Fusarium wilt disease complex of cotton. Phytopathology 83:645–654

Khan MR (2004) Chemical approach for managing root-knot nematode, *Meloidogyne incognita* race 2, infecting jute. Nematol Medit 32:195–199

Khan FA, Khan AM (1973) Studies on the reniform nematode, *Rotylenchulus reniformis,* I. Host range and population changes. Indian J Nematol 3:24–30

Koenning SR, Kirkpatrick TL, Starr JL et al (2004) Plant-parasitic nematodes attacking cotton in the U.S.: old and emerging problems. Plant Dis 88:100–113

Mukhopadhyay MC, Haque MS (1980) Influence of soil types on *Rotylenchulus reniformis* on castor. Indian J Nematol 10:89–91

Mukhopadhyay AK, Roy K (2005) Efficacy of neem cake and some chemicals to manage the *Meloidogyne incognita - Macrophomina phaseolina* disease complex in jute. Int J Nematol 15 (2):203–206

Neal DC (1954) The reniform nematode and its relationship to the incidence of Fusarium wilt in cotton at Baton Rouge, Louisiana. Phytopathology 44:447–450

Ogallo JL, Goodell PB, Eckert J, Roberts PA (1999) Management of root-knot nematodes with resistant cotton cv. NemX. Crop Sci 39:418–421

Palanisamy S, Balasubramanian P (1981) Assessment of avoidable yield loss in cotton variety 'Suvin' (*Gossypium barbadense* L) by fumigation with metham sodium (Abstract). In: Second all India nematology symposium, Coimbatore, p 52

Phukan PN, Roy AK (1983) Infestation level of *Meloidogyne incognita* and cultivar reduction of jute. Indian J Nematol 13(1):118–121

Robinson AF, Inserra RN, Caswell-Chen EP et al (1997) *Rotylenchulus* species: identification, distribution, host ranges, and crop plant resistance. Nematropica 27:127–180

Schmidt LM, Hewlett TE, Green A et al (2010) Molecular and morphological characterization and biological control capabilities of a *Pasteuria* spp. parasitizing *Rotylenchulus reniformis*, the reniform nematode. J Nematol 42:207–217

Seshadri AR, Sivakumar C (1963) A preliminary note on the reniform nematode (*Rotylenchulus reniformis* Linford and Oliveira, 1940) on a number of cultivated crops in South India. *Madras.* Agric J 50:134–137

Sivakumar CV, Seshadri AR (1972) Effect of soil texture on the reniform nematode, *Rotylenchulus* reniformis. Indian J Nematol 2:83–86

Sonavane PS (2010) Plant parasitic nematodes associated with Bt cotton (*Gossypium* spp.). M. Sc. (Agriculture) thesis, Department of Plant Pathology , University of Agricultural Sciences, Dharwad, Karnataka

Walia RK, Chakraborty PK (eds) (2018) Nematodes problems of crops in India. ICAR-All India Coordinated Research Project on Nematodes in Cropping Systems, New Delhi. 402 pp

# Sugar Crops

# 6

**Abstract**

Plant-parasitic nematodes of sugar crops like sugar beet and sugarcane have been variously reported to be associated with decline in sugar production. This chapter focuses on the parasitic nematodes of sugarcane (species of *Pratylenchus, Meloidogyne, Helicotylenchus, Tylenchorhynchus,* and *Hoplolaimus*) and sugar beet (*Heterodera schachtii, Meloidogyne* spp.). The symptoms of damage, their economic importance, and measures for their management are discussed. The effective nematode management strategies that can improve yield in sugar crops include cultural and chemical control methods, host plant resistance, and integrated nematode management.

**Keywords**

*Pratylenchus* spp. · *Meloidogyne* spp. · *Heterodera* spp. · Symptoms · Management

## 6.1 Sugar Beet, *Beta vulgaris*

### 6.1.1 Cyst Nematode, *Heterodera schachtii*

The sugar beet nematode cyst nematode (SBCN), *Heterodera schachtii* is a major parasite of sugar beets. Wherever sugar beets are grown, the nematode is responsible for serious stand and yield reductions.

#### 6.1.1.1 Symptoms

In sugar beet fields, SBCN infestation initially appears as circular to oval areas of stunted plants. Symptoms of malnutrition or poor growth are evident. Infested plants are usually stunted, show reduced leaf growth, older outer leaves become yellow and


P. P. Reddy, *Nematode Diseases of Crops and their Management*,
https://doi.org/10.1007/978-981-16-3242-6_6

**Fig. 6.1** Sugar beet plants parasitized by *Heterodera schachtii* showing severe stunting and yellowing of infected plants

wilted, and overall growth in the field is very uneven (Fig. 6.1). Plants often tend to become pale yellow and wilt at midday, even when the field has just been irrigated and soil moisture is at field capacity. Some plants may show nutrient deficiency symptoms. The taproot tends to be stunted with fibrous "bearded roots." Seedling wilt and death can result when the nematode infestation is very severe.

When affected plants are carefully dug and the roots examined, the most important confirmation of SBCN infection is the presence of tiny white to yellow lemon-shaped females attached to feeder roots or yellow-brown cysts (dead mature females) (Fig. 6.2) in soil.

#### 6.1.1.2 Life Cycle

Eggs and second stage larvae of nematodes within the cysts survive in soil (Fig. 6.3). Under favorable conditions (warm temperatures (70–81 °F) and sufficient soil moisture) and the presence of root exudate from hosts, second-stage juveniles hatch from eggs, enter the root tissue, and move to cortical tissues where they feed and develop into adult females after undergoing third and fourth molts. Adult males are thread-like in appearance. The adult males stop feeding and migrate from roots to the soil, while mature swollen females remain attached to the roots as a sedentary endoparasite with the posterior end outside the root (Fig. 6.3). A single female can produce about 200–300 eggs during the growing season. The eggs are mostly embedded within the female, but few eggs may be deposited in a gelatinous matrix outside the nematode body. When eggs are fully developed, the female dies and its body hardens to form a brown or yellow-brown cyst that protects the eggs (Fig. 6.3). The nematode completes its life cycle in about 4–6 weeks based on soil temperature.

#### 6.1.1.3 Host Range

The sugar beet cyst nematode has a very wide host range (more than 200 plant species): some cultivated species (beets, spinach, cabbage, and canola/rapeseed),

**Fig. 6.2** *Left*—White lemon-shaped cyst nematode females feeding on sugar beet roots. *Right*—Yellow-brown cysts from infected plant

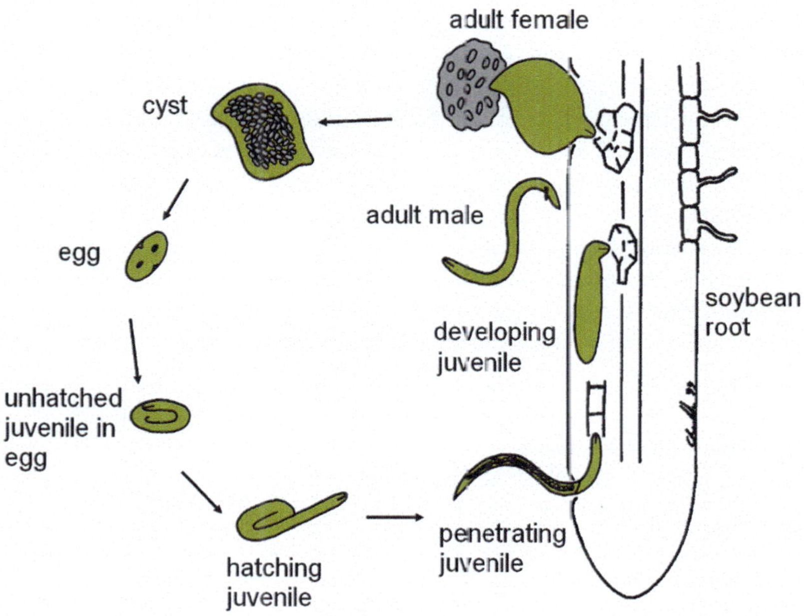


**Fig. 6.3** Life cycle of sugar beet cyst nematode

some intermediate crop species (white mustard, fodder radish, and some leguminous plants). It can also survive on common weeds, such as wild mustard, pigweed, lambs quarters, shepherds purse, and purslane.

#### 6.1.1.4 Spread and Survival

SBCN is a soil-borne pest, so anything that can move soil will move the nematode. Farm machinery, irrigation water, animals, and tare soil from harvested beets are the agents of cyst dissemination. In the soil profile, cysts can be found from the surface to 60 cm deep, but the highest numbers are found in the root zone (5–25 cm soil depth).

Some nematodes in the form of eggs within the cysts can survive as long as 12 years in fallow soils.

#### 6.1.1.5 Management

##### 6.1.1.5.1 Cultural Methods

Rotation with nonhost crops, including wheat, barley, corn, bean, potato, and alfalfa. In severely infected fields, reduction in cyst nematode population by 40–60% takes place in a year by following three- to 4-year rotation.

Growing green manure cover crops like fodder radish (*Raphanus sativus*) and white mustard (*Sinapis alba*) preceding sugar beet crops also gave effective management of the cyst nematode (Lelivelt and Hoogendoorn 1993).

Planting trap crops attract SBCN, but do not allow them to develop and reproduce. The sustainable production of sugar beet in fields infested with the sugar beet cyst nematode was achieved by using Brassicaceous green manure cover crops which act as trap crops for nematodes (Matthiessen and Kirkegaard 2006). Some SBCN-tolerant cultivars of oil seed radish (Defender, Image, and Colonel) and white mustard are effective (Fig. 6.4).

Early planting when soil temperatures are not favorable (<59 °F) for infection by SBCN, controlling weeds that are hosts for SBCN in sugar beet and rotation crops, and avoiding returning of tare soil with SBCN to fields in which sugar beet is grown are some of the other cultural control measures.

##### 6.1.1.5.2 Biological Methods

Biological seed treatment which utilizes spores of *Pasteuria nishizawae* may help to manage SBCN on tolerant sugar beet varieties.

##### 6.1.1.5.3 Host Resistance

Plant SBCN-tolerant cultivars which are available.

##### 6.1.1.5.4 Integrated Methods

This will include avoidance of, or limited use of, host crops in the rotation; good control of weed hosts; avoiding use of machinery and equipment from infested fields; and thorough cleaning of farm machinery used in infested fields. Use proper

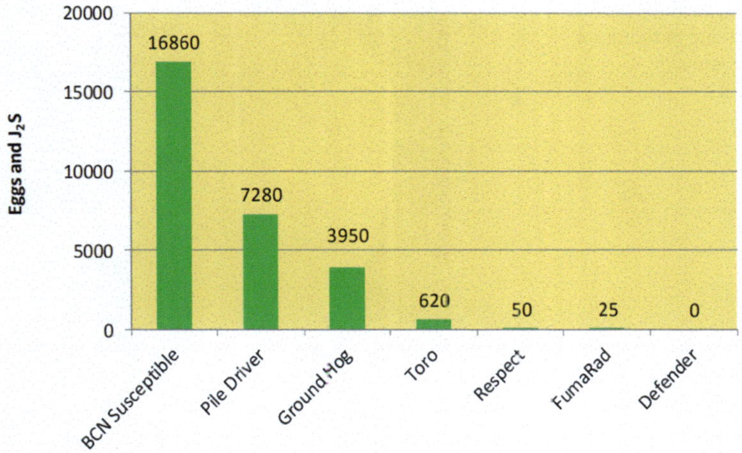


**Fig. 6.4** Sugar beet cyst nematode multiplication on oilseed radish varieties

sanitation/sterilization measures after visiting areas with known SBCN problems and then visiting fields that do not have this pest.

## 6.1.2 Root-Knot Nematodes, *Meloidogyne* spp.

### 6.1.2.1 Crop Losses

Root-knot nematodes are responsible for yield loss to the extent of over 50%.

### 6.1.2.2 Symptoms

Beets are stunted in small, conspicuous patches of the field. The nematode-infected beets are reduced in size as compared to healthy beets. The infested area can be diagnosed by excessive leaf wilting in hot weather in severely infested fields. The tap root is severely galled if infected early in the season (Fig. 6.5). Later infections occur principally on secondary roots. Weed patches in the beet field is an indication of cyst nematode infection later in the season. The yield is drastically reduced in infested field.

### 6.1.2.3 Ecology

Galls can appear as quickly as a month after planting; nematodes prefer sandy soils and damage in areas of field with this type of soil is most likely.

### 6.1.2.4 Management

Following a 2- to 3-year crop rotation with fescue, marigold, or small grains can slowly decrease the number of nematodes present in the soil. Incorporating rotting organic matter into the soil can help to decrease nematode activity by enhancing the activity of nematode-destroying microorganisms.

**Fig. 6.5** Galled roots of sugar beet due to root-knot nematode infection

In order to reduce the nematode populations in infested fields, use of soil solarization during peak summer and planting resistant cultivars have been recommended.

Under field conditions, after 3 months, bionema (containing *Bacillus penetrans* at a concentration of 20% at 10 L/ha) proved to be the most effective material causing significant reduction (55.6, 67.9, 78.5, and 57%) in number of juveniles in the soil, females, galls on sugar beet roots, and rate of nematode build-up, respectively. After 6 months, a combination of bionema + plant residue of mud sugar beet at 1 L/ha gave the best results (82.3, 70.8, 78.3, 84.1, and 81.1%) in reducing the nematode parameters. These results show how improved plant growth and technological characteristics help reduce the nematode *Meloidogyne incognita* (Abd Elhameed El-Nagdi and Abd El Fattah 2011).

## 6.2 Sugarcane, *Saccharum officinarum*

The major nematodes on sugarcane include lesion nematode *Pratylenchus zeae*; lance nematode *Hoplolaimus indicus*; stunt nematodes *Tylenchorhynchus* spp.; and spiral nematodes *Helicotylenchus* spp.

### 6.2.1 Lesion Nematode, *Pratylenchus zeae*

This is an important plant parasitic nematode found in sugarcane, maize, chickpea, etc.

#### 6.2.1.1 Crop Losses

Losses due to these nematodes varied from 1.06% to 62.9%. *P. zeae* showed maximum harmful effect (Nath et al. 1978; Haider and Nath 1996). Nematode populations have been correlated with deterioration in cane juice quality (Haider et al. 1993).

#### 6.2.1.2 Symptoms

Cortical parenchyma cells adjacent to the point of nematode entry become brownish and show dark round to elongated lesions (Fig. 6.6). Often roots show slight thickening at lesion zones. The dark color of root masks the nematode symptoms on older roots. Increase in population density of *P. zeae* even to one nematode/g soil and above shows a marked reduction in yield of sugarcane. Damage to sugarcane roots was similar as in other plants.

#### 6.2.1.3 Interaction with Other Pathogens

The lesion nematode is also important in inducing wilt symptoms in sugarcane due to combined action of nematode and fungi like *Fusarium moniliforme* and

**Fig. 6.6** Reddish lesions on sugarcane roots due to lesion nematode infection

*Cephalosporium sacchari*. *Pratylenchus* spp. also interacts with *Rhizoctonia solani* (Nath and Haider 1975; Nath et al. 1976a).

#### 6.2.1.4 Management

##### 6.2.1.4.1 Cultural Methods

Combination of lentil and garlic along with sole crop of sugarcane was effective in reducing total nematode population (42.5% and 39%) in field (Haider and Pathak 1998).

Paddy-chickpea sequence reduced nematode population nearly 100% followed by paddy-mustard (37%), paddy-lentil, and paddy-garlic (13.7%) while monoculture of sugarcane resulted in increase of 93% nematode population (Haider and Nath 1999).

Use of lesion nematode-resistant sugarcane varieties like Fernside soybean, A6785, Holt, and Kairi is recommend in nematode-infested fields.

A decrease (66%) in population of sugarcane nematodes was observed in sesame-mustard-sunflower sequence while there was continuous increase in maize-wheat-green gram sequence (Haider and Pathak 1998).

Inclusion of legume crops like soybean and peanut crops in rotation can reduce lesion nematode numbers by 80–90%.

Soil mulching with green sugarcane trash provides a better environment for beneficial organisms such as free living nematodes and predators of plant-parasitic nematodes. Minimum tillage systems which preserve the trash blanket between crops help to minimize populations of lesion nematodes. There were 95% fewer lesion nematodes (*Pratylenchus zeae*) in roots growing in amended soil than in roots from the unamended control (Stirling et al. 2005).

##### 6.2.1.4.2 Chemical Methods

Carbofuran and Fensulfothion were effective in minimizing nematode population of sugarcane in field and increasing growth characters significantly (Nath et al. 1976b).

##### 6.2.1.4.3 Host Resistance

Sugarcane varieties found tolerant from Sugarcane Research Institute, Pusa Centre, include BO 91, 109, 126, COP 9501, and 9702. These are tolerant against lance, spiral, stunt, and root-knot nematodes.

### 6.2.2 Root-Knot Nematodes, *Meloidogyne incognita, M. javanica*

#### 6.2.2.1 Symptoms

The root-knot nematode enters through the root tips and cause reduced root growth and gall formation on root tips (Fig. 6.7). Root-knot nematodes reduced fresh root weight by 33%, reduced tillering by 33–70%.

**Fig. 6.7** Sugarcane roots infected with root-knot nematodes

#### 6.2.2.2 Management

High populations of root-knot nematodes reestablish if a green cane trash cover is not maintained. It provides a better environment for beneficial organisms such as free living nematodes and predators of plant-parasitic nematodes.

Minimum tillage systems which preserve the green cane trash cover between crops help to minimize nematode populations. Tillage operations kill beneficial nematodes, allowing nematodes to quickly reestablish.

The root-knot nematode populations can be reduced by rotation of sugarcane with resistant peanut cvs, such as A6785 soybean, Kairi peanut, or Holt.

*Eragrostis curvula*, as a cover crop, succeeded in controlling an infestation consisting mainly of *Meloidogyne* and *Pratylenchus* but was too slow to be of practical value.

Nematicides Rugby 100G, Nemacur 10G (granules), and Nemacur 400 (liquid) reduce nematode populations.

The populations of root-knot nematode *M. javanica* will be markedly reduced in the presence of high inoculum of endospores of *P. penetrans* in soil.

*B. subtilis* provided effective biological control of root-knot and lesion nematodes.

Sugarcane lines KEN83-737 and Co945 had lowest nematode numbers with mean egg mass indices of 0.8.

### 6.2.3 Lance Nematode, *Hoplolaimus indicus*

Lance nematode, *Hoplolaimus indicus* is widely distributed in sugarcane fields and other crops of Bihar.

#### 6.2.3.1 Symptoms

An inoculum level of 1000 nematodes per 100 g soil suppressed development of healthy laterals. Both juveniles and adults penetrated the roots, damaging the cortical parenchymatous tissues parallel to long axis of the root. These nematodes were also found associated in wilt-affected plots. Association of lance nematodes with root-knot and lesion nematodes has been reported from Bihar (Haider et al. 1993).

#### 6.2.3.2 Management

Tolerant varieties of sugarcane found against *H. indicus* include BO-117, 119, 122, COP-8603, 9206, 9601, and 9702.

### 6.2.4 Stunt Nematodes, *Tylenchorhynchus mashhoodi, T. nudus*

*Tylenchorhynchus mashhoodi* and *T. nudus* are widespread in the state. This is an ectoparasitic nematode feeding on the root surface, root hairs, and epidermal cells. Its high numbers are recorded in sugarcane fields.

#### 6.2.4.1 Symptoms

The nematode has not been found causing discernable symptoms on sugarcane tops. A very high population of 10,000 and above/kg soil causes stunting of the sugarcane crop.

#### 6.2.4.2 Management

Tolerant varieties of sugarcane against *T. mashhoodi* include BO-125, 127, 131, 132, 133, 134, 135, COP-9202, 9204, 9601, 9701 and 9702.

### 6.2.5 Spiral Nematodes, *Helicotylenchus dihystera, H. indicus*

Sugarcane crop is often found associated with high population of *Helicotylenchus dihystera* and *H. indicus*. These are ectoparasites feeding on epidermal cells of roots.

#### 6.2.5.1 Symptoms

No pronounced disease symptoms are seen. The sloughing of epidermal jacket of sugarcane root is evident when population is high. The root system shows blunt malformed roots and reduction of small branches. The poor root system gives rise to stunted appearance of the sugarcane crop.

#### 6.2.5.2 Management

Tolerant varieties of sugarcane against *H. indicus* are BO-123, 126, 130, 131, 132, 134, 135, COJ-76, COP-8901, 8902, 9105, 9106, and 9102.

## 6.3 Conclusion

The impact of PPN on sugarcane cultivation and how to reduce the associated damage is widely studied. The PPN associated with sugarcane is the first step to facilitate the development of appropriate management strategies to minimize damage by PPN associated with sugarcane.

There is a need to educate local farmers on the large diversity of parasitic nematodes associated with sugar-cane and their damage potentials by creating awareness programs. Screening of seed plants and regular plant quarantine services should be employed prior to the introduction of new varieties to sugarcane farmers, and moreover, effective nematode management strategies should be considered in order to improve yield in sugarcane plantations.

Management of sugar beet nematodes begins with minimizing the transport of soil from field to field. Long rotations are very helpful in keeping populations in check. Incorporation of any practices that improve soil health and improve the populations of beneficial organisms is helpful. Utilization of trap crops such as Colonel or Defender oilseed radish anywhere in the cropping rotation is an effective means of reducing sugar beet cyst nematode populations. In order to reduce the nematode population and to enhance yields, the use of nematode-resistant varieties is recommended. No single management practice is as effective as the combination of practices when managing this pest. Effective nematode management can be achieved by early diagnosis and adoption of management strategies immediately.

## References


Abd Elhameed El-Nagdi WM, Abd El Fattah AI (2011) Controlling root-knot nematode, *Meloidogyne incognita* infecting sugar beet using some plant residues, a biofertilizer, compost and biocides. J Plant Prot Res 51(2):107–113

Haider MG, Nath RP (1996) Pathogenicity of plant parasitic nematodes in sugarcane in Bihar, crop losses under semi-controlled conditions. Indian Sugar 16(8):669–678

Haider MG, Nath RP (1999) Integrated control of sugarcane nematodes in infested field (Abstract). In: National Seminar on Nematological Research in India, C.S. Azad University of Agriculture and Technology, Kanpur, p 45

Haider MG, Pathak KN (1998) Nematode population as influenced by cropping sequence with oilseed crops (Abstract). In: National symposium on rational approaches in nematode management for sustainable agriculture, held at GAU, Anand, Gujarat, p 56

Haider MG, Samed A, Waris ML (1993) Efficacy of organic soil amendments and carbofuran on plant parasitic nematodes on yield and juice quality of sugarcane. Indian Sugar 3:25–29

Lelivelt CLC, Hoogendoorn J (1993) The development of juveniles of *Heterodera schachtii* in roots of resistant and susceptible genotypes of *Sinapsis alba*, *Brassica napus*, *Raphanus sativus* and hybrids. Neth J Pl Path 99:13–22

Matthiessen JN, Kirkegaard JA (2006) Biofumigation and enhanced biodegradation: opportunity and challenge in soil-borne pest and disease management. Crit Rev Plant Sci 25:235–265

Nath RP, Haider MG (1975) Combined effect of *Fusarium moniliforme* and *Hoplolaimus indicus* on sugarcane plants (Abstract). In: 45th Session of National Academy of Sciences of India

Nath RP, Singh RK, Haider MG (1976a) Effect of *Fusarium moniliforme* and *Hoplolaimus indicus* on sugarcane. Sugarcane Pathol Newsl 17:24–25

Nath RP, Singh RK, Haider MG, Ashraf MH (1976b) Effect of some nematicides on sugarcane in nematode infested field. Sugarcane Pathol Newsl 17:23–24

Nath RP, Haider MG, Sinha BK, Sinha AC (1978) Growth response of sugarcane in lesion nematode, *Pratylenchus zeae*. Indian J Nematol 8(2):165–167

Stirling GR, Wilson EJ, Stirling AM et al (2005) Amendments of sugarcane trash induce suppressiveness to plant-parasitic nematodes in a sugarcane soil. Aust J Plant Pathol 34:203–211

# Part V

# Fruit Crops

# Tropical Fruit Crops

# 7

**Abstract**

The most important nematode pests that are potentially damaging to tropical fruit crops like banana, citrus, papaya, and pineapple include root-knot (*Meloidogyne* spp.), reniform (*Rotylenchulus reniformis*), lesion (*Pratylenchus* spp.), and citrus (*Tylenchulus semipenetrans*) nematodes. Distribution, crop losses, biology, survival and spread, and management methods for tropical fruit crops such as physical, cultural, chemical, biological, host resistance, and integrated nematode management are discussed in this chapter.

**Keywords**

*Meloidogyne* spp. · *Rotylenchulus reniformis* · *Pratylenchus* spp. · Symptoms · Management

## 7.1 Banana, *Musa* spp.

The major nematode pest of banana is the burrowing nematode *Radopholus similis*. *Pratylenchus coffeae, Meloidogyne incognita*, and *Helicotylenchus multicinctus* are the other nematode parasites of banana that are of some economic importance.

### 7.1.1 Burrowing Nematode, *Radopholus similis*

Cobb (1893) first described the burrowing nematode from necrotic root lesions of banana growing in Fiji. The burrowing nematode is responsible for decline, blackhead, *Radopholus* root rot, and blackhead toppling disease of bananas. Nair et al. (1966) first reported this nematode on banana from Kerala state, India.



P. P. Reddy, *Nematode Diseases of Crops and their Management*,
https://doi.org/10.1007/978-981-16-3242-6_7

#### 7.1.1.1 Distribution

Throughout India.

#### 7.1.1.2 Crop Losses

The burrowing nematode is responsible for 30.76–41.00% yield loss in banana (Rajagopalan and Naganathan 1977; Nair 1979; Parvatha Reddy et al. 1996c; Vadivelu et al. 1987). Root population of *R. similis* is indirectly correlated with the yield (Charles et al. 1985). The burrowing nematode is responsible for 30–60% loss in yield.

#### 7.1.1.3 Symptoms

Wounding of banana roots by the burrowing nematode usually induces reddish-brown cortical lesions which are diagnostic of the disease (Fig. 7.1). The cortical lesions can be clearly observed when the infected root is examined immediately after it is longitudinally split open. Root damage due to *R. similis* is responsible for various above-ground symptoms like stunting of plants, yellowing of leaves, and uprooting of mature plants.

With the increase in nematode population, feeding roots are invaded and destroyed as fast as they are formed. The resulting setback in the uptake of plant nutrients leads to debility and smaller fruit. The lesioning of the primary roots together with the girdling and death of those roots which anchor the plant to the ground makes the plant prone to tip over under wind pressure (Fig. 7.1).

#### 7.1.1.4 Life Cycle

The burrowing nematode has a migratory endoparasitic habit and although the stages remain vermiform throughout, sexual dimorphism is apparent with adult males being somewhat degenerate and probably nonparasitic. Eggs are normally laid in infested tissue over 7–8 days at the rate of about 4 eggs per day. Eggs take 8–10 days to

**Fig. 7.1** *Left*—Toppling of banana plant due to poor root system due to *Radopholus similis* infection. *Right*—Banana roots infected with *R. similis* showing reddish-brown lesions

hatch, while larvae take 10–13 days to mature. The nematode completes its life cycle by about 20–25 days.

#### 7.1.1.5 Host Range

Poucher et al. (1967) list 244 hosts for the burrowing nematode. It is known to infect *Musa* spp., *Ipomoea batatas, Pueraria phaseoloides javanica,* pepper, areca nut, coconut, ginger, turmeric, sugarcane, coffee, maize, avocado, tea vegetables, ornamentals, trees grasses, and weeds.

#### 7.1.1.6 Spread and Survival

The most important means whereby the burrowing nematode is spread from one locality to another is through planting material (corms). Trimming sets (paring), which is often done before they are planted, is usually not sufficiently rigorous to remove infection that extend deeply into a set.

*R. similis* may be disseminated by water that drains from infested areas and reenters either irrigation channels or streams from which water is being taken to irrigate bananas. Soil that adheres to implements, tires of motor vehicles, and shoes of plantation workers may also spread *R. similis*.

After a plantation is destroyed, the corms may remain succulent for several months and nematodes in these corms probably survive for as long as the corms remain succulent. The nematode cannot survive for more than 12 weeks without its host.

#### 7.1.1.7 Physiological Races

Venkitesan (1976) reported four host types of *R similis*. The population of *R similis* isolated from pepper roots did not infect *Citrus sinensis, C. reticulata, C aurantifolia, Coffea arabica,* coconut, and areca nut, but infected banana only. Banana population infected pepper, coconut, and areca nut; coconut population infected pepper and banana but not areca nut; while areca nut population infected pepper and banana but not coconut.

#### 7.1.1.8 Management

##### 7.1.1.8.1 Regulatory Methods

The unintentional planting of banana suckers already infected with nematodes is the principal means of nematode dispersal to uninfested sites. In order to restrict the nematode spread, the first line of defense is to obtain planting stock free from nematode infestation. Certified nematode-free sets can be obtained by either paring/heat treatment/selection/tissue cultured plants. Micro-propagation from nematode-free banana materials using sterile techniques offers a good way to obtain nematode-free planting materials (Fig. 7.2).

##### 7.1.1.8.2 Physical Methods

Corms of suckers should be trimmed carefully to remove the nematode affected necrotic tissue (Paring) (Fig. 7.3) before planting.

**Fig. 7.2** Production of nematode-free banana planting material by tissue culture method

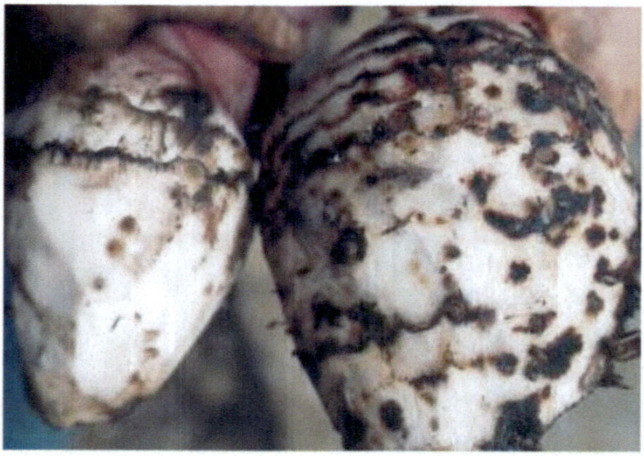

**Fig. 7.3** Pared banana suckers

The pared sets can be disinfected by dipping them in a hot water bath at 55 °C for 25 min without causing significant set losses (Fig. 7.4). Fungicides applied after treatment would probably decrease any post-treatment losses caused by the fungi.

Sun drying the rhizomes of banana suckers wrapped in plastic sheet (Fig. 7.5) for 4 h before planting has been shown to be effective in the control of nematodes, which is being commonly practiced in Kerala.

#### 7.1.1.8.3 Cultural Methods

Castor cake at 100 g/plant is effective against *R. similis* in banana.

##### Cover Cropping

After destroying an infested plantation, planting of cover crops like *Panicum maximum* var. *trichoglume*, marigold, and perennial legume (*Phaseolus atropurpureus*) did not support *R. similis* population (Fig. 7.6). Planting of

**Fig. 7.4** Hot water treatment of banana planting material

**Fig. 7.5** Solarization of banana suckers under transparent plastic sheet for management of nematodes (Control is wrapped in black plastic)

**Fig. 7.6** Intercropping banana with marigold (*Left*) and perennial legume (*Right*)

sugarcane or Pangola grass (*Digitaria decumbens*) following the destruction of bananas eliminated *R. similis* after 10 weeks (Loos 1961).

Marigold is suitable allelopathic cover crop due to its low growing habit which can be incorporated into soil at flowering for nematode control. Only living marigold root systems exhibit nematicidal properties. Farmers should aim to plant marigold at 4–5 months after banana planting to maximize marigold growth and minimize competition from banana growth.

Planting of cover crops like *T. patula* and sun hemp (*Crotalaria juncea*) prior to banana crop planting in field soil infested with plant-parasitic nematodes significantly reduced the nematode population densities after 5 months.

#### Fallowing

Fallowing for a period of 3 months after banana effectively suppressed the burrowing nematode population. Rajendran et al. (1979) reported combined management of the burrowing nematode and Panama wilt by way of 5 months flood fallowing.

#### Crop Rotation

Rotation of banana with paddy, sugarcane, green gram, coriander, cotton, Pangola grass, sorghum, tobacco, or cassava suppressed the nematode population and increased the yield (Rajendran et al. 1979). The nematode population can be reduced by rotation of banana with cereals, cotton, and vegetables in garden lands and with rice in wet lands. Rice or green gram grown after banana suppressed the population of *R. similis, P. coffeae*, and *H. multicinctus* (Rajendran et al. 1979).

#### Intercropping

The nematode population both in soil and roots were considerably reduced when banana was intercropped with nematode antagonistic plants such as marigold. *Crotalaria juncea* was most effective in reduction of soil nematodes and *Coriandrum sativum* was more effective in reduction of root nematodes. *C. juncea* gave maximum bunch weight (16.7 kg) recording 51% more than control (11.03 kg) followed by *Tagetes erecta* (16.2 kg), *Sesamum indica* (15.87 kg), *Acorus calamus* (15.67 kg), and Carbofuran (15.23 kg) (Charles and Venkitesan 1993).

Intercropping with sun hemp, coriander, marigold, radish, or lucerne reduced nematode population and increased yield (Vadivelu et al. 1987). Intercropping sun hemp in banana field and plowing the green manure 45 days after sowing was found to be effective in reducing *R. similis, P. coffeae*, and *H. multicinctus* population by 38.4%, followed by marigold and cowpea which recorded 29.0% and 22.3% reduction, respectively (Shanthi 2003). Increase in plant growth and yield and reduction in nematode population is achieved by intercropping banana with *Crotalaria* (Charles et al. 1985). Lakshmana Murthy (1983) reported that when banana was intercropped with papaya and marigold, *R. similis* was not encountered. Significant reduction in the burrowing nematode population was reported by intercropping with radish, *Crotalaria*, and marigold.

#### Mulching

Mulching with black polythene at 20% moisture depletion recorded highest yield with low population of *R. similis* (Vadivelu et al. 1987). Bhattacharya and Rao (1984) found that reduction in the population of the burrowing and lesion nematodes in banana roots by mulching soil with banana and sugarcane trash, and black polyethylene sheet.

#### Organic Amendments

Application of Neem cake at 800 g per plant (two split applications) or Neem coated urea at 110 g per plant reduced the population of the burrowing nematode and also increased the yield of banana (Vadivelu et al. 1987). Maximum reduction in the burrowing nematode population and increase in plant growth and yield were observed by soil application with neem products such as Nimbicidine and Econeem.

Application of press mud obtained from sugarcane waste treated plants has shown significant reduction in nematode populations (Sundararaju et al. 2004).

### 7.1.1.8.4 Chemical Methods

Paring and pralinage with Carbofuran at 1.2 g a.i. per sucker recorded low incidence of R. similis and highest yield (Rajendran et al. 1979). Dipping of banana suckers in 750 ppm of Carbofuran for 10 min, 750 ppm of Fensulfothion for 10 min, 0.03% solution of Parathion for 20 min, and 50 ppm of Phorate for 15–30 min increased plant height, bunch weight, fingers per bunch, induced early flowering, and effectively reduced the burrowing nematode population.

Soil application of Fenamiphos at 3 kg a.i. per ha was effective in increasing plant height, plant girth, number of leaves and yield, and in reducing the nematode population (Parvatha Reddy et al. 1991). Various other chemicals (Carbofuran, Ethoprofos, and Oxamyl) have also been found to be effective for the management of R. similis on banana under field conditions.

### 7.1.1.8.5 Biological Methods

Field application of *Pseudomonas fluorescens* (isolate PfB 13) at 10 g/plant significantly reduced nematode population in soil and roots and root lesion index (14 as compared to 54 in control); and increased plant height, pseudo stem girth, leaf area, number of leaves, and fruit yield (15.6 kg/plant as compared to 10.9 kg/plant in control).

Application of *P. fluorescens*, a biocontrol agent at 20 g/$m^2$ at the time of planting was effective in increasing the plant growth parameters (pseudo stem height and girth, number of leaves, root length and weight) and in reducing the nematode population in soil (62.34%) and roots (60.35%) (Nirmal Johnson and Devarajan 2004).

For effective control of nematodes, banana plantlets should be inoculated with *Purpureocillium lilacinum* strain 251 and re-inoculated into the soil at transplanting (Mendoza et al. 2004).

In field, increase in plant growth and bunch weight, and decrease in the burrowing and root-knot nematode population was noticed by inoculation of *G. mosseae* during planting of banana. Sosamma et al. (1998) reported that banana plants colonized with endomycorrhiza increased bunch weight by 35.7%, number of hands by 30%, and number of fingers by 39.6%.

Soil application of biofertilizer *Azospirillum* at 20 g/plant had significant effect in enhancing the plant growth and reducing the population of *R. similis* in soil to an extent of 50.1% over control (Shanthi et al. 2004).

#### 7.1.1.8.6 Host Resistance

Selections (tetraploids) from crosses between wild bananas (diploid male parent) and "Gros Michel" (triploid female parent) show considerable resistance to R. similis as well as to Panama wilt and Sigatoka (*Mycosphaerella musicola*) diseases. A new variety Bodles Altaport, a cross between Gros Michel and Pisang Lilin, has been released for growing on the hill sides of Jamaica. This variety has an extensive root system, withstands wind, and is more resistant to R. similis than Lacatan.

In American Samoa, the banana hybrid FHIA-01 has been found relatively resistant to the burrowing nematode, while the hybrid FHIA-25 is relatively resistant to all the nematodes feeding on banana.

#### 7.1.1.8.7 Integrated Methods

##### Cultural and Biological Methods

Integration of organic amendments (FYM/Neem cake/Karanj cake) with bioagents (*Trichoderma harzianum/T. viride/Pochonia chlamydosporia/Pastueria penetrans/ Pseudomonas fluorescens*) or endomycorrhiza (*Glomus mosseae*) was found to be effective against the burrowing nematode. Soil application of 2 kg FYM with *P. fluorescens* (with $1 \times 10^9$ spores/g) and *P. chlamydosporia* (with $1 \times 10^6$ spores/g) per plant at the time of planting and at an interval of 4 months significantly reduced the burrowing nematode by 64% compared to control.

Increase in plant growth and reduction in *R. similis* population by 50% was obtained by integration of soil application of *Pastueria penetrans* at 100 g soil (300 spores/g)/*T. harzianum/T. viride* at 250 g/plant while planting and repeated 4 months after planting. Increase in banana fruit yield by 45 kg/plant and early harvest of bunches by 65–75 days was achieved by combined application of FYM enriched with *T. harzianum* at 2 kg/plant and Neem cake at 500 g/plant.

Combined application of *G. mosseae* (containing 25–30 chlamydospores/g of inoculum) at 100 g/plant with Karanj cake increased banana fruit yield (16.61 kg/ plant as compared to 5.45 kg/plant in control), while application of *G. mosseae* along with Neem cake reduced the burrowing nematode population. Mycorrhizal root colonization and number of chlamydospores of *G. mosseae* were highest maximum in Neem cake amended soil (Parvatha Reddy et al. 2002).

Application of FYM enriched with bio-control agents like *Purpureocillium lilacinum* and *T. harzianum* to the banana plants along with Neem cake at the time of planting and repeating the same treatment at 2 months interval proved beneficial.

Soil application of FYM enriched with *P. lilacinum* and *P. fluorescens* at 2 kg/ plant at the time of planting and subsequent application for four times at an interval of 6 months reduced the root population of *R. similis* and *M. incognita* by 64.5% and 75%, respectively, and increased the fruit yield by 21% over control (Rao et al. 2016).

Physical and Chemical Methods

The removal of infected tissue along with some of the surrounding healthy tissue will normally disinfect sets (Loos and Loos 1960). The pared sets are dipped in mud slurry and sprinkled with 40 g Carbofuran 3G per set.

Physical, Cultural and Chemical Methods

Paring of suckers followed by hot water treatment (55 ° C for 20 min) along with application of Carbofuran at 0.5 g a.i./plant and Neem cake at 1 kg/plant, the population of *R. similis* decreased by 72–82% at 4 and 8 week interval after treatment. Maximum yield (2.7 t/ha) was recorded in the above treatment which was 97% increase over control.

Integration of paring of banana suckers, dipping in 0.5% Monocrotophos solution for 45 min along with intercropping with marigold or sun hemp gave higher fruit yield (62.838 and 61.816 MT/ha, respectively, as compared to 57.000 t/ha in control) and benefit:cost ratio (1.70 and 1.28, respectively) (Patil et al. 1999).

Physical and Biological Methods

Paring and pralinage of suckers and application of FYM/Vermicompost enriched with *T. harzianum* + *P. lilacinum* at 2 kg/plant at planting and thrice afterwards at 3 months interval was found to be effective.

Cultural and Chemical Methods

Application of Neem cake at 1 kg/plant (two split doses, first at planting and second 3 months after planting) and Carbofuran (3G) at 16.6 g/plant in the pit at planting was found to be effective.

Cultural, Chemical and Biological Methods

Reduction in the burrowing nematode population both in soil and roots and increase in plant growth, yield, and the benefit-cost ratio (2.65) was obtained by combined application of Neemark, Carbofuran, *P. penetransi*, and *G fasciculatum* (Vidya and Reddy 1998).

### 7.1.2 Interaction of Burrowing Nematode with Panama Wilt

#### 7.1.2.1 Interaction

Incidence and losses due to Panama wilt caused by *Fusarium oxysporum* f. sp. *cubense* is enhanced in association with the burrowing nematode *R. similis* under high nematode population. This clearly indicates the existence of synergistic interaction between the burrowing nematode and *Fusarium* wilt pathogen in banana (Fig. 7.7).

**Fig. 7.7** Burrowing nematode and Panama wilt disease complex in banana

Newhall (1958) showed that the incidence of Panama wilt in banana caused by *F. oxysporum* f. sp. *cubense* was doubled in the presence of *R. similis* during the experimental period of 3 months. When Gros Michel bananas infected with *R. similis* were inoculated with *F. oxysporum* f. sp. *cubense*, the period between inoculation and the onset of wilt were also considerably shortened (Loos 1959).

Risbeth (1960) suggested that nematodes breakdown resistance to Panama wilt in Lacatan bananas.

Lesions formed after inoculation with both *R. similis* and *F. oxysporum* f. sp. *cubense* were more extensively necrotic and increased in size more rapidly than when *R. similis* alone was used (Blake 1966). *F. oxysporum* f. sp. *cubense* readily establishes itself in the feeder roots of banana when they are invaded by the nematode *R. similis*, but the fungus has seldom been recovered from nematode-free roots (Blake 1966).

### 7.1.2.2 Management

#### 7.1.2.2.1 Cultural Methods

Application of compost and Neem cake was most effective in reducing the disease intensity by *Meloidogyne* sp. and *F. oxysporum* f. sp. *cubense* on banana (Roy et al. 1998).

#### 7.1.2.2.2 Integrated Methods

Soil application of Neem cake + *T. viride* + Carbendazim was found to be effective in reducing the burrowing nematode (*R. similis*) and wilt (*F. o.* f. sp. *cubense*) disease complex, lesion index (1.0 as compared to 4.0 in control), increasing the banana fruit yield by 53.55% (15.147 t/ha as compared to 9.887 t/ha in control) giving the benefit:cost ratio of 2.72 (Ravi et al. 2001).

### 7.1.3 Lesion Nematode, *Pratylenchus coffeae*

The decline in banana plantations occurred due to the infestation the lesion nematode *P. coffeae*.

#### 7.1.3.1 Crop Losses

In India, crop loss caused by the root-lesion nematode in banana cv. Nendran was reported to be 45%.

#### 7.1.3.2 Symptoms

The symptoms caused by *P. coffeae* are very similar to those caused by *R. similis*. Both *R. similis* and *P. coffeae* are associated with the blackhead toppling diseases of bananas. *P. coffeae* enter the roots and feed on cortical cells resulting in the production of dark lesions in the cortex (Fig. 7.8). The corm is similarly infected resulting in necrosis of the outer tissues.

#### 7.1.3.3 Life Cycle

The life cycle of *P. coffeae* is similar to that of *R. similis*. The egg-to-egg cycle is completed in about 27 days at 26–32 °C. The optimum temperature for reproduction is 29.5 °C.

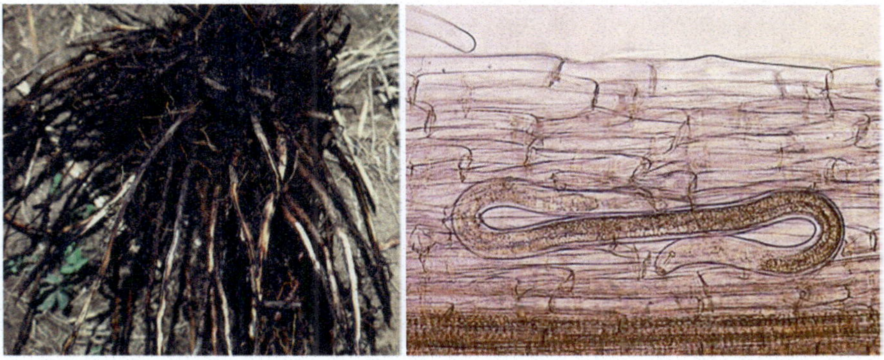

**Fig. 7.8** *Left*—Blackened and necrotic lesions on banana roots caused by *Pratylenchus coffeae*. *Right*—Lesion nematode in the cortical cells of banana root

#### 7.1.3.4 Survival and Spread

The lesion nematode can survive for 6 months in the soil, in the absence of its host. It has been observed infesting the corm, and hence spread can occur through planting material.

#### 7.1.3.5 Interaction with Other Pathogens

The lesion nematode causes most damage to banana in association with *Fusarium oxysporum.* f. sp. *cubense* (Vadivelu et al. 1987).

#### 7.1.3.6 Management

##### 7.1.3.6.1 Cultural Methods

Fallowing

A fallow period of 6 months after destruction of all banana plants eliminated *P. coffeae*.

Intercropping

A significant reduction in *Pratylenchus* sp. population (85%) was observed in the banana field where *Tagetes erecta* was grown as an intercrop and significantly increased the fruit yield (12 kg/plant) when intercropped compared with the untreated control (7 kg/plant) (Sundararaju et al. 2002). Reduction in the lesion nematode population by 38.4% was obtained with intercropping of banana with sun hemp, followed by marigold and cowpea which recorded a 29.0% and 22.3% reduction, respectively. The lesion nematode population was also reduced by rotation of banana with green gram/rice.

Organic Amendments

A significant reduction in *P. coffeae* population and yield increase was recorded in plants that have received 50% N applied through Neem cake (Sundararaju and Kumar 2003). A significant reduction in *P. coffeae* population and increased yield in plants treated with Press mud (15 t/ha) was recorded. Sundararaju et al. (2002) reported that eco-friendly and economical way of controlling nematodes is through soil application of press mud. In banana cv. Nendran, maximum reduction in lesion nematode population and significant increase in plant growth was obtained by soil application of Neem cake at 200 g/plant.

##### 7.1.3.6.2 Biological Methods

Antagonistic Fungi

Application of *Trichoderma viride* at 20 g/plant at the time of planting and repeated 3 months after planting were found to be effective in controlling *P. coffeae* as well as reducing the incidence of Panama wilt in banana.

Application of *Purpureocillium lilacinum* at 10 g along with FYM at 500 g or Neem cake at 250 g/plant effectively suppressed the lesion nematode population by 60% and 63%, respectively (Poornima et al. 2004).

Antagonistic Bacteria

Reduction in lesion nematode population in soil (67.12%) and roots (62.00%) and increasing pseudo stem height and girth, number of leaves, root length and weight were obtained by soil application of *Pseudomonas fluorescens* at 20 g/plant ($2.5 \times 10^8$ cfu/g).

#### 7.1.3.6.3 Host Resistance

Banana cvs./hybrids, viz., Kunnan, Pey Kunnan, Then Kunnan, Nattu Poovan, Pidi Monthan, Chirapunji, Singhlal, Sakkarachayam, Malai Kala, Manik Champa, Kartobium Tham, Vennettukunnan and Hybrid-74 (Matti × Pisang Lilin), H-21, H-55, H-59, H-65, H-84, H-109, and H-110, were found to be tolerant/resistant to *P. coffeae* (Vadivelu et al. 1987; Sundararaju et al. 2004).

### 7.1.4 Spiral Nematode, *Helicotylenchus multicinctus*

*H. multicinctus* is an important parasite of banana throughout the banana growing areas of the world.

#### 7.1.4.1 Crop Losses

This nematode causes serious decline of bananas in the Jordan valley, Israel, and was responsible for about 20% loss in yield. It is responsible for 33.83% loss in yield, 55.88% loss in number of fruits per bunch, and delayed fruiting by 134 days (Vadivelu et al. 1987).

#### 7.1.4.2 Symptoms

The spiral nematode incites discrete relatively shallow, necrotic lesions on banana roots (Fig. 7.9). In Israel, *H. multicinctus* causes extensive root necrosis, dieback, and dysfunction leading eventually to debility of the entire plant.

#### 7.1.4.3 Host Range

This nematode is known to infect banana, cocoa, sweet potato, citrus sugarcane, cassava, coffee, maize, mango, oil palm, rice, rubber, tea, yam, avocado, Bermuda grass, and grapevine.

#### 7.1.4.4 Life Cycle

Groups of 8–26 eggs were observed in discolored cortical tissues. Eggs hatch in 48–51 h in tap water at 30 °C and the first molt is believed to take place outside the egg shortly after hatching. The second-stage larvae are recognized by their ventral digitate tail process. In the early third-stage larvae and at the third molt, the female gonad primordium is 2 and 6 celled, respectively. During the fourth molt, the male

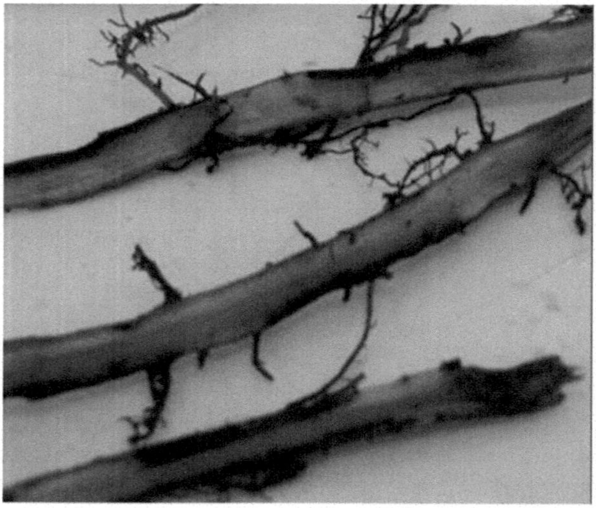

**Fig. 7.9** Banana roots showing necrotic lesions (confined to outer cortex) due to infection by *Helicotylenchus multicinctus*

and the female gonads complete their development and the vulva and vagina are seen in female which perhaps complete entire life cycle within the roots.

#### 7.1.4.5 Spread and Survival

The nematodes easily get introduced into virgin land with banana soil and suckers (rhizomes), usually brought from old infested plantations. The nematodes survived 4 months without host plants.

#### 7.1.4.6 Management

##### 7.1.4.6.1 Physical Methods

Removal of black necrotic tissue from the banana suckers and dipping them in hot water maintained at 53–55 °C for 20 min is recommended to eliminate nematodes.

##### 7.1.4.6.2 Cultural Methods

Crop Rotation

Rice or green gram after banana suppressed the population of *H. multicinctus*.

Intercropping

Intercropping of banana with sun hemp was found to be effective and reduced *H. multicinctus* population by 38.4%, followed by marigold and cowpea which recorded a 29.0% and 22.3% reduction, respectively.

Organic Amendments

Application of Distillery sludge at 2.5 kg + Vermicompost 1 kg + Neem cake at 1 kg + Poultry manure at 2.5 kg/plant at 3, 5, and 7 months after planting significantly reduced *H. multicinctus* population. It was on par with Distillery sludge at 2.5 kg + Neem cake at 1 kg treatment.

#### 7.1.4.6.3 Chemical Methods

In recent years, emphasis has been shifted to use of granular nematicides (Oxamyl/Aldicarb/Carbofuran/Fensulfothion/Ethoprophos/Fenamiphos) for the management of the spiral nematode.

#### 7.1.4.6.4 Biological Methods

Reduction in spiral nematode population in soil (58.87%) and roots (56.54%) and increasing the plant growth parameters (pseudo stem height and girth, number of leaves, root length and weight) was obtained by soil application of *Pseudomonas fluorescens* at 20 g/plant ($2.5 \times 10^8$ cfu/g).

#### 7.1.4.6.5 Integrated Methods

Jonathan and Cannayane (2002) reported significant reduction in population of *H. multicinctus* and the increase the plant growth parameters (pseudo stem height and girth, number of leaves, root length and weight) was observed in banana plants treated with *P. fluorescens* at 2 g/*Glomus mosseae* at 25 g along with Press mud at 3 kg/plant.

### 7.1.5 Root-Knot Nematodes, *Meloidogyne* spp.

Root-knot nematodes are present in all banana-growing regions and can be especially damaging to the cultivars of the Cavendish subgroup. The root-knot nematodes (*M. incognita*, *M. javanica*, *M. arenaria*, and *M. hapla*) are found to be commonly associated with banana.

#### 7.1.5.1 Crop Losses

The average yield loss due to this nematode is estimated to be 27.3% (Anon 2009). Jonathan and Rajendran (2000) reported 30.90% loss in banana fruit yield due to *M. incognita*. *M. incognita* is responsible for annual yield loss of 15% in banana amounting to ₹9710.46 million in India (Walia and Chakraborty 2018).

#### 7.1.5.2 Symptoms

The symptoms on aerial plant parts caused by *Meloidogyne* spp. include reduction in size and chlorosis of leaves, stunting of plants, yield reduction, and delay and unevenness in fruit ripening. Leaf symptoms include a progressive dieback of older leaves from the leaf tip. Symptoms exhibited by severely infected banana plants include stunting of plants, thinning of pseudo stems, and discolored greenish-yellow bands on the leaf blades.

The most characteristic symptoms of root-knot nematode include galling on both primary and secondary roots. In severe infestation, the root galls are enlarged and distorted. Galls vary in size, reaching up to 15 mm in diameter (Fig. 7.10). In

**Fig. 7.10** Banana roots infected with root-knot nematodes

bananas, *M. incognita* infection typically results in the formation of small galls on the roots or fine dilations in thicker fleshy roots. Rhizome upper cell layers are colonized by different species of root-knot nematodes in a single gall. The nematodes cause reddish-brown discoloration in the vascular system. When the infected root is longitudinally split open, females in different stages of development can be observed along with black haloes (female chambers).

#### 7.1.5.3 Biology and Life Cycle

The nematode is soil-borne and sedentary endoparasite. After penetration, the J2 larvae feed endoparasitically on conducting vessels and develop to adult females which are spherical in shape and lays egg masses in compound galls. The males are vermiform in shape which leaves roots to enter the soil. During summer, the nematode completes its life cycle in about 25–30 days, while it takes longer during winter months.

#### 7.1.5.4 Survival and Spread

Root-knot nematodes survive on dicotyledonous plants, which are usually present in most soils in which bananas are growing. Survival and spread also occurs with the planting material on infected roots and corms. The nematodes spread through infected planting material, irrigation water, and agricultural implements.

#### 7.1.5.5 Management

##### 7.1.5.5.1 Regulatory Methods

Measures to prevent introduction and spread—such as use of certified planting material, is an important management strategy.

#### 7.1.5.5.2 Physical Methods

The suckers can be pared to a depth of 1 cm and treated in hot water at 55 °C for 20 min (Inomoto and Monteiro 1989).

### 7.1.5.6 Cultural Methods

Application of Neem cake at 250 g/plant twice (once at planting and second after 4 months) helps in suppressing the nematode population.

Intercropping with Marigold at early growth stages in basins and incorporated into the soil around the plants was found to be effective.

Growing antagonistic plants, namely, *Tagetes erecta, T. patula* and *Crotalaria juncea* as intercrops in banana field drastically, reduces the root-knot nematode population.

#### 7.1.5.6.1 Chemical Methods

Application of Carbofuran 3G at 20 g/plant or Rugby at 5 g/plant or Caldan at 5 g/plant applied twice once at planting and second application after 3 months of planting is known to reduce the nematode population and increase the yield significantly.

Application of either Carbofuran or Cartap or Carbosulfan at 0.3 g a.i./$m^2$ proved beneficial (Anon 2012).

#### 7.1.5.6.2 Biological Methods

The promising biological agents such as *Purpureocillium lilacinum, Trichoderma viride, Pochonia chlamydosporia*, endomycorrhiza (*Glomus fasciculatum*) and bacterium, *Pasteuria penetrans* are recommended to control root-knot nematodes on banana.

The biocontrol agent *Pochonia chlamydosporia* parasitized the eggs of root-knot nematode and prevented the hatching significantly.

*Pastueria penetrans* at 20 g/$m^2$ (5 $\times$ $10^7$ spores/g root powder) mixed with 200 ml sand and evenly distributed per plant at planting and 45 days after planting effectively reduced *M. incognita* population in banana besides increasing the yield. Karuna et al. (2004) reported that the plants treated with *P. penetrans* (3.5 kg/plant) gave higher yield when compared to control plants (2.1 kg/plant).

#### 7.1.5.6.3 Host Resistance

In Tamil Nadu, banana cvs./hybrids such as Chakia, CO 1, Pisang Lilin, H 74, H 94, and H 110 were found to be resistant to root-knot nematodes (Jonathan 1994; Sundararaju et al. 2004).

#### 7.1.5.6.4 Integrated Methods

Infected corms can be disinfected of root-knot nematodes by paring the planting material followed by immersing it in hot water at 50–55 °C for 30 min.

Dipping of pared suckers in mud slurry mixed with Carbofuran (3G) at 20 g/L of slurry followed by drying suckers in shade before planting was found to be effective.

Paring and pralinage of suckers with Farm yard manure/Vermicompost enriched with *Trichoderma harzianum* + *Purpureocillium lilacinum* at 2 kg/plant at planting and three times afterwards at 3 months' interval proved beneficial.

In heavily nematode-infested area, paring banana suckers and dipping in hot water at 55 °C for 20 min and two split applications of Neem cake at 1 kg/plant (at planting and 3 months later)/Carbofuran 0.5 g a.i./ha was found to be effective for the control of root-knot nematode in banana.

Significant reduction in the root-knot nematode population by 76% as compared to control was obtained by integration of soil application of 2 kg FYM enriched with *Pochonia chlamydosporia* (with $1 \times 10^6$ spores/g) per plant and *Pseudomonas fluorescens* (with $1 \times 10^9$ spores/g) and at the time of planting and at an interval of 4 months.

Jonathan and Rajendran (2000) reported that application of *P. lilacinum* multiplied on Neem cake at 15–20 g/plant significantly reduced root gall index, egg masses, eggs, females, and soil population of *M. incognita* in banana.

Application of *T. harzianum* along with FYM has reduced the root-knot nematode population in soil (from 175/100 ml soil in control plot to 42/100 ml soil in treated plot) and roots (from 17/g root in control plot to 4/g root treated plot). Further, application of the bioagent also reduced *Fusarium* wilt incidence and increased the yield by 4.53 t/ha over control.

#### 7.1.5.6.5 Success Story in Odisha

Two technologies developed under AICRP on Nematodes for the management of root-knot nematode (*M. incognita*), that is, (1) application of *Pseudomonas fluorescens*, a biocontrol agent at 20 g/m$^2$ at the time of planting, and (2) paring of suckers followed by hot water treatment (55 °C for 20 min) along with application of Carbofuran at 0.5 g a.i./plant and Neem cake at 1 kg/plant were demonstrated, besides the untreated check (farmers' practice) in Shri. Basudeb Behera farmer's field located in Mangalpur Village of Puri District. The application of *P. fluorescens* and combined treatment of paring, hot water treatment, Carbofuran, and Neem cake registered average bunch weight of 24.2 kg and 25.8 kg with 17.5% and 25.0% increase in yield in comparison to untreated plants (20.5 kg), respectively. But the maximum benefit cost ratio (11.3) was noticed with application of *P. fluorescens* with a net profit of ₹52,000/– per hectare. The farmers were well convinced with the benefit of this sustainable, eco-friendly, and low-cost technology of bioagent application for nematode management.

## 7.2 Citrus, *Citrus* spp.

The major nematode problems on citrus include the citrus nematode *Tylenchulus semipenetrans* and the root-knot nematode *Meloidogyne indica*.

### 7.2.1 Citrus Nematode, *Tylenchulus semipenetrans*

Siddiqi (1961) reported *T. semipenetrans* for the first time from India in a paper presented before the Indian Science Congress and observed that about 80% of the citrus trees at Aligarh, Uttar Pradesh, were infested. In India, *T. semipenetrans* is the causal agent of slow decline disease and also involved as a component in citrus dieback disease complex.

#### 7.2.1.1 Crop Losses

The citrus nematode is responsible for annual yield loss of 27% in citrus amounting to ₹9828.22 million in India (Walia and Chakraborty 2018).

*T. semipenetrans* was responsible for 69%, 29%, and 19% loss in fruit yield of sweet orange (Baghel and Bhatti 1983), lemon (Mukhopadhyaya and Suryanarayana 1969), and sweet lime (Mukhopadhyaya and Dalal 1971), respectively.

#### 7.2.1.2 Symptoms

*T. semipenetrans* causes "slow decline" disease of citrus. Decline in tree vigor and growth, leaf yellowing, leaf drop, and small sized fruits are some of the common above-ground symptoms. The decline symptoms are often more pronounced in the uppermost portion of the trees (Fig. 7.11). In Israel, the decline symptoms are associated with a critical nematode population of approximately 40,000 larvae and males per 10 g of roots, above which the tree performance is markedly lowered.

**Fig. 7.11** Citrus trees exhibiting symptoms of citrus dieback (thinning canopy and bare branches) caused by the citrus nematode *Tylenchulus semipenetrans*. Soil particles usually cling tightly, even after washing to the gelatinous egg masses which cover the protruding part of the female nematode body (Figs. 7.12 and 7.13). In heavily infested roots, the cortex separates readily from the vascular stele

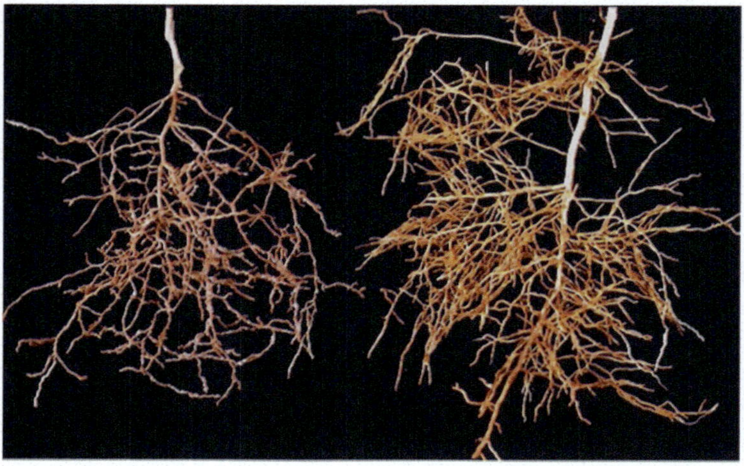

**Fig. 7.12** Soil particles clinging to the citrus roots infected with *Tylenchulus semipenetrans* (*Left*) and healthy roots (*Right*)

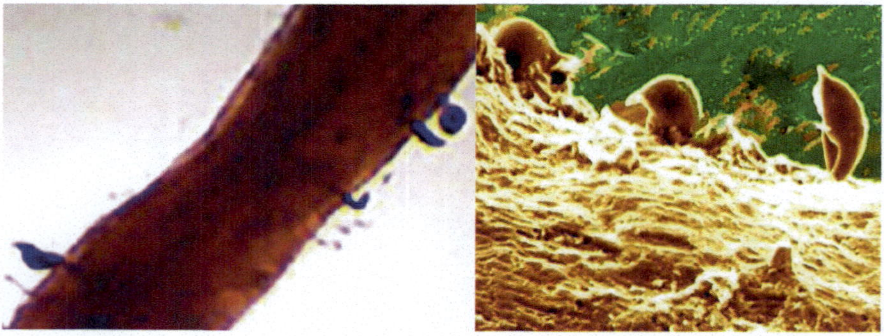

**Fig. 7.13** Citrus nematode females attached to the roots

### 7.2.1.3 Life Cycle

Van Gundy (1958) studied the life cycle in detail. Eggs hatched in 12–14 days at about 24 °C and sex differentiation was possible at the second stage. The second-stage male larvae developed to maturity within 7 days without feeding, while it takes 14 days for the second-stage female to locate the root, feed on the epidermal cells and molt. The life cycle from egg to egg required 6–8 weeks.

In experiments conducted in Israel, the minimum time required at 24 °C for completion of the life cycle was 14 weeks on *Poncirus trifoliata*, 10 weeks on *Ruta bracteosa*, and 7 weeks on *Citrus aurantium* and *C. limettoides* (Cohn 1965).

#### 7.2.1.4 Host Range

Vilardebo and Luc (1961) list 29 species of *Citrus*, 21 citrus hybrids, and 11 other species of Rutaceae as hosts from 23 countries. The citrus nematode has developed strains or biotypes to attack grasses (*Andropogon rhizomatus, Panicum* spp.), olive (*Olea* spp.), grapevines (*Vitis* spp.), persimmon (*Diospyros lotus*), pear (*Pyrus communis*), *Calodendrum capense,* climbing hemp weed (*Mikania batatifolia*), and lilac (*Syringa vulgaris*) (Stokes 1969). In India, *T. semipenetrans* is known to attack grapes and loquats (Parvatha Reddy and Singh 1978b).

#### 7.2.1.5 Interaction with Other Pathogens

In the presence of *T. semipenetrans*, significantly more citrus plants become infected by *Fusarium* spp. The debilitation of citrus plants is caused by the combined action of the citrus nematode along with *Fusarium oxysporum* and *F. solani.*

#### 7.2.1.6 Ecology

Optimal nematode reproduction occurs at soil temperatures of 28–31 °C. In California, optimum nematode reproduction occurred in soils with a clay content of 10–15%. A pH range of 5.6–7.6 has been found to be favorable for nematode reproduction.

After infested citrus seedlings are planted in virgin soils, it takes 12–17 years for the citrus nematode to attain the critical level (40,000 larvae per 10 g of feeder roots) at which stage the tree performance declines.

#### 7.2.1.7 Spread and Survival

The widespread distribution of *T. semipenetrans* throughout the citrus growing regions of the world is mainly due to the transfer of infested citrus planting material. Short distance and local spread of the free living stages of the nematode is through soil movement. Agricultural implements, animals and man, wind, and water are common agents in spreading the nematode in soil.

In soil, the nematode may remain viable in the absence of a host for as long as 10 years.

#### 7.2.1.8 Physiologic Races

Based on the pathogenicity to host differentials like sweet orange, citrange, trifoliate orange, grapevine, and lilac, Baines et al. (1974) recorded four races of *T. semipenetrans* in California. A "grass strain," which parasites *Andropogon rhizomatus* but failed to infect citrus, was reported by Stokes (1969).

#### 7.2.1.9 Management

##### 7.2.1.9.1 Physical Methods

The citrus nematode can be controlled by seedling bare root dips in hot water at 45 °C for 25 min (Bindra et al. 1967).

During summer, exposing the soil to hot sun can control the nematodes substantially (Sehgal et al. 1990).

#### 7.2.1.9.2 Cultural Methods

The citrus nematode is easily transported to new orchards through infested seedlings and budded plants which are usually carried "balled" with infested soil. Therefore, nurseries should never be established on or near old citrus orchards and nursery soil should be sterilized before planting.

Care should also be taken not to spread infection through tools, machinery, and irrigation water used in infested groves.

Neem cake at 100 g/m$^2$ is very effective in managing the citrus nematode population in the soil.

Singh and Sitaramaiah (1973) reported that castor cake was highly effective in reducing 50% of the citrus nematode population.

Effective reduction in the citrus and root-knot nematode populations was achieved by soil application of Neem cake at 20 kg per tree basin at an interval of 4 months (Mani and Murthy 1986).

Irrigation of citrus orchards with sewage water reduces the citrus nematode population to a very low level.

Castor (*Ricinus communis*), Aak (*Calotropis procera*) and Datura (*Datura strumonium*) chopped leaves applied at 20 and 40 g/kg soil in 9 m$^2$ area around tree trunks can result in reduction of nematode population and increase in yield (Baghel 1995).

Intercropping with antagonistic crops like onion, garlic, or marigold reduces the citrus nematode infection. Intercropping of acid lime with marigold and mustard markedly reduced the rate of multiplication of *T. semipenetrans* to 10.03 and 8.59, respectively (Mani 1988b).

#### 7.2.1.9.3 Chemical Methods

##### Nursery Bed Treatment

Soil application Carbofuran at 4 kg a.i./ha was effective for the control of *T. semipenetrans* (Parvatha Reddy 1988).

##### Bare Root Dip Treatment

Dipping of citrus bare roots in appropriate concentrations of Ethoprop, Fensulfothion, and Thionazin can be employed for the management of *T. semipenetrans*. Mani (1990) found that the multiplication of *T. semipenetrans* can be reduced by 90% by dipping of citrus bare roots in Chlorpyriphos/ Monocrotophos at 1000 ppm for 45 min.

##### Soil Treatment

Application of Carbofuran 3G at 13 g/m$^2$ (about 9 m$^2$ around the tree trunk) just before flowering; pulverizing the soil in the basin area and mixing the chemical thoroughly followed by irrigation was found to be effective. Application of nematicide just before flowering proved beneficial. Application of Carbofuran at 1 kg a.i./ ha in citrus soil reduced the population of *T. semipenetrans* in soil and root by 67% and thereby increased the yield of citrus by 54%.

Chhabra et al. (1977) reported increase in pumello yield by 68–76% by soil application of Fensulfothion at 3 kg a.i./ha for the management of *T. semipenetrans*. Significant reduction in the citrus nematode population (76–84%) in sweet orange and increase in yield (61–244%) was observed with soil application of Carbofuran at 6 kg a.i. per ha (Baghel and Bhatti 1983).

#### 7.2.1.9.4 Biological Methods

Application of Pseudomonas fluorescens *at* 20 g/m$^2$ in the rhizosphere at the time of flowering proved effective.

The number of nematodes in roots and soil around citrus trees inoculated with *Pupureocillium lilacinum* was significantly lower than that in the nontreated or the nematicide-treated trees (Jatala 1986). Mani and Murthy (1989) reported the reduction in the multiplication of *T. semipenetrans* by 71.7–97.8% by soil application of spore suspension of *P. lilacinum*.

Arbuscular mycorrhizal fungus *Glomus mosseae* when added to the soil with *T. semipenetrans* in *Citrus jambhiri* limited the nematode development and partially neutralized the adverse effect of the nematode (Baghel et al. 1990).

*Pasteuria penetrans* can be readily multiplied on the citrus nematode. The parasite seems to offer potential for the biocontrol of *T. semipenetrans* (Fig. 7.14) (Mani 1988a).

#### 7.2.1.9.5 Host Resistance

Citrus rootstocks that are resistant or highly tolerant to the citrus nematode would offer practical and effective means of control. Trifoliate orange (Poncirus trifoliata), Troyer Citrange, and Swingle Citrumello and its hybrids (Citrumello) were highly resistant to the citrus nematode (Parvatha Reddy and Singh 1978a; Parvatha Reddy and Agarwal 1987; Mani and Reddy 1986).

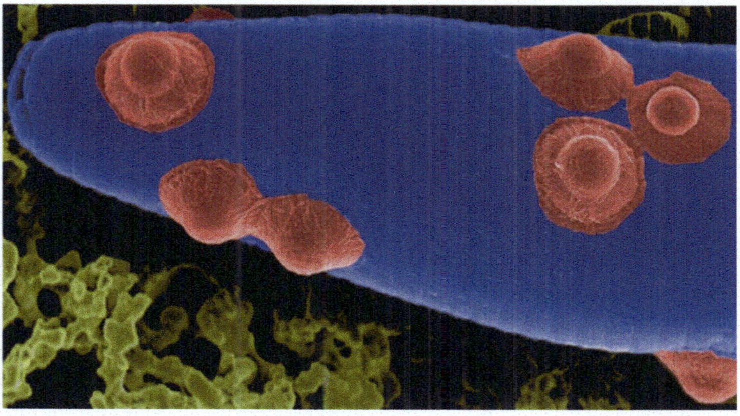

**Fig. 7.14** *Pasteuria penetrans* infecting citrus nematode larva

#### 7.2.1.9.6 Integrated Methods

Application of Carbofuran 3G at 7 g/m$^2$ + 1 kg Neem cake (around 9 m$^2$ around tree trunk) just before flowering; pulverizing the soil in the basin area; and mixing the chemical thoroughly followed by irrigation were found to be effective.

Integration of organic amendments (Neem cake and Castor cake) with biological control agents (*Pupureocillium lilacinum, Trichoderma harzianum, Verticillium lecanii, Pasteuria penetrans, Pseudomonas fluorescens*) and arbuscular mycorrhizal fungi (*Glomus fasciculatum*) was found to be effective for the management of the citrus nematode (Parvatha Reddy et al. 1993, 1995, 1996a, b, 2000).

## 7.2.2 Root-Knot Nematode, *Meloidogyne indica*

Whitehead (1968) reported *M. indica* on citrus from India. Heavy infestation of the root-knot nematode *M. indica* was especially observed on acid lime in Banaskantha district of north Gujarat. The citrus root-knot nematode is widely distributed in the states of Gujarat (Anand, Banaskantha, and Mahesana districts) and Andhra Pradesh.

### 7.2.2.1 Symptoms

The above-ground symptoms of the citrus root-knot nematode include severe stunting of trees, yellowing of leaves, poor vigor, unthrifty tree appearance, and twig blight leading to citrus decline. The plants were showing yellowing of foliage in initial stage and were completely drying at later stage. On uprooting such infected plants, numerous small to very big sized galls/knots were noticed on the roots, resulting in very poorly developed root system (Patel et al. 1999). They fail to flower and produce fruits even after several years of planting. The roots have conspicuous galls on pioneer and fibrous roots (Fig. 7.15). Big cavities were seen where galls were present earlier during advanced stage. On the surface of roots thin films can be observed which contain nematode egg masses.

### 7.2.2.2 Host Range

Cowpea and bottle gourd were recorded to be very good hosts, while cotton was poor host. Piludi, a weed, was found to be severely galled by the nematode and was considered a very good host.

### 7.2.2.3 Management

#### 7.2.2.3.1 Physical Methods

Soil solarization is one of the effective ways of suppressing nematode populations in nursery beds and can be employed mostly during hot weather days to raise nematode-free planting material. The prepared field may be covered with polythene sheet of 100 gauge thickness for about 7–10 days before sowing.

**Fig. 7.15** Acid lime roots infected with root-knot nematode *Meloidogyne indica*

#### 7.2.2.3.2 Cultural Methods

Intercropping

Growing sun hemp in tree basin reduced the population of root-knot nematodes. *Crotalaria* roots exert a toxic effect on *Meloidogyne* population. The root-knot nematode larvae which have penetrated sun hemp roots cannot survive.

Crop Rotation

Once the life of the citrus orchard comes to an end, it is highly preferable to grow cereals and millets for at least 2–3 seasons before initiating new orchard in the same field. This will help in reducing the populations of *Meloidogyne* species.

Organic Amendments

Soil application of Neem or Karanj or Mustard cake, Farm yard manure, and Poultry manure at 2.5 kg/plant were found to be more effective in increasing the plant growth and reducing root-knot index.

#### 7.2.2.3.3 Biological Methods

Use organic amendments enriched with bioagents (*Pupureocillium lilacinum*) for raising seedlings and in standing crop in the field proved beneficial.

The root-knot nematode in sweet orange can be managed by soil application of 100 kg FYM enriched with *Pseudomonas fluorescens* ($2 \times 10^6$ cfu/g) at 20 kg/ha.

#### 7.2.2.3.4 Chemical Methods

The initial seedling dip treatment with Carbosulfan at 500 ppm for 6 hr. recorded more plant height and effectively controlled root-knot nematode.

#### 7.2.2.3.5 Host Resistance

Citrus rootstocks such as Rangpur lime 8784, Sour Orange Tirupati, Citrumello 4479, Rangpur 8748, Acid Lime PDKV, Sai Sarbati, and Galgal were found to be resistant against *M. indica*. Rangpur lime Chethalli, Trifoliate Orange Chethalli, Nasnaran, Hazara Australia, Rangpur lime Kirumakki, Pramalini, and Anand Selection were moderately resistant.

## 7.3 Papaya, *Carica papaya*

Root-knot nematodes (*Meloidogyne incognita* and *M. javanica*) and the reniform nematode (*Rotylenchulus reniformis*) have been identified as proven pathogens on papaya.

### 7.3.1 Root-Knot Nematodes, *Meloidogyne incognita, M. javanica*

#### 7.3.1.1 Crop Losses

The root-knot nematode *Meloidogyne incognita* and the reniform nematode *Rotylenchulus reniformis* are responsible for 30% annual yield losses in papaya amounting to ₹2516.64 million in India (Walia and Chakraborty 2018).

#### 7.3.1.2 Symptoms

Orchard show patches of poor growth with many plants missing in rows. Symptoms include poor growth, yellowing of foliage, dropping of leaves (Fig. 7.16), reduction in leaf production, weak vigor, and premature dropping of fruits. Roots exhibit galls of varying sizes (Fig. 7.17). In heavily infected old roots, adjacent galls join together and form large galls.

#### 7.3.1.3 Biology and Life Cycle

The root-knot nematode is soil-borne, sedentary endoparasite. After penetration, the J2 larvae feed endoparasitically on conducting vessels and develop to adult females which are spherical in shape and lays egg masses in compound galls. The males are vermiform in shape which leaves roots to enter the soil. Females lay eggs in masses on root surface on galls, or in advanced stages inside the compound galls. During summer, the nematode completes its life cycle in about 25–30 days, while it takes longer during winter months.

#### 7.3.1.4 Spread

Root-knot nematodes are disseminated through infected planting material, irrigation water, and agricultural implements.

**Fig. 7.16** Field infestation of papaya with root-knot nematodes showing stunted plant growth and leaf yellowing symptoms

**Fig. 7.17** Papaya seedling (*Left*) and roots (*Right*) infected with root-knot nematodes

### 7.3.1.5 Management

#### 7.3.1.5.1 Physical Methods

Hot water treatment of papaya seedling roots at 50 °C for 10 min proved beneficial. Soil solarization was found to be effective to raise nematode-free seedlings.

#### 7.3.1.5.2 Cultural Methods

Neem cake at 2.5 t/ha gave reduction in root-knot nematode (*M. incognita*) population in both soil and roots and increased yield by about 200% (10.4 kg/plant compared to 5.5 kg/plant in control) (Nayak 1986). Application of Farm yard manure 8 kg and Neem cake 1 kg/plant was found beneficial.

#### 7.3.1.5.3 Chemical Methods

Application of Carbofuran and Fenamiphos each at 2 kg a.i./ha was found to be effective.

#### 7.3.1.5.4 Biological Methods

Nursery bed treatment with *P. lilacinum* and *T. harzianum* each at 10 kg per ha proved beneficial.

Soil application of *Pochonia chlamydosporia* at 5 kg/ha increased plant height (3.68 m as compared to 2.07 m in control) and fruit yield (38.1 kg/plant as compared to 20.5 kg/plant in control) of papaya and reduced nematode population both in soil and roots.

#### 7.3.1.5.5 Host Resistance

Use resistant papaya cvs. CO-2, Pusa 22-3, Pusa Delicious, Pusa Dwarf, and Pusa Giant (Nayak 1986; Laqman Khan 2001).

#### 7.3.1.5.6 Integrated Methods

Seed treatment with *P. fluorescens* ($10^8$ spores/g) combined with soil application of *T. harzianum* ($10^6$ spores/g) and *P. fluorescens* ($10^8$ spores/g) at 5 g/kg soil gave significant reduction in *M. incognita* population both in soil and roots, number of eggs per egg mass and hatching of eggs, and increased root colonization by both the bioagents in papaya under nursery conditions (Rao 2007).

The root-knot nematode can be managed by soil application of Neem cake at 2.5 tons per ha and Carbofuran at 6 kg a.i./ha. Highest yield of 11 kg per plant was recorded in Carbofuran followed by Neem cake (10.4 kg), whereas control plants yielded 5.5 kg (Nayak 1986).

Nursery bed treatment with *T. harzianum* and *P. lilacinum* each at 5 or 10 g/kg soil resulted in production of highly vigorous papaya seedlings whose roots were colonized with both the bioagents. There was significant reduction in root galling in the combination treatments (2.9–3.2) compared to control (8.9) (Rao and Naik 2003).

Simultaneous application of *P. lilacinum* and *T. harzianum* both at 1 g a.i. per plant in the main field gave maximum increase in plant growth parameters and highest reduction in reproduction factor and root galling in papaya (Khan 1991).

Integrated management strategy for root-knot nematodes (*M. incognita*) infecting papaya was standardized by nursery bed treatment with Neem-based formulation of *P. lilacinum* ($1 \times 10^6$ cfu/g) + *T. harzianum* ($1 \times 10^9$ cfu/g) (each at 5 or 10 g/kg soil) and application of 35 g each of *P. lilacinum* + *T. harzianum*/pit while transplanting. The above treatment increased plant height, plant weight and reduced root galling, number of egg masses/plant, and number of eggs/egg mass in nursery. The integration of both the bioagents also increased root colonization of bioagents, propagule density in soil, and parasitization of egg masses under field conditions.

Application of 2 kg FYM enriched with *P. fluorescens* ($10^9$ spores/g) and *P. lilacinum* ($1 \times 10^6$ cfu/g) per plant at the time of planting and at an interval of 6 months significantly reduced root-knot nematodes on roots by 70%. Significant increase in fruit yield (28%) was also observed. Benefit:cost ratio (calculated for marginal cost of biopesticides and returns accrued by application of biopesticides) was 3.2 (Anon 2012).

## 7.3.2 Interaction of Root-Knot Nematode with *Fusarium* Wilt

### 7.3.2.1 Symptoms

Increasing inoculum of *M. incognita*, whether present alone or together with *Fusarium solani*, decreased seedling emergence of papaya. The combination of highest inoculum of both pathogens (3000 nematodes and 3 g culture of fungus) caused maximum inhibition of seedling emergence and also increased post-emergence damping-off of papaya seedlings (Khan and Hussain 1990).

Inoculation of nematodes 12 days prior to fungus or simultaneous inoculation of both pathogens resulted in significant reduction in plant height, root length, and shoot and root weight of papaya. Root galling was highest in case of nematode alone followed by nematode inoculation 12 days prior to fungus, and simultaneous inoculation of nematode and fungus (Kishore et al. 2005).

### 7.3.2.2 Management

Khan et al. (1997) have reported that *P. lilacinum* gave better overall protection of papaya plants against *M. incognita* and *F. solani* disease complex than *T. harzianum*. They reported that application of both the biocontrol agents further limited the damage caused by *M. incognita* and *F. solani* and gave a 35% increase in plant growth compared to individual applications.

## 7.3.3 Reniform Nematode, *Rotylenchulus reniformis*

The reniform nematode was first reported on papaya from Northern India by Prasad et al. (1964).

### 7.3.3.1 Distribution

The reniform nematode has been found in the states of Gujarat, Madhya Pradesh, Orissa, Punjab, Tamil Nadu, and Uttar Pradesh (Khan and Khan 1998).

### 7.3.3.2 Crop Losses

*R. reniformis* caused 28% loss in fruit yield of papaya (Rajendran and Naganathan 1981).

#### 7.3.3.3 Symptoms

Nematode-affected plants are stunted in growth; foliage is slightly yellow in color with reduced leaf size. Under severe nematode infection, roots show discoloration with dark brown lesions which are prone to attack by soil pathogens. Egg masses on roots are covered with soil particles.

#### 7.3.3.4 Ecology

The incidence of *R. reniformis* is higher in wetter soils.

#### 7.3.3.5 Management

##### 7.3.3.5.1 Chemical Methods

Foliar application of Fenamiphos and Oxamyl on papaya was effective for the management of reniform nematode (Ayala et al. 1971).

Rajendran and Naganathan (1981) reported that Carbofuran at 2 kg a.i. per ha gave significant reduction in soil population of *R. reniformis* and increased papaya fruit yield by 38.4% over control.

##### 7.3.3.5.2 Host Resistance

The inhibition of development and reproduction of *M. incognita* was observed in Papaya cultivars CO-1, CO-2, CO-3, CO-4, CO-5, Thailand, and Giant (Nayak 1986). Patel et al. (1989) reported that papaya cvs. Solo and Washington were resistant, while Coorg Honey Dew was moderately resistant. Papaya cv. Mammoth was found resistant, while CO-1, Peradeneya, Sel.7, and Washington were moderately resistant (Khan and Khan 1998).

##### 7.3.3.5.3 Integrated Methods

Application of 2 kg FYM enriched with *P. fluorescens* ($10^9$ spores/g) and *P. lilacinum* ($1 \times 10^6$ cfu/g) per plant at the time of planting and at an interval of 6 months significantly reduced reniform nematodes on roots by 66%. Significant increase in fruit yield (28%) was also observed. Benefit-cost ratio (calculated for marginal cost of biopesticides and returns accrued by application of biopesticides) was 3.2 (Anon 2012).

## 7.4 Pineapple, *Ananas comosus*

`The most common nematodes found to be detrimental to crop production are *Meloidogyne* species and *Rotylenchulus reniformis*.

## 7.4.1 Root-Knot Nematodes, *Meloidogyne incognita, M. javanica*

### 7.4.1.1 Crop Losses

Losses for the first crop were greater than 40% due to the attack of root-knot nematodes.

### 7.4.1.2 Symptoms

The root growth is greatly slowed down. Reduction in pineapple plant weight, leaf weight, and fruit weight by 47.5% was observed due to root-knot nematode infection. Root-knot nematodes produce distinct terminal galls on the roots (Fig. 7.18), stopping further root development. The root infested simultaneously by several larvae stops the growth and may produce many secondary roots.

### 7.4.1.3 Life Cycle

Second-stage juveniles penetrate roots in the meristematic region of the root tip and become sedentary after 2–3 days. Vermiform adult males and saccate sedentary females were subsequently developed. Reproduction is by mitotic parthenogenesis and female nematodes produce eggs contained in a gelatinous matrix.

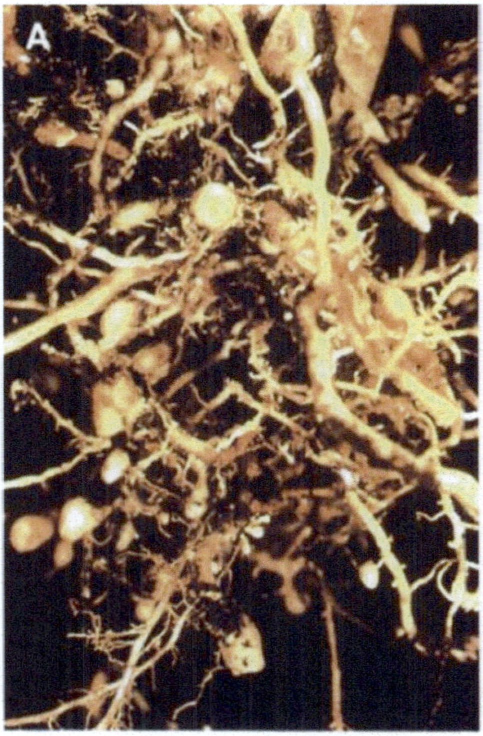


**Fig. 7.18** Root-knot nematode (*M. javanica*) galls on pineapple roots

#### 7.4.1.4 Survival and Spread

Eggs of *M. javanica* contained in galled tissue survived for 20 days at relative humidity of 90%, while the free-living forms survived for 20–24 weeks in desiccated soil without a host. *M. javanica* can survive, although at low levels, as long as 2 years in fallow field soil. The nematode spread over long distances in soil adhering to workers feet, implements and equipment that is moved from field to field. Infected planting stock also serves as a means of nematode spread.

#### 7.4.1.5 Ecology

Two seasonal *M. incognita* population peaks were observed first in April–May and second during November. Soil temperature range of 21.2–27.5 °C and soil moisture range of 14.0–18.6% were found to favor the population development.

#### 7.4.1.6 Management

##### 7.4.1.6.1 Cultural Methods

Yield increases have been observed when planting was done in a field, which was previously had pigeon pea, owing to decrease in infestation by *M. incognita*. The reduction in population of *M. incognita, Criconemoides* spp., *Helicotylenchus* spp., and *R. reniformis* and increase in fruit yield in the planted crop (23.5 tons/acre as against 15.1 tons/acre in control) and first ratoon crop (18.4 tons/acre as against 9.3 tons/acre in control) was observed when pineapple was rotated with Pangola grass (Ayala et al. 1967).

The addition of chopped pineapple material to soil at 125–375 tons/ha significantly reduced galling caused by root-knot nematode.

Maintenance of weed-free and host-free fallow period of at least for 6 months was found to be effective for significant decline in nematode population.

##### 7.4.1.6.2 Chemical Methods

Dipping crowns in solutions of Oxamyl at 2400–4800 ppm, or Fenamiphos at 300 ppm, provides a good protection against *M. incognita*.

##### 7.4.1.6.3 Biological Methods

Application of *Trichoderma harzianum/viride* and *Pseudomonas fluorescens* for treatment of seed/seedling/planting materials in the nurseries and field proved beneficial.

##### 7.4.1.6.4 Host Resistance

*Ananas ananasoides* and three of its hybrids are highly resistant to attacks by *M. incognita*.

## 7.4.2 Reniform Nematode, *Rotylenchulus reniformis*

### 7.4.2.1 Crop Losses

The reniform nematode can reduce the pineapple plant populations by more than 83% in the fields. Complete failure of ratoon crop could take place due to severe nematode infestation along with moisture stress.

### 7.4.2.2 Symptoms

Leaves of infected plants are less erect than those of healthy plants, are reddish in color and show poor growth. The reniform nematode reduces the number of lateral and fine feeder roots (Fig. 7.19). Plant collapse and death can result due to severe nematode infestations. Reniform nematode infection inhibits secondary root formation, and root systems are poorly developed.

### 7.4.2.3 Life Cycle

Egg hatch is stimulated by root exudates of certain host plants, and second-stage juveniles leave the egg and move into the soil. Once in the soil they undergo 3 molts without feeding, yielding adult males and pre-adult females. Females enter the root system, initiate a feeding site, become sedentary, and develop into swollen, mature egg-producing females. Males do not feed.

### 7.4.2.4 Survival and Spread

The reniform nematode survives extended periods (6 months to 2 years) without a host. The nematode survives fallow periods in the egg stage or as anhydrobiotic juvenile stages, depending on soil moisture.

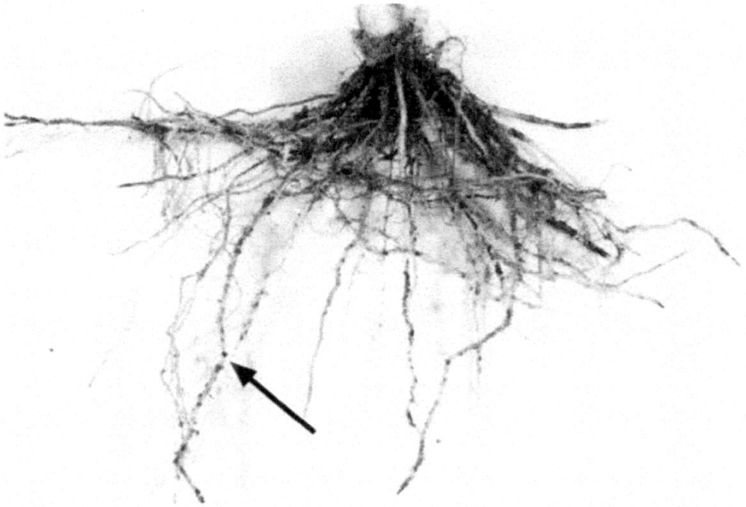

**Fig. 7.19** Lack of feeder roots and soil clinging to reniform nematode egg masses (arrow) on pineapple roots

#### 7.4.2.5 Management

##### 7.4.2.5.1 Cultural Methods

Rotation of pineapple with French marigold (*Tagetes patula*), Rhodes grass (*Chloris guyana*), sun hemp (*Crotalaria juncea*), *Desmodium unicatum*, sugarcane, or Pangola grass was effective in controlling *R. reniformis*. Sugarcane is generally considered a non-host for *R. reniformis*, so planting pineapple into sugarcane soils may decrease problems with the nematode, provided that weed hosts are not present.

Intercropping and incorporation after 3 months of *Crotalaria juncea* suppressed *R. reniformis* numbers in pineapple. *Crotalaria juncea* treatment increased bacterivorous nematodes and nematode-trapping fungal population densities in soil (Wang et al. 2002).

Reduction in the population of *R. reniformis* can be obtained by irrigated fallow, since the nematode prefers low moisture regimes in pineapple crop.

##### 7.4.2.5.2 Chemical Methods

Application of Carbofuran at 2.25 kg a.i. per ha, Oxamyl at 1 kg a.i. per ha, and Fenamiphos at 4.5 kg a.i. per ha were effective in controlling *R. reniformis* and in increasing fruit yields. Foliar application of Fenamiphos at 600–2400 ppm gave effective control of reniform nematode. Foliar application of Oxamyl may be as effective as Fenamiphos if applied at twice the rate.

##### 7.4.2.5.3 Host Resistance

Pineapple varieties/species such as Venezolana, *Ananas ananasoides*, and several of its hybrids are highly resistant to *R. reniformis* and do not allow the initial infestation to be maintained.

##### 7.4.2.5.4 Integrated Methods

Integrating sun hemp with soil solarization is promising for suppressing reniform nematodes and is more compatible with this shorter production cycle in pineapple (Wang and Sipes 2009).

## 7.5 Mulberry, *Morus rubra*

Among mulberry nematodes, root-knot nematode, *Meloidogyne incognita* is economically important as it affects the crop quantitatively and qualitatively. It is a major constraint in mulberry cultivation and plays an important role in reducing herbage yield and quality of leaves besides the life span of mulberry plants.

### 7.5.1 Root-Knot Nematode, *Meloidogyne incognita*

#### 7.5.1.1 Crop Losses

It is responsible for 10–12% herbage yield loss (Govindaiah et al. 1991).

### 7.5.1.2 Symptoms

The above-ground symptoms include typical yellowing of the foliage, stunted growth, and wilting of the young seedling plants. The uprooted plants show heavy galling on the roots, with few or no feeder roots (Fig. 7.20). The severely affected plants showed stunted growth with low water moisture in leaves, and later marginal necrosis and yellowing of leaves. The underground symptoms include the formation of characteristic knots/galls on the root (Fig. 7.20). Galls are spherical and vary in size; young galls are too small and yellowish-white in color, old galls are big and pale brown. The nematode-infected roots cannot absorb water and nutrients from soil effectively due to root injury.

### 7.5.1.3 Predisposing Factors

- Contaminated soil, farm implements, and run-off irrigation form the primary means of nematode spread.
- Secondary means of nematode spread include infected planting stock, other susceptible crops and weeds in and around the mulberry fields.
- Temperature between 27 and 30 °C, soil moisture of less than 40%, and pH of 5–7 are favorable for the development of the root-knot nematode.
- The disease outbreak is throughout the year and more common in sandy soils under irrigated conditions.

### 7.5.1.4 Interaction with Other Pathogens

The root-knot nematode *Meloidogyne incognita* and the root rot fungus *Macrophomina phaseolina* play an important role in reducing herbage yield and quality of leaves besides the life span of mulberry plants. Reduction in the nutritive value of mulberry leaves which adversely affect the growth and development of silkworm ultimately leading to poor cocoon production results due to the above disease complex. Besides causing direct damage, nematodes are also involved in

**Fig. 7.20** Mulberry roots infected with root-knot nematodes

disease complex with other microorganisms and provide easier entrance to weak pathogenic fungi and break the host plant resistance.

Inoculation of *M. phaseolina* prior to *M. incognita* significantly reduced the nematode population. Significant reduction in plant growth occurs due to inoculation of *M. phaseolina* 15 days after nematode inoculation or simultaneous inoculation of both the pathogens (Muthulakshmi et al. 2010b).

### 7.5.1.5 Management

#### 7.5.1.5.1 Physical Methods

Hot water treatment of mulberry seedlings is effective at 50 °C for 15 min. Hot water immersion must be followed up with a cold water treatment.

#### 7.5.1.5.2 Cultural Methods

Govindaiah et al. (1990) recorded increased leaf moisture, leaf yield of mulberry, and management of *M. incognita* by intercropping of mulberry with marigold at wider spacing of 30–45 cm.

Marigold and sun hemp plants at 10 per square meter can be grown in between mulberry rows. Intercropping of these plants and mulching of the same in soil after vegetative growth provides an additional organic matter to the soil which enriches soil fertility in addition to control of the nematode disease.

The root-knot disease can be easily and effectively controlled by the use of amaranth plants intercropped with mulberry plants in the naturally root-knot infested field which acts as a trap crop for root-knot nematodes increasing silk production. In order to keep the nematode population to a low level, amaranth could be grown as intercrop that can be cut and marketed at frequent intervals as an additional benefit (Datta 2005; Datta and Datta 2007).

In mulberry, application of Neem oil-cake at1 ton/ha/year in four equal split dosages was found to be effective and safe to silkworm.

#### 7.5.1.5.3 Integrated Methods

Use of bioagents is most preferred for the management of *M. incognita* since it is not advisable to use chemicals as the leaves are fed to the silkworms. Soil application of *Purpureocillium lilacinum* at 25 g + 100 g Neem cake/plant is recommended for the management of *M. incognita* infesting mulberry.

Combined soil application of *Pseudomonas fluorescens* + *Trichoderma viride* each at 10 g/plant as soil application was effective to check the root-knot nematode disease and to improve growth of mulberry with increased leaf yield and reduced nematode population (Muthulakshmi et al. 2010a).

Combined inoculation of AMF and *Pseudomonas fluorescens* had positive effect on root-knot nematode control on mulberry (Kumutha 2001).

Maximum reduction of root galling (85.30) and egg mass production (86.45%) was achieved by soil application of 24 g of Neem cake +200 g FYM + 1 g of Bionema +100 g of AMF mixed in 20 kg of FYM applied in three split doses/year (Vijaya Kumari and Sujathamma 2014).

#### 7.5.1.5.4 Success Story in Karnataka

Management of root-knot nematode (*Meloidogyne incognita*) in mulberry was demonstrated in Shri. Ravi's farm located in Uppinahalli Village, Hosakote Taluk, Bangalore North District, using soil application of talc formulation of *Purpureocillium lilacinum* at 20 kg/ha ($10^8$ spores/g talc) along with Neem cake at 1 ton/ha. Use of bioagents is most preferred for the management of *M. incognita* since it is not advisable to use chemicals as the leaves are fed to the silkworms. Individual farmer's gain was 60% more leaf yield which amounts to nearly ₹1,00,000/ha as compared to the poor crop which fetches ₹50,000 to ₹67,500/ha.

## 7.6 Conclusion

Although many species of plant parasitic nematodes are associated with tropical fruit crops, root-knot nematode *Meloidogyne incognita*, burrowing nematode *Radopholus similis*, and citrus nematode *Tylenchulus semipenetrans* are the major nematode pests that infect these fruit crops. Damage due to nematodes is documented for relatively few host-parasite combinations. These should be recognized by growers of tropical fruit crops, and such combinations be avoided in selecting planting sites. Good management practices and sanitation are essential in all phases of nursery and orchard operations, because once a serious nematode problem is established on a fruit tree, it is very difficult to restore tree health and production.

Knowledge of the host-parasite associations of nematodes on tropical fruit species is essential in deciding what fruit species to plant in sites which are infested by one or more nematode species. Future studies are needed to determine the nematode population levels which are damaging to the various fruit species, and control measures should be developed for the appropriate cases. Further studies needs to be directed in order to identify resistant genes and nematode effectors to raise up new management approaches. The strategy for nematode management could be shifted from the concept of control to the concept of management by combining several methods in an integrated manner at keeping the nematode population below economic injury level.

## References

Anon (2009) Nematological research in Odisha - the tiny feet towards excellence. All India coordinated research project on plant parasitic nematodes with integrated approach for their control, Department of Nematology, College of Agriculture, Orissa University of Agriculture and Technology, Bhubaneswar

Anon (2012) Three decades of nematological research in Odisha (1977–2007). All India coordinated research project on plant parasitic nematodes with integrated approach for their control, Department of Nematology, College of Agriculture, Orissa University of Agriculture and Technology, Bhubaneswar

Ayala A, Roman J, Gonzalez J (1967) Pangola grass as a rotation crop for pineapple nematode control. J Agri Sci Univ Puerto Rico 51:94–96

Ayala A, Acosta N, Adsuar JA (1971) A preliminary report on the response of *Carica papaya* to foliar application of two systemic nematicides. Nematropica 1:10

Baghel PPS (1995) Nematode problems in citrus. In: Swarup G, Dasgupta DR, Gill JS (eds) Nematode pest management-an appraisal of eco-friendly approach. Indian Agricultural Research Institute, New Delhi, pp 203–210

Baghel PPS, Bhatti DS (1983) Evaluation of pesticides for the control of phytonematodes on citrus. In: Third all India nematology symposium, Himachal Pradesh Agricultural University, Solan, pp 38–39

Baghel PPS, Bhatti DS, Jalali BL (1990) Interaction of VA mycorrhizal fungus and *Tylenchulus semipenetrans* on citrus. In: Proceedings of the National conference on Mycorrhiza, Haryana Agricultural University, Hisar, p 118

Baines RC, Cameron JW, Soost RK (1974) Four biotypes of *Tylenchulus semipenetrans* in California identified, and their importance in the development of resistant citrus rootstocks. J Nematol 6:63–66

Bhattacharya RK, Rao VNM (1984) Effect of soil covers and soil moisture regimes on nematode population in soil and in root of banana. J Res Assam Agri Univ 5:206–209

Bindra OS, Chhabra HK, Chadha KL, Mehrotra NK (1967) A study on the correlation of citrus nematode population with decline of citrus. J Res Punjab Agric Univ 4:543–546

Blake CD (1966) The histological changes in banana roots caused by *Radopholus similis* and *Helicotylenchus multicinctus*. Nematologica 12:129–132

Charles JSK, Venkitesan TS (1993) Status report on the nematological investigations on Banana in Kerala Agricultural University. Kerala Agricultural University, Vellanikkara, Kerala, 43 pp

Charles JSK, Venkitesan TS, Thomas Y, Varkey PA (1985) Correlation of plant growth components to bunch weight in banana infested with burrowing nematode, *Radopholus similis* (Cobb). Indian J Nematol 15:186–190

Chhabra HK, Bindra OS, Singh I (1977) Further studies on the control of the citrus nematode, *Tylenchulus semipenetrans* Cobb, 1913 infecting *Citrus maxima* Merrill (Abstract). In: Proceedings of the International symposium on citriculture, Bangalore, p 38

Cobb NA (1893) Nematodes, mostly Australian and Fijian. Linnean Society New South Wales, Macleay Memorial, 59 pp

Cohn E (1965) The development of the citrus nematode on some of its hosts. Nematologica 11:593–600

Datta SC (2005) Possible use of amaranth as catch crop for root-knot nematodes intercropped with mulberry. J Environ Sociobiol 2:61–65

Datta SC, Datta R (2007) Intercropping amaranth with mulberry for managing root-knot nematodes and improving sericulture. Sericologia 47:297–306

Govindaiah, Dandin SB, Philip T, Datia RK (1990) Effect of marigold (*Tagetes patula*) intercropping against *Meloidogyne incognita* infecting mulberry. Indian J Nematol 20:96–99

Govindaiah, Dandin SB, Sharma DD (1991) Pathogenicity and avoidable leaf yield loss due to *Meloidogyne incognita* in mulberry (*Morus alba* L.). Indian J Nematol 21:52–57

Inomoto MM, Monteiro AR (1989) Thermal treatment of suckers of ‘Giant Cavendish’ banana for the eradication of plant-parasitic nematodes. Nematol Bras 13:139–150

Jatala P (1986) Biological control of nematodes. Annu Rev Phytopathol 24:453–489

Jonathan EI (1994) Studies on the root-knot nematodes (*Meloidogyne incognita*) (Kofoid and White, 1919) Chitwood, 1949, in Banana (*Musa* sp.) cv. Poovan (AAB). Ph. D. Thesis, Tamil Nadu Agricultural University, Coimbatore, Tamil Nadu

Jonathan EI, Cannayane I (2002) Field application of biocontrol agents for management of spiral nematode, *Helicotylenchus multicinctus* in banana (Abstract). Global conference on banana and plantain, Bangalore, p 193

Jonathan EI, Rajendran G (2000) Assessment of avoidable yield loss in banana due to root-knot nematode, *Meloidogyne incognita*. Indian J Nematol 30:162–164

Karuna K, Ravichandra NG, Krishnappa K, Sreenivasa KR (2004) Management of root-knot nematode, *Meloidogyne incognita* on banana by biocontrol agents. National symposium on paradigms in nematological research for biodynamic farming, University of Agricultural Sciences, Bangalore, p 70

Khan TA (1991) Studies on a disease complex of papaya caused by *Meloidogyne incognita* and *Fusarium solani*. Final technical report, DST project. Aligarh Muslim University, Aligarh, 47 pp

Khan TA, Hussain SI (1990) Effect of interaction of various inoculum levels of *Meloidogyne incognita* and *Fusarium solani* on seedling emergence and post-emergence damping-off of papaya seedlings. New Agric 1:17–20

Khan TA, Khan ST (1998) Nematode diseases of papaya. In: Trivedi PC (ed) Nematode diseases in plants. CBS Publishers and Distributors, New Delhi, pp 344–360

Khan T, Khan ST, Fazal M, Siddiqi ZA (1997) Biological control of *Meloidogyne incognita* and Fusarium solani disease complex in papaya using *Paecilomyces lilacinus* and *Trichoderma harzianum*. Int J Nematol 7:127–132

Kishore R, Kamalwanshi RS, Pandey M (2005) Effect of root-knot nematode with *Fusarium solani* on damping-off of papaya. Indian J Nematols 35:96–97

Kumutha K (2001) Symbiotic influence of AM fungi and rhizobacteria on biochemical and nutritional changes in mulberry (*Morus alba*). Ph. D. thesis, Tamil Nadu Agricultural University, Coimbatore

Lakshmana Murthy KV (1983) Studies on burrowing nematode, *Radopholus similis* (Cobb, 1933) Thorne, 1949 on banana in Karnataka. M. Sc. (Agriculture) thesis, University of Agricultural Sciences, Bangalore

Laqman Khan M (2001) Dr. Y.S. Paramar University of Horticulture and Forestry, Solan, Himachal Pradesh. In: Dhawan SC et al (eds) Indian nematology-progress and prospectives. Division of Nematology, Indian Agricultural Research Institute, New Delhi, pp 153–158

Loos CA (1959) Symptom expression of *Fusarium* wilt disease of the Gros Michel banana in the presence of *Radopholus similis* (Cobb, 1893) Thorne, 1949 and *Meloidogyne incognita acrita* Chitwood, 1949. Proc Helminth Soc Wash 26:103–111

Loos CA (1961) Eradication of the burrowing nematode, *Radopholus similis*, from bananas. Pl Dis Reptr 45:457–461

Loos CA, Loos SB (1960) Preparing nematode free banana "seed". Phytopathology 50:383–386

Mani A (1988a) Studies on the bacterial parasite *Pasteuria penetrans:* I. Spore viability after storage. II. Culture on citrus nematode *Tylenchulus semipenetrans*. Int Nematol Network Newsl 5(2):24–25

Mani A (1988b) Effect of interculture of marigold and mustard with acid lime on citrus nematode, *Tylenchulus semipenetrans*. Int Nematol Netw Newsl 5(4):14–15

Mani A (1990) Control of citrus nematode, *Tylenchulus semipenetrans*, by bare-root dip treatment of acid lime seedlings in pesticides. Int Nematol Netw Newsl 6(3):20–21

Mani A, Murthy VD (1986) Citrus nematode and root-knot nematodes: distribution, crop loss, symptoms and management. In: Swarup G, Dasgupta DR (eds) Plant parasitic nematodes - problems and progress. Indian Agricultural Research Institute, New Delhi, pp 249–260

Mani A, Murthy VD (1989) National symposium on insect pests and disease management on citrus. Jawaharlal Nehru Krishi Viswa Vidyalaya, Chindwara

Mani A, Reddy GS (1986) Studies on the varietal reaction of some species of *Citrus* and *Poncirus* to *Tylenchulus semipenetrans* Cobb. Indian J Nematol 16:267–268

Mendoza A, Sikora RA, Kiewnick S (2004) Efficacy of *Paecilomyces lilacinus* strain 251 for the control of *Radopholus similis* in banana. Commun Agric Appl Biol Sci 69:365–372

Mukhopadhyaya MC, Dalal MR (1971) Effect of two nematicides on *Tylenchulus semipenetrans* and on sweet lime yield. Indian J Nematol 1:95–97

Mukhopadhyaya MC, Suryanarayana D (1969) Citrus decline in Haryana: role of *Tylenchulus semipenetrans* and its control. Indian Phytopathol 22:495–497

Muthulakshmi M, Devrajan K, Jonathan EI (2010a) Biocontrol of root knot nematode, *Meloidogyne incognita* (Kofoid and White) Chitwood in mulberry (*Morus alba* L.). J Biopest 3(2):479–482

Muthulakshmi M, Jonathan EI, Devrajan K (2010b) Interaction of *Meloidogyne incognita* and *Macrophomina phaseolina* on Mulberry (*Morus alba* L.), National conference on Innovations in nematological research for agricultural sustainability - challenges and a roadmap ahead. Nematological Society of India & Tamil Nadu Agricultural University, Coimbatore, p 108

Nair KKR (1979) Studies on the chemical control of banana nematodes. Agric Res J Kerala 17:232–235

Nair MRGK, Das NM, Menon MR (1966) On the occurrence of the burrowing nematode *Radopholus similis* (Cobb, 1893) Thorne, 1949 on banana in Kerala. Indian J Entomol 28:553–554

Nayak GM (1986) Studies on papaya in relation to certain plant diseases. M. Sc. Thesis, University of Agricultural Sciences, Bangalore, 106 pp

Newhall AG (1958) The incidence of Panama disease of banana in the presence of the root-knot and burrowing nematodes (*Meloidogyne* and *Radopholus*). Pl Dis Reptr 42:853–856

Nirmal Johnson SB, Devarajan K (2004) Effect of *Pseudomonas fluorescens* on banana nematodes. In: National symposium on paradigms in nematological research for biodynamic farming, University of Agricultural Sciences, Bangalore, pp 81–82

Parvatha Reddy P (1988) Nematode management in horticultural crops. Technical Bulletin No 4, Indian Institute of Horticultural Research, Bangalore, 52 pp

Parvatha Reddy P, Agarwal PK (1987) Resistance in citrus rootstocks to the citrus nematode, *Tylenchulus semipenetrans*. Indian J Hort 44:111–114

Parvatha Reddy P, Singh DB (1978a) Evaluating the reaction of some species and varieties of *Citrus* and *Poncirus* to the citrus nematode. Indian J Nematol 8:82–84

Parvatha Reddy P, Singh DB (1978b) The association of the citrus nematode with grape roots in a commercial orchard. Curr Sci 47:640–641

Parvatha Reddy P, Khan RM, Rao MS (1991) Evaluation of chemicals for the management of the burrowing nematode, *Radopholus similis* infecting banana. Curr Nematol 2:1–2

Parvatha Reddy P, Khan RM, Rao MS (1993) Integrated management of the citrus nematode, Tylenchulus semipenetrans using neem cake and the parasitic fungus, *Paecilomyces lilacinus*. In: Proceedings of world neem conference, Bangalore, pp 647–650

Parvatha Reddy P, Rao MS, Mohandas S, Nagesh M (1995) Integrated management of the citrus nematode, *Tylenchulus semipenetrans* Cobb using VA mycorrhiza, *Glomus fasciculatum* (Thaxt.) Gerd & Trappe and oil cakes. Pest Mangmt Hort Ecosyst 1:37–41

Parvatha Reddy P, Rao MS, Nagesh M (1996a) Management of the citrus nematode, *Tylenchulus semipenetrans* by integration of *Trichoderma harzianum* with oil cakes. Nematol Medit 24:265–267

Parvatha Reddy P, Rao MS, Nagesh M (1996b) Management of the citrus nematode on acid lime by integration of parasitic fungi and oil cakes. Pest Mangmt Hort Ecosyst 2:15–18

Parvatha Reddy P, Rao MS, Nagesh M (1996c) Crop loss estimation in banana due to the burrowing nematode, *Radopholus similis*. Pest Mangmt Hort Ecosyst 2:85–86

Parvatha Reddy P, Nagesh M, Rao MS, Rama N (2000) Management of *Tylenchulus semipenetrans* by integration of Pseudomonas fluorescens with oil cakes. In: Proceedings of the international symposium on citriculture, Nagpur, pp 830–833

Parvatha Reddy P, Rao MS, Nagesh M (2002) Integrated management of burrowing nematode (*Radopholus similis)* using endomycorrhiza (*Glomus mosseae)* and oil cakes. In: Singh HP, Chadha KL (eds) Banana – improvement, production and utilization. AIPUB, Trichy, pp 344–348

Patel HV, Patel DJ, Patel BA (1989) Reaction of papaya varieties to *Rotylenchulus reniformis*. Int Nematol Netw Newsl 6(3):24

Patel DJ, Patel BA, Patel SK et al (1999) Root-knot nematode, *Meloidogyne indica* on kagzi lime in North Gujarat. Indian J Nematol 29(2):185–205

Patil NM, Dharne PK, Deshmukh SS (1999) Sucker dip treatment in monocrotophos with intercrops for the management of plant parasitic nematodes in banana. In: Seminar on technological advancement in banana production, handling and processing management, Jalgaon, Maharashtra, pp 126–128

Poornima K, Kumar N, Balamohan TN (2004) Biological control of lesion nematode in banana. In: National symposium on paradigms in nematological research for biodynamic farming, University of Agricultural Sciences, Bangalore, p 99

Poucher C, Ford HW, Suit RF, DuCharme EP (1967). Burrowing nematode in citrus. Florida Department of Agriculture, Division of Plant Industry, Bulletin No 7, 63 pp

Prasad SK, Dasgupta DR, Mukhopadhyaya MC (1964) Nematodes associated with commercial crops in northern India. I. Uttar Pradesh. Indian J Nematol 3:8–23

Rajagopalan P, Naganathan TG (1977) Studies on nematode parasites of banana. Tamil Nadu Agric Univ Ann Rep 6:131

Rajendran G, Naganathan TG (1981) Evaluation of systemic chemicals for the control of *Rotylenchulus reniformis* on papaya. Indian J Nematol 11:130

Rajendran G, Naganathan TG, Sivagami V (1979) Studies on banana nematodes. Indian J Nematol 9:54

Rao MS (2007) Papaya seedlings colonized by the bio-agents *Trichoderma harzianum* and *Pseudomonas fluorescens* to control root-knot nematodes. Nematol Medit 35:199–203

Rao MS, Naik D (2003) Effect of *Trichoderma harzianum* and *Paecilomyces lilacinus* on *Meloidogyne incognita* on papaya (Carica papaya L.) nursery seedlings. Pest Mangmt Hort Ecosyst 9:155–160

Rao MS, Umamaheswari R, Priti K et al (2016) Role of biopesticides in the management of nematodes and associated diseases in horticultural crops. In: Hakeem K, Akhtar M, Abdullah S (eds) Plant, soil and microbes. Springer, Cham

Ravi K, Nanje Gowda DN, Reddy PP (2001) Integrated management of disease complex in banana involving borrowing nematode and Panama wilt—an ecofriendly approach. In: Verghese A, Parvatha Reddy P (eds) IPM in horticultural crops: emerging trends in the new millennium. Association for Advancement of Pest Management in Horticultural Ecosystems, Bangalore, p 154

Risbeth J (1960) Factors affecting the incidence of banana with Panama disease. Emp J Expt Agric 28:109–113

Roy S. Ojha, P.K., Ojha, K.L. *et al.* 1998.Integrated management of wilt complex diseases of banana (*Musa* sp.). J Appl Biol. 8:46–49

Sehgal M, Chawla ML, Singh S, Balakrishnan K (1990) Studies on interactive effect of moisture and temperature on *Tylenchulus semipenetrans*. Curr Nematol 1:35–36

Shanthi A (2003) Management of *Radopholus similis*, *Helicotylenchus multicinctus* and *Pratylenchus coffeae* in Banana. Ph. D. thesis, Tamil Nadu Agricultural University, Coimbatore

Shanthi A, Rajendran G, Sivakumar M (2004) Influence of biofertilizers on lesion nematodes in banana. In: National symposium on paradigms in nematological research for biodynamic farming, University of Agricultural Sciences, Bangalore, p 97

Siddiqi MR (1961) Occurrence of the citrus nematode, *Tylenchulus semipenetrans* Cobb, 1913, and the reniform nematode, *Rotylenchulus reniformis* in India (Abstract). In: Proceedings of the 48th Indian science congress Part III, p 504

Singh RS, Sitaramaiah K (1973) Control of plant parasitic nematodes with organic amendments of soil. Final technical report on effect of organic amendments, green manuring and inorganic fertilizers on root-knot of vegetable crops, G.B. Pant University Of Agriculture And Technology, Bulletin No 6, 289 pp

Sosamma VK, Koshy PK, Samuel R (1998) Effect of vesicular arbuscular mycorrhizae, *Glomus mosseae* on nematodes infecting banana under coconut based high density multispecies cropping system. In: Mehta UK (ed) Nematology – challenges and opportunities in 21st century. Sugarcane Breeding Institute, Coimbatore, pp 94–97

Stokes DE (1969) *Andropogon rhizomatus* parasitized by a strain of *Tylenchulus semipenetrans* not parasitic to four citrus rootstocks. Pl Dis Reptr 53:882–885

Sundararaju P, Kumar V (2003) Management of root-lesion nematode, *Pratylenchus coffeae* in six commercial cultivars of banana through organic and inorganic amendments. Infomusa 12:21–25

Sundararaju P, Cannayane I, Thangavelu R (2002) Biomanagement of *Pratylenchus coffeae* and *Meloidogyne incognita* on banana by using endophytic fungi. Global conference on banana and plantain, Bangalore, p 191

Sundararaju P, Cannayane I, Sathiamoorthy S (2004) Nematode management in banana and plantains. Technical Bulletin No 10, National Research Centre for Banana, Thiruchirapalli, Tamil Nadu, India, 14 pp

Vadivelu RS, Rajendran G, Naganathan TG, Jayaraj S (1987) Studies on the nematode pests of banana and their management - a review. In: Third group discussion on the nematological problems of plantation crops, Sugarcane Breeding Institute, Coimbatore, pp 32–33

Van Gundy SD (1958) The life history of the citrus nematode *Tylenchulus semipenetrans*. Nematologica 3:283–294

Venkitesan TS (1976) Studies on the burrowing nematode *Radopholus similis* (Cobb, 1893). Thorne, 1949 on pepper (*Piper nigrum* L.) and its role in slow wilt disease. Ph.D. thesis, University of Agricultural Sciences, Bangalore, 122 pp

Vidya K, Reddy BMR (1998) Integrated management of *Radopholus similis* infecting banana using plant product, nematicide and biocontrol agents. Mysore J Agric Sci 32:186–190

Vijaya Kumari N, Sujathamma P (2014) Integrated management of root knot nematode *Meloidogyne incognita* Chitwood in mulberry (*Morus* spp.). J Pharmacy Biol Sci 9(1):30–33

Vilardebo A, Luc M (1961) Slow decline of citrus caused by the nematode, *Tylenchulus semipenetrans* (In French). Fruits 16:445–454

Walia RK, Chakraborty PK (eds) (2018) Nematodes problems of crops in India. ICAR-All India Coordinated Research Project on Nematodes in Agriculture, New Delhi. 402 pp

Wang KH, Sipes BS (2009) Solarization and cover cropping as alternatives to soil fumigants for nematode management in Hawaii's pineapple fields. University of Hawaii, Manoa, SCM 29, 4 pp

Wang KH, Sipes BS, Schmitt DP (2002) Management of *Rotylenchulus reniformis* in pineapple, *Ananas comosus*, by intercycle cover crops. J Nematol 34(2):106–114

Whitehead AG (1968) Taxonomy of *Meloidogyne* (Nematodea: Heteroderidae) with descriptions of four new species. Trans Zool Soc Lond 31:263–401

# Subtropical Fruit Crops

# 8

**Abstract**

There has been an outbreak of serious root-knot nematode problems in the recent years associated with guava (*Meloidogyne enterolobii*) resulting in death of trees within 3 months once the symptoms are initiated. The nematode is disseminated through infected planting stock. Besides causing direct damage, the root-knot nematode species is also responsible for causing wilt disease complexes in association with soil-borne fungus *Fusarium solani*. The major nematode problems on grapevine include root-knot nematode *Meloidogyne incognita*; reniform nematode *Rotylenchulus reniformis*; and dagger nematodes *Xiphinema* spp. The distribution, crop losses, symptoms, interaction with other soil-borne pathogens, survival and spread, and management of subtropical fruit crops are discussed in this chapter.

**Keywords**

*Meloidogyne* spp. · Rotylenchulus reniformis · Xiphinema spp. · Symptoms · Management

## 8.1 Guava, *Psidium guajava*

### 8.1.1 Root-Knot Nematode, *Meloidogyne enterolobii*

The worst nematode problem affecting guava is the root-knot nematode *Meloidogyne enterolobii* (=*M. mayaguensis*) which is recognized as a limiting factor in commercial guava production. Recently, *M. enterolobii* has been reported from Tamil Nadu and some other states also.



P. P. Reddy, *Nematode Diseases of Crops and their Management*,
https://doi.org/10.1007/978-981-16-3242-6_8

#### 8.1.1.1 Crop Losses

Root-knot nematodes are responsible for 28% annual yield loss in guava amounting to Rs. 2350.88 million in India (Walia and Chakraborty 2018).

#### 8.1.1.2 Symptoms

*Meloidogyne enterolobii* affects growth, yield, life span, and tolerance to environmental stresses of infected plants. Young infected guava trees exhibit symptoms of yellowing, shedding of leaves, drying of branches, reduced vigor, and less productivity. Rapid decline in tree health and death of trees results under heavy nematode infestation. Moderate infestations are associated with general chlorosis, nutrient deficiency symptoms, and reduced flowering and fruiting (Fig. 8.1). Roots of infected trees show multiple galls and root necrosis, causing a drastic decrease in feeder roots (Fig. 8.2) (Cetintas et al. 2007). The root-knot nematode infects the whole root system, from small rhizoids to more lignified roots. The fruit yield and quality are affected by the nematode. Carneiro et al. (2007) reported that *M. enterolobii* causing serious economic problems to growers and the economy of the Brazil region since this nematode problem is making guava cultivation unfeasible in the infested areas.

#### 8.1.1.3 Host Range

Similar to *M. incognita*, *M. enterolobii* also has very broad host range that include economically important crops such as bell pepper, (watermelon), coffee, soybean, sweet potato, tomato, tobacco, bean, guava, aborigine, ornamental plants (e.g., *Ajuga, Brugmansia, Clerodendron, Tibouchina*), and wild plants (e.g., *Bidens*

**Fig. 8.1** Symptoms of root-knot nematode on guava trees showing stunted growth and leaf yellowing (*Left*) and bronzing of leaves (*Right*)

**Fig. 8.2** Root galling on guava seedling (*Left*) and roots (*Right*)

*pilosa*) (Humphreys et al. 2012). Additional host ranges of *M. enterolobii* include species of *Cactus, Ficus, Syngonium, Rosa*, and *Vitis* in the Netherlands.

#### 8.1.1.4 Biology and Life Cycle

he root-knot nematode is sedentary endoparasite and soil-borne. After penetration, the J2 larvae feed endoparasitically on conducting vessels and develop to adult females which are spherical in shape and lays egg masses in compound galls. The males are vermiform in shape which leaves roots to enter the soil. Under north Indian conditions, the nematode completes its life cycle in 25–30 days during summer months; while it takes 60–90 days in winter.

#### 8.1.1.5 Interaction with Other Pathogens

The decline and death of guava trees infected by *M. enterolobii* is a complex disease involving other microorganisms like *Fusarium solani* (Gomes et al. 2010), *Verticillium dahliae, Pythium aphanidermatum*, and *Trichoderma* sp. (Avelar-Mejia et al. 2001).

#### 8.1.1.6 Spread

The root-knot nematode is disseminated through infected planting material (grafted material, ground layers, etc.). As is the case for other plant-parasitic nematodes, *M. enterolobii*'s own movement is limited at most to a few tens of centimeters in the soil. The nematode is mainly spread through infested planting stock and soil attached to equipment and machinery and irrigation water (Castagnone-Sereno 2012).

#### 8.1.1.7 Management

##### 8.1.1.7.1 Physical Methods

Deep plowing exposes nematodes to sunlight.

##### 8.1.1.7.2 Cultural Methods

- Use nematode-free planting material.
- Use sterilized potting media (e.g., sterilized by steam or roasting).
- Use clean implements and tools.
- Do not use previously infested fields for planting.
- Use recommended fertilizers to improve crop vigor.
- Use trap crops like marigold.
- Green manuring with repellent crop before planting sun hemp and mustard.
- Add burnt paddy husk to planting hole.
- Restrict flow of irrigation water from infected fields to healthy fields.
- Add compost to the soil to reduce the effect of nematodes (after infestation).
- Remove weeds and infested crop residues to protect nearby healthy trees.
- Suppress nematodes by biofumigation with crucifers, intercropping with cabbage and mustard and their soil incorporation.
- Add organic soil amendments like neem cake, green manure, FYM, compost, etc. to the planting pit.
- Remove infested plants from the field.
- Use of nonhost cover crops like peanut (cvs. IAC Oirã, IAC Poitã, IAC Tatuí and Cavalo vermelho), rye (cv. IPR 89), green gram, oat (cv. IAPAR 126).
- Growing nonhost crops or bare fallow is the most efficient method for reducing *M. enterolobii* populations. Fourteen fruit trees are nonhosts to *M. enterolobii* (atemoya, avocado, cashew, citrus rootstocks, coconut, and mango) or poor hosts (assai palm, common mulberry, jaboticaba, grape rootstocks, passion fruit, sapodilla, and sour sop). Crops like cabbage, garlic, grapefruit, maize, peanut, sour orange, and welsh onion are nonhosts.
- Growing marigold around guava plants.

##### 8.1.1.7.3 Chemical Methods

Effective management of guava root-knot nematode can be achieved by soil application of Carbofuran at 15 g a.i./ $m^2$ followed by irrigation.

##### 8.1.1.7.4 Biological Methods

Incorporation of egg parasitic fungi, *Pochonia chlamydosporia* and *Purpureocillium lilacinum* mixed in FYM into the pits before planting.

Application of *Purpureocillium lilacinum/Psudomonas fluorescens* at 20 g/$m^2$ (about 9 $m^2$ around the tree trunk) in the rhizosphere region at the time of flowering gave effective control of *M. enterolobii.*

#### 8.1.1.7.5 Host Resistance

Use resistant varieties and rootstocks such as Horana White, Horana Giant, Pubudu, and Kanthi (all are moderately resistant).

Only Costa Rican wild guava, *P. friedrichsthalianium* was considered resistant to *M enterolobii*. Bogantes-Arias and Mora-Newcomer (2010) reported the absence of nematodes in the root system of "Criollo" and "Arrayán" rootstocks, confirming the resistance of *P. friedrichsthalianium*.

The resistance was observed only in *P. guajava* accessions S3 K-10, A-06, J-16, and B-12 (Casassa et al. 1997; Milan 2007, 2010).

Under field conditions, the compatible and resistant rootstocks for *P. guajava* cv. "Paluma" include one accession of *P. friedrichstalianium* (Carneiro et al. 2012).

Milan (2007) found that only *P. guajava* Acc. B-12 has the potential to control the root-knot nematode and graft-compatible with commercial clone of guava because they are from the same species.

#### 8.1.1.7.6 Integrated Methods

The guava growers are advised to avoid nematode-infected saplings, removal and destruction of nematode-affected trees, application of farm yard manure at 100 kg, Neem cake at 250 g, and nematode egg parasitic fungus *Purpureocillium lilacinum* at 25 g, and Carbofuran 3G at 60 g/tree at early stage of nematode infestation to overcome the nematode problem and to enhance fruit yield.

Effective management of guava root-knot nematode can be achieved by soil application of 1 kg Neem cake enriched with bioagents – Carbofuran at 7 g a.i. / $m^2$ just before flowering.

### 8.1.2 Interaction of Root-Knot Nematode with *Fusarium* Root-Rot

The root-knot nematode *M. enterolobii* interacts with the soil-borne fungus, *Fusarium solani* in causing root rot disease complex in guava. The nematode predisposes the roots for fungal infection.

#### 8.1.2.1 Interaction

*M. enterolobii* and *F. solani* have a synergistic effect on disease complex development. Whereas *M. enterolobii* alone was able to induce galling and a limited amount of root rot, *F. solani* did not cause root rot when in isolation. When combined, nematode and fungus led to drastic root rot and typical guava decline symptoms, which suggests that guava decline is a disease complex between *M. enterolobii* and *F. solani* (Gomes et al. 2010, 2012).

#### 8.1.2.2 Symptoms

Declining trees show chlorosis, scorching of margin, wilting, and shedding of leaves (Anon. 2011). Gomes et al. (2008) have shown that these symptoms are associated with foliar deficiency of nitrogen, potassium, and phosphorus, a near deficiency of calcium and magnesium, and an imbalanced accumulation of manganese, chlorine, and sodium. Declining trees develop progressive rotting of the entire root system, eventually succumbing to the disease within few months.

#### 8.1.2.3 Management

##### 8.1.2.3.1 Chemical Methods

Propiconazole (0.2%) was found to be significantly superior as compared to other chemicals. Trees treated with Propiconazole (0.2%) took 105 days to recover, while trees treated with Difenconazole (0.2%) took 120 days (Sonyal 2010).

##### 8.1.2.3.2 Biological Methods

The most effective biological control agent for the management of the disease complex includes *Aspergillus niger* followed by *Trichoderma viride*. Misra et al. (2000) reported that *Aspergillus niger* was found to be very fast growing, easy to propagate, and most effective in controlling the wilt disease in field.

##### 8.1.2.3.3 Host Resistance

In wilt sick plot that was artificially inoculated repeatedly with *Gliocladium roseum*, *Fusarium solani* and *Fusarium oxysporum*, F1 population of *Psidium molle* x *P. guajava* was free from the above wilt causing pathogens. The rootstock besides being resistant to wilt pathogens was also found to be graft compatible with *P. guajava* (Misra et al. 2003).

##### 8.1.2.3.4 Integrated Methods

Integrated eco-friendly approach for the control of guava wilt was suggested by Misra et al. (2004) and Misra (2005) utilizing bioagent *Aspergillus niger* strain AN17, using resistant rootstock (*P. molle x P. guajava*), intercropping of guava with turmeric or marigold as well proper cultural practices to minimize losses due to the disease complex.

Integration of *T. harzianum* + *P. fluorescens* each at $1 \times 10^6$ gave effective management of the disease complex (Sonyal 2010).

## 8.2 Grapevine, *Vitis vinifera*

The root-knot (*Meloidogyne incognita*), the reniform (*Rotylenchulus reniformis*), and dagger (*Xiphinema index*) nematodes are the most important grapevine pests.

### 8.2.1 Root-Knot Nematode *Meloidogyne incognita*

#### 8.2.1.1 Crop Losses

The root-knot nematode is responsible for 30% annual yield loss in established vineyards of grapevine amounting to Rs. 3940.44 million in India (Walia and Chakraborty 2018).

#### 8.2.1.2 Symptoms

The affected vines show decline in vigor with inadequate sprouting after pruning, reduction in leaf size, marginal drying of leaves, and perceptible reduction in yield (Fig. 8.3). The nematode produces small galls on the roots (Fig. 8.4). Heavy infections may completely destroy the root systems of young plants. The plant roots cannot absorb water and nutrients from soil due to disruption of conducting tissues resulting in death of roots beyond site of nematode infection.

#### 8.2.1.3 Biology and Life Cycle

Root-knot nematodes are sedentary endoparasites and soil-borne. Second-stage juveniles infect the roots; the nematode develops endoparasitically feeding on the vascular tissues. The adult females are spherical in shape and lay egg masses in compound galls. The males are vermiform in shape which leaves roots to enter the

**Fig. 8.3** Grapevine plants infected with root-knot nematodes showing stunting (*Left*) and leaf yellowing (*Right*) symptoms in field

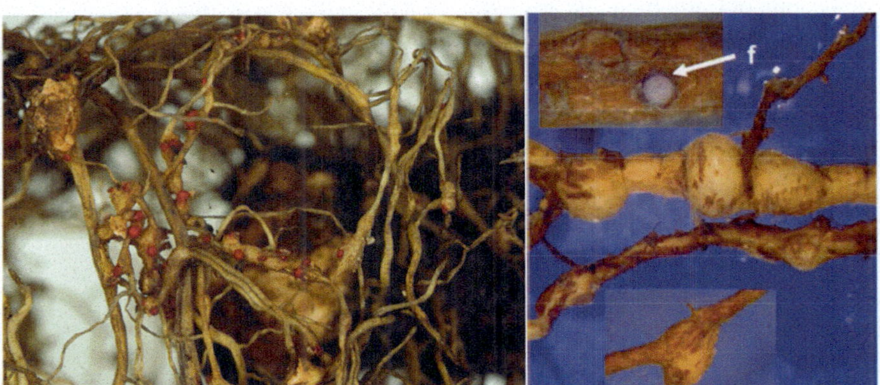


**Fig. 8.4** Grapevine roots infected with root-knot nematodes (*Left*), close-up root-knot galls and female (f) (*Right*)

soil. During summer, the nematode completes its life cycle in about 25 to 30 days, while it takes longer during winter months.

#### 8.2.1.4 Ecology

Grapevines are attacked and seriously injured by root-knot nematodes only when the vines are grown in light, sandy, or sandy loam soils. Vines in fertile loams and heavier soils are seldom seriously damaged.

#### 8.2.1.5 Spread

Root-knot nematodes are disseminated through infected transplanting material, irrigation water, and agricultural implements.

#### 8.2.1.6 Management

A well-planned control program should include the use of nematode-free planting material, nematode-resistant rootstocks, and the selection of the most effective soil treatment for the root-knot nematode infestation.

##### 8.2.1.6.1 Physical Methods

Hot water treatment for root-knot nematode control is necessary if roots are lightly infested with nematodes but otherwise have made good growth. The roots are submerged in hot water maintained at 51.7 °C for 5 minutes and cooled immediately before planting.

##### 8.2.1.6.2 Cultural Methods

Use of nematode-free planting materials and application of Neem cake at 200 g/vine was found effective.

##### 8.2.1.6.3 Chemical Methods

Application of Carbofuran 3G at 13 g/m$^2$ in 9 m$^2$ basin area of vine after 3 weeks of pruning followed by flood irrigation proved beneficial.

##### 8.2.1.6.4 Biological Methods

Sundarababu et al. (1999) observed that commercial formulation of *Pseudomonas fluorescens* reduced root-knot nematode population effectively and enhanced yield in grapevine.

##### 8.2.1.6.5 Host Resistance

In soils where the root-knot nematode infestations are heavy, it may be impossible to grow grapevines without the use of nematode-resistant rootstocks.

The grapevine rootstocks 1613, Dogridge and Saltcreek are resistant, while 1616 and Talaki 5-A are moderately resistant to the root-knot nematode (Parvatha Reddy and Singh 1984) (Fig. 8.5).

**Fig. 8.5** Root-knot nematode resistant (healthy and vigorous vines on either side) and susceptible (weak and missing vines in the middle row) grapevine rootstocks

#### 8.2.1.6.6 Integrated Methods

Application of 100 kg moist FYM enriched with *Pseudomonas fluorescens* ($2 \times 10^6$ cfu/g) or *Trichoderma viride* each at 20 g/m$^2$ at October pruning proved beneficial.

Soil application of Carbofuran 3G at 7 g/m$^2$ at flowering and growing garlic as an intercrop was found effective.

Pruning (during July) and soil application of 4 g talc formulation of *P. fluorescens* (containing 15 x $10^8$ cfu/g) around each root-knot infested grapevine at 15 cm depth in the basin at the time of pruning (during July) significantly reduced root galling due to *M. incognita* (39%), number of egg masses (250%), and increased fruit yield (166%) (Shanthi et al. 1998).

The combined application of two bioagents *Purpureocillium lilacinum* and *Pasteuria penetrans* was most effective in increasing the fruit yield of grapevine besides checking *M. incognita* with benefit: cost ratio of 3.3 (Sivakumar and Vadivelu 1999).

Application of Neem cake (200 g/plant) + Carbofuran (10 g/plant) reduced the nematode populations in the soil and roots. Combination of intercropping garlic and application of Carbofuran at 0.1 g a.i./m$^2$ reduced nematode population (44.2–47.6%) and increased yield (88.3–105.6%) over control with benefit: cost ratio of 4–5 (Anon 1993).

Application of Neem cake at 200 g and *Purpureocillium lilacinum* at 50 g/vine helps in improving vine growth and yield. Application of FYM enriched with *Trichoderma harzianum* and *P. lilacinum* to the field at the rate of 2 kg per plant for 5–6 times at an interval of 2 months reduced the nematode problems significantly and improved the yield levels of grapes.

The nematode can be effectively managed by integration of reduced doses of Neem cake and nematicide followed by intercropping with onion and garlic which is profitable.

## 8.2.2 Reniform Nematode *Rotylenchulus reniformis*

The reniform nematode is a serious problem on Anab-e-Shahi variety of grapevine.

#### 8.2.2.1 Symptoms

Feed on and damage secondary and feeder roots (Fig. 8.6). Affected roots show brownish discoloration. As a result, nutrient uptake is affected and vine appears sick.

#### 8.2.2.2 Management

##### 8.2.2.2.1 Chemical Methods

Soil application of Metham Sodium at 40 liters per ha is effective and increased the bunch weight by 63%.

##### 8.2.2.2.2 Host Resistance

Grapevine cultivars such as Dakshi, Joazbeli, and Mukchilani are resistant to the nematode.

## 8.2.3 Dagger Nematode *Xiphinema index*

#### 8.2.3.1 Symptoms

The dagger nematode *X. index* feeds on root tips causing swelling and bending as well as many dead feeder roots. If population levels are high, multiple prolonged attacks can result in darkened, necrotic spots that spread through the root tip (Fig. 8.7). The above-ground symptoms include poor growth, stunting of the plant, yellowing or wilting of the foliage, and reduced root systems (root necrosis, lack of feeder or secondary roots, and occasional tufts of stubby rootlets).

**Fig. 8.6** Grapevine root infected with reniform nematodes

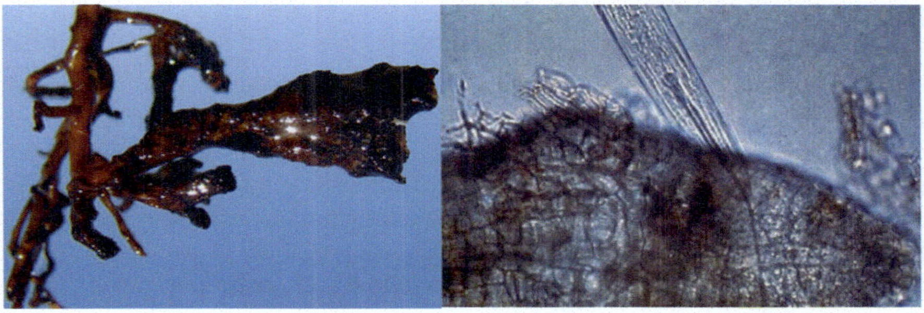

**Fig. 8.7** *Left*—Terminal galling of grapevine root caused by a dagger nematode *Xiphinema index*. *Right*—A dagger nematode *Xiphinema index* feeding on root tip by piercing its stylet

**Fig. 8.8** Grapevine infected with fan leaf virus disease transmitted by *Xiphinema index*

#### 8.2.3.2 Interaction with Other Pathogens

The dagger nematode *X index* acts as a vector for transmission of grapevine fan leaf virus. The symptoms of dagger nematode transmitted virus in grapevine plants include yellowing of leaf veins, mosaic, and malformation. In warmer regions, symptom expression was less pronounced in white varieties. *Xiphinema index* is especially important because of its ability to transmit Grapevine fan leaf virus (Fig. 8.8) when feeding on grapevine roots.

#### 8.2.3.3 Management

The most effective control of both *Xiphinema index* and *Meloidogyne javanica* was obtained with Carbofuran at 0.2 g/L. Integration of Neem seed powder at 4.08 g/pot + Fertinemakil fertilizer at 2.5 g/pot and Marigold + Fertinemakil is effective against *X. index* and *M. javanica*.

## 8.3 Conclusion

Currently nematodes have become a major problem to the commercial subtropical fruit growers. Diverse nematode pathogens which cause diseases of subtropical fruit crops can be limiting factors in the production and marketing of these commodities, and are often difficult to manage. It is clear from the above information that there is adequate information available on nematode infestation in subtropical fruit crops. Resistance, the most effective tool with which many of these nematodes are managed, is usually available in coevolved pathosystems but may be uncommon in new encounter situations. Inadequate host resistance can be a significant barrier in the management of new encounter nematode diseases, as well as those that are caused by generalist pathogens. Development of integrated nematode and disease management in subtropical fruit crops is the need of the hour. Further there is an urgent need to do further studies on host parasite relationship and epidemiology of the diseases in order to have a sound integrated management approach.

## References


Anon (1993) Biennial report (1991–93) of AICRP on plant parasitic nematodes with integrated approach for their control. Dept. of Nematology, Haryana Agri. Univ., Hisar, p 80

Anon. (2011) Meloidogyne enterolobii. EPPO Bull 41:329–339

Avelar-Mejia JJ, Teliz-Ortiz D, Zavaleta-Mejia E (2001) Patógenos asociados con el "declinamiento del guayabo". Revista Mexicana de Fitopatologia 19:223–229

Bogantes-Arias A, Mora-Newcomer E (2010) Evaluation of four rootstocks for graft in guava (*Psidium guajava* L.). Agronomia Mesoamericana 21:103–111

Carneiro RMDG, Castro JMC, Carneiro RG et al (2012) Major guava nematodes and control prospects using resistance on *Psidium* spp. and non-host crops. In C.a.F. Santos *et al.* (eds.) *proc. 3rd Int. Symp. On guava and other Myrtaceae.* Acta Hort 959:41–49

Carneiro RMDG, Cirotto PA, Silva DB, Carneiro RG (2007) Resistance to *Meloidogyne mayaguensis* in *Psidium* spp. accessions and their grafting compatibility with *P. guajava* cv. 'Paluma'. Fitopatol Bras 32:281–284

Casassa AM, Matheus J, Crozzoli R et al (1997) Response of some selections of guava to *Meloidogyne incognita*, in Mara County, Zulia state, Venezuela. Fitopatologia Venezolana 10:5–8

Castagnone-Sereno P (2012) Meloidogyne enterolobii (=M. mayaguensis): profile of an emerging, highly pathogenic, root-knot nematode species. Nematology 14:133–138

Cetintas R, Kaur R, Brito JA et al (2007) Pathogenicity and reproductive potential of *Meloidogyne mayaguensis* and *M. floridensis* compared with three common *Meloidogyne* spp. Nematropica 37:21–31

Gomes VM, Souza RM, Corrêa FM, Dolinski C (2010) Management of *Meloidogyne mayaguensis* in commercial guava orchards with chemical fertilization and organic amendments. Nematol. Bras. 34:23–30

Gomes VM, Souza RM, Midorikawa G et al (2012) Guava decline: Evidence of nationwide incidence in Brazil. Nematropica 42:153–162

Gomes VM, Souza RM, Silva MM, Dolinski C (2008) Caracterizac ̧a˜o do estado nutricional de goiabeiras em declı'nio parasitadas por *Meloidogyne mayaguensis*. Nematol Bras 32:154–160

Humphreys DA, Williamson VM, Salazar L et al (2012) Presence of *Meloidogyne enterolobii* Yang & Eisenback (=*M. mayaguensis*) in guava and acerola from Costa Rica. Nematology 14:199–207

Milan AR (2007) Breeding of *Psidium* species for root knot nematode resistance in Malaysia. Acta Hort. 735:61–69

Milan AR (2010) Collection and evaluation of guava (*Psidium guajava* L.) for nematode resistance in Malaysia. Acta Hort 849:357–362

Misra AK (2005) Important diseases of guava in India with special reference to wilt. Acta Hort. 735:507–523

Misra AK, Prakash O, Sen B (2000) Biological control of guava wilt by *Aspergillus Niger* strain AN17 (Pusa Mrida). Proc. National Seminar on Hi-Tech Hort, Bangalore, pp 149–149

Misra AK, Prasad D, Prasad B, Shukla SK (2004) Effective management of wilt disease of guava. In: Proc. of the National Symp. On crop surveillance: disease forecasting and management. Div. of Plant Pathology, IARI, New Delhi, pp 92–93

Misra AK, Rajan S, Prasad B et al (2003) Resistant source of wilt disease of guava. In: *Seminar on plant diseases of National Importance with special reference to guava wilt and mango malformation* (souvenir and abstract), Zonal conference (east zone), Indian Society of Mycology and Plant Pathology. CISH and IISR, Lucknow, pp 51–52

Parvatha Reddy P, Singh DB (1984) Reaction of selected cultivars/rootstocks of grapevine to *Meloidogyne incognita*. Indian J Nematol 14:73–74

Shanthi A, Rajeswari S, Sivakumar CV (1998) Soil application of *Pseudomonas fluorescens* (Migula) for the control of root-knot nematode (*Meloidogyne incognita*) on grapevine (*Vitis vinefera* Linn.). In: Mehta UK (ed) Nematology – challenges and opportunities in 21st century. Sugarcane Breeding Institute, Coimbatore, pp 203–206

Sivakumar M, Vadivelu S (1999) Management of *Meloidogyne incognita* on grapevine using biocontrol agents, botanicals and biofertilizers. Pest Mangmt Hort Ecosystems 5:127–131

Sonyal S (2010) *Studies on Pomegranate Wilt Complex*. M. Sc. (Agri.) thesis, Dept. of Plant Path., Univ. of Agri. Sci., Dharwad, India

Sundarababu R, Mani MP, Sivakumar CV (1999) A new biocontrol agent for plant parasitic nematodes. Indian Hort 12:25

Walia RK, Chakraborty PK (eds) (2018) Nematodes problems of crops in India. ICAR-All India Coordinated Research Project on Nematodes in Agriculture, New Delhi, 402 pp

# Temperate Fruit Crops

# 9

**Abstract**

Plant parasitic nematodes such as lesion nematodes (*Pratylenchus vulnus* and *P. penetrans*), root-knot nematodes (*Meloidogyne* spp.), ring nematode (*Mesocriconema xenoplax*), and bud and leaf nematodes (*Aphelenchoides* spp.) represent a worldwide concern on temperate fruit crops. For food and nutritional security, management of nematodes is important. Integrated nematode management strategies have been developed using the components like seed certification, hot water treatment of planting stock, soil application of organic amendments (including oil cakes), use of nematicides both in nursery and field, use of biological control agents, and resistant rootstocks/cultivars.

**Keywords**

*Pratylenchus* spp. · *Meloidogyne* spp. · *Mesocriconema xenoplax* · *Aphelenchoides* spp. · Symptoms · Management

## 9.1 Peach, *Prunus persica*

Root-knot, lesion, and ring nematodes have been recognized as pathogens causing economic damage to peach crop.

### 9.1.1 Root-Knot Nematodes, *Meloidogyne* spp.

Of the many nematodes associated with peach trees, the root-knot nematode was the first known to cause damage in the form of knotted, swollen roots, reduced root and top growth, and leaf chlorosis.


P. P. Reddy, *Nematode Diseases of Crops and their Management*, https://doi.org/10.1007/978-981-16-3242-6_9

#### 9.1.1.1 Distribution

Infestation of *Meloidogyne* spp. has been reported from Delhi, Punjab, Haryana, and Himachal Pradesh. *M. javanica* and *M. incognita* occur commonly in warmer peach-growing areas, whereas *M. hapla* is restricted to colder regions only.

#### 9.1.1.2 Symptoms

Above-ground symptoms on peach plants infected with *Meloidogyne* spp. include poorly developed foliage, lack of vigor, dieback of twigs, and yield reduction (Fig. 9.1). In periods of drought, infected trees wilt more quickly than noninfected ones. The peach grove infested with root-knot nematode is usually composed of unevenly sized trees due to uneven distribution of nematodes, and their population variation within one field.

The root system of nematode-infested trees has poorly developed secondary and tertiary branches bearing galls of varying sizes. When many juveniles penetrate a root, root tip is devitalized and elongation of the root ceases resulting in the production of many fine tertiary branches forming brush-like appearance (Fig. 9.1). The nematode prevents the availability of water and nutrients and desired energy for the production of fruit resulting into yield loss.

#### 9.1.1.3 Histopathology

The larvae of *M. javanica* readily penetrate susceptible peach rootstock (Sharbati) and develop into mature females. The production of giant cells was detected as early as fourth day after nematode penetration. Giant cell formation takes place in pericycle area, a few cells away from nematode head. The cytoplasm of affected cells becomes dense, followed by enlargement of nuclei and nucleoli. The affected cells undergo fast multiplication of nuclei without cytoplasmic division, thus forming syncytium. In most cases, giant cells are located within vascular tissues.

**Fig. 9.1** Root galling on peach due to root-knot nematode infection

#### 9.1.1.4 Interaction with Other Pathogens

The root-knot nematode *M. javanica* interacts with soil-borne bacterium *Agrobacterium tumefaciens* in causing disease complex in peach. This interaction has a synergistic effect resulting in increasing the incidence of both the pathogens.

#### 9.1.1.5 Management

##### 9.1.1.5.1 Physical Methods

Hot water treatment of peach seedling roots at 50–52 °C for 5–10 min eliminates the nematode infection.

##### 9.1.1.5.2 Cultural Methods

Nematode-free planting stock is essential for transplanting in field. Planting of cover crops like Max Q and tall fescues before taking up peach helps in reducing the root-knot nematode population in soil and enhance plant growth.

##### 9.1.1.5.3 Chemical Methods

Treatment of nursery beds with Carbofuran 3G at 3 kg a.i. /ha; root dip treatment of nursery plants in 0.1% solution of Fenamiphos for 6 h before planting in the main field; and soil application of Carbofuran 3G at 100 and 150 g/plant twice or thrice a year in the main field proved beneficial.

##### 9.1.1.5.4 Biological Methods

*G. mosseae* at 400 g/pot (50 spores/g soil) colonized 67% peach roots and significantly enhanced plant growth as indicated by increased shoot and root weight and length and suppressed the multiplication of *M. javanica* on peach cv. Floridasun (Baghel and Bhatti 1990). Strobel et al. (1982) reported that AM fungus *Gigaspora margarita* decreased the number of *M. incognita* eggs produced per gram of peach root growth at a low P level.

##### 9.1.1.5.5 Host Resistance

Nemagaurd, Nemared, Flordaguard (Fig. 9.2), Okinawa, and Rancho are the most popular rootstocks resistant to *M. incognita, M. javanica,* and *M. arenaria* (Sharpe et al. 1969). Yunnan, Shalil, and Bokhara have been found resistant to *M. incognita* but susceptible to *M. javanica*. Root-knot nematode-resistant peach cultivars include Chinaflat, Bidbillos Early, Japanese Double, and Elberta (Verma 1987).

### 9.1.2 Ring Nematode, *Mesocriconema xenoplax*

The ectoparasitic ring nematode causes economic damage to peach and is considered as one of the important components in peach tree short life (PTSL) disease complex.

**Fig. 9.2** Root-knot susceptible (*left*—Sharbati) and resistant (*right*—Flordaguard) peach rootstocks

#### 9.1.2.1 Crop Losses

The estimated losses that occurred in peach orchards suffering from PTSL in Georgia were 30–70% after 5 years (Wehunt et al. 1980). *M. xenoplax* was responsible for 33% loss in fruit yield of peach (Anon. 1990).

#### 9.1.2.2 Distribution

*M. xenoplax* was first observed on peach grown in Northern India. It is distributed in Himachal Pradesh and Haryana.

#### 9.1.2.3 Symptoms

The infected plant roots are darkened and often contain longitudinal cracks. The most serious infection appears in the form of the death of finer roots due to direct feeding by nematodes. Due to excessive killing of feeder roots, plants are stunted, showing mineral deficiency syndrome, and more susceptible to water stress. Parasitism of *M. xenoplax* causes chlorosis, leaf drop, stunting, and loss of vigor on top of peach trees (Fig. 9.3). Peach roots showed lesions, pits, and fewer feeder roots (Fig. 9.4). The roots are predisposed to bacterial canker caused by secondary infection of *Pseudomonas syringae*.

#### 9.1.2.4 Life Cycle

According to Seshadri (1964), eggs are deposited singly, 8–15 in number per female in 2–3 days egg laying period. Total life cycle was completed in 25–34 days at 22–26 °C in sandy soil at pH 7.0 and 15.5% moisture. It takes 11–13 days for egg stage, 3–5 days for second stage, 4–7 days for third stage, and 5–6 days for fourth stage development and preoviposition period of adult. The nematode is multivoltine and completes several generations in a year.

**Fig. 9.3** Field infestation of ring nematode showing decline of peach trees (three plants on right side)

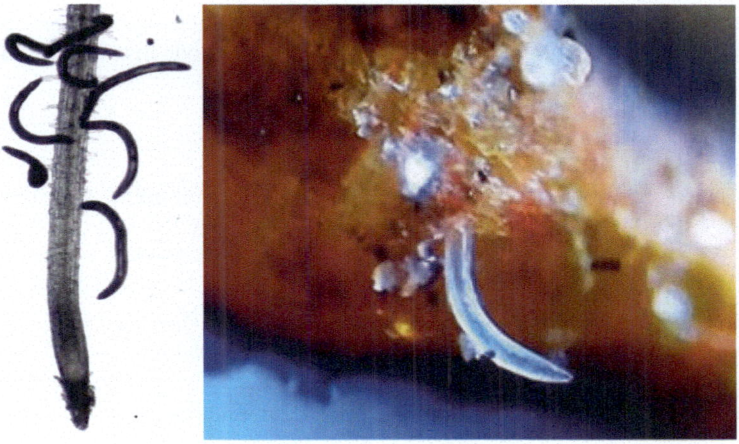

**Fig. 9.4** Ring nematodes feeding on peach roots (*left*) and close up of nematode feeding (*right*)

#### 9.1.2.5 Interaction with Other Pathogens

Parasitization by *M. xenoplax* predisposes peach trees to bacterial canker caused by *Pseudomonas syringae* (Lownsbery et al. 1977). *M. xenoplax* has been strongly implicated in the peach tree short life (PTSL) condition in some parts of the USA. PTSL disease syndrome is characterized by a sudden collapse of the new growth above soil line and death of trees in late winter or early spring shortly after foliation (Taylor et al. 1970).

### 9.1.2.6 Management

#### 9.1.2.6.1 Physical Methods

Hot water treatment of peach seedlings at 50–52 °C for 5–10 min. Eliminates the nematode infection.

A 37–100% reduction in population of *M. xenoplax* and other nematodes has been achieved by soil solarization in South Africa (Barbercheck and von Broembsen 1986).

#### 9.1.2.6.2 Cultural Methods

Sorghum as a green manure with and without plastic did suppress the population of *M. xenoplax* in the early stages (Nyczepir and Rodriguez-Kabana 2004).

#### 9.1.2.6.3 Biological Methods

*Hirsutella rhossiliensis* reduced nematode population by 50%. Rate of nematode decay was much faster for parasitized second- or third-stage juveniles (Fig. 9.5) (Jaffee et al. 1988).

Treatment of peach seedlings with a strain of *Pseudomonas aureofaciens* isolated from suppressive soil inhibited infection by the ring nematode in soil (McInnis et al. 1990).

#### 9.1.2.6.4 Host Resistance

Lorell rootstock of peach has been recorded as most tolerant to *M. xenoplax* (Rom et al. 1985)

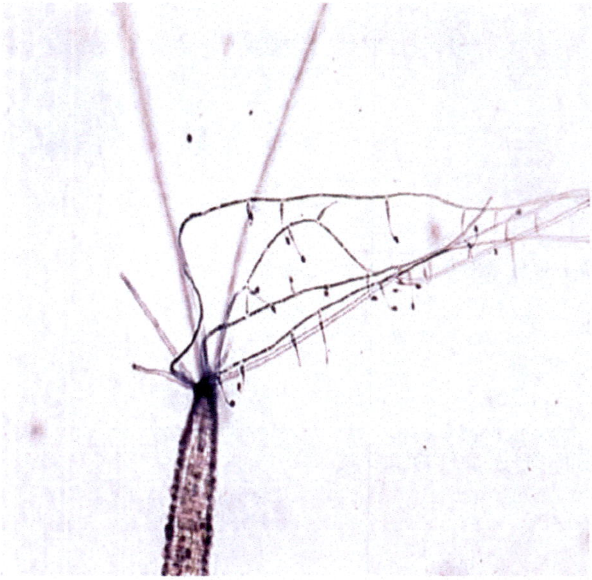

**Fig. 9.5** Parasitic fungus *Hirsutella rhossiliensis* infecting ring nematode

### 9.1.3 Lesion Nematodes, *Pratylenchus vulnus* and *P. Penetrans*

In Canada, the lesion nematode causes "peach replant problem."

#### 9.1.3.1 Symptoms

Above-ground symptoms include tree decline with general nutrition deficiency symptoms and reduced growth (Fig. 9.6). Under-ground symptoms are root lesions or discoloration of roots usually reddish brown later turning dark and black (Fig. 9.7). The track of lesions on root becomes visible as the nematodes feed and migrate from one place to another. Rapid deterioration of feeder roots and reduced vigor are symptoms caused by *P. vulnus*. The nematode poses threat to replanting of peach.

**Fig. 9.6** *Left*—Healthy peach tree. *Right*—Root lesion nematode-infected peach trees (white circle)

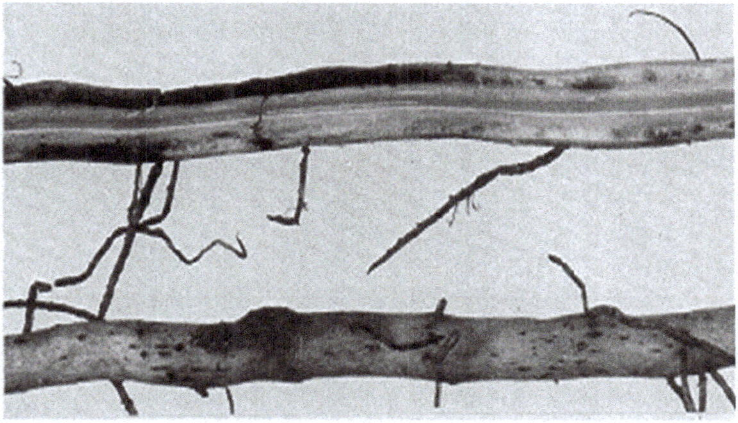

**Fig. 9.7** Peach roots infected with *Pratylenchus vulnus*

#### 9.1.3.2 Life Cycle

The nematode enters roots and moves inter- and intracellularly and feed on cortical cells; the nematode larvae and adults both have capacity to enter or leave the roots. Eggs are laid by the adult female nematode in the soil or in roots. Hatching begins to take place immediately after egg laying. After hatching, different stages of larvae feed on roots and develop into adults. Under unfavorable conditions, nematodes leave the roots and migrate to new favorable roots for penetration and feeding.

#### 9.1.3.3 Management

##### 9.1.3.3.1 Physical Methods

Hot water treatment of peach seedlings at 52 °C for 15 min eliminates *P. penetrans* infection.

##### 9.1.3.3.2 Chemical Methods

Combining preplanting treatment with Carbofuran and postplanting treatment with Phorate annually to standing crop proved beneficial.

##### 9.1.3.3.3 Biological Methods

Pinochet et al. (1995) found that precolonization of peach roots with *Glomus mosseae* was effective in reducing the *P. vulnus* population.

## 9.2 Strawberry, *Fragaria* × *ananassa*

The economically important nematodes infecting strawberry include bud and leaf, lesion, and root-knot nematodes.

### 9.2.1 Bud and Leaf Nematode, *Aphelenchoides fragariae*

Strawberry dwarf is caused by at least three species of *Aphelenchoides*. The nematode is responsible for a cool weather disease of strawberry called "spring dwarf." The summer dwarf is caused later in the season by *A. besseyi.* Susceptible varieties showed 60–82% reduction in yield after 2 years of infestation.

#### 9.2.1.1 Symptoms

Spring dwarf (caused by *A. fragariae*) symptoms appear in the early spring as soon as plant growth starts. Bud and leaf development is retarded and plants fail to develop normal foliage. Leaves and petioles in the crown are modified into a compact mass of crinkled, twisted, distorted, and twisted structure. When the apical bud is killed, numerous adventitious buds develop and give rise to multiple-crowned plants (Fig. 9.8). Generally, fruit buds are killed. The petioles of certain cultivars develop a prominent red coloration, and hence the name "red plant."

**Fig. 9.8** *Top*—Foliar nematode-infected strawberry plant showing twisting and puckering of leaves, undersized leaves with crinkled edges (*left*), and reddening of petioles (*right*)

#### 9.2.1.2 Interaction with Other Pathogens

*A. fragariae* and/or *A. ritzemabosi* occasionally interact with the bacterium *Rhodococcus fascians* causing "cauliflower disease." Both nematode and bacterium must be present to produce the disease. Diseased plants are stunted with deformed stems, leaves, and flowers which in advanced stages resemble tiny cauliflowers.

#### 9.2.1.3 Favorable Conditions

Wet weather during the strawberry-growing season is most favorable for spread of foliar nematodes.

#### 9.2.1.4 Survival and Spread

- Nematodes are found to overwinter as juveniles and adults in soil, on dead leaves and in dormant buds.
- Spread by contact between plants in the presence of water.
- They migrate in the film of water on plant leaves.

#### 9.2.1.5 Management

##### 9.2.1.5.1 Regulatory Methods

The bud and leaf nematode can be managed by selecting nematode-free propagation stock. Whenever possible, certified stocks should be used.

##### 9.2.1.5.2 Physical Methods

Hot water treatment of strawberry runners at 50 °C for 5 min or 48 °C for 15 min or 46 °C for 20 min eliminate the nematode infection.

Tissue culture technology can be employed to regenerate nematode-free strawberry plants from clean, dormant, excised axillary buds.

#### 9.2.1.5.3 Cultural Methods

Uprooting and destroying infected plants and propagation through healthy planting stock are recommended for the control of nematodes.

Crop rotation with grain crops such as barley and rye proved beneficial.

#### 9.2.1.5.4 Chemical Methods

Foliar sprays with Fenamiphos at 4–6 ml/3.8 liters of water gave good control of *A. fragariae.* Chemicals used for disinfestation of strawberry plants include Parathion as a foliar spray and soil application of Carbofuran in granular form.

#### 9.2.1.5.5 Host Resistance

In Poland, the strawberry varieties "George Soltwedel," "Regina," and "Talizman" are comparatively resistant as are "Saksonka" and "Festival'naya" in USSR.

## 9.2.2 Lesion Nematode, *Pratylenchus penetrans*

### 9.2.2.1 Symptoms

Damage is frequently seen as patches in the field. The above-ground symptoms include stunting of plants, pale green leaves, reduced yields, and eventually death of plants. The below-ground symptoms include distinct brown necrotic lesions on roots in early or light infestation. Shortened crop cycle, reduction in crop yield, and decreased production of runners are also noticed. Fewer fine feeder roots and a brushy appearance of the plant can be observed (Fig. 9.9). Roots have individual reddish-brown lesions that can be so numerous on the feeder roots (Fig. 9.9). The root is girdled. The entire root system eventually becomes black and necrotic as populations increase and secondary organisms invade (Fig. 9.9).

**Fig. 9.9** *Left*—Healthy root system on the right and necrotic lesions (red arrows) affecting the root systems in the middle and left. *Right*—Lesion nematodes feeding inside strawberry roots (stained red)

#### 9.2.2.2 Interaction with Other Pathogens

The presence of the lesion nematode makes the plant susceptible to soil-borne disease pathogens like species of *Verticillium, Fusarium, Cylindrocarpon, Rhizoctonia,* and *Pythium.*

#### 9.2.2.3 Management

##### 9.2.2.3.1 Regulatory Methods

Use certified plants.

##### 9.2.2.3.2 Physical Methods

Hot water treatment of dormant strawberry plants at 54.4 °C for 1 min or 52.8 °C for 3 min or 51 °C for 7.5 min or 49.4 °C for 17.5 min eliminates the nematode infection.

##### 9.2.2.3.3 Cultural Methods

Crop rotation with potato (moderately susceptible and causing moderate nematode reproduction), oats or barley (not susceptible but causing nematode reproduction), and beet (not susceptible and suppressing nematode reproduction) was found effective.

Use cover crops like yellow mustard (*Brassica juncea*) which contains allyl isothiocyanate which is effective against nematodes. For effective nematode management, *Brassica carinata* also de used.

Fallow periods can disrupt the nematode life cycle. Practice good weed control using herbicides or tillage.

##### 9.2.2.3.4 Biological Methods

MeloCon WG (*Purpureocillium lilacinum* strain 251) at 2–4 kg/ha plus a soil wetting agent can be applied to established plants, although it might be better used when applied to plants just before planting.

##### 9.2.2.3.5 Integrated Methods

Integration of cultural (crop rotation) and chemical (nematicides) methods can be employed to manage the lesion nematode:

- Soil application of nematicide reduces nematode population to a very low level.
- Rotation of strawberry with potato (which is moderately susceptible and causing moderate nematode reproduction), beet root (which is not susceptible and suppressing nematode reproduction), and oats or barley (which are not susceptible but causing nematode reproduction).

## 9.2.3 Root-Knot Nematodes, *Meloidogyne* spp.

### 9.2.3.1 Symptoms

Leaf yellowing, stunting of plant growth, wilting of plants during hot sun, and yield reduction are some of the symptoms that appear on the above-ground plant parts. The main below-ground symptoms produced by the root-knot nematodes include galling at the tips of roots and abundant branching of roots above the root galls (Fig. 9.10). Shortened crop cycle, reduction in crop yield, and decreased production of runners are also noticed.

In warm, sandy soils, attack by *M. hapla* can cause severe stunting. When the nematode interacts with other soil-borne plant pathogens, death of plants is the possibility. In the presence of nematodes, starvation of plants for water is very severe during drought season.

### 9.2.3.2 Management

#### 9.2.3.2.1 Cultural Methods

Planting nematode-free runners certified nematode-free is imperative. Crop rotation with nonhost or poor crops can be effectively employed to reduce the northern root-knot nematode population. Typically, 2 years out of strawberries with corn or small grains is recommended. Cover crops like yellow mustard (*Brassica juncea*), which contains allyl isothiocyanate, and *Brassica carinata* are an excellent way to disrupt the nematode life cycle.

#### 9.2.3.2.2 Biological Methods

Application of *Purpureocillium lilacinum* at 2–4 kg a.i. /ha at the time of planting is beneficial.

**Fig. 9.10** *Left*—Strawberry roots infected with *M. incognita*. *Right*—Severe root galling on strawberry due to *M. hapla* infection

#### 9.2.3.2.3 Host Resistance

Strawberry cvs. Tiago, Chandler, and Torrey were found resistant to root-knot nematodes (Laqman Khan 2001).

#### 9.2.3.2.4 Integrated Methods

Application of neem cake in combination with Carbofuran, applied at the time of planting, increased the yield by 35.2% over control (Laqman Khan 2001).

## 9.2.4 Cauliflower Disease Complex

### 9.2.4.1 Symptoms

Both nematode (*Aphelenchoides fragariae*, *A. ritzemabosi*) and bacterium (*Rhodococcus fascians*) must be present to produce the disease. The disease manifests in the form of malformation of the inflorescence, shortening, and thickening of the aerial parts of strawberry plants (Fig. 9.11).

Cauliflower disease complex on strawberry is vectored by the nematodes like *A. fragariae* and *A. ritzemabosi* and the bacterium, *R. fascians* which play an

**Fig. 9.11** Symptoms of cauliflower disease on strawberry showing shortening and thickening of the aerial plant parts

obligatory role. Strawberry plants infected with both *A. ritzemabosi* and certain strains of *R. fascians* resemble small cauliflowers.

In extreme cases, plants are stunted, petals are small and greenish in color or fail to develop, and flowers are deformed. The sepal becomes much enlarged. Development of fruits is prevented due to the malformation of stamens and receptacles. Auxiliary buds are continually produced in the crowns. Therefore, both the organisms actively contribute to the cauliflower syndrome. The bacterium attaches to the cuticle of the nematodes and is transported to the crown tissue of strawberry plants.

#### 9.2.4.2 Management

- Use certified nematode-free propagation material.
- Root dip treatment of strawberry runners in hot water maintained at 50 °C for 5 min or 48 °C for 15 min or 46 °C for 20 min.
- Regenerating plants from clean, dormant, excised axillary buds.
- Uprooting and destroying infected plants.
- Foliar sprays with Fenamiphos at 4–6 ml/3.8 liters of water.
- Strawberry varieties "George Soltwedel," "Regina," "Talizman," "Saksonka," and "Festival'naya" are comparatively resistant.

## 9.3 Apple, *Malus domestica*

The lesion nematode is the major limiting factor in successful production of apple crop.

### 9.3.1 Lesion Nematodes, *Pratylenchus penetrans* and *P. vulnus*

#### 9.3.1.1 Symptoms

Root-lesion nematodes penetrate into roots and cause damage by feeding and migrating through the cortical tissues. Interaction of root-lesion nematodes with other soil-borne organisms can increase injury to roots. The lesion nematodes are sometimes responsible for the apple replant disease (characterized by young tree decline after transplanting). Apple trees (especially younger trees) may exhibit poor growth, stunting (Fig. 9.12), lack of vigor, twig dieback, and a gradual decline in yields. Severely infected root systems may lack feeder roots (Fig. 9.12). The economic threshold of *P. penetrans* is 15 nematodes per 100 g soil which cause significant growth reduction in apple trees (Crossa Raynand and Audergon 1987).

#### 9.3.1.2 Life Cycle

The juveniles of *Pratylenchus* sp. enter the roots wherever the tissue is immature so that the penetration may become easy. Walia and Bajaj (2003) observed that the larvae feed on root cortex and multiply by moving inter- and intracellularly. Migration through host tissues and feeding activities result in destruction of host cells

**Fig. 9.12** Root-lesion nematode-infected young apple trees showing drastic reduction in root system (*left*) and healthy (*right*) apple seedlings

which leads to the formation of necrotic lesions. Sexual reproduction is common in *P. penetrans.* The adult female laid eggs in soil or in roots (Parvatha Reddy 2008). Among four molts in nematode life cycle, the first one is inside the egg and subsequent three molts are outside the roots. All the life stages except $J_1$ are parasitic. Entire life cycle completes in 30–45 days. The population of the lesion nematode is much higher within the roots as compared to the surrounding soil, since it is an endoparasite.

#### 9.3.1.3 Interaction with Other Pathogens

The addition of *Phytophthora cactorum* to apple replant disease soil containing the nematode significantly reduced plant growth compared with the corresponding individual treatments. *Phytophthora parasitica* did not by itself reduce plant height, but in the presence of nematode, it reduced plant height to a greater extent than did the nematode alone (Utkhede et al. 1992).

#### 9.3.1.4 Management

##### 9.3.1.4.1 Physical Methods

In order to eliminate the nematode infection, the bare roots of apple rootstocks should be dipped in hot water maintained at 50 °C for 15 min or at 47 °C for 30 min. The material should first be stored in a cold place (at 1–3.3 °C), the treatment conducted in January, and the plants subsequently restored in a cold place until planting. Immersion in hot water should be followed immediately by a cold bath.

#### 9.3.1.4.2 Cultural Methods

Use certified (nematode-free) planting material. Remove old roots and plant green manure cover crops (tall fescue, red fescue, oat cultivar Saia, or perennial ryegrass as orchard ground cover) resistant to root-lesion nematodes (*Pratylenchus vulnus*) for 1–2 years or fallow the site for 4 years.

#### 9.3.1.4.3 Chemical Methods

Preplant soil fumigation in fall with Metam sodium is highly recommended.

#### 9.3.1.4.4 Host Resistance

The Geneva series rootstocks G11 and G30 support lower populations than Malling or Malling-Merton rootstocks in eastern Washington.

## 9.4 Conclusion

The major nematode problems on temperate fruit crops worldwide include species of *Pratylenchus*, *Meloidogyne*, and *Mesocriconema*. Deeper and broader understanding of nematode diseases is essential to develop effective, economical, and eco-friendly nematode management strategies. Work on disease complexes where nematodes and soil microorganisms are involved needs to be intensified with adequate collaboration between nematologists and plant pathologists. A prior knowledge of host-parasite relationship of major nematode pests and techniques for precise determination of damage threshold limit of nematode population is a must for a successful nematode management program. Also while adopting a management strategy, economy and ecology must be taken into consideration.

For food and nutrition security, management of nematodes is very essential. There is a need to develop bio-intensive integrated nematode management strategies using nematode-free planting stock by seed certification, heat treatment for denmatizing planting stock, use of crop rotation and intercropping with nonhosts or antagonistic crops, soil application of organic amendments enriched with biological control agents, treatment of nurseries and main field with safer chemicals, and development of resistant rootstocks.

In the recent years, some cultivars/rootstocks resistant to plant parasitic nematodes have been developed. However, for an encouraging result, more emphasis is required in this field by identifying new sources of plant resistance and its incorporation into crops by conventional breeding or genetic engineering biotechnology.

## References

Anon. (1990) Final report of research scheme on nematode problems in some selected temperate fruit crops and their management. Y.S. Parmar Univ. of Hort. & Forestry, Solan, 54 pp

Baghel PPS, Bhatti DS (1990) Interaction of vesicular arbuscular mycorrhizal fungus *Glomus mosseae* and *Meloidogyne javanica* on peach. *Proc. of seminar on bioagents in nematode management*. Indian Agri. Res. Inst, New Delhi, p 41

Barbercheck ME, von Broembsen SL (1986) Effects of soil solarization on plant-parasitic nematodes and *Phytophthora cinnamomi* in South Africa. Plant Dis 70:945–950

Crossa Raynand P, Audergon J (1987) Apricot rootstocks. In: Rom RC, Carlson RF (eds) Rootstocks for fruit crops. Wiley, New York, pp 295–320

Jaffee BA, Gaspard JT, Ferris H, Muldoon A (1988) Quantification of parasitism of the soilborne nematode *Criconemella xenoplax* by the nematophagous fungus *Hirsutella rhossiliensis*. Soil Biol Biochem 20:631–636

Laqman Khan M (2001) Dr. Y.S. Paramar University of Horticulture and Forestry, Solan, Himachal Pradesh. In: Dhawan SC et al (eds) Indian nematology-progress and prospectives. Division of Nematology, Indian Agri. Res. Inst, New Delhi, pp 153–158

Lownsberr BF, English H, Noel GR, Schick (1977) Influence of Nemaguard and Lovell rootstocks and *Macroposthonia xenoplax* on bacterial canker of peach. J Nematol 9:221–224

McInnis T, Klupfel D, Zehr E (1990) Suppression of *Criconemella xenoplax* on peach by rhizosphere bacteria. In: *Proc. of the 2nd Int. Nematol. Cong.*, Veldhoven

Nyczepir AP, Rodriguez-Kabana R (2004) Effectiveness of biofumigation for ring nematode control in a young peach orchard. In: *First Int. biofumigation Symp.*, Florence, Italy

Parvatha Reddy P (2008) Diseases of horticultural crops: nematode problems and their management. Scientific Publishers, Jodhpur, 379 pp

Pinochet J, Calvet C, Camprubi A, Fernandez C (1995) Growth and nutritional response of Nemared peach rootstock infected with *Pratylenchus vulnus* and the mycorrhizal fungus, *Glomus mosseae*. Fundam Appl Nematol 18(3):205–210

Rom RC, Dozier WA, Knowles JW et al (1985) Compact Fruit Tree 18:85–91

Seshadri AR (1964) Investigations on the biology and life cycle of *Criconemoides xenoplax* Raski, 1952 (Nematoda: Criconematidae). Nematologica 10:540–562

Sharpe RH, Hesse CO, Lownsbery BF et al (1969) Breeding peaches for root-knot nematode resistance. J Amer Soc Hort Sci 94:209–212

Strobel NE, Hussey RS, Roncadori RS (1982) Interaction of vesicular arbuscular fungi, *Meloidogyne incognita* and soil fertility on peach. Phytopathology 72:690–684

Taylor J, Briesbrock JA, Hendrix FF Jr et al (1970) Peach tree decline in Georgia. Georgia Agri Expt Stn Res Bulletin 77

Utkhede RS, Smith EM, Palmer R (1992) Effect of root rot fungi and root-lesion nematodes on the growth of young apple trees grown in apple replant disease soil. Zeitschrift fur pflanzenkrankheiten und pflanzenschutz 99:414–419

Verma RR (1987) Reaction of some peach cultivars to *Meloidogyne incognita*. Indian J Nematol 17:123–124

Walia RK, Bajaj HK (2003) Text book on introductory plant nematology. ICAR, New Delhi, 237 pp

Wehunt EJ, Horton BD, Prince VE (1980) Effect of nematicides, lime and herbicides on peach tree short life in Georgia, U.S.a. J Nematol 12:183–189

# Semiarid Fruit Crops

# 10

**Abstract**

The root-knot nematode *Meloidogyne incognita* is a serious pest of semiarid fruit crops like pomegranate and fig which primarily cause galls on root system that eventually are seriously hampered in their main functions of uptake and transport of water and nutrients, reduced plant growth, and productivity. In recent years, pomegranate wilt complex caused by the association of the root-knot nematode *Meloidogyne incognita* and the fungus *Ceratocystis fimbriata* is becoming an important disease which results in complete wilting of plant. The nematode distribution, crop losses, symptoms, biology and life cycle, spread, and management using cultural, chemical, biological, and integrated methods are discussed in this chapter.

**Keywords**

Pomegranate · Ber · *Meloidogyne incognita* · Symptoms · Management

## 10.1 Pomegranate, *Punica granatum*

### 10.1.1 Root-Knot Nematode, *Meloidogyne incognita*

The root-knot nematode *Meloidogyne incognita* Race 2 is a serious pest of pomegranate. The nematode-infected plant is not in a position to absorb water and nutrients from soil due to disruption of conduction vessels resulting in reduced plant growth and productivity of pomegranate (Khan et al. 2005; Nasira et al. 2011).

#### 10.1.1.1 Distribution

Mainly in Maharashtra, Gujarat, Tamil Nadu, Karnataka.





P. P. Reddy, *Nematode Diseases of Crops and their Management*,
https://doi.org/10.1007/978-981-16-3242-6_10

#### 10.1.1.2 Crop Losses

*Meloidogyne incognita* is responsible for annual yield loss of 23% in pomegranate amounting to Rs. 3023.44 million in India (Walia and Chakraborty 2018).

#### 10.1.1.3 Symptoms

Severely infested plants by root-knot nematode exhibited yellowing of foliage resulting in stunted plant growth. These plants produced less number of fruits or no fruits which might be due to the nematode-induced nutritional deficiency. In severe cases the galls were predominantly found on the entire root system (Fig. 10.1). The galls were white in color turned to light brown and hardy when they became old. The intensity of root-knot nematode damage increased with increase in age of the plant. In general, more than five-year old plants were severely affected by root-knot nematode. The adult female laid its eggs within the root galls and outside. More numbers of females were found in a single compound gall.

#### 10.1.1.4 Biology and Life Cycle

The root-knot nematode is a sedentary endoparasite which is soil-borne. After penetration, the J2 larvae feed endoparasitically on conducting vessels and develop to adult females which are spherical in shape and lays egg masses in compound galls. The males are vermiform in shape which leaves roots to enter the soil. During summer, the nematode completes its life cycle in about 25–30 days, while it takes longer during winter months.

**Fig. 10.1.** Symptoms of root-knot nematode-infected pomegranate plant showing leaf yellowing (*Left*) and roots showing galling (*Right*)

#### 10.1.1.5 Spread

The root-knot nematode is disseminated through infected planting material.

#### 10.1.1.6 Management

##### 10.1.1.6.1 Cultural Methods

Raising healthy plants (ground layers, air layers) in nematode-free soil for rooting.

The root-knot nematode can be effectively managed by intercropping pomegranate with onion, and garlic, mustard, sun hemp, fenugreek, and marigold during crop rest period.

Application of marigold flowers gave the highest reduction percentage in root galling with value of 98.8%, followed by colocynth fruits extract which averaged 96.4% without significant differences between them. The lowest egg-mass index (EI) was obtained from plants receiving marigold extract (0.0) with reduction percentage value of 100% (Nour El-Deen et al. 2016).

Soil application with castor cake at 2 t/ha is effective for the management of *M. incognita* infecting pomegranate.

##### 10.1.1.6.2 Chemical Methods

The root-knot nematode in pomegranate can be effectively managed by soil application of Carbofuran at 4 kg a.i./ha in nursery, and 2 kg a.i./ha in the main field at transplanting and at flowering.

##### 10.1.1.6.3 Biological Methods

Reduction in the number of root galls/5 g roots (29.28%) and *M. incognita* population in soil (31.28%), increasing the fruit yield of pomegranate (18.99%) with 2.37 B:C ratio can be achieved by soil application of *Pseudomonas fluorescens* at 20 kg/ha; followed by *Trichoderma viride* at 20 kg/ha (reduction in *M. incognita* population, 28.3%; number of root galls/5 g roots, 23.9%; increasing the fruit yield, 16.78%; and B:C ratio, 2.33) (Pawar et al. 2013).

Soil application of 100 kg FYM enriched with *P. fluorescens* or *T. viride* at 20 kg/ha at "Bahar" (blossom).

Apply *Purpureocillium lilacinum* or *P. fluorescens* 10–20 g/$m^2$ for reduction of root-knot nematode population.

The root-knot nematode in pomegranate can be effectively reduced by 97.9% through soil application of the symbiotic bacterium of entomopathogenic nematode, *Xenorhabdus* treatment, followed by those treated with *Photorhabdus* with value of 96%. Regarding egg-masses numbers, *M. incognita* was barely reproduced on plants treated with *Xenorhabdus* filtrate with reduction percentage value of 95.8%, followed by *Photorhabdus* with value of 91.7% (Nour El-Deen et al. 2016).

##### 10.1.1.6.4 Integrated Methods

Application of Farmyard manure at 20 tons/ha, Neem cake at 250 g/tree followed by bioagent, and *P. lilacinum* at 25 g/plant was found to be effective to manage root-knot nematode problem in pomegranate and to increase crop productivity.

Application of *P. lilacinum* at 20 kg/ha + castor cake at 2 t/ha in root zone, 30–45 cm away from tree trunk approximately 22.5 cm deep in soil at 6-monthly intervals proved beneficial.

Two applications of FYM enriched with *P. lilacinum* at 15–25 kg around each plant at 3 months interval along with application of liquid formulation of *P. lilacinum* through drip; showed recovery from symptoms, and putting forth new flushes. Saprophytic microbial population in soil has multiplied many folds, and reduced egg masses and females in roots and juveniles of root-knot nematodes were not seen in soil.

For the effective management of root-knot nematodes, the farmers of North Gujarat region cultivating pomegranate are advised to apply talc-based formulation of *Pochonia chlamydosporia* ($10^8$ spores/g) or *P. lilacinum* ($10^8$ spores/g) at 100 g/plant along with Mustard cake at 1.5, 2.0, and 2.25 kg/plant in first, second, and third year, respectively, for two times (during November and April).

Application of bioagents (*T. viride, P. lilacinum, Ps. fluorescens*) at 10–20 g/$m^2$ in nursery and 5 kg/plant of FYM/compost enriched with bioagents in the main field at transplanting and at flowering was found effective against *M. incognita*.

The following INM measures have been recommended for effective management of pomegranate root-knot nematodes:

- Intercropping with sun hemp, marigold, mustard, fenugreek, onion, and garlic during crop rest period
- Use of biocontrol agents such as *P. fluorescens, P. lilacinum, P. chlamydosporia, T. harzianum, T. viride* at 20 kg/ha
- Use of entomopathogenic nematodes (*Steinernema* spp., *Heterorhabditis* spp.) and their symbiotic bacteria (*Xenorhabdus, Photorhabdus*)
- INM using enriched Neem cake/Mustard cake/FYM/Compost/Vermicompost with biocontrol agents (*Pseudomonas fluorescens, Purpureocillium lilacinum, Pochonia chlamydosporia, Trichoderma harzianum, T. viride*)
- Enforcement of domestic quarantine.

### 10.1.2 Interaction of Root-Knot Nematode with *Ceratocystis* Wilt

The root-knot nematode *Meloidogyne incognita* interacts with soil-borne fungal pathogen *Ceratocystis fimbriata* and cause wilt disease complex.

#### 10.1.2.1 Distribution

The disease has been reported from Karnataka (Somasekhara and Wali 1999), Maharashtra (Jadhav and Sharma 2009), Andhra Pradesh, Tamil Nadu, and North Gujarat states.

#### 10.1.2.2 Symptoms

When the roots and lower portion of the stems were split open, the dark bluish-black/gray/brown discoloration of the wood is observed if the pathogen is *Ceratocytis fimbriata*. Within 2–3 months after infection, plants show symptoms like leaf yellowing in one or more branches, followed by drooping of foliage, and complete wilting of plant. Plants may reveal drooping of leaves of the entire plant all of a sudden (within 1–2 days) resulting in complete wilting. *C. fimbriata* infection causes discoloration of vascular bundles and adjoining tissues.

Root galls are observed on fine roots in early stage and other roots in advance stage if it is the root-knot nematode infestation.

#### 10.1.2.3 Interaction

There was no conspicuous gall production in nematode alone. In combination with fungi, the affected roots showed brownish discoloration in fungus alone and fungus + nematode treatments. In uninoculated treatment there was no brownish discoloration.

The two pathogens in combination (*M. incognita, C. fimbriata*) adversely affected plant growth parameters like shoot and root length, and fresh and dry shoot and root weight. Among the two pathogens inoculated individually, maximum reduction in plant growth parameters was noticed in fungus alone (*C. fimbriata*) treatment. However, the treatment receiving the inoculum of both pathogens (*M. incognita* + *C. fimbriata*) showed a highest reduction in plant growth parameters in comparison with other treatments.

#### 10.1.2.4 Survival and Spread

- The fungus survived in thick walled Aleurioconidia of *C. fimbriata*, in soil, host, and plant debris and served as a source of primary inoculum. The pathogen can also survive for very long time in infected wood.
- Infected planting material or the soil used in nursery for raising the planting material generally is the primary source of infection for wilt organisms.
- Pruning wounds are common entry points for the pathogen. *C. fimbriata* can spread through by wind and pruning knives.
- Avoid close planting to prevent the nematode spread to adjacent trees via root grafts.
- Prevent movement of irrigation water from infected field to healthy field.
- The nematodes predisposed plants to wilt pathogens by way of root wounding.
- The major sources for the spread of disease complex include implements, budding knife, secateurs, rain water, infected seedlings, irrigation water, root contact, and wind (Somasekhara et al. 2009).

#### 10.1.2.5 Management

##### 10.1.2.5.1 Biological Methods

Biological formulations if used should be reliable, fresh, and used during rest period when no other fungicides/bactericides are used. The soil application of *Bacillus subtilis, Purpureocillium lilacinum, Trichoderma harzianum*, and *Aspergillus niger* at 10–15 g/plant along with well-decomposed farm yard manure around the trunk of pomegranate trees helps to prevent wilt infections. Neem cake at 2–3 kg/plant effectively checks incidence of wilt disease complex.

Biofertilizer – Kalisena SA having *Aspergillus niger* at 2.5 kg/ha + Mychorrhizal preparation "Josh" at 12.5 kg/ha or "Josh Ultra" 2.5 kg/ha gives effective control of wilt if used from beginning or before the incidence of disease. These two bioagents should be applied twice a year along with sufficient organic matter for effective wilt management. This treatment controls several soil-borne pathogens, improves nutrient uptake, and gives disease resistance and also improves yields.

#### 10.1.2.5.2 Integrated Methods

The following practices are recommended for the integrated management of wilt disease complex in pomegranate:

- Locate disease-free orchards for selection of disease-free cuttings/air layers to raise seedlings in sterilized soil.
- The planting material (sapling as well as soil in which it is planted) should be free from all wilt organisms.
- Before planting, sterilize soil with solarization or formalin at 50 ml/L.
- Avoid water stagnation and provide proper drainage.
- Follow recommended spacing of 4.5 m × 3.0 m in the orchard.
- The nursery beds or soil used for potting mixtures should be sterilized by soil solarization for 6 weeks using 50–100 μ thick linear low density polyethylene (LLDPE) sheet during peak summer or formalin at 2.5–5.0%. If formalin is used, ensure that the soil is free from any formalin fumes before transplanting in the bags. Soil solarization is beneficial as it kills harmful pests and pathogens and also increases population of beneficial microorganisms which are present in the soil and are thermotolerant. However, formalin treatment kills both harmful and beneficial organisms.
- The above formalin treatment can also be used for sterilizing soil after removing infected dead plant.
- On observing first symptoms of wilt due to fungal pathogens in the orchard, drench soil immediately with Propiconazole 25EC (2 ml/L) + Chlorpyriphos 20EC (2 ml/L) or Carbendazim 50WP (2.0 g/L) + Chlorpyriphos 20EC (2 ml/L). Use 5–10 liters solution/plant depending on plant growth to drench soil up to 30 cm depth. Drench at least 3–4 healthy plants all around the infected plant/s, and repeat drenching 3–4 times at 20–25 days interval. Propiconazole was significantly superior over remaining treatments. Propiconazole-treated plants were completely recovered from wilting after 105th day, followed by Difenconazole the plants were recovered after 120th day. The control showed maximum percentage of wilting (97.61%).
- Basin soil application of Carbofuran 3G at 40 g/plant or Phorate 10G at 25 g/plant to manage wilting due to root-knot nematodes. Drenching with Azadirachtin (1%) at 2 ml/L is also recommended. Application of Neem cake at 1–3 kg/plant depending on age is advisable twice a year. Plant *Tagetes erecta* (African marigold varieties best followed by French marigold) in between row space or

in a ring, on the border of plant basin. For effective results, these should be grown for more than 4–5 months and incorporated in the soil. Crops like onion, tomato, chili, potato, bell pepper, gram, legumes, cucurbits, gerbera, gladiolus, etc. aggravate nematode infestation and hence should be avoided as intercrops. Green manuring with sesbania is beneficial.

## 10.2 Fig, *Ficus carica*

### 10.2.1 Root-Knot Nematodes, *Meloidogyne incognita, M. javanica*

The root-knot nematode is probably the most severe nematode problems in fig cultivation. The problem is recognized as a major limiting factor in commercial fig production in the India, the United States, France, and Brazil.

#### 10.2.1.1 Distribution

Root-knot nematodes are distributed in Mediterranean, North and South American countries, India, as well as in southern Africa.

#### 10.2.1.2 Crop Losses

Average avoidable loss in the fruit yield of fig was recorded to be 24.92%, when the crop was treated with Carbofuran 3 G at 2 kg a.i./ha (Jagdev et al. 2020a).

#### 10.2.1.3 Symptoms

Root knot nematodes on fig trees often show themselves with bumps or "knots" on the roots of the tree (Fig. 10.2). Above ground, the tree has a generally stunted and unhealthy look. The above-ground symptoms are somewhat similar to nutrition deficiency. Under severe nematode infestation, severe root galling occurs which

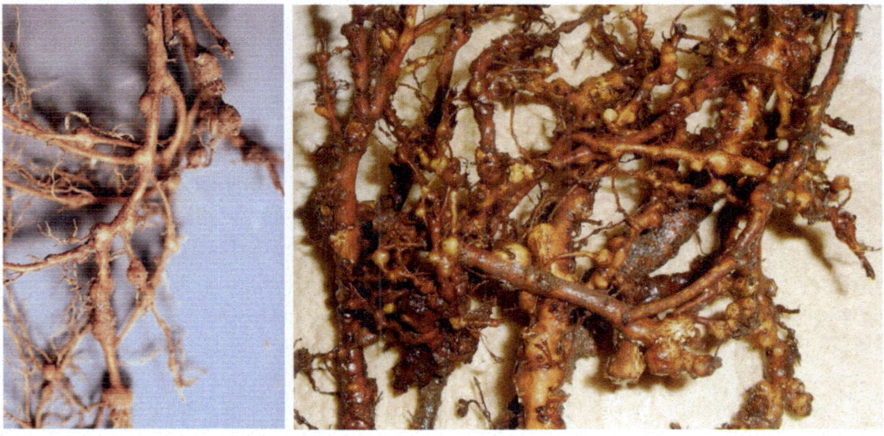

**Fig. 10.2.** Symptoms of root-knot nematode on fig roots

blocks the conducting tissues, deprive the plants from taking up nutrients leading to death of trees.

#### 10.2.1.4 Management

##### 10.2.1.4.1 Chemical Methods

Soil application of Carbofuran 3G at 2 kg a.i. /ha was found to be most effective in reducing root-knot nematode population (53.66%), number of root galls (50.19%), and number of egg masses (45.36%) at intermediate stage and increasing the fruit yield of fig (37.85%) with 1:17.79 ICBR at termination (Jagdev and Mhase 2019). Nematode populations were considerably reduced in young trees by stem treatments with an experimental paste formulation of Fenamiphos (Inserra and O'Bannon 1974).

##### 10.2.1.4.2 Biological Methods

At termination stage, the treatment of Phule *Trichoderma* plus at 20 kg/ha was found to be most effective in reducing nematode population (40.36%), number of root galls (37.08%), and number of egg masses (41.02%) and increasing the fruit yield of fig (13.54%) with 1: 11.20 ICBR (Jagdev et al. 2020b).

##### 10.2.1.4.3 Host Resistance

Four *Ficus* species (*F. racemosa*, *F. cocculifolia*, *F. gnaphalocarpa,* and *F. palmata*) showed a high degree of resistance to species of root knot nematodes, as well as good graft compatibility with *F. carica* (Krezdorn and Glasgow 1970). O'Rourke (1992) recommended that the commercial nematode-resistant fig cv. "LSU Purple" (from "Hunt" an unknown California Capri fig) for planting. The rootstock "Zidi" performed well in a number of field trials (Hosomi et al. 2002). In Israel, the fig varieties "Celeste" and "Poulette" were considered resistant to the nematode (Gur 1955). "Ficus Echo" (*Ficus sycomorus*) seems to be effective as a nematode-resistant rootstock for figs for South Florida. "Ficus Echo" propagates readily by cuttings.

##### 10.2.1.4.4 Integrated Methods

The most effective management combinations in suppressing the nematode densities in the rhizosphere of fig trees were *Purpureocillium lilacinum* + *Pastueria penetrans* + cow manure, *Trichoderma harzianum* + *P. penetrans* + cow manure, *P. lilacinum* + *P. penetrans* + urea 46%, and *Pochonia chlamydosporia* + *P. penetrans* + urea 46% (54.68–57.17% nematode reductions) (Dawabah et al. 2019).

## 10.3 Conclusion

The most commonly encountered plant-parasitic nematodes in semiarid fruit crops like pomegranate and fig include root-knot nematode *Meloidogyne incognita*. Complete death of the plant occurs due to wilt disease complex in pomegranate caused by the association of the root-knot nematode *Meloidogyne incognita* and the fungus *Ceratocystis fimbriata*. Cover crops can be important tools for sustainable nematode management in arid and semiarid fruit crops. Using appropriate cover crops can prevent or reduce nematode population and enhance soil health. The successful system of cover cropping is limited by availability of water especially during frequent droughts. Choosing the right cover crops and managing them appropriately can provide long-term benefits for many nematode problems.

There is a need for development of suitable root-knot and wilt management schedules involving cultural practices, sanitation measures, and chemical methods. Emphasis needs to be given to develop root-knot-wilt resistant rootstock to evolve suitable wilt-resistant variety to ensure more economical, feasible, and effective wilt management schedule.

## References


Dawabah AAM, Al-Yahya FA, Lafi HA (2019) Integrated management of plant-parasitic nematodes on guava and fig trees under tropical field conditions. Egypt J Biol Pest Control 29:29

Gur A (1955) The Fig. Depts. of Hort. and Agri. Edn., Ministry of Agriculture, Israel, Bulletin No. 57 (in Hebrew)

Hosomi A, Dan M, Kato A (2002) Screening of fig varieties for rootstocks resistant to soil sickness. J Japanese Soc Hort Sci 71:171–176

Inserra RN, O'Bannon JH (1974) Systemic activity of phenamiphos for control of Meloidogyne arenaria on Gardenia jasminoides and Ficus carica. Pl Dis Reptr 58:1075–1077

Jadhav VT, Sharma KK (2009) Integrated management of diseases in pomegranate (Lead Paper). In: Int. Symp. on Pomegranate and other Minor Fruits including Mediterranean Fruits. Univ. of Agri. Sci., Dharwad, pp 48–52

Jagdev GH, Hole UB, Kulkarni SR, Mhase NL (2020a) Assessment of avoidable yield losses due to root-knot nematode, *Meloidogyne incognita* infesting fig under field conditions. Pharma Innov J 9(2):13–16

Jagdev GH, Hole UB, Kulkarni SR, Mhase NL (2020b) Management of root-knot nematode, *Meloidogyne incognita* infesting fig under field conditions. Int J Chem Stud 8(1):1592–1597

Jagdev GH, Mhase NL (2019) Management of root-knot nematode, *Meloidogyne incognita* infesting fig under field conditions. J Entomol Zool Stud 7(6):193–197

Khan A, Shaukat SS, Siddiqui IA (2005) A survey of nematodes of pomegranate orchards in Balochistan province, Pakistan. Nematol Medit 33:25–28

Krezdorn AH, Glasgow SK (1970) Propagation of *Ficus carica* on tropical species of *Ficus*. Proc Trop Reg Am Soc Hort Sci 14:156–164

Nasira K, Shaheen N, Shahina F (2011) Root-knot nematode *Meloidogyne incognita wartellei* on pomegranate in swat KPK. Pakistan J Nematol 29(1):117–118

Nour El-Deen AH, Al-Barty Amal F, Darwesh Hadeer Y, Al-Ghamdi AS (2016) Eco-friendly management of root-knot nematode, *Meloidogyne incognita* infecting pomegranate at Taif governorate, KSA. Res J Pharma Biol Chem Sci 7(1):1070–1076

O'Rourke EN (1992) LSU purple, a new fig for the Gulf south. Louisiana Agri 35:19–20

Pawar SA, Mhase NL, Kadam DB, Chandele AG (2013) Management of root-knot nematode, *Meloidogyne incognita,* race-II infesting pomegranate by using bioinoculants. Indian J Nematol 43(1):92–94

Somasekhara YM, Wali SY (1999) New record of *Ceratocystis fimbriata* causing wilt of pomegranate. Plant Dis 83(4):400

Somasekhara YM, Wali SY, Shaik MK (2009) Studies and the management of pomegranate *(Punica granatum)* wilt *(Ceratocystis fimbriata,* Latin American group*)*. In: Paper presented in: *2nd Intn. Symp. Pomegranate and minor fruits including Mediterranean fruits*. Univ. Agric. Sci., Dharwad, pp 132–133

Walia RK, Chakraborty PK (eds) (2018) Nematodes problems of crops in India. ICAR-All India Coordinated Research Project on Nematodes in Agriculture, New Delhi, 402 pp

# Part VI

# Vegetable Crops

# Solanaceous Vegetable Crops 11

**Abstract**

Root-knot (*Meloidogyne* spp.) and cyst (*Globodera* spp.) nematodes are the major nematode problems in some major solanaceous vegetable crops like potato, tomato, brinjal, chilli, and bell pepper. Nematodes interact with soil-borne pathogens in causing disease complexes. In this chapter, economic importance and losses, symptoms/damage, pre-disposing factors, epidemiology, survival and spread, and integrated management using physical methods, cultural practices, botanical pesticides, chemical pesticides, bioagents, arbuscular mycorrhizal fungi, and host resistance are discussed. For the effective and eco-friendly management of nematodes on solanaceous vegetable crops, strategies like crop sanitation, crop rotation, destruction of crop residues, use of safe chemicals, biological control agents, development of resistant cultivars, and integrated management have been adopted.

**Keywords**

*Meloidogyne* spp. · *Globodera* spp. · Symptoms · Management

## 11.1 Potato, *Solanum tuberosum*

The cyst nematodes (*Globodera rostochiensis* and *G. pallida*) and the root-knot nematodes (*Meloidogyne incognita, M. javanica*, and *M. hapla*) are the major problems on potato in India.



P. P. Reddy, *Nematode Diseases of Crops and their Management*,
https://doi.org/10.1007/978-981-16-3242-6_11

### 11.1.1 Cyst Nematodes, *Globodera rostochiensis, G. pallida*

The potato cyst nematode (PCN) is the most important quarantine pest that is causing concern to potato production and trade throughout the world. In India, the potato cyst nematode *Globodera rostochiensis* was first reported by Jones (1961) who detected the cysts from the roots of potato plants which exhibited the symptoms of yellowing of leaves from Nilgiri Hills of Tamil Nadu state. This report initiated a great awareness among administrators in the country resulting in establishment of Nematology Unit in Tamil Nadu as well as initiating an Indo-German Collaborative Program of intensive chemical application (Dasanit) in the infested area to avoid further nematode spread. The Government of Tamil Nadu amended the Destructive Insect Pest Act 1919 during 1971 in order to prevent the potential damage due to potato cyst nematodes by ensuring strict prohibition of movement of seed potatoes from infested area to any other parts of the state or the country.

#### 11.1.1.1 Distribution

The earlier surveys have indicated the presence of both species of potato cyst nematode (*Globodera rostochiensis* and *G. pallida*) restricted to potato growing regions of Nilgiri, Palani, and Kodaikanal Hills of Tamil Nadu (Krishna Prasad 1986) and Kerala (Ramana and Mohandas 1988).

The presence of potato cyst nematodes was reported by Ganguly et al. (2010) from the Potato Breeding Farms of Central Potato Research Institute located in Kufri, Shimla District of Himachal Pradesh. The nematode infestation to the extent of 58–548 cysts per 100 g soil was recorded. The European and Mediterranean Plant Protection Organization (EPPO) had also reported potato cyst nematode *G. pallida* from Himachal Pradesh earlier to Ganguly et al. (2010). Recently it was recorded from the states of Uttarakhand and Jammu and Kashmir also. These reports have a serious concern with ramifications on export and quarantine issues.

#### 11.1.1.2 Crop Losses

Potato cyst nematodes inflict huge quantitative and qualitative losses in potato. While 26% losses are normal, the damage may be as high as 53–66% in badly infested fields. In severely infested fields, the yields may be even less than the tubers used for planting.

An average loss of about 9% of global potato is accounted to the cyst nematodes amounting to about 45 million tons (Prasad 1989).

Potato cyst nematodes are responsible for 26% annual yield loss amounting to ₹127.04 million in Nilgiri Hills, Tamil Nadu, India (Walia and Chakraborty 2018).

#### 11.1.1.3 Symptoms

Under conditions of severe infestations, plant growth is stunted and wilting occurs during hot parts of the day (Fig. 11.1). The disease usually occurs in patches (Fig. 11.1). Plants show tufting of leaves on the top as the outer leaves turn yellow and die. Root system is smothered, secondary roots are induced at the collar region, and the plants can be easily pulled out. Tubers formed are less in number and

**Fig. 11.1** *Left*—Patchy appearance of potato field damaged by cyst nematodes. *Right*—Healthy and cyst nematode infected roots of potato plants

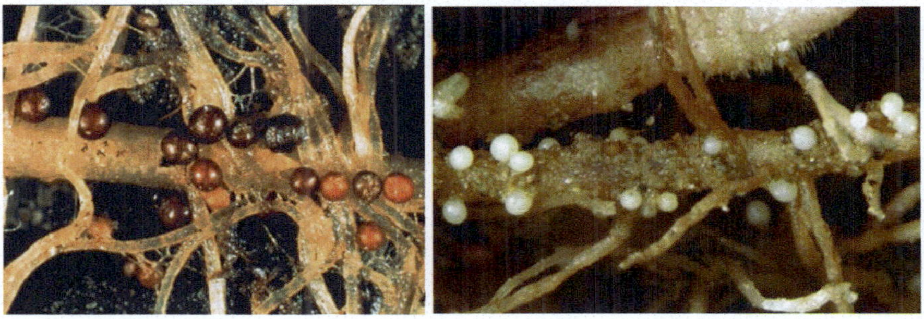

**Fig. 11.2** Potato roots infected with cyst nematodes (*Left—Globodera rostochiensis*, Right—*G. pallida*)

reduced in size. In extreme cases tuber formation is arrested. Outside the root surface, very small (about the size of a pin head, 0.5 mm) white females that turn to yellow or brown cysts after death can be observed at flowering or later (Fig. 11.2).

#### 11.1.1.4 Host Range

All known varieties of potato, tomato, and brinjal are severely attacked by the cyst nematode. A number of other Solanaceae are recorded as hosts, but the majority of hosts are species and varieties of tuberous *Solanum*. The only susceptible plant outside the family is snap dragon, *Antirrhinum majus*.

#### 11.1.1.5 Biology and Life Cycle

Cysts contain 200–500 eggs that remain viable for several years. Hatching commences with the planting of host crop; potato root diffusates strongly stimulate the emergence of juveniles from cysts. Only 30–40% eggs hatch in a season; the remaining eggs hatch over a period of several years. Under Nilgiri Hill conditions, the nematode completes its life cycle in about 5–7 weeks.

#### 11.1.1.6 Ecology

Well-drained, aerated sands, silts, and peat soils with a moisture content of 50–75% water capacity are suitable for survival and movement of the free-living stages of the potato cyst nematode. A minimum soil temperature of 10 °C is necessary for larval activity.

#### 11.1.1.7 Spread and Survival

The cysts can be spread by the movement of soil through seed material for a very long distance. They can also be spread through field tools, implements, seedlings grown in infested soil for transplanting elsewhere, labors' feet, etc. from one field to another. The cysts can remain viable in the soil without a host for 30 years.

#### 11.1.1.8 Interaction with Other Pathogens

The potato cyst nematode and the black-scurf fungus *Rhizoctonia solani* together caused much greater reduction in yield than did the organisms separately. The severity of the Verticillium wilt *Verticillium dahlia* is also increased in the presence of potato cyst nematodes. The leaf symptoms appeared earlier and the tuber yields were depressed in the presence of the nematode.

The potato cyst nematodes *Globodera rostochiensis* and *G. pallida* in association with the wilt bacterium *Ralstonia solanacearum* cause wilt disease complex in potatoes.

#### 11.1.1.9 Physiological Races

In India, potato cyst nematodes are restricted to Nilgiri and Palani hills in south India, where five pathotypes—Ro 1, Ro 2 of *G. rostochiensis*, and Pa 1, Pa 2, and Pa 3 of *G. pallida*—have been detected (Amalraj and Rao 1992).

#### 11.1.1.10 Management

##### 11.1.1.10.1 Regulatory Methods

The infested area should be quarantined and cultivation of host plants to be discontinued indefinitely. Volunteer seedlings in the infested area should be destroyed. The seed material should not be allowed to move out of the infested area. The infested area can be put to nonagricultural use like construction of buildings, etc.

Potato cyst nematodes mainly spread through infected planting stock. Hence, all seed producers and growers selling seeds of commercial crops should be registered under certified schemes in all states to produce nematode-free seed tubers.

##### 11.1.1.10.2 Cultural Methods

###### Crop Rotation

It is the most practical method of control. It is recommended to grow the host crop once in every 3 years if the infestation is confined to a few patches, but only once in 4 or 5 years if the infestation is extensive. Grow nonhost crops like cabbage, cauliflower, beetroot, radish, carrot, beans, strawberry, and turnip in rotation. In Nilgiri Hills, potato is rotated with wheat and strawberry to control the cyst nematode. Growing of resistant potato cultivar Kufri Swarna in rotation is beneficial.

Trap Cropping

Trap cropping should only be considered where PCN infestations are very large and need to be reduced quickly. The trap crop *Solanum sisymbriifolium* generates high mortality in PCN but requires a full growing cycle to be effective. The technique is costly (at least as costly as soil fumigation) and needs very careful management.

Intercropping

Cyst nematode in Tamil Nadu hills is managed by intercropping potato with beans or wheat.

Disease Escape

Early varieties of potatoes like Irish Cobbler usually suffer less than do late varieties. The potato crop should be planted during spring when the soil temperatures are very low for nematode activity, and the crop can mature before the nematode can multiply to sufficient level to cause damage to the crop.

#### 11.1.1.10.3 Chemical Methods

The infested area can be treated with an eradication dosage of Fesnulfothion or Carbofuran both at 3 kg a.i. per ha in two doses at an interval of 15–20 days. Dazomet, which liberates methyl isothiocyanate, is more nematicidal in pearls and clay soils and often gives large yield increases.

#### 11.1.1.10.4 Biological Methods

Application of *Pseudomonas fluorescens* at 10 g/m$^2$ gives good nematode control (Mani et al. 1998).

*Pseudomonas fluorescens* at 10 kg/ha gave maximum reduction in cyst nematode multiplication (Reproduction factor, 1.12) and improvement in tuber yield (19.40 t/ha) (Umamaheswari et al. 2010).

*Purpureocillium lilacinum* treated plots had higher marketable yield (20.58 t/ha as compared to 15.72 t/ha in control) and significantly lower numbers of cysts (58.88 as compared to 70.35 cysts/50 ml soil) (Jatala 1985).

#### 11.1.1.10.5 Host Resistance

*Solanum vernei* (multiple gene resistance) and *S. tuberosum* s. sp. *andigena* (single gene resistance) were found to have marked resistance to the cyst nematode of potato. *S. andigena* roots produce a hatching factor and are readily invaded by the larvae, which, however, fail to mature and reproduce. The failure is because the root tissues are not induced to produce giant cells, but instead the cells near the nematode head become necrotic and there is inadequate nutrition for the female to develop.

In the United States, both the USDA and Cornell University scientists have evolved two commercial potato varieties, viz., "Peconic" and "Wauseon" which were resistant to the cyst nematode. "Peconic" derives its resistance from the subspecies andigena. "Wauseon" is a selection from a cross between B 4159-8 and Katahdin.

Potato cv. Kufri Swarna developed by Ooty Station of Central Potato Research Institute, Shimla, is found to be resistant to PCN.

#### 11.1.1.10.6 Integrated Methods

The integrated management of cyst nematodes can be achieved by combining practices like planting early maturing cultivars (nematode escape) and soil application nematicides. Other practices like use of certified seed, crop hygiene, along with rotation of potato with nonhost crops can also be used for the nematode management in seed production regions. Use of a resistant variety followed by a nematicide would kill 99% of the nematode population. The most effective measure combines crop rotation, use of nematicides and resistant varieties to keep the nematodes at an economically acceptable level (Jones 1961).

An integrated system utilizing resistant cultivars, nonhost crops and susceptible cultivars with chemical treatment (Carbofuran) will successfully manage densities of *Globodera rostochiensis* at desired levels.

Application of Neem cake at 1 t/ha or FYM at 20 t/ha or compost at 10 t/ha enriched with *Trichoderma harzianum/Purpureocillium lilacinum* gave effective control of cyst nematodes.

Umamaheswari et al. (2012) reported that reduction in nematode reproduction (Reproduction factor, 1.09) and increase in tuber yield (23.14 t/ha) was achieved by integrating soil incorporation of Neem cake at 5 t/ha along with *Trichoderma viride* at 0.5 kg/ha.

Sitaramaiah and Naidu (2003) found that soil solarization during peak summer along with soil incorporation with *Pasteuria penetrans* and karanj cake/farmyard manure gave good control of *G. rostochiensis*.

A well-known integration of methods can control *G. rostochiensis* in potato in a 4-year rotation combining soil fumigation (causes 30% reduction in nematode population), cultivation of nonhost (50% reduction), cultivation of resistant potato variety (30% reduction), cultivation of nonhost (50% reduction), and then cultivation of susceptible variety of potato (Oostenbrink 1972).

### 11.1.2 Root-Knot Nematodes, *Meloidogyne* spp.

Thirumalachar (1951) noticed root-knot galls on potato tubers from Shimla during 1950 for the first time and since then it has been recorded from all the potato growing regions of the country. Root-knot nematodes *Meloidogyne* spp. are by far the most important nematode parasites of potatoes in India.

#### 11.1.2.1 Crop Losses

An initial level of two root-knot larvae per gram of soil resulted in 100% tuber infection with an overall yield reduction of 42.5% (Prasad 1989).

Severe infestation of root-knot nematodes leading to crop failure have been noticed in Mahasu District of Himachal Pradesh and Hassan in Karnataka where farmers did not get even as much as planted by them.

#### 11.1.2.2 Symptoms

Plants affected by root-knot nematodes are generally stunted, slightly yellow, and may wilt during hot weather. The diagnostic symptoms of a *Meloidogyne* attack on potatoes are the galls found on both roots and tubers (Fig. 11.3). Tuber surface becomes uneven and warty because of numerous blisters like galls. When an infested tuber is transversely cut, the white glistening swollen females of the size of pin head can be easily seen embedded in the potato tissue. Tubers infected with blisters are responsible for significant crop losses due to their very small size, quickly rot, and not marketable.

#### 11.1.2.3 Life Cycle

Species of *Meloidogyne* are sedentary root and tuber endoparasites, the life cycle being similar to that of *G. rostochiensis*, with some important differences. The white, swollen females do not develop into persistent, brown cysts, but deposit all their eggs in an external egg mass. Hatching occurs when the physical conditions are suitable, and is not dependent on a hatching factor from host roots. Root-knot nematodes reproduce parthenogenetically, and males even though present are not required for reproduction.

#### 11.1.2.4 Host Range

Species of *Meloidogyne* usually have wide host range among crops and weeds in many botanical families.

**Fig. 11.3** *Left*—Potato roots infected with root-knot nematodes. *Right*—Blisters on potato tubers caused by *Meloidogyne incognita*

#### 11.1.2.5 Interaction with Other Pathogens

*Meloidogyne* species are associated with the fungi *Rhizoctonia solani* and *Verticillium* sp. and the bacterium *Ralstonia solanacearum* in disease complexes.

#### 11.1.2.6 Ecology

Root-knot damage to potatoes is usually associated with light soil. Spread of nematodes through irrigation water enhances the nematode attack. Soil temperature requirements of these nematodes vary with species. *M. hapla* requires cool climate, whereas *M. incognita* and *M. javanica* require higher temperatures for their development and reproduction.

#### 11.1.2.7 Physiological Races

In India, all four races of *M. incognita* have been reported and the race 1 is predominant (Sharma and Gill 1992).

#### 11.1.2.8 Survival and Spread

Root-knot nematodes survive on several weeds and volunteer crops present in the field after harvest of the main crop. They can also survive in the form of eggs and anhydrobiotic second-stage larvae. Infected tubers, plant parts, and planting material as well as movement of infested soil by farm machinery and irrigation water are the main avenues of disseminating *Meloidogyne* spp.

#### 11.1.2.9 Management

##### 11.1.2.9.1 Physical Methods

*Meloidogyne* spp. in seed tubers can be controlled by hot water treatment at 46–47.5 °C for 2 h and by dry heat treatment at 45 °C for 48 h (Nirula and Bassi 1965).

##### 11.1.2.9.2 Cultural Methods

###### Hygiene

Tubers from *Meloidogyne*-infested crops should not be used as seed, as an even outwardly clean seed may be infested.

###### Early Planting

In long season, warm climate regions, potatoes may be planted during the winter months and harvested before injury occurs in the spring without visible infection. Under Shimla Hill conditions, both root and tuber infestation of *M. incognita* can be prevented by early planting of potato during March third or fourth week of without affecting the yield (Ahuja 1983; Prasad et al. 1983). Late planting of autumn crop and early planting of spring crop in North Western Plains; while in hills, early planting of summer crop in fourth week of March is ideal. At Jalandhar, early planting of spring crop in first week of January and late planting of autumn crop in second and third week of October reduced the root-knot infection in potato.

Crop Rotation

Crop rotation with maize, wheat, and barley effectively check the buildup of *Meloidogyne* population. Follow crop rotation with vegetables (radish, garlic, cruciferous crops) and green manuring with lupin.

Trap Crops

*Tagetes patula* planted in alternate rows with potato gives good control of root-knot nematodes.

Growing nematode antagonistic crops such as crotalaria, asparagus, and mustard as cover crops prior to potato planting significantly brought down soil population of plant parasitic nematodes by 41.6%.

Intercropping

Potatoes grown with white mustard in a pot of infested soil were less heavily attacked by root-knot nematodes than potatoes growing alone. Potato root diffusate was ineffective in the presence of leachings from the roots of mustard seedlings. Mustard oil increased the yield of potatoes by reducing the severity of nematode attack. In mustard, the nematode toxic substance involved is allyl isothiocyanate. The intercropping of marigold with potato has resulted in a 7% increase in tuber yield.

Intercropping of onion and maize reduce galling by *M. incognita* on potato roots. Potatoes intercropped with maize have been found to be extremely vigorous with well- developed root systems.

Organic Amendments

Application of organic amendments such as FYM, compost, green manures, and crop residues suppress the nematode population. This may promote fungal and other natural enemies of nematodes or release toxic compounds.

#### 11.1.2.9.3 Chemical Methods

Treatment of the tubers with Carbosulfan is beneficial.

Application of Carbofuran 3G or Phorate 10G (1 kg a.i. /ha) in furrows while seeding was found to be effective.

#### 11.1.2.9.4 Biological Methods

Seed treatment with *Pseudomonas fluorescens* at 20 g/kg seed and its soil application at 2.5 kg/ha was found to be effective for managing potato cyst nematode.

Soil application of *Purpureocillium lilacinum* significantly lower root galling index (*M. incognita*), enhanced egg parasitization (86% egg masses were infected with *P. lilacinum*, and over 54% of the eggs within these egg masses were destroyed), and marketable yield of potatoes (302.59 kg/400 $m^2$ as compared to 147.12 kg/400 $m^2$ in control).

#### 11.1.2.9.5 Host Resistance

Potato cv. Kufri Dewa was found resistant to root-knot nematodes.

#### 11.1.2.9.6 Integrated Methods

Application of Neem cake/FYM/compost enriched with *Trichoderma harzianum*/*Purpureocillium lilacinum* gave effective control of root-knot nematodes.

Following 2 years crop rotational sequence of maize-wheat-potato-wheat coupled with summer fallow after two or three deep plowings in North Western Plains reduced root-knot damage significantly.

## 11.2 Tomato, *Solanum lycopersicum*

### 11.2.1 Root-Knot Nematodes, *Meloidogyne incognita, M. javanica*

#### 11.2.1.1 Crop Losses

Root-knot nematodes are responsible for 23% annual yield loss in tomato worth ₹6035.2 million in India (Walia and Chakraborty 2018).

*M. incognita* was responsible for 30.57% to 46.92% loss in fruit yield of tomato (Bhatti and Jain 1977; Reddy 1985; Darekar and Mahse 1988), while *M. javanica* caused 77.5% loss in yield (Anon 1993).

#### 11.2.1.2 Symptoms

Infected seedlings exhibit reduced growth, yellowing, wilting, and moderate to large number of galls on the root system (Fig. 11.4). Similarly transplanted seedlings in infested field show reduced growth. The infested plants yield less, exhibit yellowing similar to nutrient deficiency, and galls can be seen on the root system upon uprooting. The infested plants succumb to wilt and viral diseases quickly.

**Fig. 11.4** Tomato roots infected with root-knot nematode

#### 11.2.1.3 Management

##### 11.2.1.3.1 In Nursery

Physical Methods

For the management of root-knot nematodes and weeds in tomato nursery, soil solarization through tarping with 100 gauge (25μm) LLDPE clear plastic film for 15 days during May is recommended. It increased number of transplantable seedlings by 61.5% and decreased root-knot disease and weeds by 66% and 93%, respectively.

Rabbing root-knot infested tomato nursery with *bajra* husk at 7 kg/m$^2$ a week prior to seeding was found to be effective for the management of root-knot disease.

Cultural Methods

Incorporation of green materials of Naffatia and Neem leaves each at 2 kg/1.44 m$^2$ or Parthenium and neem leaves each at 2 kg/1.44 m$^2$ or Calotropis, Naffatia, Parthenium, and neem leaves each at 1 kg/1.44 m$^2$ 15–20 days prior to tomato seeding was found to be effective.

Green manuring with naffatia (*Ipomoea fistulosa*) leaves or water hyacinth (*Eichhornia crassipes*) or *aak* (*Calotropis* sp.) or congress grass (*Parthenium hysterophorus*) each at 3 kg/m$^2$ 15 days prior to seeding proved beneficial for the management of root-knot disease and production of healthier transplantable seedlings in tomato nursery during *kharif* season.

Application of poultry manure at 2 t/ha to 3 t/ha 15 days prior to seeding proved beneficial for the management of root-knot disease and getting more numbers of transplantable tomato seedlings.

Use of Neem cake at 750 g per m$^2$ in nursery beds of tomato proved beneficial.

Planting of *Tagetes* sp. (dwarf variety) in the nursery beds was found to be effective.

Chemical Methods

Nursery bed infested with root-knot nematode *Meloidogyne incognita* treated with Carbofuran at 0.3 g a.i./m$^2$ was found to be effective in reducing the gall index and thereby increasing the yield of tomato.

Application of Metham sodium 32 EC at 60 ml/m$^2$ or Carbofuran at 0.2 g a.i./m$^2$ to tomato nursery controlled the root-knot nematode *M. incognita* and improved the performance of transplants by 80 and 60%, respectively.

In susceptible variety, the nursery treatment with Carbofuran 3.3 g a.i./m$^2$ gave 39.2% increase in yield whereas in resistant variety it was only 10.5%.

Biological Methods

Nursery treatment with either *Trichoderma viride* formulation (FYM incubated) at 2.5 g/m$^2$ or *Pseudomonas fluorescens* formulation at 20 g/m$^2$ was found to be effective.

Nursery bed treatment with *Pasteuria penetrans* at 10 g culture (1 × $10^8$ spores/g)/$m^2$ before sowing proved effective for managing root-knot nematode *Meloidogyne incognita* infecting tomato and increasing the yield.

#### Host Resistance

Root galling caused by *M. incognita* was lower (2.95 as compared to 8.08 in control) and the fruit yield was higher (1.785 kg $plant^{-1}$ as compared to 1.118 kg $plant^{-1}$ in control) in tomato plants grafted on the rootstock Beaufort (Kapkavalci et al. 2009).

#### Integrated Methods

Soil application of poultry manure at 2 t/ha followed by soil solarization (with 100 gauze LLDPE clear film for 15 days in hot summer) is recommended for the control of root-knot nematodes and higher production of tomato transplants in tomato nursery.

In nursery, integration of *Pasteuria penetrans* (at 28 × $10^4$ spores/$m^2$), *P. lilacinum* (at 10 g/$m^2$ with 19 × $10^9$ spores/g), and Neem cake (at 0.5 kg/$m^2$) gave maximum increase in plant growth and number of seedlings/bed. Integration of Neem cake with *P. lilacinum* resulted in maximum parasitization of eggs, while integration of Neem cake with *P. penetrans* gave highest parasitization of *M. incognita* females. In field, planting of tomato seedlings (raised in nursery beds amended with Neem cake + *P. penetrans*) in pits incorporated with *P. lilacinum* (at 0.5 g/plant) gave least root galling and nematode multiplication rate and increased fruit weight and yield of tomato (Parvatha Reddy et al. 1997).

*Pseudomonas fluorescens* applied in the nursery bed at 20 g/$m^2$ along with Carbofuran at 1.5 g a.i./$m^2$ has shown promising results against root-knot nematode in bringing down the population and increasing the yield of tomato. Similarly, application of *Glomus fasciculatum* at 200 spores/kg soil in nursery bed of tomato provided more than 50% reduction in root-knot population and 36% increase in yield of tomato.

Soil application of *Trichoderma viride* at 20 g/$m^2$, mixed with Neem cake/FYM/Vermicompost at 100 g/$m^2$ in the beds was found promising for the management of root-knot nematodes (*Meloidogyne* spp.) in tomato.

For the management of root-knot nematode in vegetables like tomato, nursery should be covered with polythene sheet for 15 days during May/June after giving light irrigation. At the time of sowing, apply Carbofuran 3G at 10 g/$m^2$ in the nursery prior to sowing.

### 11.2.1.3.2 In Field

#### Physical Methods

Polythene mulching in the main field for 3 weeks prior to transplanting recorded a reduction of nematode population by 52.3% in susceptible variety and 50% in the resistant variety. Polythene mulching gave 32.9% increase in fruit yield over no tillage, followed by plowing and exposure (16.4%).

### Cultural Methods

Cultivation of cowpea as a cover crop, later removing and burning proved beneficial for the management of *M. incognita* in tomato.

Mixed cropping of tomato along with onion/garlic reduced the population of root-knot nematode and the fruit yield was also improved by 22% in 2 years.

Application of Neem cake at 500 kg/ha in the main field before transplanting was found to be effective.

Rotation of tomato with bajra was found beneficial for the management of root-knot nematode problem.

Sowing of nonhost crop (cauliflower) in *kharif* and disease-resistant variety Hisar Lalit of tomato in *rabi* in fields infested with root-knot nematode was found to be effective.

Biofumigation with broccoli, *Brassica oleracе* var. *italica* at 14 t $ha^{-1}$ (all plant parts were added to the soil 1 week before planting and incorporated to a depth of 20 cm with a rotary tiller), could be used efficiently against *M. incognita* in organic tomato production. Biofumigation with broccoli gave lower root-knot index (4.33 as compared to 8.08 in control) and higher yield (2.315 kg $plant^{-1}$ as compared to 1.118 kg $plant^{-1}$ in control) (Kapkavalci et al. 2009).

### Chemical Methods

Bare root dip treatment of root-knot nematode-infested seedlings of tomato was done using Carbosulfan and Triazophos at 0.1% and 0.5% for 6 h increased the yield and reduced the root-knot index.

Seed treatment of tomato with Abamectin 40% FS at 0.4 g a.i./kg seed caused minimum root-knot index (3.75) and 36.9% reduction in soil population of *M. incognita* which reflected in 21.4% increase in yield of tomato.

### Biological Methods

The root-knot nematode *M. incognita* population can be effectively reduced by 62% and the fruit yield increased by 42% by bare root dip treatment ($10 \times 10^9$ cfu/100 ml water) and soil incorporation of SkNAn5 isolate of *Aspergillus niger* ($2 \times 10^9$ cfu/kg soil) (Anwar and Khan 2010).

Application of *P. penetrans* at $1 \times 10^9$ spores/$m^2$ was found to be effective in reducing the nematode population (*M. incognita*) in the main field by 72.3% over control and improved plant growth, fruit yield (7 800 t/ha compared to 3.325 t/ha in control), and nutritional status of tomato fruits (Ravichandra and Reddy 2008).

### Host Resistance

Use of resistant varieties of tomato, for example, Hisar Lalit, Punjab NR-7, SL-120, Mangala, Utkal Pallavi (BT-1), Utkal Dipti (BT-2), Utkal Urbasi (BT-12), and BT-17 gave effective control.

Indian Institute of Vegetable Research, Varanasi, has developed grafting technology to mitigate the problem of root-knot nematode infesting tomato. *Solanum torvum* (wild brinjal germplasm) which is identified as resistant against root-knot nematode was used as rootstock to graft with scions of promising tomato varieties cv. Kashi Aman, Kashi Vishesh, and Hisar Lalit. Grafted plants were compatible between rootstock and scion and also showed significant resistance against root-knot nematode by reducing soil population, reproduction, and gall index (Gowda et al. 2017).

### Integrated Methods

Application of Carbofuran 3G at 10 g/m$^2$ at sowing and 1 kg a.i. /ha in the main field 1 week after transplanting proved beneficial.

Application of Carbofuran 3G at 10 g or Phorate 10G at 3 g/m$^2$ at seeding in the nursery beds and spot application of Carbofuran 3G at 33 kg or Phorate 10G at 10 kg + Neem cake at 400 kg/ha at the base of plant 10 days after transplanting was found to be effective for the control of root-knot nematode.

Treated nursery + polythene mulching in main field + Neem cake at 200 kg/ha reduced the population of *M. incognita* by 70.9% and increased tomato yield by 39.9%.

Soil solarization along with application of Neem cake at 200 kg/ha in the main field increased fruit yield of tomato by 180.6% by reducing infestation by *M. incognita*.

Solarization coupled with treated nursery (Carbofuran at 0.3 g a.i./m$^2$) or resistant variety (NR-7) along with Neem cake at 200 kg/ha effectively controlled the root-knot nematode and increased the yield. This package has been demonstrated at farmer's field at Benikheda village of Panagar, where average increase in yield was 43%.

In tomato soil solarization along with the use of resistant variety Hisar Lalit, significantly reduced *M. incognita* infestation.

Application of *Pupureocillium lilacinum* ($2 \times 10^6$ cfu/g) at 50 g/m$^2$ in nursery bed (0.5 t/ha) + *P. lilacinum* at 5 kg along with 2.5 t of FYM/ha in the main field prior to planting tomato proved effective for suppressing soil nematode population and enhancing yield by 17.2–19.3%.

Adoption of INM technology involving summer plowing and fallowing, growing nonhost crops like wheat, maize, sorghum, bajra; intercrops like marigold, sun hemp; green manuring with sun hemp; application of FYM and Neem cake; and use of nematicides like Phorate 10G and Carbofuran 3G proved beneficial.

Application of *Aspergillus niger* and *P. lilacinum* along with Mustard cake gave maximum reduction in nematode population both in root and soil with enhanced plant vigor. *P. lilacinum* parasitize eggs, while *A. niger* being toxic agent kills the second-stage juveniles of *M. incognita* present in the rhizosphere. As a result, there is an overall reduction in root and soil population. Goswami et al. (1998) reported that the general plant health of tomato is maintained by soil incorporation of mustard cake which also has nematicidal properties.

The integration of castor leaves with *P. lilacinum* increased the efficiency of nematode control (*M. javanica*) in tomato. The residual effect on the second crop of tomato revealed the reduction in root-knot index (29.6–356.8%) and $J_2$ population in soil (26.4–57.9%) (Zaki 1998). *P. lilacinum* in combination with castor leaves reduced the nematode population up to 89% and increased plant growth and yield in tomato (Zaki and Bhatti 1991).

Tomato seedlings raised in nursery beds treated with *P. chlamydosopria* and *P. penetrans* when transplanted in the main field significantly lowered root-knot index, number of eggs per egg mass, nematode population in roots and soil; increased root colonization and egg parasitization with *P. chlamydosporia,* infection of *M. incognita* females with *P. penetrans*; and increased tomato fruit yield (Rao et al. 1998).

Seed treatment with *Pseudomonas fluorescens* at 10 g/kg seed 24 h before sowing and seed treatment with *Azospirillum* at 100 g/kg just before sowing in nursery and main field treatment with *P. fluorescens*, *Azospirillum*, and *Phosphobacterium* each at 2 kg/ha was found to be effective in reducing nematode population in soil and root by 48.9% and 57.8%, respectively, over control. Shanthi (2010) found that there was 58.9% increase in fruit yield and increased benefit-cost ratio to 4.2 by the above treatment.

### 11.2.2 Interaction of Root-Knot Nematode with *Fusarium* Wilt

The root-knot nematode and *Fusarium* wilt are frequently associated with tomato resulting in considerable damage to the crop.

#### 11.2.2.1 Interaction

Bhagawati and Goswami (2000) reported the interaction of *M. incognita* and *Fusarium oxysporum* f. sp. *lycopersici* on tomato (Fig. 11.5). It was found that when either or both pathogens were inoculated simultaneously or nematode was

**Fig. 11.5** Interaction of root-knot nematode (*Meloidogyne* spp.) with Fusarium (*Fusarium oxysporum*) causing wilt disease complex

inoculated 10 days prior to inoculation of the fungus, the symptoms were visible by 20 days as against no such symptoms in plants where fungus was inoculated 10 days prior to nematodes. Nematode inoculation prior to fungus and simultaneous inoculation of both the pathogens resulted in significantly higher intensity of wilt after 40–60 days. The number of galls and egg masses and nematode population in soil was significantly reduced when simultaneously inoculated or prior inoculation of nematodes.

Young plants inoculated with both the pathogens collapsed earlier before crop maturity. The nematode, by impairing water and nutrient availability, disrupts root function and plant growth processes. These effects combined with the vascular blocking due to the wilt fungi can be particularly severe, and if widespread, result in total crop failure. Giant cells in root galls produced by the root-knot nematode provide the substrate suitable for rapid colonization of the fungal pathogen. Root-knot nematodes can thus significantly enhance disease development and yield loss, elevating primary or secondary pathogens to major pest status.

#### 11.2.2.2 Management

##### 11.2.2.2.1 Biological Methods

Antagonistic Bacteria

Root infection caused by *M. javanica* and the soil-borne root-infecting fungi *F. oxysporum* and *F. solani* was effectively suppressed following application of *Pseudomonas aeruginosa*. Biocontrol and growth promoting potential of the bacterium was enhanced when soil was kept at 50% or 75% moisture holding capacity. Siddiqui and Ehteshamul-Haque (2001) found that increase in plant growth was optimal at an inoculum level $2.5 \times 10^8$ cfu/ml of the bioagent, *P. aeruginosa*.

Antagonistic Fungi

Soil application of *T. harzianum* at 50 kg/ha ($2 \times 10^8$ cfu/g) was effective in improving the plant growth parameters and fruit weight followed by *P. fluorescens*, Carbofuran, *A. niger*, *T. virens*, and Neem seed powder. Carbofuran was found highly effective against *M. incognita*, Topsin-M against *F. solani*, and *T. harzianum* was effective against both the pathogens (Haseeb et al. 2006).

*Paenibacillus polymyxa* GBR-462, GBR-508, and *P. lentimorbus* GBR-158 strains showed the strongest antifungal and nematicidal activities. These three strains used in pot experiment reduced the symptom development of the disease complex caused by *M. incognita* and *F. oxysporum* f. sp. *lycopersici* on tomato (wilting and plant death), and increased plant growth. The control effects were estimated to be 90–98%, and also reduced root gall formation by 64–88% compared to the untreated control (Son et al. 2009).

##### 11.2.2.2.2 Chemical Methods

Application of Metham sodium through drip irrigation was found to be effective against root-knot and *Fusarium* sp. complex of tomato (Roberts et al. 1988).

#### 11.2.2.2.3 Host Resistance

Abawi and Barker (1984) found that the monogenic resistance in tomato cvs. (Florida, MH-1, and Manapal) against *F. oxysporum* f. sp. *lycopersici* race 1 remained unaffected in the presence of *M. incognita.*

#### 11.2.2.2.4 Integrated Methods

##### Bioagents and Botanicals

*T. harzianum* controlled wilt in the presence of *M. incognita* for 6 weeks after transplanting. Tomato roots that received *P. lilacinum* along with *T. harzianum* and Neem cake were free from root-knot nematodes (*M. incognita*) and did not wilt due to *F. o.* f. sp. *lycopersici* till harvest. The roots were also free from *Fusarium* infection. The above treatment also reduced the % of wilt (10% compared to 90% in control), root-knot index (1.8 compared to 4.4 in control) and increased root colonization with bioagents (74% compared to 0% in control), parasitization of egg masses (56% compared to 0% in control), and eggs/egg mass (44% compared to 0% in control) (Nagesh et al. 2006).

##### Bioagents, Cultural Methods, and Host Resistance

Deep plowing and exposing soil to hot sun in summer, removal and burning of crop debris, soil application of *T. viride* and *P. lilacinum*, use of wilt-resistant varieties like Utkal Pallavi, Utkal Deepti, Utkal Kumari, Utkal Urbasi, etc. help in controlling the disease complex.

##### Arbuscular Mycorrhizal Fungi and Botanicals

Combined application of the endomycorrhiza, *Glomus etunicatum* along with mustard cake was found more effective in managing the disease complex caused by *M. incognita* and *F. oxysporum* f. sp. *lycopersici* on tomato as compared their individual application (Bhagawati et al. 2000).

##### Two Bioagents

Efficacy of *Pseudomonas aeruginosa* alone or in combination with *Purpureocillium lilacinum* was evaluated for the control of root-knot nematode and root-infecting fungi under field conditions. The efficacy of *Pseudomonas aeruginosa* alone or in combination with *Purpureocillium lilacinum* significantly suppressed soil-borne root-infecting fungi such as *Macrophomina phaseolina, Fusarium oxysporum, Fusarium solani, Rhizoctonia solani*, and the root-knot nematode *Meloidogyne javanica*. Siddiqui et al. (2000) reported that the parasitization of eggs and female of *M. javanica* was significantly better with *P. lilacinum* than *P. aeruginosa.*

### 11.2.3 Interaction of Root-Knot Nematode with Bacterial Wilt

#### 11.2.3.1 Interaction

Pani and Das (1972) have reported the association of root-knot nematode with bacterial wilt of tomato. Haider et al. (1987) reported the significant reduction in the root-knot index and larval development of *M. incognita* in soil where *Ralstonia solanacearum* was present. *R. solanacearum* and *M. incognita* alone as well as in different combinations reduced plant growth and yield significantly with the nematode followed by the bacterium combination showing the maximum reduction in growth (Fig. 11.6).

#### 11.2.3.2 Management

Treatment of nursery bed with *Pseudomonas fluorescens* ($10^9$ cfu/g) and *Trichoderma harzianum* ($10^6$ cfu/g) each at the rate of 20 g/m$^2$ and subsequent application of 5 t of farm yard manure enriched with 5 kg each of *P. fluorescens* ($10^9$ cfu/g) and *P. lilacinum* ($10^6$ cfu/g) per hectare significantly reduced *R. reniformis* and *M. incognita* in tomato roots by 74% and 70%, respectively; reduced the incidence of bacterial wilt; and increased the yield by 24.2%. Rao et al. (2009) reported that the benefit:cost ratio worked out to be 4.4 when both the bioagents were used for the management of disease complex.

**Fig. 11.6** Root-knot nematode and bacterial wilt disease complex

## 11.3 Brinjal, *Solanum melongena*

Root-knot nematodes are the major nematode problems in successful cultivation of brinjal crop.

### 11.3.1 Root-Knot Nematodes, *Meloidogyne incognita, M. javanica*

#### 11.3.1.1 Crop Losses

The estimated brinjal fruit yield losses ascribed to *M. incognita* was found to be 27.30–48.55% (Bhatti and Jain 1977; Darekar and Mahse 1988; Parvatha Reddy and Singh 1981).

Root-knot nematodes are responsible for annual yield loss of 21% in brinjal amounting to ₹3499.12 million in India (Walia and Chakraborty 2018).

#### 11.3.1.2 Symptoms

- Under severe infection, stunted growth and unthrifty development of plants occurred.
- The leaves are yellowish green to yellow, tend to drooping (Fig. 11.7) and sudden wilting if wilt organisms are present.
- Leaf margin was scorched inwards from the margin.
- The most characteristic symptom of the disease is the formation of knots or galls on the root system (Fig. 11.7).
- The characteristic symptom of root-knot infection is the formation galls of varying size on main and feeder roots.
- In advanced stages the galled tissues decay and are invaded by other pathogenic and saprophytic organisms.

**Fig. 11.7** *Left*—Healthy (one plant) and root-knot nematode infected brinjal seedlings (three plants). *Right*—Severely galled roots of brinjal

### 11.3.1.3 Management

#### 11.3.1.3.1 In Nursery Beds

Physical Methods

Solarization of nursery beds with 25µm transparent plastic film for 15 days during hot summer to raise healthy seedlings.

Chemical Methods

Application of Metham sodium 32 EC at 80 ml/m$^2$ or Carbofuran at 0.2 g/m$^2$ controlled *M. incognita* in brinjal nursery and increased the fruit yield.

Treatment of nursery bed of brinjal with Dazomet/Phorate/Carbofuran at 0.6 g a. i./m$^2$ reduced the root-knot nematode.

Biological Methods

Nursery bed treatment with *Glomus fasciculatum* at 50 g culture (200 spores/g)/m$^2$ before sowing proved beneficial for managing root-knot nematode *Meloidogyne incognita* infecting brinjal and increasing the yield.

Nursery treatment with *Bacillus macerans* at 25 g/m$^2$ + drenching with 2% solution of formulated product of *B. macerans* 7 days after sowing was found to be effective for managing root-knot nematode.

Nursery bed treatment with either *Trichoderma viride* enriched with FYM at 2.5 g/m$^2$ or *Pseudomonas fluorescens* at 20 g/m$^2$ gave effective control.

Nursery bed treatment with *Pasteruia penetrans* at 10 g culture (1 × 10$^8$ spores/g)/m$^2$ before sowing was found to be effective for managing root-knot nematode *Meloidogyne incognita* infecting brinjal and increasing the yield.

Integrated Methods

Application of Neem cake at 500 kg/ha 15 days before sowing and Carbofuran at 1 kg a.i./ha at sowing is effective in brinjal (Saikia et al. 2007).

#### 11.3.1.3.2 In Field

Cultural Methods

**Crop Rotation** Crop rotation with non-solanaceous crops such as paddy, pigeon pea, gingelly, marigold, and groundnut helps in reducing the buildup of nematodes in soil. Sheela et al. (2002) recorded 47% reduction in the root-knot nematode population and 22% increase in fruit yield by rotating brinjal with sweet potato (cv. Sree Bhadra). Crop rotation with sorghum, wheat, and chili reduced the root-knot nematode population. The population of *M. incognita* larvae decreased under brinjal-wheat, brinjal-chili-wheat, and red gourd-brinjal-mustard. The root-knot larvae were completely eliminated in marigold-marigold-fallow and chili-cauliflower-cauliflower. Garlic following brinjal also brought about considerable reduction in larvae of root-knot nematode. The population levels of *M. incognita* on highly infested land dropped to zero after 18 months of continuous cultivation of *Panicum* *maximum*. Growing of brinjal after *P. maximum* was completely free from root-knot nematodes (Netscher 1983).

**Intercropping** Intercropping brinjal with marigold, onion, garlic, or resistant tomato (SL-12) improved plant growth and reduced number of galls and final soil population of nematode (Jain et al. 1990). Marigold (cv. Calcutta Yellow)-brinjal cropping sequence has been recommended to the farming community for management of *M. javanica* affecting brinjal (Singh 1991).

**Trap Cropping** Growing of trap crop like knol-khol with brinjal but removed and destroyed before new generation of nematode could emerge helps in reducing the root-knot nematode population (Ayyar 1926).

**Influence of Fertilizers** Row application of fly ash at 0.6 kg/ $m^2$ enhanced the fruit yield by 27.7% and inhibited the reproduction of *M. incognita* (Khan and Ghadipur 2004).

**Organic Amendments** Nursery bed treatment with Neem cake at 200 g/$m^2$ gave maximum reduction in nematode population and resulted in yield of 2.075 t/ha (Sheela and Nisha 2004).

Enhancement of fruit yield and reduction in the nematode infestation was effectively achieved by row incorporation of saw dust at 210 g/3 m row and spot application of mustard cake at 25 g/spot. Similarly, Hazarika (1990) found that spot application of Castor/Mustard/Neem cake at 15 g/spot and furrow application of the above cakes at 100 g/3 m furrow reduced the infestation of *M. incognita* and increased the yield of brinjal. The yield was maximum in spot application of Neem cake along with maximum reduction of *M. incognita* population in soil.

**Time of Planting** In early transplanted (April) brinjal, the rate of multiplication of *M. incognita* was low, in comparison to late transplanted crop (May–June); the latter significantly reduced yield in comparison to early crops (April). Galling was also comparatively higher in the late sown crops as compared to crop sown in April. The study revealed that the crop sown at the right time is more tolerant to the *Meloidogyne* population as compared to the late crop.

#### Chemical Methods

Bare root dip treatment of brinjal seedlings with Zolone (Phosalone 35 EC) or Monocrotophos (Monocil 35 EC) or Carbosulfon (Marshall 25 ST) at 0.05% for 6 h before transplanting reduced the population of *M. incognita* and increased the yields substantially.

#### Host Resistance

Use of resistant varieties, for example, Utkal Madhuri (BB-44), Utkal Jyoti (BB-13), Utkal Anushree (BB-45C), Ghatikia white, and KS-224 (for Bihar), gave effective control.

Cultivar identified from Bangalore Centre as resistant to *M. incognita* race-1 includes Matti Gulla (Brinjal).

#### Integrated Methods

Integration of soil solarization for 15 days in summer or deep summer plowing with application of Neem cake at 200 kg/ha followed by transplanting of brinjal seedlings raised in nursery bed treated with Carbofuran 3G at 0.3 g a.i./$m^2$ was effective in management of *M. incognita*.

Application of *P. lilacinum* ($2 \times 10^6$ cfu/g) at 50 g/$m^2$ in nursery bed + main field application of *P. lilacinum* ($2 \times 10^6$ cfu/g) at 5 kg along with 2.5 t of farm yard manure/ha prior to transplanting was found to be effective for the management of root-knot nematode in brinjal.

Application of Carbofuran 3G at 3 kg a.i./ha in the nursery beds prior to seeding and Neem cake at 500 kg/ha 10 days before transplanting in the main field was found to be effective.

Application of *Purpureocillium lilacinum* at 25 kg spore dust ($10^9$ conidia/g) with carrier/ha at the time of transplanting + Poultry manure at 10 t/ha 1 week prior to transplanting or Mustard cake at 2 t/ha 1 week prior to transplanting or *P. lilacinum* at 25 kg spore dust ($10^9$ conidia/g) with carrier/ha at the time of transplanting + Neem cake at 2 t/ha 1 week prior to transplanting proved beneficial.

Application of Carbofuran 3G at 10 g or Phorate 10G at 3 g/$m^2$ at seeding in the nursery beds and spot application of Carbofuran 3G at 33 kg or Phorate 10G at 10 kg + Neem cake at 400 kg/ha at the base of plant 10 days after transplanting proved beneficial for the control of root-knot nematode.

At harvest, enhancement of brinjal plant growth and yield (7.4 kg/6 $m^2$ as compared to 5.2 kg/6 $m^2$ in control) and reduction in number of juveniles in 100 ml soil (67 compared to 152 in control), number of juveniles in 5 g root (62 compared to 162 in control), root galling index (5.2 compared to 7.8 in control), and number of egg masses in 5 g root (20 compared to 49 in control) were observed when brinjal seedlings raised in nursery treated with *P. chlamydosporia* at 50 g/$m^2$ were transplanted in field (Rao et al. 2003).

Application of *Pseudomonas fluorescens* at 10 g/$m^2$ + Carbofuran at 0.15 g a.i./$m^2$ effectively controlled initial root-knot nematode (*M. incognita*) population in the nursery which in turn protected the brinjal crop in the main field also and increased the yield by 17–22% (Sheela et al. 2002).

In field, transplanting of brinjal seedlings raised in nursery treated with Neem cake at 400 g/$m^2$ along with *G. mosseae* (at 500 g/$m^2$ containing 26–32 chlamydospores/g) and *P. lilacinum* (at 2 L/$m^2$ containing $6 \times 10^5$ spores/ml) gave maximum reduction in root galling, egg mass production, fecundity, and final nematode population in soil and roots; and enhanced fruit yield (7.8 kg/6 $m^2$ as 5.3 kg/6 $m^2$ in control), root colonization with *G. mosseae* and *P. lilacinum* and egg parasitization with *P. lilacinum* (Rao and Parvatha Reddy 2001).

## 11.3.2 Interaction of Root-Knot Nematode with Bacterial Wilt

### 11.3.2.1 Interaction

Eggplant is prone to many soil-borne diseases among which the bacterial wilt (*Ralstonia solanacearum*) in combination with root-knot nematode (*M. incognita*) takes heavy toll every year all over the world (Naik et al 2003). Combined inoculation of nematode and the bacterium had a greater pathogenic effect than pathogens inoculated individually (Fig. 11.8) (Ravichandra et al. 1990; Savitha and Lingaraju 1996).

### 11.3.2.2 Management

#### 11.3.2.2.1 Biological Methods

Significant reduction in the final soil nematode population and wilt disease incidence in brinjal was obtained by soil application of biological control agents like *Pseudomonas fluorescens* at 5 g/kg soil and *T. harzianum* at 3 g/kg soil. The highest reduction in bacterial population in soil was observed in the treatment *P. fluorescens* (Barua and Bora 2008). Further, significant increase in the antagonist population in soil and reduction in root galling, egg mass production, and final population of *M. incognita* and *R. solanacearum* in soil were observed with combined application of wheat bran formulations of *P. fluorescens* and *T. harzianum* (Barua and Bora 2009).

Root dip of brinjal seedlings in *P. fluorescens* (5 g/L of water for 30 min) gave maximum suppression of root-knot nematode population in soil followed by *P. chlamydosporia* (soil application at 10 g/plant + root dip at 5 g/L of water for 30 min), Neem cake, *P. fluorescens* (soil application at 10 g/plant + root dip at 5 g/L of water for 30 min), *P. chlamydosporia* as soil application (10 g/plant), *P. fluorescens* as soil application (10 g/plant), and *P. chlamydosporia* as root dip (5 g/L of water for 30 min) (Dhawan et al. 2008).

**Fig. 11.8** Root-knot nematode and bacterial wilt disease complex in brinjal

Nursery bed treatment with *P. chlamydosporia* at 100 g/m$^2$ gave reduced root galling; and increased seedling height, seedling weight, and root colonization by *P. chlamydosporia* (Naik et al. 2003). The seedlings colonized by *P. chlamydosporia* were highly vigorous and reached main field when transplanted.

#### 11.3.2.2.2 Integrated Methods

The nursery seedling stand was good in seed treatment with *P. fluorescens* at 50 g/kg + nursery bed treatment with *P. fluorescens* at 50 g/m$^2$ followed by nursery bed treatment with *T. harzianum* + *P. fluorescens*, and Neem-based *P. fluorescens* (Naik et al. 2003).

Integration of seedling dip for 15 min in talc-based formulation of *P. fluorescens* + *P. chlamydosporia* (5 + 5 g/L) with soil application of the above bioagents gave maximum survival of brinjal plants under field conditions (53.33%) compared to 33.50% survival in control (Naik et al. 2003).

Seed treatment with talc-based formulation of *P. fluorescens* at 50 g/kg gave least mortality of brinjal plants (13.18%) followed by seed treatment with Neem-based formulation of *P. fluorescens* at 50 g/kg (16.66%) compared to 45.29% mortality in control plants (Naik et al. 2003).

Combined application of *T. harzianum* and *P. chlamydosporia* gave effective management of the disease complex and gave least root gall index (RGI), least wilt index (WI), and maximum fruit yield (1.8 RGI, 22.18% WI, and 2.024 kg/3 m$^2$ yield) followed by *P. fluorescens* + *P. chlamydosporia* (1.81 RGI, 25.13 WI, and 1.782 kg/3 m$^2$ yield) and *T. harzianum* + *P. fluorescens* (1.93 RGI, 27.28 WI, and 1.680 kg/3 m$^2$ yield) (Naik 2004).

The highest reduction of *M. incognita* and *R. solanacearum* population in soil was observed in combined application of *T. harzianum* and *P. fluorescens*. *P. fluorescens* was proved to be more promising followed by *T. harzianum* in suppressing the population of *R. solanacearum*, while *T. harzianum* proved more effective than *P. fluorescens* in reducing root galling. The population of both the bioagents increased in the soil (Barua and Bora 2009).

Integration of two summer plowings during May–June, half recommended dose each of Carbofuran (0.75 kg a.i. /ha), Neem cake (7.5 g/spot), Streptocycline (250 ppm at 30 ml/spot), and full dose of *T. harzianum* (150 g/spot) improved plant growth parameters and fruit yield with corresponding decrease in the nematode reproduction rate (*M. incognita*), bacterial population in the soil, and bacterial wilt incidence (*R. solanacearum*) in brinjal (Hussain and Bora 2008).

## 11.4 Chilli and Bell Pepper, *Capsicum annuum*

The root-knot nematode is the major limiting factor in successful production of chilli and bell pepper crops.

### 11.4.1 Root-Knot Nematodes, *Meloidogyne incognita, M. javanica*

#### 11.4.1.1 Crop Losses

Root-knot nematodes are responsible for annual yield loss of 15% in chilli and 10% in bell pepper amounting to ₹744.90 million and ₹52.92 million, respectively (Walia and Chakraborty 2018).

#### 11.4.1.2 Symptoms

- Presence of galls on the roots (very small galls in chilli) (Fig. 11.9).
- Stunting of plant growth and rapid wilting of plants especially under dry growing conditions.
- Growth may be retarded and leaves may be chlorotic (Fig. 11.9).
- Death of seedling in nursery bed and non-survival of transplants in main field when seedlings are infected.
- Significant reduction in survival of plants and drastic reduction in flowering and fruit yield.

#### 11.4.1.3 Interaction with Other Pathogens

Bell pepper is prone to many soil-borne diseases among which the bacterial wilt (*Ralstonia solanacearum*) in combination with root-knot nematode (*M. incognita*) takes heavy toll every year all over the world (Naik 2004).

#### 11.4.1.4 Management

##### 11.4.1.4.1 Physical Methods

Rabbing nursery beds with castor husk at 7 kg/m$^2$ or soil solarization with 25μm LLDPE clear film for 15 days during summer proved beneficial.

**Fig. 11.9** *Left*—Chilli seedlings infected with root-knot nematode. *Right*—Severe galling on chilli roots

#### 11.4.1.4.2 Cultural Methods

Khan et al. (1975) found that rotation of chilli with bottle gourd, marigold, and spinach or fallowing of field reduced *M. incognita* population on chili. Growing of chili along with marigold gave highest degree of reduction in root-knot nematode population on chili roots, followed by onion, garlic, and asparagus (Trivedi and Tiagi 1984).

Khan and Ghadipur (2004) reported that inhibition in the reproduction of *M. incognita* and 21.3% increased fruit yield was obtained by in row application of fly ash at 0.6 kg/m$^2$.

Soil application of decaffeinated tea leaves effectively reduced root galling on chilli plant (Trivedi and Tiagi 1984).

#### 11.4.1.4.3 Chemical Methods

Bare root dip of chili seedlings in 0.05% Carbosulfan solution for 1 h was found to be effective.

Application of Carbofuran 0.2 g/m$^2$ to chili nursery improved plant stand and increased the seedling weight by controlling *M. incognita.*

#### 11.4.1.4.4 Biological Methods

Nursery treatment with either *Trichoderma viride* formulation (FYM incubated) at 2.5 g/m$^2$ or *Pseudomonas fluorescens* formulation at 20 g/m$^2$.

#### 11.4.1.4.5 Host Resistance

Recommended resistant varieties for root-knot nematode on Chilli include BC-21-2, Pusa Jwala, Agnirekha, Suryamukhi, Utkal Ava (BC- 14-2), CA- 2057, CA-2123, TC-52-18, CA-2123, C-26-4-5, and BR-31 (Anon 2015).

Bell pepper grafted on Graffito rootstock of *Capsicum annuum* had good compatibility, does not affect scion growth or production, and is resistant to *Phytophthora capsici* and root-knot nematodes (Morra and Billoto 2006).

#### 11.4.1.4.6 Integrated Methods

For effective management of root-knot nematode in chilli, nursery should be covered with polythene sheet during May/June for 15 days after giving light irrigation. At the time of sowing, application of Carbofuran 3G at 10 g/m$^2$ in the nursery prior to sowing was found to be effective. Also before transplanting, application of Neem cake at 500 kg/ha in the main field proved beneficial.

Effective management of root-knot nematode (*Meloidogyne* spp.) in chilli was obtained by solarization of soil under turf for 2 weeks during peak summer, soil incorporation of Carbofuran at 3.3 g/m$^2$ before sowing followed by bare root dip treatment of transplantable seedlings with Carbosulfan at 500 ppm at transplanting in the main field.

The demonstration trials on chili showed that nursery application of Carbofuran at 0.3 g a.i. /m$^2$ and root dip with Carbosulfan 500 ppm significantly reduced *M. incognita.*

Enhancement of the yield of the bell pepper crop and reduction in the number of nematodes in the roots and soil and root-galling index (*M. incognita*) under field conditions was achieved by integration of *P. fluorescens* and *P. chlamydosporia* in the nursery bed (Naik 2004).

In capsicum, the seedling stand was good where the combination of Neem-based *P. fluorescens* and *T. harzianum* was used (4.4), followed by seed treatment with *P. fluorescens* + soil application of *T. harzianum* (Naik 2004).

In greenhouse and field studies, Neem cake at 50 g/m$^2$ + Carbofuran 3G at 10 g/m$^2$ application in the pots as well as in the field was most effective in reducing nematode population as well as increasing the plant growth and yield in chilli (Thakur 2016).

Naik (2004) reported that combination of *T. harzianum* along with *P. fluorescens* increased bell pepper yield (3.320 kg/3 m$^2$ plot) followed by combination of *T. harzianum* and *P. chlamydosporia* (2.960 kg/3 m$^2$). The least gall index was present where the combination of *T. harzianum* and *P. fluorescens* was used (1.5) followed by *T. harzianum* + *P. chlamydosporia* (1.7) and *P. fluorescens* + *P. chlamydosporia* (1.7).

Capsicum seedlings raised in nursery treated with *P. fluorescens* at 50 g/kg seed (seed treatment) combined with nursery bed treatment with *P. chlamydosporia* at 50 g/m$^2$ when transplanted in field significantly reduced nematode population both in soil and roots, root galling (4.4 as compared to 8.1 in control), and increased root colonization by bioagents, propagule density in soil, parasitization of eggs, and fruit yield (5.5 kg/6 m$^2$ as compared to 4.0 kg/6 m$^2$ in control) (Rao et al. 2004).

### 11.4.2 Interaction of Root-Knot Nematode with Bacterial Wilt

#### 11.4.2.1 Interaction

Capsicum is prone to many soil-borne diseases among which the bacterial wilt (*Ralstonia solanacearum*) in combination with root-knot nematode (*Meloidogyne incognita*) takes heavy toll every year all over the world (Fig. 11.10) (Naik et al. 2003).

#### 11.4.2.2 Management

Naik et al. (2003) found that treatment of the nursery beds with Neem-based formulations of *T. harzianum* + *P. fluorescens* followed by seed treatment with talc-based *P. fluorescens* + nursery bed treatment with Neem-based *P. fluorescens* enhanced the seedling stand.

Enhancement of the bell pepper crop yield by 37% and significant reduction in bacterial wilt disease complex to the extent of 70% was obtained by integration of Neem-based formulations of *P. fluorescens* and *P. chlamydosporia/T. harzianum* at 40 g/m$^2$ in nursery beds and transplanting these seedlings in the main field (Rao et al. 2002). Combinations of the bioagents did not affect the colonization of the individual bioagents on the roots and hence the transplants carried the bioagents to the main field which controlled the wilt disease complex.

**Fig. 11.10** Root-knot nematode-bacterial wilt disease complex in sweet pepper

Integration of seed treatment with *P. fluorescens* and nursery bed treatment with *P. chlamydosporia* at 25 or 50 g/m$^2$ increased plant height, weight of seedlings, and root colonization with bioagents; and reduced root galling in bell pepper (Naik et al. 2003). Transplanting of such seedlings colonized with the bioagents carried the inoculum to the main field.

The seedlings raised in nursery beds treated with *P. chlamydosporia* at 25 or 50 g/m$^2$ + seed treatment with *P. fluorescens* when transplanted in the main field gave least root galling and increased fruit yield, colonization of roots with the bioagents, and parasitization of eggs by *P. chlamydosporia* (Naik et al. 2003).

Naik (2004) reported that enhancement of bell pepper fruit yield (3.320 kg/3 m$^2$ plot) was achieved by integration of *T. harzianum* along with *P. fluorescens* followed by integration of *T. harzianum* and *P. chlamydosporia* (2.960 kg/3 m$^2$). Integration of *T. harzianum* and *P. fluorescens*, followed by *T. harzianum* + *P. chlamydosporia*, and *P. fluorescens* + *P. chlamydosporia* gave the least root galling of 1.5, 1.7, and 1.7, respectively. Integration of *T. harzianum* + *P. fluorescens* followed by *T. harzianum* + *P. chlamydosporia* showed the percent bacterial wilt disease (*R. solanacearum)* incidence of 16.90 and 17.20, respectively.

Naik et al. (2003) reported that soil application of organically developed *Purpureocillium lilacinum* (50 g/m$^2$ of formulated product containing $10^6$ spores/g) along with pure culture of *Bacillus pumilis* ($10^8$ cfu/ml) to the nursery beds of bell pepper increased plant growth, root galling, and root colonization by the bioagents. Transplanting of such biological control agents colonized plants in the main field resulted in significant reduction of root galling and wilt incidence, and increased root colonization, propagule density in soil, egg parasitization, and fruit yield.

## 11.5 Conclusion

Plant parasitic nematodes constitute one of the major limiting factors for cultivation of solanaceous vegetable crops. Several nematode species are associated with solanaceous vegetable crops and few of them negatively impact yield. Severe yield losses have been reported in most regions where solanaceous vegetable crops are grown. In order to select suitable nematode management practices, correct diagnosis of the causal agent is prime importance. Overall, these nematode problems in solanaceous vegetable crops are better managed when integrated management practices are used, that is, regulatory methods (plant quarantine and seed certification), cultural practices (crop rotation, cover crops), genetic control, and ultimately by the use of nematicides. Therefore, for a sustainable cropping of solanaceous vegetable crops, growers, extension services, and researchers must consider these nematodes holistically, the impact they cause and whether these management practices are economically, environmentally, and technically sound.

Integrated nematode management could be tackled with the intelligent planning of nematode-suppressive crop sequences, summer plowing, organic manuring, clean cultivation, adjusting sowing time, water and irrigation management, and sensible use of nematicides. The increasing concerns on the ill effects of chemical pesticides on the environment have driven recent research interests on the use of several alternative strategies like botanicals, biopesticides, and cropping system research for management of nematodes. Some of the successes have been obtained for managing plant parasitic nematodes with neem-based formulations, fungal formulation of *Purpureocillium lilacinum* and *Trichoderma* spp., and bacterial formulations of *Pasteuria penetrans* and *Pseudomonas fluorescens*. Inconsistent efficacy of the biological control agents in different agro-climatic regions is the main reason for not adopting them widely under field scale. Integration of more than one option can be explored on the basis of their compatibility, economic viability and availability to the farmers. The major challenge ahead for the greater benefit of the farmers is the need for development of holistic approaches for managing field problems including nematodes induced disease complexes with other soil-borne pathogens. There is a need for development of economic, eco-friendly, and sustainable bio-intensive integrated nematode management strategies using all the available components.

## References

Abawi GS, Barker KR (1984) Effects of cultivar, soil temperature, and population levels of *Meloidogyne incognita* on root necrosis and *Fusarium* wilt of tomatoes. Phytopathology 74:433–438

Ahuja S (1983). Studies on the variability in infection of root-knot nematode *Meloidogyne incognita* on potato. In: Third All India nematology symposium, Dr. Yashwant Singh Parmar University of Horticulture and Forestry, Solan, p 4

Amalraj SFA, Rao GR (1992) Potato cyst nematode pathogens in the southern hills of India. Indian J Nematol 22:82–85

Anon (1993) Biennial report (1991–1993) of AICRP on plant parasitic nematodes with integrated approach for their control. Department of Nematology, Haryana Agricultural University, Hisar, 80 pp

Anon (2015) Biennial report, AICRP (nematodes). Palampur Centre, Himachal Pradesh. 47 pp

Anwar MA, Khan MR (2010) Biological control of root-knot of tomato with the application of Aspergillus niger isolates. National conference on innovations in nematological research for agricultural sustainability – challenges and a roadmap ahead. Tamil Nadu Agricultural University, Coimbatore, p 100

Ayyar PNK (1926) A preliminary note on the root nematode, *Heterodera radicicola* Mitter and its economic importance in South India. Sci Cult 22:391–393

Barua L, Bora BC (2008) Comparative efficacy of *Trichoderma harzianum* and *Pseudomonas fluorescens* against *Meloidogyne incognita* and Ralstonia solanacearum complex in brinjal. Indian J Nematol 38:86–89

Barua L, Bora BC (2009) Compatibility of *Trichoderma harzianum* and *Pseudomonas fluorescens* against *Meloidogyne incognita* and Ralstonia solanacearum complex on brinjal. Indian J. Nematol. 39:29–34

Bhagawati B, Goswami BK (2000) Interaction of *M. incognita* and Fusarium oxysporum f. sp. lycopersici on tomato. Indian J Nematol 30:93–94

Bhagawati B, Goswami BK, Singh CS (2000) Management of disease complex of tomato caused by *M. incognita* and *Fusarium oxysporum* f. sp. *lycopersici* through bioagents. Indian J Nematol 30:16–22

Bhatti DS, Jain RK (1977) Estimation of loss in okra, tomato and brinjal yield due to *Meloidogyne incognita*. Indian J Nematol 7:37–41

Darekar KS, Mahse NL (1988) Assessment of yield losses due to root-knot nematode, *Meloidogyne incognita* Race 3 in tomato, brinjal and bitter gourd. Int Nematol Netw Newsl 5(4):7–9

Dhawan SC, Singh S, Kamra A (2008) Biomanagement of root-knot nematode, *Meloidogyne incognita* by *Pochonia chlamydosporia*. Indian J Nematol 38:119–121

Ganguly S, Singh M, Ganguly AK (2010) Record of potato cyst nematode, *Globodera rostochiensis* and *G. pallida* in Shimla, Himachal Pradesh, India. Indian J Nematol 40:96–102

Goswami BK, Uma Rao, Singh S (1998) Potentiality of some fungal bioagents against root-knot nematode, *Meloidogyne incognita* infecting tomato. In: Proceedings of the first national symposium on pest management in horticultural crops, Bangalore, pp. 304–307

Gowda MT, Rai AB, Singh B (2017) Root-knot nematode: a threat to vegetable production and its management. Technical BulletinsNo 76, Indian Institute of Vegetarian Restaurants, Varanasi, 32 pp

Haider MG, Nath RP, Thakur SC, Ojha KL (1987) Interaction of *Meloidogyne incognita* and *Pseudomonas solanacearum* on tomato plants. *Indian*. J Nematol 17:174–176

Haseeb A, Kumar V, Shukla PK, Ahmad A (2006) Effect of different bioinoculants, organic amendments and pesticides on the management of *Meloidogyne incognita – Fusarium solani* disease complex on tomato cv. K-25. Indian J Nematol 36:65–69

Hazarika K (1990) Pathogenicity and management of *Meloidogyne incognita* on Brinjal (*Solanum melongena*). M. Sc. (Agriculture) thesis, Assam Agricultural University, Jorhat

Hussain Z, Bora BC (2008) Integrated management of *Meloidogyne incognita* and *Ralstonia solanacearum* complex in brinjal. Indian J Nematol 38:159–164

Jain RK, Paruthi IJ, Gupta DC (1990) Influence of intercropping on the incidence of root-knot nematode (*Meloidogyne javanica*) in brinjal. Indian J Nematol 20:236–238

Jatala P (1985) Biological control of nematodes. Annu Rev Phytopathol 24:453–489

Jones FGW (1961) The potato root eelworm *Heterodera rostochiensis* Woll. in India. Curr Sci 30:187

Kapkavalci G, Tuzel Y, Dura O, Oztekyn GB (2009) Effects of alternative control methods against *Meloidogyne incognita* in organic tomato production. Ekoloji 18(72):23–31

Khan MR, Ghadipur MH (2004) Management of root-knot disease of some solanaceous vegetables by soil application of fly ash. In: National Symposium on Paradigms in Nematological Research for Biodynamic Farming, University of Agricultural Sciences, Bangalore, p 92

Khan AM, Saxena SK, Siddiqi ZA, Upadhyay RS (1975) Control of nematodes by crop rotation. Indian J Nematol 5:214–221

Krishna Prasad KS (1986) Potato nematodes. In: Swarup G, Dasgupta DR (eds) Plant parasitic nematodes in India. Indian Agricultural Research Institute, New Delhi

Mani MP, Rajeswari S, Sivakumar C (1998) Management of the potato cyst nematodes, *Globodera* spp. through plant rhizosphere bacterium *Pseudomonas fluorescens* Migula. J Biol Control 12:131–134

Morra L, Billoto M (2006) Evaluation of new rootstocks for resistance to soil borne pathogens and productive behavior of pepper (*Capsicum annuum* L.). J Hort Sci Biotechnol 81:518–524

Nagesh M, Hussaini SS, Ramanujam B, Chidanandaswamy BS (2006) Management of *Meloidogyne incognita* and *Fusarium oxysporum* f. sp. *lycopersici* wilt complex using antagonistic fungi in tomato. Nematol Medit 34:63–68

Naik D (2004) Biotechnological approaches for the management of wilt disease complex in Capsicum (*Capsicum annum* L.) and Eggplant (*Solanum melongena*) with special emphasis on biological control. Ph. D. thesis, Kuvempu University, Shimoga

Naik D, Rao MS, Shylaja M et al (2003). Management of wilt disease complex in capsicum using *Bacillus pumilus* and *Paecilomyces lilacinus*. In: Proceedings of Int. plant growth promoting rhizobacteria workshop, Calicut, India, pp 276–281

Netscher C (1983) Control of *Meloidogyne incognita* in vegetable production by crop rotation in Ivory Coast. Acta Hort 152:219–225

Nirula KK, Bassi KK (1965) Thermotherapy for root-knot nematode, *Meloidogyne incognita* in potato tubers. Indian Potato J 7:9–11

Oostenbrink M (1972) Evaluation and integration of nematode control methods. In: Webster JM (ed) Economic nematology. Academic Press, New York, pp 497–514

Pani AK, Das SN (1972) Studies on etiological complexes in plant disease. I. Association of root-knot nematodes in bacterial wilt of tomato. J Res Orissa Univ Agri Technol 2:54–59

Parvatha Reddy P, Singh DB (1981) Assessment of avoidable yield loss in okra, brinjal, French bean and cowpea due to root-knot nematodes. In: Third international symposium on plant pathology, New Delhi, pp 93–94

Parvatha Reddy P, Nagesh M, Devappa V (1997) Effect of integration of *Pasteuria penetrans*, *Paecilomyces lilacinus* and neem cake for the management of root-knot nematodes infecting tomato. Pest Mangmt Hort Ecosyst 3:100–104

Prasad KSK (1989) Nematological problems and progress of research on potato in India. In: Fourth group meeting on nematology problems of plantation crops, University of Agricultural Sciences, Bangalore

Prasad KSK, Raj D, Sharma RK (1983) Effect of planting time on *Meloidogyne incognita* infesting potato at Shimla Hills. In: Third all India nematology symposium, Yashwant Singh Parmar University of Horticulture and Forestry, Solan, pp 5–6

Ramana KV, Mohandas C (1988) Occurrence of potato cyst nematode *Globodera pallida* in Kerala. Indian J Nematol 18:141

Rao MS, Parvatha Reddy P (2001) Control of *Meloidogyne incognita* on eggplant using *Glomus mosseae* integrated with *Paecilomyces lilacinus* and neem cake. Nematol Medit 29:153–157

Rao MS, Parvatha Reddy P, Nagesh M (1998) Integrated management of *Meloidogyne incognita* on tomato using *Verticillium chlamydosporium* and *Pasteuria penetrans*. Pest Mangmt Hort Ecosyst 4:32–35

Rao MS, Naik D, Shylaja M, Parvatha Reddy P (2002) Prospects for the management of the root-knot nematode and bacterial wilt disease complex in capsicum by integration of fungal and bacterial biocontrol agents. In: Proceedings of international conference on vegetables, Bangalore, pp 347–351

Rao MS, Naik D, Shylaja M (2003) Management of *Meloidogyne incognita* on egg-plant using a formulation of Pochonia chlamydosporia Zare et al. (Verticillium chlamydosporium Goddard). Pest Mangmt Hort Ecosyst 9:71–76

Rao MS, Naik D, Shylaja M (2004) Bio-intensive management of root-knot nematodes on bell pepper using *Pochonia chlamydosporia* and *Pseudomonas fluorescens*. Nematol Medit 32:159–163

Rao MS, Ramachandran N, Sowmya DS et al (2009) Biological control of nematode induced disease complex in certain vegetable crops. In: International Conference on Horticultural, Bangalore

Ravichandra NG, Reddy BMR (2008) Efficacy of *Pasteuria penetrans* in the management of *Meloidogyne incognita* infecting tomato. Indian J Nematol 38:172–175

Ravichandra NG, Krishnappa K, Setty KGH (1990) Interaction of *Meloidogyne javanica* and race 1, race 2, race 3 of *Meloidogyne incognita* with Pseudomonas solanacearum on a few brinjal lines. Indian J Nematol 20:138–147

Reddy DDR (1985) Analysis of crop losses in tomato due to *Meloidogyne incognita*. Indian J Nematol 15:55–59

Roberts PA, Magyarosy AC, Mathews W, May DW (1988) Effect of metham sodium applied by drip irrigation on root-knot nematodes, *Pythium ultimum* and *Fusarium* sp. in soil and carrot and tomato roots. Plant Dis 72:213–217

Saikia J, Borah A, Bora BC (2007) Efficacy of organic amendments in combination with the nematicide for management of *Meloidogyne incognita* in brinjal. Indian J Nematol 37 (1):179–181

Savitha RM, Lingaraju S (1996) Interaction of root-knot nematode and wilt bacterium on brinjal. Karnataka J Agric Sci 9:657–660

Shanthi A (2010) Management of root-knot nematode, *Meloidogyne incognita* in tomato using bioinoculants. In: National conference on innovations in nematological research for agri. sustainability – challenges and a roadmap ahead. Tamil Nadu Agricultural University, Coimbatore, p 93

Sharma HK, Gill JS (1992) Prevalence of races in root-knot nematode, *Meloidogyne incognita* in India. Indian J Nematol 22:43–49

Sheela MS, Nisha MS (2004) Impact of biological control agents for the management of root-knot nematode in brinjal. In: National Symposium on Paradigms in Nematological Research for Biodynamic Farming, University of Agricultural Sciences, Bangalore, pp 63–64

Sheela MS, Nisha MS, Mohandas C (2002) Impact of trap cropping and crop rotation for the management of root-knot nematode in vegetables. In: Proceedings of the international conference on vegetables, Bangalore, pp 344–346

Siddiqui IA, Ehteshamul-Haque S (2001) Suppression of the root rot–root knot disease complex by *Pseudomonas aeruginosa* in tomato: the influence of inoculum density, nematode populations, moisture and other plant-associated bacteria. Plant Soil 237:81–89

Siddiqui IA, Qureshi SA, Sultana V et al (2000) Biological control of root rot-root knot disease complex of tomato. Plant Soil 227:163–169

Singh D (1991) Studies on the effect of tagetes species on *Meloidogyne javanica* infectivity in brinjal (*Solanum melongena*). Ph. D. thesis, Haryana Agricultural University, Hisar, 83 pp

Sitaramaiah K, Naidu H (2003) Plant parasitic nematodes-management opportunities and challenges. In: Trivedi PC (ed) Nematode management in plants. Scientific Publishers, Jodhpur, pp 57–80

Son SH, Khan Z, Kim SG, Kim YH (2009) Plant growth-promoting rhizobacteria, *Paenibacillus polymyxa* and *Paenibacillus lentimorbus* suppress disease complex caused by root-knot nematode and Fusarium wilt fungus. J Appl Microbiol 107:524–532

Thakur R (2016) Management of root-knot nematode (*Meloidogyne incognita* Kofoid & White) in Chilli (*Capsicum annuum* L.). M. Sc. Thesis, Dr. Yashwant Singh Parmar University of Horticulture and Forestry, Solan, Himachal Pradesh, 60 pp

Thirumalachar MJ (1951) Root-knot nematode on potato tubers in Shimla. Curr Sci 20:104

Trivedi PC, Tiagi A (1984) Effect of some plants on *Meloidogyne incognita* population in chilli soil. Indian J Bot Sci 63:416–419

Umamaheswari, R., Somasekhar, N. and Joseph, T.A. 2010. Evaluation of biocontrol agents on potato cyst nematodes, *Globodera rostochiensis* and *G. pallida* infesting potato. In: National conference on innovations in nematological research for agricultural sustainability – challenges and a roadmap ahead. Tamil Nadu Agricultural University, Coimbatore, p 86

Umamaheswari R, Somasekhar N, Manorama K, Joseph TA (2012) Eco-friendly management of potato cyst nematodes in the Nilgiris of Tamil Nadu. Potato J 39(2):185–190

Walia RK, Chakraborty PK (eds) (2018) Nematodes problems of crops in India. ICAR-All India Coordinated Research Project on Nematodes in Agriculture, New Delhi. 402 pp

Zaki FA (1998) Biological control of *Meloidogyne javanica* in tomato by *Paecilomyces lilacinus* and castor leaves. *Indian.* J Nematol 28:132–139

Zaki FA, Bhatti DS (1991) Effect of castor *Ricinus communis* and the biocontrol fungus, *Paecilomyces lilacinus* on *Meloidogyne javanica*. Nematologica 36:114–122

# Malvaceous Vegetable Crops 12

**Abstract**

The production of okra throughout the globe is threatened by root-knot nematodes (*Meloidogyne incognita* and *M. javanica*). Nematodes interact with *Fusarium* wilt and *Rhizoctonia* root rot pathogens in causing disease complexes. In this chapter, crop losses, symptoms, biology and life cycle, spread, and management using physical, cultural, chemical, biological methods, host resistance, and integrated methods for root-knot nematodes are discussed.

**Keywords**

*Meloidogyne incognita* · M. javanica · Symptoms · Biology · Management

## 12.1 Okra, *Abelmoschus esculentus*

Root-knot nematodes *Meloidogyne* spp. are the major nematode problems on okra which interact with *Fusarium* wilt and *Rhizoctonia* root rot in causing disease complexes.

### 12.1.1 Root-Knot Nematodes, *Meloidogyne incognita, M. javanica*

#### 12.1.1.1 Crop Losses

*M. incognita* was responsible for 28.08–90.90% loss in fruit yield of okra (Bhatti and Jain 1977; Parvatha Reddy and Singh 1981), while *M. javanica* caused 20.20–41.20% loss in yield (Jain et al. 1986).

Root-knot nematodes are responsible for 19.5% annual yield loss in okra amounting to Rs. 2480.86 million in India (Walia and Chakraborty 2018).



P. P. Reddy, *Nematode Diseases of Crops and their Management*,
https://doi.org/10.1007/978-981-16-3242-6_12

#### 12.1.1.2 Symptoms

*Meloidogyne* species cause similar upper ground symptoms (stunting, leaf yellowing, and wilting) as in other vegetable crops. On roots prominent galls are formed which are large in size (Fig. 12.1) and easily attacked by secondary pathogens to form disease complexes.

#### 12.1.1.3 Biology and Life Cycle

Root-knot nematodes are soil-borne and sedentary endoparasites. After penetration of the second-stage larvae in the root, they feed on the vascular tissues and develop endoparasitically. Adult females are sac-like and lay eggs in masses on the surface of galls, or in advanced stages inside the compound galls, while the vermiform males leave the roots and come out in soil. During summer months, the nematode takes about 25–30 days to complete its life cycle.

#### 12.1.1.4 Spread

Nematodes disseminate through infected transplanting material, irrigation water, and agricultural implements.

#### 12.1.1.5 Management

##### 12.1.1.5.1 Physical Methods

There was 71.8% reduction in nematode population in okra by soil solarization alone while fallowing alone led to 56.2% reduction.

##### 12.1.1.5.2 Cultural Methods

Organic Amendments

Application of Neem/Eupatorium leaves at 15 t/ha 2 weeks before sowing proved beneficial for controlling root-knot and reniform nematodes.Application of saw dust/

**Fig. 12.1** Heavy galling on okra roots due to root-knot nematode infestation

paddy husk at 500 g per plant/Neem leaves/Eupatorium leaves @ 250 g/plant in the basins 3 weeks prior to planting and watering daily proved beneficial for managing root-knot nematode.Application of Neem cake at 1 t/ha at sowing in the main field was found to be effective.

In *kharif* okra, application of Press mud at 15 t/ ha (broadcast) 1 week prior to seeding proved effective.

#### Green Manuring

Use of *toria* crop as green manure after okra and before tomato was observed to keep the nematode infestation low in the okra-*toria*-tomato rotation.

#### Crop Rotation

Rotation with nonhost crops like cereals was found to be effective.

Okra-sweet potato cv. Shree Bhadra-chili sequence is effective in minimizing the population of *M. incognita* and increasing the okra yield to a tune of 21–22%.

The cropping sequence of okra-wheat-cluster bean decreased soil population of root-knot nematode by 52–56.8% in comparison to 57.3% increase in susceptible check (okra-brinjal-okra). There was a 72.3% decrease in final nematode population by growing a combination of onion/cluster bean/resistant tomato variety, Hisar Lalit for 3 years.

Okra-rice-fallow or okra-cucumber-mustard cropping sequence suppressed soil population of root-knot nematode (*M. incognita*) in vegetables.

Cauliflower-onion-okra has been recommended in root-knot nematode infested fields and is economically beneficial.

Follow the following rotations:

- First year: Planting of okra in *kharif* and tomato in *rabi* season.
- Second year: Root-knot resistant cowpea variety "Anand Vegetable Cowpea 1" in *kharif* and root-knot resistant tomato variety "Hisar Lalit" in *rabi*.
- Third year: okra in *kharif* and tomato in *rabi*.

#### Biofumigation

Biofumigation using crop residues of cabbage and cauliflower at 2.5 kg/m$^2$ reduced the nematode population and increased fruit yield in okra.

Incorporation of cabbage leaves at 5 t/ha 15 days before sowing is effective against root-knot nematode in okra (Anon. 2017).

### 12.1.1.5.3 Chemical Methods

Seed treatment with Carbosulfan 25 ST at 3% w/w proved beneficial.

Seed dressing of okra with Carbosulfan, Benfuracarb (Oncol 50 WP), and Triazophos at 1.5 and 3% inhibited the gall index and increased the yield.

Seed treatment with Abamectin 40% FS at 0.4 g a.i. /kg seed achieved highest yield of okra (6.87 t/ha) and proved better that Carbosulfan 25DS at 3% a.i. w/w with reference to the gall index and population of *M. incognita*.

Fenamiphos, Quinalphos, Carbosulfan, and Monocrotophos, each at 500 and 1000 ppm as seed dip for 12 h lowered the root-knot index in okra.

Soaking of seeds of okra cv. MDU 1 for 12 h in 0.1% solutions of Monocrotophos and Phosphamidon caused higher shoot weight of 30.71 and 30.69 g/plant and lower root-knot index of 2.33 and 2.40, respectively, when compared to control which recorded a shoot weight of 19.12 g/plant and root-knot index of 4.83.

Root-knot nematodes in okra can be effectively managed by soil application of Carbofuran 3G at 1 kg a.i. /ha (Choudhury et al. 2011).

Application of Carbofuran 3G in two split doses of 1.5 kg +1.5 kg a.i./ha at 30 day interval in okra was observed to be more effective in reducing root-knot infestation as compared to single dose application of Carbofuran 3G at the time of sowing.

#### 12.1.1.5.4 Biological Methods

Seed treatment with *Bacillus macerans* at 3% w/w of seed was found to be effective for the management of root-knot nematodes.

Application of *Purpureocillium lilacinum at* 20 g/L as soil drench proved beneficial for managing root-knot and reniform nematodes.

Soil application of *Trichoderma viride* at 2.5 kg/ha at the time of sowing successfully managed root-knot nematode *Meloidogyne* spp. and increased the okra yield by 16.4–18.7%.

#### 12.1.1.5.5 Host Resistance

Varieties found resistant against *Meloidogyne* spp. are Okra - D-1-87-5, HOE-101, and Red Wonder.

#### 12.1.1.5.6 Integrated Methods

Soaking of okra seeds in 10% Castor cake suspension mixed with spores of *P. lilacinum* ($1.5 \times 10^6$ spores/ml) for 30 min. and sowing in soil drenched with 10% Castor cake suspension at 20 liters/6 $m^2$ was effective in reducing root galling (4.7 as compared to 9.2 in control), final nematode population of *M. incognita* and increasing the fruit yield (6.8 kg/6 $m^2$ as compared to 4.2 kg/6 $m^2$ in control), root colonization, propagule density in soil, and parasitization of eggs by *P. lilacinum* (Rao et al. 1997).

*Bacillus circulans* and *B. macerans* and their combinations were effective for the management of root-knot nematodes and increased the fruit yield ranging from 27 to 31%.

Increase in fruit yield and reduction in nematode population is obtained by seed treatment with the biopesticide formulation of *P. lilacinum/Trichoderma harzianum* ($2 \times 10^6$ cfu/g) at 20 g/kg seed + soil application of *P. lilacinum/T. harzianum* ($2 \times 10^6$ cfu/g) at 2.5 kg along with 2.5 tons of FYM/ha prior to sowing of okra.

Seed dressing of okra with Carbosulfan (25ST) 3% w/w with summer plowing effectively control root-knot nematode and increased yield.

Seed treatment with *B. macerans* at 3% w/w of seeds + drenching pits with 2% bioagent solution may be used as sticking agents for seed treatment in the endemic areas of root-knot infestation.

Increase in okra fruit yield and reduction in root-knot nematode population (*M. incognita*) is achieved by soil incorporation of *P. lilacinum* at 2.5 kg/ha in 2.5 t FYM.

In okra, summer plowing + seed treatment with Carbosulfan at 3% a.i. + Carbofuran application at 1 kg a.i./ha led to 76–79% decrease in nematode population and 35.1% higher yield over untreated check.

Solarization of nematode infested field + seed dressing of direct sown crops with Carbosulfan 25 DS at 3% a.i. (w/w) was found to be effective for management of root-knot (*Meloidogyne* spp.), reniform (*Rotylenchulus reniformis*), and lesion nematodes (*Pratylenchus* spp.) in okra.

Soil solarization for 15 days in summer and application of Neem cake at 200 kg/ha or seed dressing of okra with Carbosulfon (25ST) 3% w/w is effective in the management of *M. incognita* and in getting higher yield.

Seed treatment with *T. harzianum* at 20 g/kg of seed + main field application of *T. harzianum* at 2.5 kg along with 2.5 tons of FYM/ha showed 18.8% increase in okra yield.

Seed treatment with *P. lilacinum* at 20 g/kg ($2 \times 10^6$ cfu/g) seed in combination with its application at 2.5 kg along with 2.5 tons of FYM/ha prior to sowing.

Reduction in nematode population in okra was obtained with seed treatment of *P. lilacinum* at 5 ml/kg followed by soil application of Vermicompost at 2.5 t/ha enriched with *P. lilacinum* at 10 ml/kg.

In okra cv. Pusa Sawani the integration of seed treatment with Carbosulfan 3% with either mulching with polythene sheet after plowing and exposure for 20 days or plowing and exposure for 20 days reduced the nematode population by 57.8 and 48.6% and also increased the fruit yield by 46% and 34%, respectively.

The highest yield with greater suppression of soil nematode population was obtained in the treatment with *P. lilacinum* at 2.5 kg in 2.5 tons FYM and followed by soil application with *Pseudomonas fluorescens* at 2.5 kg/ha, *Pochonia chlamydosporia* at 2.5 kg/ha, and *Trichoderma harzianum* at 2.5 kg/ha.

### 12.1.2 Interaction of Root-Knot Nematode with *Fusarium* Wilt

#### 12.1.2.1 Interaction

*Meloidogyne incognita-Fusarium oxysporum* f. sp. *vasinfectum* disease complex has been considered important on many crops including okra leading to reduction in its productivity.

#### 12.1.2.2 Management

Highest suppression of root colonization of the fungus (*F. o.* f. sp. *vasinfectum*) was achieved in plants treated with Carbendazim (Bavistin) (5.0) followed by Neem seed powder (7.5), and Neem cake (12.5). The greatest suppression of root-knot development was achieved by Carbofuran (0.25) followed by *T. harzianum* (0.75), and Neem seed powder (0.90). In general, Carbofuran and *T. harzianum* were highly effective against the nematodes, Carbendazim against the fungus, and Neem seed powder and Neem cake against both the pathogens (Abuzar and Haseeb 2006).

The damage caused by *M. incognita* and *F. oxysporum* f. sp. *vasinfectum* was reduced to the tune of 68 and 57%, respectively, was achieved by soil incorporation of deoiled Neem cake enriched with *Pseudomonas fluorescens* ($2 \times 10^6$ cfu/g) or *Pochonia chlamydosporia* ($2 \times 10^6$ cfu/g) at 25 g/m$^2$. This treatment also increased the yield of okra fruits by 24% under field conditions (Chaya et al. 2010).

### 12.1.3 Interaction of Root-Knot Nematode with *Rhizoctonia* Root Rot

#### 12.1.3.1 Interaction

*M. incognita* and *R. bataticola* when inoculated simultaneously in soil reduce the germination of seeds in okra (Chhabra and Sharma 1981). Combined attack of both these pathogens cause significantly greater damage to the crop than that of the damage caused by either pathogen alone (Bhagawati et al. 2007). If the nematode preceded the fungus by 3 weeks, the root rot was more severe and appeared within 14–21 days. *M. incognita* predisposed roots to *R. solani* which resulted in severe root rot and subsequent plant death.

#### 12.1.3.2 Management

Seed treatment of okra with Carbofuran at 3% a.i. (w/w) or Carbendazim 0.2% a.i. (w/w) + Difolaton 0.25% a.i. (w/w) was effective in the control of reniform nematode/root-knot nematode-Rhizoctonia wilt complex.

Seed treatment with Carbofuran at 3% a.i. (w/w) + Difolaton 0.25% a.i. (w/w) controlled reniform nematode/root-knot nematode *Rhizoctonia* wilt complex.

*Purpureocillium lilacinum* can be effectively used against *R. solani-M. incognita* complex of okra (Shahzad and Gaffer 1989).

Chaitali et al. (2003) observed that *Trichoderma viride* combined with Neem cake controlled the disease complex better than *T. viride* combined with Groundnut cake in okra.

Bhagawati et al. (2009) reported that seed treatment with both Carbosulfan 25 SD + Carbendazim 50 WP and *T. harzianum* + *Pseudomonas fluorescens* reduced the pre-emergence damping-off by 5.01% and 6.26% and post-emergence damping-off by 7.73% and 7.73%, respectively.

## 12.2 Conclusion

Of all the pathogens, root-knot nematodes (*Meloidogyne* spp.) are among the most serious sedentary endoparasitic nematodes worldwide. Although chemical usage is a common and popular practice to manage nematodes, search for better options has been emphasized. A viable approach for the management of plant parasitic nematodes is the adoption of sequential cropping system using suppressive crops that have been tested for host status under local nematode populations. There is a need for development of most effective and economical method, that is, the use of resistant cultivars. An economical practice to be adopted by the farmers for gradual decrease in nematode populations in the infested fields is by growing resistant cultivars of crops in crop rotations. Integrated approaches to root-knot nematodes management be further developed using resistant/moderately resistant crop varieties combined with sequential cropping. They are more likely to be adopted by smallholder farmers due to the multiple benefits associated with them.

## References


Abuzar S, Haseeb A (2006) Efficacy of carbofuran, bavistin, neem, *Trichoderma harzianum* and *Aspergillus Niger* against *Meloidogyne incognita* and *Fusarium oxysporum* f. sp. *vasinfectum* disease complex on okra. Indian J Nematol 36:299–301

Anon. (2017) Biennial report (2015–2016). AICRP on Nematode pests of crops & their control, Assam Agri. Univ., Jorhat

Bhagawati B, Choudhury BN, Sinha AK (2009) Management of *Meloidogyne incognita-Rhizoctonia solani* complex on okra through bioagents. Indian J. Nematol 39:156–161

Bhagawati B, Das BC, Sinha AK (2007) Interaction of *Meloidogyne incognita* and *Rhizoctonia solani* on okra. Ann Plant Prot Sci 15:533–535

Bhatti DS, Jain RK (1977) Estimation of loss in okra, tomato and brinjal yield due to *Meloidogyne incognita*. Indian J Nematol 7:37–41

Chaitali L, Singh S, Goswami BK (2003) Effect of cakes with *Trichoderma viride* for the management of disease complex caused by *Rhizoctonia bataticola* and *Meloidogyne incognita* on okra. Ann Plant Prot Sci 11:178–180

Chaya MK, Rao MS, Ramachandran N (2010) Biomanagement of disease complex in okra (*Abelmoschus esculentus* L. Moench) using microbial antagonists. In: National Conf. On innovations in Nematol. Res. For Agri. Sustainability – challenges and a roadmap ahead. Tamil Nadu Agri Univ, Coimbatore, pp 96–97

Chhabra HK, Sharma JK (1981) Combined effect of *Meloidogyne incognita* and *Rhizoctonia bataticola* on pre-emergence damping-off of okra and brinjal. Sci Cult 47:256–257

Choudhury BN, Bhagawati B, Bora BC (2011) Three decades of nematology in Assam (1977–2011). AICRP on Plant Parasitic Nematodes with Integrated Approach for their Control, Dept. of Nematol., Assam Agri. Univ., Jorhat, 49 pp

Jain RK, Paruthi IJ, Gupta DC, Dhankar BS (1986) Appraisal of losses due to root-knot nematode, *Meloidogyne javanica* on okra under field conditions. Trop Pest Mangmt 32:341–342

Parvatha Reddy P, Singh DB (1981) Assessment of avoidable yield loss in okra, brinjal, French bean and cowpea due to root-knot nematodes. In: *Third Int. Symp. Plant Pathol.*, New Delhi, pp 93–94

Rao MS, Parvatha Reddy P, Nagesh M (1997) Integrated management of *Meloidogyne incognita* on okra by castor cake suspension and *Paecilomyces lilacinus*. Nematol Medit 25:17–19

Shahzad S, Gaffer A (1989) Use of *Paecilomyces lilacinus* in the control of root rot and root-knot complex in okra and mungbean. Pakistan J Nematol 7:47–53

Walia RK, Chakraborty PK (eds) (2018) Nematodes problems of crops in India. ICAR-All India Coordinated Research Project on Nematodes in Agriculture, New Delhi, 402 pp

# 13 Leguminous Vegetable Crops

**Abstract**

Root-knot (*Meloidogyne incognita* and *M. javanica*) and the reniform (*Rotylenchulus reniformis*) nematodes are the major problems on leguminous vegetables like French bean, cowpea, and pea. Nematodes interact with soil-borne fungal pathogens like *Fusarium solani* and *Rhizoctonia solani* in causing wilt/root rot disease complexes in leguminous vegetables. In this chapter, Crop losses, symptoms and management, using physical, cultural, chemical, biological methods, host resistance, and integrated methods for root-knot nematodes are discussed.

**Keywords**

*Meloidogyne* spp. · *Fusarium solani* · *Rhizoctonia solani* · Symptoms · Management

## 13.1 French Bean, *Phaseolus vulgaris*

Root-knot and reniform nematodes are the major limiting factors in successful production of French bean crop.

### 13.1.1 Root-Knot Nematodes, *Meloidogyne incognita, M. javanica*

The root-knot nematode *M. arenaria* was recorded on beans for the first time from Bihar, India (Lal and Ansari 1960).


P. P. Reddy, *Nematode Diseases of Crops and their Management*,
https://doi.org/10.1007/978-981-16-3242-6_13

#### 13.1.1.1 Crop Losses

*M. incognita* was responsible for 18.77% avoidable yield loss in French bean cv. contender (Mohapatra 2015). Crop losses have been estimated by various workers varying from 19.38% to 43.48% loss in pod yield (Parvatha Reddy 2008). *M. incognita* was responsible for 19.38% to 43.48% loss in pod yield of French bean (Das 1994; Parvatha Reddy and Singh 1981), while *M. javanica* caused 30–40% loss in yield (Sharma et al. 2002).

#### 13.1.1.2 Symptoms

Above-ground symptoms of severe root-knot infestation include patches of chlorotic, stunted, necrotic, or wilted plants. Earlier wilting was noticed when infested plants were under temperature or moisture stress compared to other plants.

The characteristic under-ground symptom of infected plants is the formation of root galls stimulated by feeding of nematodes. Root-galling and nematode reproduction harms the plant because they both take away nutrients and photosynthates from the developing top growth and harvestable bean crop. Severely galled roots can be short and thick (Fig. 13.1).

#### 13.1.1.3 Interaction with Other Pathogens

*M. incognita* in association with soil-borne fungi such as *Rhizoctonia solani* and *Fusarium solani* was responsible for root-rot/wilt disease complex in French bean. Considerable increase in root-rot incidence due to *R. solani* occurred due to root-knot nematode infection (Parvatha Reddy et al. 1979; Singh et al. 1981b).

**Fig. 13.1** French bean roots severely infected with the root-knot nematode, *Meloidogyne incognita*

#### 13.1.1.4 Management

##### 13.1.1.4.1 Physical Methods

Root-knot nematodes can be effectively controlled by a 4–8 weeks solarization, assuming that the land will remain non-cropped during summer.

Rabbing by rice husk or sawdust before 10–15 days of sowing is effective in reducing root-knot nematode infestation in French bean (Ahmed and Choudhury 2004c).

##### 13.1.1.4.2 Cultural Methods

Cropping sequence of French bean-*ahu* rice-sesame is effective in controlling *M. incognita* in French bean (Ahmed and Choudhury 2004a).

The root-knot nematode damage in bean crop was reduced by rotation with certain cover/green manure crops like crotalaria, hairy indigo, joint vetch, pigeon pea, and velvet bean.

Rotation of root-knot nematode resistant processing tomatoes with common bean was successful in preventing root-knot damage. This rotation worked because root-knot nematodes were unable to reproduce on the resistant tomatoes, thereby reducing their numbers to below damaging levels. A similar reduction in root-knot nematodes can be achieved by planting beans after a crop of root-knot nematode-resistant cotton.

Reduction in root-knot nematode, *M. incoginita* reproduction was obtained when French bean was intercropped with Chenopodium, bell pepper, maize, marigold, and sesame. Maize, marigold, and sesame did not allow multiplication of the nematode and reduced the population by 0.26, 0.61, and 0.68 times, respectively.

Intercropping of French bean with finger millet, chili, and groundnut helps in the management of *M. incognita* Race 2 (Ramappa 1988). Growing of chili or groundnut or finger millet in rotation with French bean significantly reduced the root-knot nematode population (Ramappa 1988).

Early planting during the end of winter or early in the spring (temperatures below 15 °C) would help bean plants to escape from nematode infection and establish well. Yield would increase because the large root system can withstand the damage caused by delayed nematode invasion. Moreover, the beans would be harvested by the end of spring or early in summer, thus limiting the number of generations produced (often to only one) and overall population densities.

Ahmed and Choudhury (2004b) reported that the reduction in root galling (64.4 compared to 122.9 in control), number of egg masses (16.0 compared to 44.4 in control), final nematode population (253 compared to 437 in control), and in increasing plant growth, and yield (3.375 t/ha compared to 0.715 t/ha in control) by spot application of neem cake at 16.6 g/spot 15 days before sowing.

##### 13.1.1.4.3 Chemical Methods

Parvatha Reddy (1984) reported that the effective management of the root-knot nematode, *M. incognita* infecting French bean and enhancement of pod yield was obtained by seed treatment with Carbofuran and Fenamiphos each at 1%

concentration was effective in controlling and in increasing their pod yields. Seed treatment with Oxamyl at 3–10% prevented development of *M. incognita*.

Root-knot nematodes on French bean can be effectively managed by soil application of Carbofuran 3G at 1 kg a.i./ha (Choudhury et al. 2011). Soil application of Carbofuran 3G at 2 kg a.i./ha was found to be most effective in reducing the root galls and egg masses by 76.50% and increasing the pod yield by 50.81% with benefit:cost ratio of 10.74 (Mahse 2012).

Effective management of the population of root-knot nematode and French bean pod yield enhancement to the extent of 100% to 112% was achieved by soil incorporation of Fenamiphos, Ethoprophos, and Carbofuran each at 1 to 2 kg a.i. per ha.

##### 13.1.1.4.4 Host Resistance

French bean cvs. Banat, Blue Lake Stringless, Bountiful Flat, Brown Beauty, Cambridge Countess, Gallaroy, Kenya-3, Pinto W5-114, Seafarer, and Sutton's Masterpiece were reported to be resistant to *M. incognita* (Singh et al. 1981a). French bean cv Contender showed moderately resistant reaction to *M. javanica*.

### 13.1.2 Reniform Nematode, *Rotylenchulus reniformis*

#### 13.1.2.1 Symptoms

General symptoms include reduced root systems, leaf chlorosis, overall stunting of host plants, and reduced yields and plant longevity. When viewed under a dissecting microscope, the female nematodes and their egg masses are seen on the plant roots.

#### 13.1.2.2 Interaction with Other Pathogens

Vadhera et al. (1995) demonstrated that the incidence of root-rot was highest in simultaneous inoculation of *R. reniformis* and *F. solani* on beans. The combination treatment resulted in maximum reduction in plant growth parameters.

#### 13.1.2.3 Management

##### 13.1.2.3.1 Cultural Methods

Effective management of the reniform nematode in French bean was obtained with soil application of Karanj and Neem oil cakes at 2 t/ha.

##### 13.1.2.3.2 Chemical Methods

Satisfactory nematode control has been obtained with integration of six foliar sprays of Oxamyl at 0.56 kg a.i./ha along with soil application of 2.24 kg a.i./ha of the same chemical. Furrow application of 2.5 kg a.i./ha of Carbofuran was also found to be effective.

## 13.2 Cowpea, *Vigna unguiculata*

The major limiting factors in successful production of cowpea crop include root-knot, reniform, and pigeon pea cyst nematodes.

### 13.2.1 Root-Knot Nematodes, *Meloidogyne incognita, M. javanica*

#### 13.2.1.1 Crop Losses

Parvatha Reddy and Singh (1981) reported that *M. incognita* was responsible for 28.60% loss in pod yield of cowpea.

#### 13.2.1.2 Symptoms

- Root-knot infected field showed stunting of plants, yellowing of leaves, and patchy growth (Fig. 13.2).
- Heavy galling on roots (Fig. 13.2).
- Poor growth of seedlings in nursery.
- Yield considerably reduced.

#### 13.2.1.3 Interaction with Other Pathogens

The presence of heavy infestations of *M. javanica* on a cowpea cv. tolerant to wilt incited by *Fusarium oxysporum* f. sp. *tracheiphilum* caused increased wilting when compared to the susceptible cultivar.

#### 13.2.1.4 Management

##### 13.2.1.4.1 Cultural Methods

Application of Neem or Eupatorium leaves at 15 t/ha, 2 weeks before sowing was found to be effective for the management of root-knot and reniform nematodes.

**Fig. 13.2** *Left*—Field infestation of cowpea with root-knot nematodes showing leaf yellowing symptoms. *Right*—Cowpea roots severely infected with root-knot nematode

The crop rotation sequence recommended for the management of root-knot nematode includes cowpea-graminaceous crops/*Crotalaria spectabilis*. Populations of root-knot decreased greatly when compared to fallowed plots when *C. spectabilis* was grown as a weed free cover crop. Effective management of root-knot nematode was obtained by mulching of soil with cowpea foliage. One crop of rice was sufficient to effectively reduce root-knot nematode infestations in succeeding susceptible cowpea crop which is superior over rotation with non-host crops.

*M. javanica* populations were lower when cowpea and maize were grown under mixed rather than under sole crop cropping systems.

#### 13.2.1.4.2 Chemical Methods

Seed treatment of cowpea with Carbofuran at 3% w/w gave 35% reduction in gall index and 68% increase in yield over control.

Seed soaking of cowpea cv. CO 1 for 12 h in Monocrotophos at 0.1% gave a reduction of 92% of juvenile and adult nematode population in roots.

Soaking of seeds of cowpea cv. CO 2 for 12 h with 0.1% solutions of Phosphamidon and Methyl Demeton caused higher shoot weight of 46.20 and 44.57 g/plant and lower root- knot index of 2.59 and 2.69, respectively, as against the shoot weight of 36.85 g/plant and the root-knot index of 5.00 in the untreated control.

#### 13.2.1.4.3 Biological Methods

Cowpea seed treatment with *Purpureocillium lilacinum* spores was reported to be effective in reducing the root-knot infestation level. Application of certain organic materials, viz., Subabool leaves, Neem cake, or Saw dust, prior to the application of *P. lilacinum* not only enhanced the growth and multiplication of the fungus resulting into high percentage of egg mass infection but also effectively reduced the soil incidence in cowpea (Hasan and Jain 1998). Application of *P. lilacinum* at 15 kg/ha along with certain organic materials effectively reduced the root-knot nematode incidence in cowpea (Hasan and Jain 1992).

#### 13.2.1.4.4 Host Resistance

For the effective management of root-knot nematode, the farmers of middle Gujarat growing vegetables are advised to grow root-knot resistant varieties, that is, cowpea (Anand Vegetable Cowpea-1) in *kharif* and tomato (Hisar Lalit) in *rabi* for 3 years.

Cowpea genotypes, viz., 82-1-B (Gujarat-1) and C-152, tested against *M. incognita* recorded resistant reaction.

#### 13.2.1.4.5 Integrated Methods

Solarization of nematode infested field + seed dressing with Carbosulfan 25 DS at 3% a.i. (w/w) in cowpea proved effective for management of root-knot (*Meloidogyne* spp.), reniform (*Rotylenchulus reniformis*), and lesion nematodes (*Pratylenchus* spp.).

Hasan and Jain (1992) reported that soil application of *P. lilacinum* cultured on sorghum seeds together with certain organic matter effectively reduced the incidence of *M. incognita* and increased the crop yield of cowpea.

Summer plowing along with seed treatment with Carbosulfan 3% w/w or seed soaking in Monocrotophos at 0.1% for 6 h gave effective control of root-knot nematodes and increased the cowpea yield.

Combined application of *G. fasciculatum* and Achook, a Neem product was very effective in reducing root-knot nematode population in cowpea (Jain and Hasan 1995).

Reduction in *M. incognita* population was obtained in cowpea by combined soil application of chopped leaves of *Prosopis fuliflora* along with *G. fasciculatum* that increased the spore production and root colonization of *G. fasciculatum*.

### 13.2.2 Reniform Nematode, *Rotylenchulus reniformis*

The emergence of cowpea seedlings was delayed by 7 days and seedling stand was reduced to the tune of 6% to 11% due to *R. reniformis* at one nematode/g of soil (Nanjappa et al. 1978).

#### 13.2.2.1 Crop Losses

*R. reniformis* was responsible for 13.20% to 32.00% loss in pod yield of cowpea (Palanisamy and Sivakumar 1981; Hasan and Jain 1998).

#### 13.2.2.2 Symptoms

The symptoms include leaf chlorosis, delayed emergence of seedlings, stunted growth, wilting, and collapse of plant.

#### 13.2.2.3 Physiological Races

The nematode has two races based on their ability to parasitize cowpea, castor, or cotton with “race A” reproducing on all three hosts and “race B” only on cowpea (Dasgupta and Seshadri 1971).

#### 13.2.2.4 Management

##### 13.2.2.4.1 Physical Methods

Enhanced cowpea yield and significant reduction of soil population of the reniform nematode up to 15 cm depth was achieved by soil solarization under field conditions.

Summer plowing was found to be effective for the management of reniform nematodes and increased the cowpea yield over non-plowed field.

##### 13.2.2.4.2 Cultural Methods

Soil application of Neem seed kernel at 2.5 t/ha was most effective in reducing the number of females/plant (46.2 compared to 80.7 in check), number of egg masses/plant (26.0 compared to 45.5 in check), number of eggs and larvae/egg mass (46.2

compared to 80.7 in check), and final soil population/200 ml soil (143.5 compared to 831.0 in check) of the reniform nematode (Ram and Baheti 2004).

The reniform nematodes associated with cowpea can be effectively managed by the application of Neem or Eupatorium leaves at 15 t/ha two weeks before sowing.

The reniform nematode densities were suppressed when cowpea was grown intercropped with maize.

#### 13.2.2.4.3 Host Resistance

Cowpea cvs. V-16 and Pusa Phalguni are reported to be resistant to *R. reniformis* (Thakar and Patel 1984).

#### 13.2.2.4.4 Integrated Methods

Summer plowing along with seed treatment with Carbosulfan 3% w/w or seed soaking in Monocrotophos at 0.1% for 6 h gave effective control of reniform nematodes and increased the cowpea yield.

Soil application of Neem cake at 100 g/m$^2$ + *Trichoderma viride* at 2.5 kg/ha is effective against reniform nematode and enhancing yield of cowpea.

## 13.3 Pea, *Pisum sativum*

Root-knot and reniform nematodes are the important limiting factors in successful cultivation of pea crop.

### 13.3.1 Root-Knot Nematodes, *Meloidogyne* spp.

Pea pod yield loss of 20.00–50.61% was attributed to the root-knot nematode *Meloidogyne incognita* (Sharma 1989; Upadhyay and Dwivedi 1987; Parvatha Reddy 1985b).

#### 13.3.1.1 Symptoms

The infested fields show patches in which peas are stunted, chlorotic, and have few flowers which produce small and often empty pods. The root systems are reduced in size, exhibit large galls, and poor nodulation (Fig. 13.3). Senescence also tends to occur earlier.

#### 13.3.1.2 Interaction with Other Pathogens

The root-knot nematode *Meloidogyne incognita* along with wilt fungus *Fusarium oxysporum* causes wilting of pea plant that leads to death (Abuzar 2012). Maximum reduction in plant height was observed in nematode inoculated 1 week prior to fungus. Both the organisms suppressed the plant growth characters in general and plant height in particular. Maximum nematode population was found to be in nematode alone and nematode followed by fungus inoculation. Fifty days after

**Fig. 13.3** *Left*—Field infestation of pea with root-knot nematodes. *Right*—Pea roots infected with root-knot nematodes

inoculation of *Fusarium* alone, *Fusarium* prior to nematode, and nematode prior to *Fusarium* resulted in wilt index of 3.5, 4.0, and 4.2, respectively.

### 13.3.1.3 Management

#### 13.3.1.3.1 Cultural Methods

Intercropping with marigold or mustard with pea reduced the damage of root-knot nematodes.

Peas escape nematode attack, in the subtropics when sowing is postponed to mid-autumn, when temperatures drop.

#### 13.3.1.3.2 Chemical Methods

##### Seed Treatment

Bhagawati and Phukan (Bhagawati and Phukan 1990) found that Carbofuran at 3% w/w as seed treatment was effective in reducing galls and egg masses in roots of pea and increased yields. Seed treatment with Fenamiphos and Carbofuran both at 1% concentration was effective in controlling the root-knot nematode (*M. incognita*) infecting peas and in increasing their pod yields (Parvatha Reddy 1984).

##### Main Field Treatment

Enhancement of pea pod yields to the extent of 58–85% and reduction in the root-knot nematode population occurred following soil incorporation of Carbofuran, Ethoprophos, and Fenamiphos each at 1 to 2 kg a.i. per ha (Parvatha Reddy 1985a).

Decrease in the root-knot nematode, *M. incognita* in pea was achieved by drill application of Carbofuran at 1.5 kg a.i. /ha below the seed level.

#### 13.3.1.3.3 Host Resistance

Pea cv. Wando is reported to be highly resistant to *M. incognita* and moderately resistant to *M. javanica.*

Pea selection "Pant P-42" showed highest tolerance among all the selections tested and can be recommended for field infested with both pathogens (Abuzar 2012).

### 13.3.2 Reniform Nematode, *Rotylenchulus reniformis*

Dalal and Vats (Dalal and Vats 1998) reported that *R. reniformis* was responsible for 15.80% loss in pod yield of peas.

#### 13.3.2.1 Interaction with Other Pathogens

Vats and Dalal (1997) reported maximum and early wilting in pea due to prior inoculation of *R. reniformis* followed by *F. o.* f. sp. *pisi.*

#### 13.3.2.2 Management

Intercropping with marigold or mustard with pea reduced the damage of reniform nematodes.

## 13.4 Conclusion

So far, limited importance has been placed on the management of leguminous vegetable nematodes, while much attention has been devoted to the use of different nematicides. But as eco-friendly approaches become more popular to maintain a pollution-free environment, integrated pest management practices for control of nematodes on leguminous vegetable crops are gaining priority, and the use of bio-agents has recently become an important aspect in control of nematode diseases.

Research on the impact of plant-parasitic nematodes and the development of management strategies to limit their damage on leguminous vegetable crops need to be pursued. Various gaps in terms of the generation of knowledge need to be addressed concerning the distribution and impact of nematode pests on several leguminous vegetable crops. Initiatives to reduce population levels of plant-parasitic nematodes in leguminous vegetable crops have to be made available to the farmers which are practical and cost-effective to ensure sustainable crop production.

There is a need for collection of information on the distribution and yield losses due to major nematode species on a regional basis. Research on cropping patterns and systems is another important aspect that will help in devising nematode disease management programs. Only when these research results are acquired can we then go forward in this field and expect to gain the desired results.

## References


Abuzar S (2012) Response of field pea (*Pisum sativum*) selections/cultivars to root-knot nematode (*Meloidogyne incognita*) – wilt fungus (*Fusarium oxysporum*) disease complex under greenhouse conditions. Int J Appl Res Stud 1(2):1–9

Ahmed FA, Choudhury BN (2004a) Management of *Meloidogyne incognita* in French bean through crop rotation. Ann Plant Prot Sci 12:118–120

Ahmed FA, Choudhury BN (2004b) Management of root-knot nematode, *Meloidogyne incognita* using organic amendments in French bean. Indian J Nematol 24:137–139

Ahmed FA, Choudhury BN (2004c) Effect of rabbing on *Meloidogyne incognita* in French bean. Ann Plant Prot Sci 12:156–158

Bhagawati B, Phukan PN (1990) Chemical control of *Meloidogyne incognita* on pea. Indian J Nematol 20:73–83

Choudhury BN, Bhagawati B, Bora BC (2011) Three decades of nematology in Assam (1977–2011). AICRP on plant parasitic nematodes with integrated approach for their control. Department of Nematology, Assam Agricultural University, Jorhat, 49 pp

Dalal MR, Vats R (1998) Estimation of loss in pea (*Pisum sativum* L.) due to *Rotylenchulus reniformis*. In: Mehta UK (ed) Nematology – challenges and opportunities in 21st century. Sugarcane Breeding Institute, Coimbatore, pp 8–9

Das AK (1994) Assessment of yield loss due to meloidogyne incognita on French bean and its management. M. Sc. (Agriculture) thesis, Assam Agricultural University, Jorhat

Dasgupta DR, Seshadri AR (1971) Races of the reniform nematode, *Rotylenchulus reniformis* Linford and Oliveira, 1940. Indian J Nematol 1(1):21–24

Hasan N, Jain RK (1992) Field application of Paecilomyces lilacinus in combination with certain organic matters for controlling Meloidogyne incognita infecting cowpea followed by lucerne. First Afro-Asian Asian Nematology. Symposium Aligarh Muslim University, Aligarh, pp 10–11

Hasan N, Jain RK (1998) Nematological research in Uttar Pradesh: an overview. In: Trivedi PC (ed) Phytonematology in India. C.B.S. Publishers and Distributors, New Delhi, pp 150–169

Jain RK, Hasan N (1995) Nematode problems in forage crops. In: Swarup G, Dasgupta DR, Gill JS (eds) Nematode pest management- appraisal of eco-friendly approaches. Nematological Society of India, Indian Agricultural Research Institute, New Delhi, pp 217–227

Lal BS, Ansari MNA (1960) Field studies on the root-knot nematodes (*Meloidogyne* spp.) (Nematoda: Heteroderidae). Sci Cult 26:279–281

Mahse NL (2012) Studies on the root-knot nematode, *Meloidogyne incognita* (Kofoid & White, 1919) Chitwood, 1949 infesting French Bean, Phaseolus vulgaris L. M. Sc. thesis, Mahatma Phule Agriculture University, Rahuri

Mohapatra S (2015) Pathogenicity and avoidable yield loss due to *Meloidogyne incognita* in French bean. M. Sc. (Nematology) thesis, Orissa University of Agriculture and Technology, Bhubaneswar

Nanjappa CK, Panda M, Seshadri AR (1978) Effects of *Meloidogyne incognita* and *Rotylenchulus reniformis* on the emergence and survival of seedlings. In: Third International Congress on Plant Pathology, Germany

Palanisamy S, Sivakumar CV (1981) Assessment of avoidable yield loss in cowpea, black gram, maize and finger millet. In Second all India nematology symposium. Tamil Nadu Agricultural University, Coimbatore, p 58

Parvatha Reddy P (1984) Efficacy of seed treatment with three nematicides for the control of *Meloidogyne incognita* infecting cowpea, French bean and peas. Indian J Nematol 14:39–40

Parvatha Reddy P (1985a) Chemical control of root-knot nematodes infecting peas. Indian J Nematol 15:120–121

Parvatha Reddy P (1985b) Estimation of crop losses in peas due to *Meloidogyne incognita*. Indian JNematol 15:226

Parvatha Reddy P (2008) Disease of horticultures crops, nematode problems and their management. Scientific Publishers, Jodhpur, p 131

Parvatha Reddy P, Singh DB (1981) Assessment of avoidable yield loss in okra, brinjal, French bean and cowpea due to root-knot nematodes. In: The third international symposium plant pathology, New Delhi, pp 93–94

Parvatha Reddy P, Singh DB, Sharma SR (1979) Interaction of *Meloidogyne incognita* and *Rhizoctonia solani* in a root rot disease complex of French bean. Indian Phytopath 32:651–652

Ram B, Baheti BL (2004) Management of reniform nematode, *Rotylenchulus reniformis* on cowpea through seed and soil treatment with plant products. Indian J Nematol 34:193–195

Ramappa HK (1988) Studies on nematode parasites of french bean and their control. M. Sc. (Agriculture) thesis, University of Agricultural Sciences, Bangalore

Sharma GL (1989) Estimated losses due to root-knot nematodes, *Meloidogyne incognita* and *M. javanica* in pea crop. Int Nematol Netw Newsl 6(1):28–29

Sharma SB, Sharma HK, Pankaj (2002) Nematode problems in India. In: Prasad D, Puri SN (eds) Crop Pest and disease management: challenges for the millennium. Jyoti Publishers, New Delhi, pp 267–275

Singh DB, Parvatha Reddy P, Rao VR, Rajendran R (1981a) Cultivars of French bean resistant to root-knot nematode, *Meloidogyne incognita,* in India. Trop Pest Mangmt 27:29–31

Singh DB, Parvatha Reddy P, Sharma SR (1981b) Effect of the root-knot nematode, *Meloidogyne incognita* on *Fusarium* wilt of French beans. Indian J Nematol 11:84–85

Thakar NA, Patel CC (1984) Screening of few cowpea varieties against reniform nematode, *Rotylenchulus reniformis* Linford and Oliveira, 1940. Indian J Nematol 14:204

Upadhyay KD, Dwivedi K (1987) Analysis of crop losses in pea and gram due to *Meloidogyne incognita*. Int Nematol Netw Newsl 4(4):6–7

Vadhera I, Sheila BN, Bhatt J (1995) Interaction between reniform nematode (*R. reniformis*) and *Fusarium solani* causing root rot of French bean. Indian J Agric Sci 65:774–777

Vats R, Dalal MR (1997) Interaction between *R. reniformis* and *Fusarium oxysporum* f. sp. pisi on pea. Ann Biol Ludhiana 13:239–242

# Cucurbitaceous Vegetable Crops 14

**Abstract**

Root-knot (*Meloidogyne incognita* and *M. javanica*) and the reniform (*Rotylenchulus reniformis*) nematodes are the major problems on cucurbitaceous vegetables like cucumber, water melon, pointed gourd, bottle gourd, and bitter gourd. Crop losses, symptoms, biology and life cycle, spread, management, using physical methods, cultural methods, chemical methods, biological methods, host resistance, and integrated methods for root-knot and reniform nematodes are discussed.

**Keywords**

*Meloidogyne* spp. · *Rotylenchulus reniformis* · Biology · Symptoms · Management

## 14.1 Cucumber, *Cucumis sativus*

The root-knot nematode is the most important pest on cucumber that is of some economic importance.

### 14.1.1 Root-Knot Nematodes, *Meloidogyne incognita, M. javanica*

#### 14.1.1.1 Crop Losses

Root-knot nematodes are responsible for annual yield loss of 12% in cucumber amounting to Rs. 110.46 million in India (Walia and Chakraborty 2018).


P. P. Reddy, *Nematode Diseases of Crops and their Management*,
https://doi.org/10.1007/978-981-16-3242-6_14

#### 14.1.1.2 Symptoms

The infected plants exhibit stunted plant growth, leaf yellowing, and reduced yield. Large number of galls on root system and multiple galling are common. In case of heavy attacks, galls can become very large, the root system being reduced to a swollen stump without hairs (Fig. 14.1). It restricts the uptake of nutrients from the root system to the foliage, resulting in a leaf yellowing and stunted plant growth. Ultimately plants wilt and die.

#### 14.1.1.3 Biology and Life Cycle

Root-knot nematodes are soil-borne, sedentary endoparasites. After penetration the juvenile feed on conducting vessels and develops endoparasitically. Adult females are sac-like, vermiform males leave the roots and come out in soil. Females lay eggs in masses on root surface on galls, or in advanced stages inside the compound galls. During summer months, the nematode takes about 25–30 days to complete its life cycle.

#### 14.1.1.4 Spread

Root-knot nematodes are disseminated through infected transplanting material, irrigation water, and agricultural implements.

#### 14.1.1.5 Management

##### 14.1.1.5.1 Cultural Methods

Application of Neem cake at 30 g/plant 10 days before sowing is effective in cucumber (Anon. 2012).

In polyhouse grown cucumber, Neem cake at 320 g/m$^2$ proved most effective and resulted in highest yield followed by Mustard cake (Hema 2014). The benefit-cost ratio with Neem cake was found to be 3.07 in cucumber.

**Fig. 14.1** Symptoms of root-knot nematode infestation on cucumber plants (*Left*) and roots (*Right*)

#### 14.1.1.5.2 Chemical Methods

Spot application of Carbofuran 3G at 10 g/plant at sowing is effective in cucumber (Anon. 2012).

The efficacy of Metham sodium (40 ml/m$^2$) and Dazomet (40 g/m$^2$) was evaluated in polyhouse cucumber. Both the treatments proved promising. The benefit:cost ratios with Metham sodium and Dazomet were found to be 3.84 and 3.76, respectively, in cucumber (Anon. 2015; Chandel et al. 2014).

#### 14.1.1.5.3 Host Resistance

Cucumbers grafted on the bur cucumber (*Sicyos angulatus*) rootstock (good compatibility with cucumbers) exhibited increased root-knot nematode resistance (Zhang et al. 2006).

Cucumber cv. "Adrian" plants grafted onto interspecific hybrid rootstock "Strong Tosa" (*Cucurbita maxima* × *C. moschata*) had higher plant growth and total number of fruits (19.9/plant as compared to 10.1 in non-grafted plant) and yield (5.38 kg/plant as compared to 2.56 in non-grafted plant) in first season, and the same result was found in the second season also (6.96 kg/plant and more than 28.9 fruits/ plant as compared to 4.5 kg/plant and 20.4 fruits/ plant in non-grafted plant) (Ban et al. 2014).

#### 14.1.1.5.4 Integrated Methods

Pre-plant treatment with *Purpureocillium lilacinum* (20 kg/ha) in combination with organic amendments effectively reduced root-knot infection in gherkin.

Combined inoculation of AMF and *Pseudomonas fluorescens* had positive effect on root-knot nematode control on cucumber.

Enhancement of cucumber fruit yield, reduction in root galling, and second-stage larvae of *M. javanica* in soil were achieved by soil amendments (using alfalfa, cauliflower, olive-cake residues, poultry manure, and tomato) or soil solarization. Organic amendments reduced densities of *Fusarium* spp., generally increased *Aspergillus* spp., while *Trichoderma* spp. was not affected. Combinations of solarization and addition of organic amendments substantially augmented each other, particularly with poultry manure, alfalfa hay, and to a lesser extent with cauliflower and tomato residues.

Bello (1998) reported that high initial very levels of *M. incognita* were reduced to nil in susceptible cucumber crop by integrating the cultivation of short-cycle vegetables acting as trap crops and biofumigation with mushroom residue and sheep manure in commercial greenhouse trials in Spain.

In a cucumber crop in glasshouse trial, the use of solarization and *Pasteuria penetrans* had an additive detrimental effect on *M. javanica* populations (Tzortzakakis and Gowen 1994).

Oxamyl increased the efficacy of *P. penetrans* in trials against *M. javanica* infection of cucumber crop and the effects on nematode control were additive (Tzortzakakis and Gowen 1994).

In greenhouse and field studies conducted by Tempta (2017), Neem cake at 50 g/m$^2$ + Carbofuran 3G at 10 g/m$^2$ application in the field was found to be most effective in reducing nematode population and root gall index as well as in increasing the plant growth and yield in cucumber.

## 14.2 Pointed Gourd, *Trichosanthes dioica*

The root-knot and reniform nematodes are the major limiting factors in successful production of pointed gourd.

### 14.2.1 Root-Knot Nematode, *Meloidogyne incognita*

#### 14.2.1.1 Crop Losses

*M. incognita* is responsible for 30–40% loss in yield of pointed gourd (Verma 2001).

#### 14.2.1.2 Symptoms

The common symptoms of root-knot nematode on plants include general stunting, low vigor, chlorosis, necrosis, defoliation, and twig dieback (Fig. 14.2). Infected plants show fewer, small rootlets, reduction in aerial growth and highly deformed and longer size galls are formed which are easily attacked by secondary pathogens to form disease complex.

In pointed gourd, very large galls are commonly seen outside the soil level, which is a major characteristic symptom of this nematode (Fig. 14.2).

#### 14.2.1.3 Management

##### 14.2.1.3.1 Cultural Methods

Selection of healthy looking roots for planting is recommended. The root-knot nematodes on pointed gourd was controlled by planting of marigold, *Tagetes patula* at 3–5 plants/pit.

**Fig. 14.2** *Left*—Field infestation of pointed gourd with root-knot nematode showing leaf yellowing. *Right*—Severe root galling due to root-knot nematode infection

Application of Neem cake at 100 g/pit during mid-February to mid-March when rootlets start to initiate from the nodes of the vine cuttings in pointed gourd prevent the attack of *M. incognita* and increase the yield by 5000 kg/ha. In other studies, application of Neem cake at 30 g/pit in pointed gourd as spot application at 10 days prior to planting achieved maximum average yield (750 kg/ha) along with simultaneous reduction (52%) of J2 population *M. incognita* in soil.

#### 14.2.1.3.2 Chemical Methods

Dipping of planting material of *parval* (pointed gourd) in 0.05% Carbosulfan solution for 6 hours was found to be effective.

#### 14.2.1.3.3 Integrated Methods

Dipping vines in Carbosulfan 25EC at 500 ppm for 6 hours + Carbofuran 3G at 2.5 g/pit 45 days after planting was found to be effective to enhance fruit yield and the greatest reduction of root-knot nematode population in soil (Mukhopadhyay et al. 2006).

Vine dipping in Monocrotophos at 1000 ppm for 6 h + soil application of *T. viride* at 10 g/pit in two split doses (once at planting and another at 40 DAP) reduced root galling caused by *M. incognita*, nematode population in soil and gave fruit yield almost double of the untreated plots (Khan et al. 2009).

*Purpureocillium lilacinum* (PL) formulation (talc based) enriched with FYM at 15 kg in 375 kg/ha (formulation of PL mixed with FYM and incubated for 2 weeks in shade and the mixture gets ready for pit application during planting of cuttings/roots of pointed gourd) reduced nematode infestation and increased yield to the extent of 16%.

Neem cake at 50 g/pit + Carbofuran 3G at 0.3 g a.i./pit or drenching of rhizospheric soil with Carbosulfan 25EC at 500 ppm in the infested field was found to be effective in reducing root-knot nematode population as well as to increase yield of pointed gourd.

Increase in pointed gourd plant growth and yield (6.6 kg/plant as compared to 1.4 kg in control), and reduced final nematode population in soil (66.6/100 g as compared to 86,660/100 g in control), root galling (14.3/5 g as compared to 75.0/5 g in control), number of egg masses (40/5 g roots as compared to 150/5 g in control) by intercropping with marigold (3 plants/pit) combined with soil incorporation of Neem cake (250 g) + *Trichoderma harzianum* (100 g) + *Purpureocillium lilacinum* (50 g) (Verma et al. 2005).

Integrated approaches, that is, cultural control (deep plowing followed by soil solarization for 15 days and stubble burning) with or without minimum synthetic chemical (Carbofuran at 2.0 or 3.0 kg a.i./ha) and with the inclusion of Neem components like leaf, cake, and Azadirachtin, were highly effective in checking the nematode population build-up and infection. The eco-friendly integrated treatments recorded very low final nematode population and good yield (Chakraborti 2000).

## 14.2.2 Reniform Nematode, *Rotylenchulus reniformis*

### 14.2.2.1 Symptoms

Pointed gourd (*parwal*) infested by *R. reniformis* shows slower growth, reduced plant size, thinner veins, reduced fruit size resulting in substantial loss of yield. The symptoms produced on different hosts are not easily discernable. Root discoloration, leaf shedding, and malformation of fruits and seed can also occur.

### 14.2.2.2 Bio-Ecology

Reniform nematodes can survive in air-dried soil for a long period. Young females, males, and fourth-stage juveniles can survive in soil in a coiled anhydrobiotic state formed within the encrusted cuticles retained from previous molting. Survival of reniform nematode is positively correlated with decreasing soil moisture. Alternate drying and wetting of soil result in a sharp decline of nematode population density. It takes 3–4 weeks to complete life cycle depending upon temperature and host suitability. The pathogenic potential of *R. reniformis* in pointed gourd was considered to be two nematodes per ml of soil (Roy et al. 2008).

### 14.2.2.3 Interaction with Other Pathogens

*R. reniformis* can interact with other pathogens like *Fusarium* spp., *Verticillium* spp., *Sclerotium rolfsii*, and *Rhizoctonia solani* to generate disease complexes.

### 14.2.2.4 Management

#### 14.2.2.4.1 Cultural Methods

Crop rotation with nonhost crops like chili, mustard, maize, sugarcane, marigold, and restricting susceptible crops in winter seasons in the multiple cropping systems proved beneficial.Application of oilcakes like Neem, Karanj, Mustard, FYM, etc. and two to three deep summer plowings during hot months (May–June) helps to reduce soil nematode populations.

#### 14.2.2.4.2 Chemical Methods

Seed dressing with Carbosulfan 25DS at 3% a.i. w/w has been found quite effective.

#### 14.2.2.4.3 Integrated Methods

Use of organic amendments including Neem and Castor cakes at 1 t/ha 10 days before sowing in infested fields and their combinations with seed treatment with Carbosulfan at 3.0% a.i. (w/w) has been found to reduce nematode damage.Among various bioagents, seed treatment with *P. fluorescens* + *Trichoderma viride* each at 5 g/kg seed was found to be most effective for the management of reniform nematode infecting pointed gourd.

Applying biocontrol formulations like *Purpureocillium lilacinum* either as seed treatment (10 g/kg seed) or FYM-enriched formulation (1 kg in 25 kg FYM) in nursery bed and field treatment was found to be effective.

## 14.3 Bottlegourd, *Lagenaria siceraria* and Bittergourd, *Momordica charantia*

The root-knot nematode is an important limiting factor in successful cultivation of bottle gourd and bitter gourd.

### 14.3.1 Root-Knot Nematodes, *Meloidogyne incognita, M. javanica*

#### 14.3.1.1 Crop Losses

Root-knot nematodes are responsible for annual yield loss of 13.5% in bitter gourd and 22% in bottle gourd amounting to Rs. 252.82 and Rs. 403.78 million, respectively, in India (Walia and Chakraborty 2018).

#### 14.3.1.2 Symptoms

The common symptoms of root-knot nematode on plants include general stunting, low vigor, chlorosis, necrosis, defoliation, and twig dieback. Infected plants show fewer, small rootlets, reduction in aerial growth and highly deformed and longer size galls are formed (Fig. 14.3) which can be easily attacked by secondary pathogens to form disease complex.

#### 14.3.1.3 Management

##### 14.3.1.3.1 Cultural Methods

Application of Neem cake at 10 g/pit 10 days before sowing in bottle gourd proved effective.Neem, Castor and Jatropa cakes were evaluated at 30 g/m$^2$ in bottle gourd; yield was higher in all the treatments (1.186–1.224 t/ha) compared to 1.149 t/ha in untreated plots. The benefit: cost ratio was favorable in all the treatments ranging from 2.2 to 2.5 (Anon 2011).

**Fig. 14.3** Root-knot nematode-infested bottle gourd (*Left*) and bitter gourd (*Right*) plants

Effective management of root-knot nematode and enhanced yield of bottle gourd was obtained by soil incorporation of Neem cake at 4.286 t/ha or Mustard cake at 3.333 t/ha (ICBR 1: 16.10) or Press mud at 10 t/ha (ICBR 1: 18.95) 1 week before sowing (Dahiya et al. 1998).

##### 14.3.1.3.2 Chemical Methods

Seed treatment with Carbosulfan 25DS or Carbofuran 25ST at 3% a.i. w/w (120 g/kg seed) was found to be effective for the control of root-knot nematodes.Seed dressing with Carbosulfan or Benfurocarb each at 1.5 and 3% increased yield and reduced root galling (*M. javanica*) in bottle gourd.Fenamiphos, Quinalphos, Carbosulfan, and Monocrotophos, each at 500 and 1000 ppm as seed dip for 12 hours lowered the root-knot index in bottle gourd.

##### 14.3.1.3.3 Integrated Methods

Application of *P. lilacinum* ($2 \times 10^6$ cfu/g) at 2.5 kg along with 2.5 tons of FYM/ha + *P. fluorescens* ($2 \times 10^6$ cfu/g) at 2.5 kg along with 2.5 tons of FYM/ha reduced nematode population and increased yield in bitter gourd.

Seed treatment of bottle gourd with *Gluconoacetobacter diazotrophicus* strain 35–47 at 50 ml/2 kg seed and soil incorporation combined with soil incorporation of Neem cake at 30 g/spot was found to be effective for the management of root-knot nematodes *Meloidogyne* spp.

Solarization of nematode infested field + seed dressing of direct sown crops with Carbosulfan 25 DS at 3% a.i. (w/w) was found to be effective for management of root-knot nematodes.

## 14.4 Watermelon, *Citrullus lanatus*

### 14.4.1 Root-Knot Nematode, *Meloidogyne incognita*

#### 14.4.1.1 Symptoms

In watermelons with nematode damage, leaf chlorosis is evident and leaves may be stunted and wilted. Yellowing of plants, stunted growth, and general decline in watermelon are early symptoms. Formation of root galls is the characteristic symptom of root-knot nematode infestation (Fig. 14.4). Infested roots fail to absorb water and nutrients resulting in stunted growth, yellowing, and wilt of above-ground plants.

#### 14.4.1.2 Survival and Spread

The primary source of inoculum include collateral and other hosts like Solanaceous, Malvaceous, and Leguminaceous plants and egg masses that survive in infected plant debris and soil. Secondary spread through autonomous second-stage juveniles that may also be water dispersed.

**Fig. 14.4** Galling on watermelon roots

#### 14.4.1.3 Favorable Conditions

Sandy Loam light soils are favorable.

#### 14.4.1.4 Management

Dry fallowing (weed-free) and cultivation to bring as much soil as possible to the surface are effective in killing root-knot nematode through starvation and desiccation. In order to restore the soil fertility after fallowing, application of compost/FYM/other organic amendments should be added to the soil. The nematicidal effects due to release of toxic compounds after breakdown were observed in several *Brassica* species. Soil application of neem cake/pongamia cake at 250 kg/ha or press mud at 5 t/ha at the time of last plowing helps in reducing nematode population.

Watermelon grafted onto cucurbit rootstocks grown in a field infested with southern root-knot nematode *Meloidogyne incognita* reveal that "Emphasis" bottle gourd (*Lagenaria siceraria*) and "Strong Tosa" squash hybrid (*Cucurbita moschata* × *C. maxima*) rootstocks exhibited severe root galling caused by root-knot nematodes (Fig. 14.5), while RKVL 301 and RKVL 302 wild watermelon (*C. lanatus* var. *citroides*) rootstocks were found to be resistant to root-knot nematodes (Fig. 14.6).

Percentage of root galling was significantly higher ($P = 0.05$) on the bottle gourd (Fig. 14.5-1A) and squash hybrid (Fig. 14.5) rootstocks than on the watermelon "Fiesta" (non-grafted) and the *C. lanatus* var. *citroides* lines (Fig. 14.6).

*Citrullus lanatus* var. *citroides* rootstock "Carolina Strongback" exhibited resistance to both root-knot nematode (*M. incognita*) and Fusarium wilt (*Fusarium oxysporum* f. sp. *niveum*), and could be used for grafting watermelon by growers to control these two pathogens in co-infected fields.

**Fig. 14.5** Watermelon grafted onto rootstocks "Emphasis" bottle gourd (*Left*) and "Strong Tosa" squash hybrid (*Right*) showing severe root galling

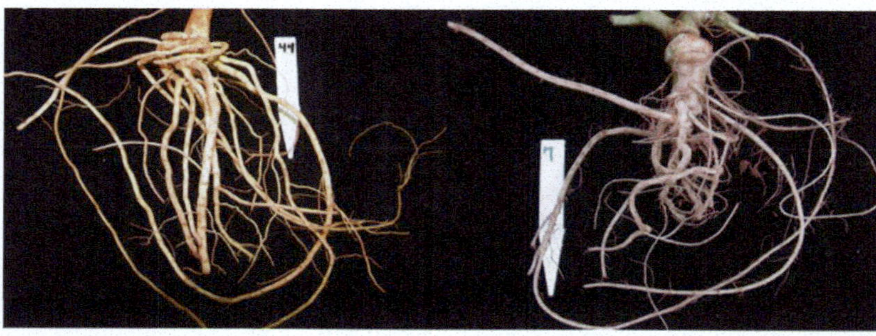

**Fig. 14.6** Watermelon grafted onto RKVL 301 (*Left*) and RKVL 302 (*Right*) wild watermelon (*C. lanatus* var. *citroides*) rootstocks showing resistant reaction to root-knot nematodes

## 14.5 Conclusion

Several species of nematodes can cause economic damage to cucurbit crops, including root-knot nematodes (*Meloidogyne* spp.) and reniform nematodes (*Rotylenchulus reniformis*). The *Citrullus lanatus* var. *citroides* accessions can be used not only as rootstocks but also in plant breeding programs to enhance cucurbit cultivars for resistance to root-knot nematodes. The development of valuable crop management strategy in order to reduce the deleterious effect of root-knot nematodes can be achieved by grafting bitter gourd on three rootstocks such as kumatikai (*Citrulus colocynthis*), African horned cucumber (*Cucumis metuliferus*), and pumpkin (*Cucurbita moschata*) which exhibited resistant reaction and sponge gourd (*Luffa cylindrica*) and mithipakal (*Momordica charantia* var. *muricata*) which showed moderately resistant reaction.

Use of chemicals cannot be advocated due to safety issues. Hence other methods, viz., biological, organic amendments, resistant cultivars, and integrated methods, should be adopted. Current efforts have been focused on using cultural and chemical controls as well as grafted watermelons on appropriate disease-resistant rootstocks. Using a combination of these management tactics is recommended when growers have serious problems with root-knot nematodes.

## References


Anon (2011) Quinquennial report. AICRP on Nematode pests of crops & their Control, Palampur centre, Himachal Pradesh, 98 pp

Anon. (2012) Biennial report (2010-2011). AICRP on Nematode Pests of Crops & Their Control, Jorhat Centre, AAU, Jorhat

Anon. (2015) Biennial Report. AICRP on Nematode pests of crops & their Control, Palampur centre, Himachal Pradesh, 47 pp

Ban SG, Žanić K, Dumičić G et al (2014) Growth and yield of grafted cucumbers in soil infested with root-knot nematodes. Chilean J Agri Res 74(1):29–34

Bello A (1998) Biofumigation and integrated pest management. In: Bello A, González JA, Arias M, Rodriguez-Kábana R (eds) Alternatives to methyl bromide for the southern European countries. DG XI EU, CSIC, Valencia, pp 99–126

Chakraborti S (2000) Management approaches for root-knot nematode in pointed gourd. Indian J Nematol 30:136–140

Chandel YS, Ravichandra NG, Mhase NL et al (2014) Integrated management of root-knot nematodes (*Meloidogyne spp.*) under protected cultivation. Indian J Nematol 44(1):92–96

Dahiya RS, Bansal RK, Paruthi IJ (1998) Efficacy of mustard cake and poultry manure for the management of *Meloidogyne javanica* in bottle gourd and okra. In: Proc. of National Symp. On rational approaches in nematode Management for Sustainable Agric. Nematol. Soc. of India, New Delhi, pp 100–102

Hema (2014) Eco-friendly Management of Nematodes in Cucumber (Cucumis sativus L.) under Protected Conditions. M. Sc. Thesis. Dr. Y.S. Parmar Univ. of Hort. and Forestry, Solan, Himachal Pradesh, 57 pp.

Khan MR, Bhattacharya I, Chattopadhyay SB, Ghosh S (2009) Integrated approach for managing root-knot nematode (*Meloidogyne incognita*) in pointed gourd (*Trichosanthes dioica* Roxb ). Indian J Nematol 39(1):25–28

Mukhopadhyay AK, Roy K, Panda TR (2006) Evaluation of nematicides and neem cake for managing root-knot nematodes in pointed gourd, *Trichosanthes dioica*. Nematol Medit 34:27–31

Roy K, Mukhopadhyay AK, Pramanik A (2008) Pathogenicity of *Rotylenchulus reniformis* Linford & Oliveira, 1940 in brinjal, *Solanum melongena* L. Ann Plant Prot Sci 16(1):179–181

Tempta N (2017) Management of Root-Knot Nematode Meloidogyne incognita (Kofoid & White, 1919) Chitwood, 1949 in Cucumber, Cucumis sativus L. M. Sc. Thesis, Dr. Y S Parmar Univ. of Hort. and Forestry, Solan, Himachal Pradesh, 70 pp.

Tzortzakakis EA, Gowen SR (1994) Evaluation of *Pasteuria penetrans* alone and in combination with oxamyl, plant resistance and solarization for control of *Meloidogyne* spp. on vegetables grown in greenhouses in Crete. Crop Prot 13:455–462

Verma AC (2001) N.D. University of Agriculture and Technology, Kumarganj, Faizabad, Uttar Pradesh. In: Dhawan SC et al (eds) Indian nematology-progress and perspectives. Divn. of Nematol., Indian Agri. Res. Inst, New Delhi, pp 121–125

Verma AC, Singh HK, Khan N (2005) Management of root-knot nematode, *Meloidogyne incognita* through antagonistic approaches in pointed gourd. Indian J Nematol 35:78–79

Walia RK, Chakraborty PK (eds) (2018) Nematodes problems of crops in India. ICAR-All India Coordinated Research Project on Nematodes in Agriculture, New Delhi, 402 pp

Zhang S, Gun X, Wang Y (2006) Effect of bur cucumber (*Sicyos angulatus* L) as rootstock on growth physiology and stress resistance of cucumber plants (in Chinese). Acta Hort Sinica 33:1231–1236

# Root Vegetable Crops 15

**Abstract**

The production of root vegetables like carrot and beetroot worldwide is threatened by root-knot nematodes (*Meloidogyne incognita* and *M. javanica*). In this chapter, crop losses, symptoms, biology and life cycle, spread, and management, using physical methods, cultural methods, chemical methods, biological methods, host resistance, and integrated methods, for root-knot nematodes on root crops are discussed.

**Keywords**

*Meloidogyne incognita* · M. javanica · Symptoms · Biology · Management

## 15.1 Carrot, *Daucus carota*

The root-knot nematodes are the major limiting factors in successful cultivation of carrot crop.

### 15.1.1 Root-Knot Nematodes, *Meloidogyne incognita* and *M. javanica*

#### 15.1.1.1 Crop Losses

The crop loss caused by root-knot nematode *M. incognita* to carrots is to the extent of 56.64% (Devi 1993).

Root-knot nematodes are responsible for annual yield loss of 34% in carrot amounting to ₹ 754.12 million in India (Walia and Chakraborty 2018).



P. P. Reddy, *Nematode Diseases of Crops and their Management*,
https://doi.org/10.1007/978-981-16-3242-6_15

#### 15.1.1.2 Symptoms

Nematode infestation causes qualitative loss in carrot along with quantitative loss. The infested plant exhibits reduced growth, yellowing of the leaves, and wilting symptoms. Galls are seen on the secondary roots of main tuberous tap root (Fig. 15.1). Under severe infestation, the tuberous roots exhibit branching/splitting/spiking. The infestation paves way for secondary degrading organisms like *Erwinia*, which are responsible for further rotting of carrots.

Root-knot nematodes caused significant reduction in the yield and reduction in consumer preference due to forking of carrots (Fig. 15.2).

#### 15.1.1.3 Biology and Life Cycle

Root-knot nematodes are soil-borne, sedentary endoparasites. After penetration, the juveniles feed on conducting tissues endoparasitically and develop into adults. Adult females are sac-like and lay eggs within the galls or on the surface of root galls, while

**Fig. 15.1** Carrot roots severely infected with root-knot nematodes

**Fig. 15.2** Forking of carrot roots due to root-knot nematode infection

the males are vermiform, leave the roots, and come out in soil. Under north Indian conditions, the nematode takes about 60–90 days to complete the life cycle.

#### 15.1.1.4 Spread

Root-knot nematodes are disseminated through infected transplanting material, irrigation water, and agricultural implements.

#### 15.1.1.5 Management

##### 15.1.1.5.1 Physical Methods

Soil solarization during summer months with 400 gauze polythene sheet for 3–4 weeks reduces nematode population in soil.

##### 15.1.1.5.2 Cultural Methods

Rotation with nonhost crops like cereals was found beneficial.

The population of *M. incognita* larvae decreases in combination of spinach-marigold-carrot and spinach-radish-carrot (Hasan and Jain 1998). Late sown carrots suffer less damage due to *M. incognita* in California, USA.

Application of neem cake or mustard cake at 300-400 kg/ha 15 days before sowing is effective for the management of root-knot nematode and in increasing yields of carrot (Choudhury et al. 2011).

In a field trial, saw dust, neem cake, and poultry manure at 1.0 and 1.5 t/ha and water hyacinth at 1 and 2 t/ha increased yield of carrot (28–66%) by reducing the infestation of *M. incognita*. Devi (1993) reported that soil application of neem cake gave maximum reduction of *M. incognita* population in soil (38.5%) followed by saw dust (23%), poultry manure (22%), and water hyacinth (14%) over control.

##### 15.1.1.5.3 Chemical Methods

Soil application with Carbofuran 3G at 16.6 g/plant proved effective.

##### 15.1.1.5.4 Biological Methods

Enhancement of yield (147.6%), reduction in forking of carrot roots (92.4%), reduction in nematode population in soil (45.23%), and least root-knot index (1.6 as against 5.0 in controls) was obtained by soil incorporation of *P. lilacinum* (containing $1 \times 10^6$ spores/g) at 10 g/ 2.5 $m^2$ (Sivakumar 1998).

##### 15.1.1.5.5 Host Resistance

Carrot cv. Arka Suraj is reported to be tolerant to the root-knot nematode (Swamy and Sadashiva 2004). Ten carrot genotypes namely New Kuroda, Selendro, Perfecta Elita, Zeno, Nantes Strong Top, Nantes, F-524, Danvers Half Long, Avenger, and HYDC-4 were found resistant to *M. incognita* (Veere Gowda et al. 2001).

#### 15.1.1.5.6 Integrated Methods

Carrot seed treatment with *P. lilacinum* ($2 \times 10^6$ cfu/g) at 20 g/kg seed + application of *P. lilacinum* ($2 \times 10^6$ cfu/g) at 2.5 kg along with 2.5 tons of FYM/ha prior to sowing managed *M. incognita*.

Soil application of *P. lilacinum* ($2 \times 10^6$ cfu/g) at 2.5 kg + FYM at 2.5 t/ha gave minimum number of galls and final nematode population (120 and 12.5, respectively) compared to control (440 and 235, respectively). The highest yield was also recorded in the above treatment (9.34 t/ha) over control (6.54 t/ha) (Somasekhara and Ravichandra 2010).

Integration of neem cake with *P. lilacinum* and *Trichoderma harzianum* was found effective and increased carrot yield (23.347 t/acre as compared to 18.245 t/acre in control).

Seed treatment with *Pseudomonas putida* ($10^9$ cfu/g) at 10 g/kg and subsequent application of 5 tons of FYM enriched with 5 kg each of *P. putida* ($10^9$ cfu/g) and *T. harzianum* ($10^6$ cfu/g) per ha significantly reduced root-knot nematode population in roots by 77% and increased yield (24%) and benefit:cost ratio (5.3) (Anon. 2012).

Seed treatment with *T. harzianum* at 10 g (with 1 x $10^6$ cfu/g)/kg and *P. fluorescens* at 10 g (with $1 \times 10^9$ cfu/g)/kg and subsequent field application of 5 tons of enriched FYM with *T. harzianum* (with 1 x $10^6$ cfu/g) and *P. fluorescens* (with $1 \times 10^9$ cfu/g) per ha significantly reduced root-knot nematodes in carrot roots by 79% and increased yield to the tune of 29.8% with a benefit:cost ratio of 13.6.

#### 15.1.1.5.7 Success Story

Management of root-knot nematode *Meloidogyne incognita* in carrot was demonstrated in farmer Mr. Chaannappa's field, Uppinahalli village, Hosakote taluk, Bangalore North District, Karnataka. Soil solarization during summer months with 400 gauze polythene sheet, soil application with *Purpureocillium lilacinum* at 25 g/m$^2$ along with neem cake at 100 g/m$^2$, and soil application with Carbofuran 3G at 16.6 g/m$^2$ gave 77.7% benefit amounting to ₹ 1,60,000/acre to the farmer.

## 15.2 Beetroot, *Beta vulgaris* sub sp. *vulgaris*

### 15.2.1 Root-Knot Nematodes, *Meloidogyne* spp.

#### 15.2.1.1 Symptoms

Beets with root-knot nematode can actually exhibit a variety of symptoms like stunting of plants, yellowing of foliage, and withering despite the presence of adequate water. Root-knot nematodes cause galls and swellings on the main root and any secondary roots that are present (Fig. 15.3).

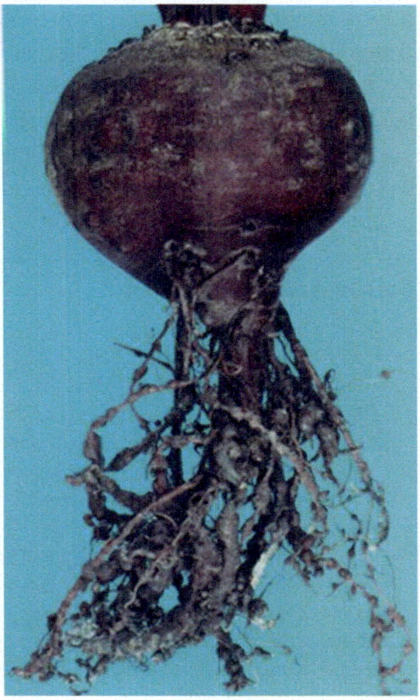

**Fig. 15.3** Beet fibrous roots infected with root-knot nematodes

#### 15.2.1.2 Management

##### 15.2.1.2.1 Cultural Methods

Since beets will mature in temperatures around 13 °C, early planting means they can escape the majority of the root-knot nematode's life cycle. Young beetroot seedlings are more susceptible than healthy, maturing beets which better tolerates root-knot nematode attack.

Maintain a 2- to 3-year crop rotation with nonhosts like fescue, marigold, or small grains which can slowly decrease the number of nematodes present in the soil.

Incorporating organic matter into the soil can help to decrease nematode activity by enhancing the activity of nematode-destroying bacteria.

##### 15.2.1.2.2 Host Resistance

Beet root cv. "Detroit Dark Red" was tolerant to *M. incognita*, whereas cv. "Crimson Globe" was resistant to *M. javanica*.

## 15.3 Conclusion

Root-knot nematodes are generally considered the most important nematode parasites of carrot and beetroot worldwide, being geographically widespread and capable of complete crop loss. Management of nematodes is heavily reliant upon the use of nematicides. With potential loss in availability of nematicides, there is a need to move back to a cropping system approach to nematode management. Alternative strategies will be required for nematode control in the future. An integrated strategy for control of nematodes is advocated involving preplant nematode counts to determine the need for nematicides, manipulation of planting date, and suitable crop rotations or break crops to reduce nematode populations in soil prior to carrot or beetroot.

## References

Anon. (2012) Annual Report (2011-2012). Indian Inst. of Hort. Res, Bangalore, pp 23–24

Choudhury BN, Bhagawati B, Bora BC (2011) Three decades of nematology in Assam (1977-2011). AICRP on Plant Parasitic Nematodes with Integrated Approach for their Control, Department of Nematology, Assam Agri. University, Jorhat, 49 pp

Devi G (1993) Pathogenicity, Crop Loss Assessment and Management of Meloidogyne incognita (Kofoid and White, 1919) Chitwood, 1949 on Carrot (Dacus carota L.). M. Sc. (Agri.) thesis, Assam Agri. Univ., Jorhat

Hasan N, Jain RK (1998) Nematological research in Uttar Pradesh: an overview. In: Trivedi PC (ed) Phytonematology in India. C.B.S. Publishers and Distributors, New Delhi, pp 150–169

Sivakumar CV (1998) Bacterial antagonists for suppression of plant parasitic nematodes. In: Singh SP, Hussaini SS (eds) Biological suppression of plant diseases, Phytoparasitic nematodes and weeds. Project Directorate of Biological Control, Bangalore, pp 128–137

Somasekhara YM, Ravichandra NG (2010) Evaluation of bio-pesticides for the management of *Meloidogyne incognita* in carrot. In: National Conf. On innovations in Nematological research for agricultural sustainability – challenges and a roadmap ahead. Tamil Nadu Agri. Univ., Coimbatore, p 73

Swamy KRM, Sadashiva AT (2004) Vegetable varieties/hybrids of IIHR. Tech. Bull. 2, Indian Inst. of Hort. Res, Bangalore, 30 pp

Veere Gowda R, Singh TH, Nagesh M, Rao MS (2001) Evaluation of carrot (*Dacus carota L.*) genotypes for resistance to root-knot nematode (*Meloidogyne incognita*). In: Proc.Second National Symp.Integrated Pest Mangmt. Horti. Crops: new molecules, biopesticides and environment, Bangalore, p 155

Walia RK, Chakraborty PK (eds) (2018) Nematode problems of crops in India. ICAR-All India Coordinated Research Project on Nematodes in Agriculture, New Delhi, 402 pp

# Bulbous Vegetable Crops

# 16

**Abstract**

Nematodes are one of the major constraints in bulb crops like onion and garlic production. The most common nematodes on onion are the stem and bulb nematode (*Ditylenchus dipsaci*) and the root-knot nematode (*Meloidogyne arenaria*). Crop losses, symptoms, biology and life cycle, spread, and management, using physical methods, cultural methods, chemical methods, biological methods, host resistance, and integrated methods, for root-knot nematodes are discussed.

**Keywords**

*Ditylenchus dipsaci* · *Meloidogyne arenaria* · Symptoms · Biology · Management

## 16.1 Onion, *Allium cepa*, and Garlic, *Allium sativum*

Stem and bulb and root-knot nematodes have been recognized as the limiting factors in successful production of onion and garlic.

### 16.1.1 Stem and Bulb Nematode, *Ditylenchus dipsaci*

The stem and bulb nematode causes great damage to the onion and garlic crops in temperate areas of many countries. This is primarily observed in heavy soils. The nematode problem becomes serious with abundant precipitation or high soil humidity. The nematode causes serious damage at inoculum level of 10 or more nematodes/500 g of soil. This is responsible for the "bloat" disease of onion.



P. P. Reddy, *Nematode Diseases of Crops and their Management*,
https://doi.org/10.1007/978-981-16-3242-6_16

#### 16.1.1.1 Symptoms

Infected young seedlings are stunted, pale, and have swollen areas along the cotyledons. Leaves can develop yellowish-brown spots, short and thickened, and stem swelling (bloating) can occur (Fig. 16.1). As the disease progresses, the foliage collapses (Fig. 16.1) and the stems and necks of bulbs soften. The bulb scales become soft and light gray. Infected bulbs (Fig. 16.1) are light in weight, may be malformed, or produce sprouts and double bulbs. The common secondary infections due to bacteria and fungi are responsible for decay and bad smell.

The leaves of garlic plants turn yellow and die but no thickening or twisting is evident (Fig. 16.2). The infected garlic bulbs become loose (the cloves separate from the base) and cannot be used for seed even if they tolerate winter storage well because the plants which form often die during the vegetative period.

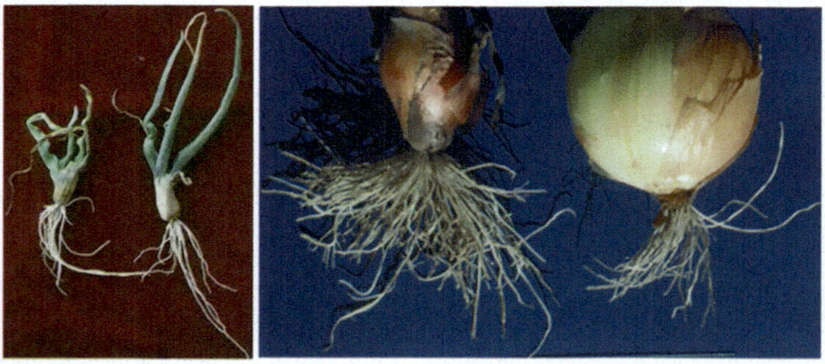

**Fig. 16.1** *Left*—Onion seedling infected with stem and bulb nematode (left side) compared to healthy onion seedling (right). *Right*—Onion bulb infected with stem and bulb nematode (left side) compared to healthy onion bulb (right)

**Fig. 16.2** *Left*—Garlic plants infected with stem and bulb nematode (three plants on left) compared to healthy onion seedlings (two plants on right). *Right*—Healthy garlic bulb (bulb on left) and garlic bulb infected with stem and bulb nematode (bulb on right)

#### 16.1.1.2 Life Cycle

The fourth-stage juveniles penetrate the stem and leaf tissue through the stomata. Egg laying begins at temperatures of 1–5 °C with the optimum at 13–18 °C. *D. dipsaci* completes one generation in 19–23 days at 15 °C. Nematode activity stops at 36 °C. The nematode prefers the cool moist climatic conditions existing in the upland tropics and wet winter seasons in the subtropics. *D. dipsaci* can parasitize plants in both heavy and light soils, although a higher incidence of infestation seems to occur on heavy soils.

#### 16.1.1.3 Spread and Survival

The nematode can survive in onion bulbs/garlic cloves used for seed as well as in the soil without a host plant for more than 1 year, and the fourth juvenile stage can survive in anabiosis for many years. The nematodes on plant residues can tide over drought for several years which becomes active with the availability of moisture.

The means by which the nematode spreads include soil adhering to bulbs, crop residues, and movement of infected bulbs. Seed-borne infections also are responsible for long-distance spread in onion. Infestations can also be maintained by volunteer host plants and weed hosts in between onion and garlic crops.

#### 16.1.1.4 Management

##### 16.1.1.4.1 Physical Methods

The nematodes can be exterminated by soaking the plants in water containing 0.5% formalin at 24 °C, followed by a hot water treatment (with addition of 0.5% formalin at 49 °C for 20 min).

Soaking of onion sets or garlic cloves in cold water for 3–4 days will also kill the nematodes.

##### 16.1.1.4.2 Cultural Methods

The diseased plants and their remnants should be gathered and destroyed.

Make every effort to ensure that seed, sets from onion, and garlic bulbs are clean to prevent contaminating the field. Sets and bulbs should be examined carefully and only plant if they appear perfectly healthy should be used for planting. The important symptoms on infected sets include lighter in weight, soft, shrunken, and discolored (dark brown) near the neck of the plant.

Rotation of onion and garlic should be avoided with susceptible crops like beans, bulbous plants, carrot, and pea. Onion and garlic may be rotated with wheat, barley, maize, rape, white clover, red clover, lucerne, potatoes, beets, crucifers, spinach, or lettuce to overcome the stem and bulb nematode damage because they are poor hosts for stem and bulb nematode. Control weeds and volunteers during rotations.

Mustard, rapeseed, oilseed, radish, and sorghum-Sudan grass can be used as a green manure which release nematode toxic chemicals when tilled into the soil. Use of green manure has been shown to reduce nematode populations.

Planting bio-fumigants such as brown mustard and Sudan grass as cover crops proved beneficial.

#### 16.1.1.4.3 Chemical Methods

Row application of Thionazin at 4.5 g/meter row gives effective control of the stem and bulb nematodes.

### 16.1.2 Root-Knot Nematodes, *Meloidogyne* spp.

Vadivelu and Rajendran (1986) reported *M. arenaria* on onion from Palladam district of Tamil Nadu. The infected plants showed patchiness and unthrifty growth with "neck twisting" symptoms.

#### 16.1.2.1 Symptoms

Affected crop show stunted growth, yellowing of leaves, smaller bulbs, delayed maturity, and wilting of the plants (Fig. 16.3). Heavy infection of older plants causes wilting of plants unexpectedly and die-off early. Small, swollen galls 1–2 mm in diameter can be found on the roots when infected plants (Fig. 16.3) are carefully lifted from the ground and freed from soil particles without damaging the roots. Depending on the species causing infection, the shape of the galls can be round or spindly, and with or without short root branches that arise from the upper part of galls. White to dark brown egg masses can be observed on the roots.

#### 16.1.2.2 Management

##### 16.1.2.2.1 Physical Methods

Hot water treatment of onion bulbs at 43.5 °C for 120 min eliminates the nematode infection. Soil solarization was also found effective.

**Fig. 16.3** Onion (*Left*) and garlic (*right*) roots infected with root-knot nematodes

#### 16.1.2.2.2 Cultural Methods

Crop rotation to a nonhost or a long fallow period helps to reduce populations of root-knot nematodes. The lowest populations of plant parasitic nematodes were recorded in onion-cowpea-fallow cropping sequence (Vetrivelkalai and Subramanian 2006).

Soil application of organic amendments enhances biological suppression of nematodes.

#### 16.1.2.2.3 Integrated Methods

Root dip treatment of onion seedlings in *P. lilacinum* + neem cake suspension for 1 h before transplanting proved effective and increased bulb yield.

Seed treatment with *P. fluorescens* ($10^9$ cfu/g) at 10 g/kg and subsequent soil application of 5 tons of FYM enriched with 5 kg each of *Pseudomonas fluorescens* ($10^9$ cfu/g) and *Pochonia chlamydosporia* ($10^6$ cfu/g) per ha significantly reduced *M. incognita* population in roots by 69% and increased bulb yield by 21% (Anon 2012).

## 16.2 Conclusion

The most common nematodes on onion are root-knot nematode (*Meloidogyne hapla*), stem nematode (*Ditylenchus dipsaci*), and root-lesion nematodes (*Pratylenchus* spp.). Considering the costs and environmental concerns of application of nematicide, planting resistant cultivars is an economically sound and environmentally viable option for sustainable nematode management in onion. Integrated management of nematodes using neem cake, trap crops, crop rotation, soil solarization, turmeric (*Curcuma longa*) grown as nonhost antagonistic crop, crop sanitation, and application of carbofuran can be explored for effective nematode management.

## References

Anon (2012) Annual report 2011–12. Indian Institute of Horticultural Research, Bangalore, pp 23–24

Vadivelu RS, Rajendran G (1986) Damage potential of the root-knot nematode, Meloidogyne arenaria to onion. Int Nematol Netw Newsl 4(2):5–8

Vetrivelkalai P, Subramanian S (2006) Plant parasitic nematode populations as affected by onion based coping sequences. Indian J Nematol 36:10–14

# Cruciferous Vegetable Crops 17

**Abstract**

The production of cruciferous vegetable crops such as cabbage and cauliflower worldwide is threatened by stunt nematode *Tylenchorhynchus brassicae,* cyst nematode *Heterodera cruciferae,* and root-knot nematode *Meloidogyne incognita*. The root-knot nematodes interact with *Fusarium* wilt and the stunt nematode interacts with *Rhizoctonia* root-rot pathogens in causing disease complexes. In this chapter, crop losses, symptoms, biology and life cycle, spread, and management, using physical methods, cultural methods, chemical methods, biological methods, host resistance, and integrated methods, for root-knot nematodes are discussed.

**Keywords**

*Tylenchorhynchus brassicae* · *Heterodera cruciferae* · *Meloidogyne incognita* · Symptoms · Biology · Management

## 17.1 Cabbage, *Brassica oleracea* var. *capitata*, and Cauliflower, *Brassica oleracea* var. *botrytis*

Stunt and root-knot nematodes are the major limiting problems in successful cultivation of cabbage and cauliflower crops.

### 17.1.1 Stunt Nematode, *Tylenchorhynchus brassicae*

Siddiqui et al. (1972) encountered *T. brassicae* as the most prevalent species around the roots of cabbage and cauliflower in Uttar Pradesh.



P. P. Reddy, *Nematode Diseases of Crops and their Management*,
https://doi.org/10.1007/978-981-16-3242-6_17

#### 17.1.1.1 Distribution and Hosts

The most efficient hosts of the stunt nematode include cabbage, cauliflower, and tomato, although all crucifers are susceptible. The populations of *T. brassicae* were exceptionally high in Aligarh, Kanpur, Ghazipur, and Rampur districts of UP. The multiplication rate on cabbage (cv. Suttons Earliest), cauliflower (cv. Patna), and Knol-khol (cv. Purple Vienna) was 13, 18, and 29, respectively. Common weeds such as *Chenopodium album, Trianthema monogyna,* and *Portulaca oleracea* were found to be hosts of *T. brassicae*.

#### 17.1.1.2 Symptoms

*T. brassicae* caused reduction in fresh weight of cabbage (Fig. 17.1) and cauliflower (Fig. 17.2) plants and suppressed root growth, resulting in reduced water absorption.

#### 17.1.1.3 Life Cycle

The optimum soil temperature (30 °C) and soil moisture (25–30%) for reproduction of *T. brassicae* was same for both cabbage and cauliflower. Further, high temperature (35–40 °C) and low moisture are highly unfavorable for survival of *T. brassicae* in the absence of host. However, at moderate temperature and moisture conditions, the nematode is capable of surviving even up to 240 days.

#### 17.1.1.4 Interaction with Other Pathogens

The stunt nematode *T. brassicae* interacts with the root rot fungus *Rhizoctonia solani* and adversely affects the emergence of crucifer seedlings when occur together (Khan et al. 1971).

**Fig. 17.1** *Left*—Cauliflower plants infected with stunt nematode. *Right*—Healthy cauliflower plants

**Fig. 17.2** *Left*—cauliflower plant infected with stunt nematode. *right*—healthy cauliflower plant

**Fig. 17.3** Cabbage cyst nematode (*left*—infected; *right*—healthy)

#### 17.1.1.5 Histopathology and Histochemistry

The histopathology of cabbage and cauliflower roots shows that *T. brassicae* penetrated throughout the intercellular region of cortex. The nematodes were mostly situated 1–3 cell layer deep the cortex and were parallel to the longitudinal axis of the roots (Fig. 17.3).

Alam et al. (1976) reported that the roots of cabbage and cauliflower infected with 10,000 *T. brassicae* enhanced total soluble phenols, ortho-dihydroxy phenols, free amino acids, and proteins, but not in total carbohydrates. Total phenols, ortho-dihydroxy phenols, and free amino acids increased with increasing seedling age at inoculation (Khan et al. 1986).

#### 17.1.1.6 Management

##### 17.1.1.6.1 Cultural Methods

Siddiqui et al. (1973) reported that rotation of cabbage and cauliflower with wheat resulted in sudden decline in population of *T. brassicae*. Enhanced crop growth and reduction in reproduction of *T. brassicae* was obtained by soil incorporation of neem cake (Siddiqi and Saxena 1987a, b). The mustard (*rabi*), radish (summer), sesame (*kharif*) cakes sequence considerably reduced the population of *Tylenchorhynchus* spp. (Haque and Gaur 1985).

Botanicals

Seed dressing with latex of *Calotropis gigantea*, *C. procera*, *Euphorbia milii*, *E. nerifolia*, and *E. tirucalli* significantly inhibited *T. brassicae* multiplication on cabbage and cauliflower and improved plant growth (Siddiqi and Alam 1989a).

Bare-root dip of cabbage and cauliflower seedlings in leaf extracts of neem and Persian lilac (Siddiqi and Alam 1989b), intercropping of marigold, neem, and Persian lilac, reduced multiplication rate of *T. brassicae* and improved plant growth.

Amendment of soil with oil cakes of neem, groundnut, mustard, and castor (equivalent to 100 kg N/ha) was effective in reducing the population of *T. brassicae* around the roots of cabbage and cauliflower (Siddiqui et al. 1976). Neem cake was the best.

##### 17.1.1.6.2 Chemical Methods

Carbofuran, Ethoprophos, and Fenamiphos each at 1–2 kg a.i. per ha were found effective in reducing the root-knot nematode population and in increasing flower yields of cauliflower.

Carbofuran at 1.5 kg a.i. per ha was effective in reducing *T. brassicae* and *T. dubius* population and in increasing yields of cabbage (Varma et al. 1978).

##### 17.1.1.6.3 Host Resistance

*Brassica rapa* cv. Sarson (yellow mustard) was found to be least susceptible to *T. brassicae* and could be grown on infected soil in preference to *B. nigra* (black mustard) (Khan et al. 1986).

### 17.1.2 Cyst Nematode, *Heterodera cruciferae*

#### 17.1.2.1 Symptoms

*Heterodera cruciferae* only infects crucifers. Nematode significantly reduces the size of heads (Fig. 17.3). As the disease progresses, leaves may wilt or curl, especially during hot weather, tap root remains small, roots are discolored and branched profusely, produce loose and small heads. Invasion of infected roots by fungi is common. A characteristic sign of this pathogen is the appearance of lemon-shaped cysts on the root surface, which are white, tan, or reddish in color (Fig. 17.4). Plants often die prematurely.

**Fig. 17.4** Cabbage roots infected with cyst nematode (*left*). Cabbage roots infected with female pearly white cyst nematode (*middle*) and brown cysts (*right*)

#### 17.1.2.2 Favorable Conditions

Loamy soils favor disease development. The nematode spreads through rains and irrigation water.

#### 17.1.2.3 Survival and Spread

- The means of survival is through cysts which overwinter in soil. After planting of the crop, eggs from cysts hatch and the second-stage larvae infect the roots.
- The means of nematode spread include infected planting stock, farm implements, and contaminated soil.

#### 17.1.2.4 Management

- Crop rotation with nonhosts (barley resistant, wheat and sugar beet moderately resistant) for a period of 3–5 years helps to reduce nematode populations.
- Incorporation of crop residues immediately after harvest proved beneficial.
- Eradication of weeds and volunteers helps control this disease.
- Application of nematicides.
- Use of resistant varieties.

### 17.1.3 Root-Knot Nematode, *Meloidogyne incognita*

#### 17.1.3.1 Symptoms

Root-knot nematodes on crucifers induce prolific root branching and galling above the point of infection (Fig. 17.5). Invasion of infected roots by fungi may occur. Above-ground symptoms include stunting, chlorosis, and wilting. Small and unmarketable heads are produced by infected plants which survive in a growing season.

#### 17.1.3.2 Interaction with Other Pathogens

Pathak and Keshari (2004) found that 2 weeks prior inoculation of root-knot nematode *M. incognita* before the soil-borne fungus *F. oxysporum* f. sp.

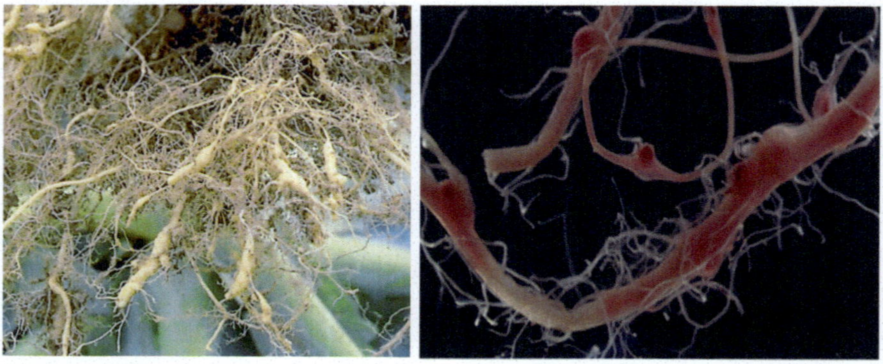

**Fig. 17.5** Symptoms of root-knot nematodes on cabbage (severe galling on roots) (*left*) and close up of root-knot nematode-infected root stained with acid fuchsine (*right*)

*conglutinans* resulted in maximum wilt score (4.0) and wilt symptom index (80.0) in cauliflower plants.

#### 17.1.3.3 Favorable Conditions

The most severe damage can occur in sandy soil with moderate moisture. Infection can occur at temperatures ranging from 10 to 35 °C. Freezing temperatures kill all life cycle stages of *Meloidogyne* species.

#### 17.1.3.4 Survival

These nematodes survive in infected root debris.

#### 17.1.3.5 Management

- Cabbage cultivar "Oxylus" is reported to be a poor host of root-knot nematodes and recommended for cultivation by farmers.
- Soil fumigation with Dazomet.
- Reduction in root-knot nematode population by fallowing or flooding.
- Crop rotation with grass crops for at least 2 years effectively reduces nematode population below damaging levels.

## 17.2 Conclusion

The stunt nematode (*Tylenchorhynchus brassicae*), the cyst nematode (Heterodera cruciferae), and root-knot nematodes (Meloidogyne incognita, M. javanica) are the major nematode pests of cole crops. Cabbage cultivar "Oxylus" is reported to be a poor host of root-knot nematodes and recommended for cultivation by farmers. Integrated nematode management strategies should be developed using crop sanitation, crop rotation, nematicide application, and biological control agents.

## References


Alam MM, Ali QG, Masood A, Khan AM (1976) Studies on the chemical changes by the infection of the root-knot nematode (*Tylenchorhynchus brassicae*) in cabbage and cauliflower roots. Indian J Expt Biol 14:517–518

Haque MM, Gaur HS (1985) Effect of multiple cropping sequences on the dynamics of nematode population and crop performance. Indian J Nematol 15:262–263

Khan AM, Saxena SK, Khan MW (1971) Interaction of *Rhizoctonia solani* Kuhn and *Tylenchorhynchus brassicae* Siddiqi, 1961 in pre-emergence damping-off of cauliflower seedlings. Indian J Nematol 1:85–86

Khan AH, Mahmood I, Saxena SK (1986) Interaction of certain crucifers to *Tylenchorhynchus brassicae*. Int Nematol Netw Newsl 3:12

Pathak KN, Keshari N (2004) Interaction of *Meloidogyne incognita* with Fusarium oxysporum f. sp. conglutinans on cauliflower. Indian J Nematol 34:85–87

Siddiqi MA, Alam MM (1989a) Seed treatment with azadirachtin for the control of the stunt nematode attacking cabbage and cauliflower. Ann Appl Biol 113:4–5

Siddiqi MA, Alam MM (1989b) Control of stunt nematode by bare-root dip in leaf extracts of margosa and Persian lilac. Pakistan J Nematol 7:33–38

Siddiqi MA, Saxena SK (1987a) Effect of interculture of margosa and Persian lilac with tomato and eggplant on root-knot and reniform nematodes. Int Nematol Netw Newsl 4(2):5–8

Siddiqi MA, Saxena SK (1987b) Studies on the control of the stunt nematode, *Tylenchorhynchus brassicae* by interculture of margosa and Persian lilac. Int Nematol Netw Newsl 4(4):27–29

Siddiqui ZA, Khan AM, Saxena SK (1972) Host range and varietal resistance of certain crucifers against *Tylenchorhynchus brassicae*. Indian Phytopath 25:275–281

Siddiqui ZA, Khan AM, Saxena SK (1973) Studies on *Tylenchorhynchus brassicae* II - effect of temperature and moisture on multiplication of nematode. Indian J Entomol 26:139–147

Siddiqui ZA, Khan AM, Saxena SK (1976) Nematode population and yield of certain vegetables as influenced by oil-cake amendments. Indian J Nematol 6:179–182

Varma MK, Sharma HC, Pathak VN (1978) Evaluation of systemic granular nematicides against *Tylenchorhynchus* spp. parasitic on cabbage. Indian J Nematol 8:59–60

# Leafy Vegetable Crops

# 18

**Abstract**

Nematodes are one of the major constraints in leafy vegetable crops like lettuce and celery production. Root-knot nematodes (*Meloidogyne* spp.) are the major problems on leafy vegetable crops. Symptoms and management using physical, cultural, chemical, biological, and integrated methods for root-knot nematodes are discussed.

**Keywords**

*Meloidogyne* spp. · Symptoms · Management

## 18.1 Lettuce, *Lactuca sativa*

Root-knot and reniform nematodes are recognized as the major limiting factors in successful cultivation of lettuce crop.

### 18.1.1 Root-Knot Nematodes, *Meloidogyne* spp.

The root-knot nematodes cause large decreases in yield of lettuce.

#### 18.1.1.1 Symptoms

Root-knot nematodes feed within the roots and cause characteristic galls on roots. The northern root-knot nematode *M. hapla* generally occurs in cooler regions than the other three *Meloidogyne* species that prefer hot summer climates. Galls formed by *M. hapla* are spherical, distinct, and generally smaller than those caused by the three "warm-climate" species (Fig. 18.1). Other symptoms include patchy



P. P. Reddy, *Nematode Diseases of Crops and their Management*,
https://doi.org/10.1007/978-981-16-3242-6_18

**Fig. 18.1** Lettuce plants infected with *Meloidogyne hapla* (*left*) and *M. incognita* (*right*)

appearance of infected fields during middle of the crop season, stunted growth of seedlings, and drastic reduction in lettuce yields.

#### 18.1.1.2 Management

##### 18.1.1.2.1 Physical Methods

Steaming of greenhouse soil is used for the management of root-knot nematodes in lettuce crop.

Integration of soil solarization along with chicken manure was found most effective against root-knot nematode, while solarization alone gave partial control.

##### 18.1.1.2.2 Cultural Methods

Planting of 4 and 5 seedlings of *Tagetes erecta* in a pot around a lettuce plant reduced root galling by 56 and 72% and nematode population by 73% and 94%, respectively (Perwez et al. 1988).

Strawberries may be a suitable rotation crop. Bio-fumigation with mustard under black polythene for 30 days followed by soil application of *Pseudomonas fluorescens* at 2.5 kg/ha at the time of planting.

##### 18.1.1.2.3 Chemical Methods

Soil application of Carbofuran at 1 kg a.i. /ha was found effective.

#### 18.1.1.2.4 Biological Methods

The number of root galls was significantly decreased in *Pasteuria penetrans*-treated pots with 75 (77.2%) and 150/root (54.4%) in 200,000 and 100,000 *P. penetrans* endospores/1 g medium, respectively, compared to the 330/root in inoculated control when harvested at 10 weeks after planting.

Effective reduction of number of eggs of *M. hapla* on lettuce was obtained by soil incorporation of biological control agents like *Streptomyces costaricus* and *Bacillus thuringiensis* (Chen et al. 2000).

The fungi *Hirsutella rhossiliensis* and *Pochonia chlamydosporia* applied against *M. hapla* on lettuce were effective in reducing second-stage juveniles in seedlings when applied alone or in combination (Viaene and Abawi 2000).

The bioagent *Paecilomyces margaundii* reduced egg numbers of *M. hapla* on lettuce as compared to control (Chen et al. 2000).

#### 18.1.1.2.5 Integrated Methods

Bio-fumigation with mustard under black polythene for 30 days followed by soil application of *Pseudomonas fluorescens* at 2.5 kg/ha at the time of planting was effective in enhancing the lettuce leaf yield by 23.29% and decreasing the population of root-knot nematodes by 76.39% which was on par with Carbofuran at 1 kg a.i./ha.

## 18.2 Celery, *Apium graveolens*

### 18.2.1 Root-Knot Nematodes, *Meloidogyne hapla, M. incognita,* and *M. javanica*

#### 18.2.1.1 Symptoms

Celery infected with root-knot nematodes cause stunting (Fig. 18.2), uneven stand of plants, and characteristic galls (numerous knobby growths) on roots (Fig. 18.3). Small feeder roots will have been destroyed, with irregular galls growing in their place. Even in the presence of enough moisture in soil, the nematode-infected plants are likely to wilt earlier as compared to healthy plants. Plants begin to wilt, when nematodes interfere with moisture uptake. New growth comes to a standstill. Since the deranged roots are not in a position to absorb enough water and nutrients from the soil, the plant leaves become chlorotic. Under severe nematode infestation during extended warm growing season, the plant stand is reduced.

#### 18.2.1.2 Management

- Infected plants to be uprooted and burnt.
- Rotating with nonhost crops can reduce nematode population levels.
- Farm equipment and machinery operated in infested fields to be thoroughly cleaned before moving those to noninfested fields.
- Celery grown in flooded soil has lower nematode damage (Fig. 18.4).

**Fig. 18.2** *Background*—field symptoms of root-knot nematodes on celery. *Front*—uprooted root-knot-infected plant showing stunting and severe root galling (*left*) and healthy (*right*) celery plants

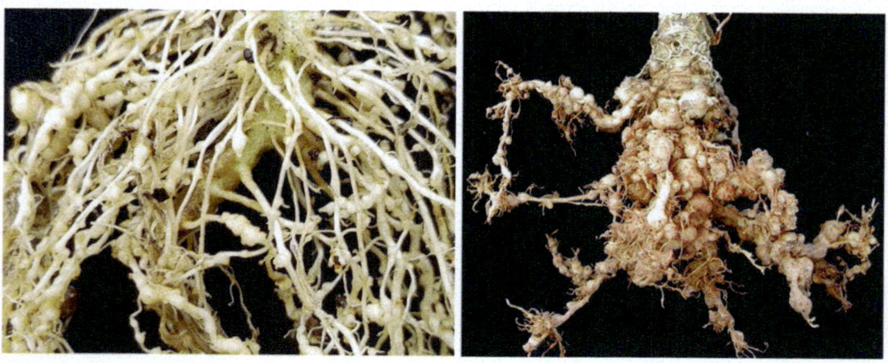

**Fig. 18.3** Severe galling on celery roots

## 18.3 Conclusion

The root-knot nematodes *Meloidogyne* spp. are so far the most important pests of leafy vegetables. In view of the short life span of the leafy vegetable crops and the systemic nature of the available nematicides, field sanitation and other cultural control methods are given preference over chemicals. Identification of resistant indigenous leafy vegetable varieties would contribute greatly to the management of the pests in the cropping systems. The identified resistant varieties can be used as rotation crops in agricultural production systems as a component of root-knot nematode suppression in the soil. Amaranth, sun hemp, and *Cleome* spp. (spider

**Fig. 18.4** Flooding in celery field for root-knot nematode management

plant) were poor hosts with some level of resistance resulting in reduced incidence and prevalence of root-knot nematodes and could be cultivated in areas where the pathogenic nematodes are endemic. The farmers are advised to grow *Crotalaria* spp. in rotational and mixed cropping systems to reduce nematode populations for subsequent and companion susceptible crops.

## References


Chen J, Abawi GS, Zuckerman BM (2000) Efficacy of *Bacillus thuringiensis, Paecilomyces marguandii* and *Streptomyces costaricanus* with and without organic amendments against *Meloidogyne hapla* infecting lettuce. J Nematol 32:70–77

Perwez MS, Rehman MF, Haider SR (1988) Effect of *Tagetes erecta* on *Meloidogyne javanica* infecting lettuce. Int Nematol Netw Newsl 5(3):18–19

Viaene NM, Abawi GS (2000) *Hirsutella rhossiliensis* and *Verticillium chlamydosporium* as biocontrol of root-knot nematode *Meloidogyne hapla* on lettuce. J Nematol 32:85–100

# Mushrooms

# 19

**Abstract**

Nematodes are one of the most dangerous pests of button mushroom, which once enter the beds cannot be eradicated completely, until and unless crop beds are destroyed and disposed-off completely. Mycelium of the fungi is a favorable source of food for nematodes. Total failure of crop or drastic reduction in yield occurs in the presence of nematodes in the beds. The various factors responsible for easy access of nematodes in cropping beds include lack of awareness about cultivation practices, improper pasteurization in short method, compost preparation using long method where no pasteurization is done, and seasonal cultivation under unhygienic conditions. It is better to prevent entry of nematodes into the beds (sterilized production of substrates, containers, implements, gloves, etc.) rather than controlling them afterward. The nematode infestation and damage can be significantly reduced by using the botanicals like oil cakes and leaves of neem, *Pongamia*, and castor and providing hygienic production conditions.

**Keywords**

*Aphelenchoides composticola* · *Ditylenchus myceliophagus* · Symptoms · Biology · Management

## 19.1 Introduction

The two commonly grown mushrooms in India include button mushroom *Agaricus bisporus* and paddy straw mushroom *Volvariella volvacea*. The first report about nematode damage to mushroom in India was made from district Solan of Himachal Pradesh by Bhardwaj et al. (1972) when they reported poor sporophore yield having brownish white, stunted mushrooms from beds infested with *Aphelenchoides* and *Ditylenchus*. Mushroom nematodes comprising mainly *Aphelenchoides agarici*,





P. P. Reddy, *Nematode Diseases of Crops and their Management*,
https://doi.org/10.1007/978-981-16-3242-6_19

*A. composticola, A. myceliophagus, A. neocomposticola, A. sacchari, A. swarupi, Aphelenchus avenae, Ditylenchus myceliophagus,* and *Rhabditis* on button mushroom (*Agaricus bisporus*) have now been well recognized as the serious constraint in achieving higher production. *Aphelenchoides composticola* and *Ditylenchus myceliophagus* were most common and were recorded from compost as well as spent compost that is used as casing material; the latter is the main source of contamination in mushroom houses. Population of fungal feeding nematodes, especially of *Aphelenchus avenae*, was highest during midcrop season and came down drastically at the end of crop season.

## 19.2 Mushroom Nematodes

### 19.2.1 Myceliophagous Nematodes

In all, 21 nematode species representing two orders—Aphelenchida and Tylenchida—have been reported to be harmfully associated with mushroom cultivation from various parts of the world. Among these, 20 species belong to 4 genera (*Aphelenchoides, Aphelenchus, Paraphelenchus,* and *Seinura*) of order Aphelenchida and the only Tylenchid is *Ditylenchus myceliophagus* (Table 19.1). *Aphelenchoides composticola* and *D. myceliophagus* are the major parasitic nematode problems on mushrooms beds in India.

#### 19.2.1.1 *Ditylenchus myceliophagus*

This nematode (Fig. 19.1) is widely distributed in almost all mushroom-growing belts of the world.

**Table 19.1** Pathogenicity of myceliophagous nematodes associated with mushrooms

| Nematode order | Nematode species | Pathogenicity |
|---|---|---|
| Tylenchida | *Ditylenchus myceliophagus*[a] | +++ |
| Aphelenchida | *Aphelencoides agarici*[a] | +++ |
| | *A. bicaudatus* | +++ |
| | *A. composticola*[a] | +++ |
| | *A. dactylocerus* | ++ |
| | *A. limberi* | ++ |
| | *A. neocomposticola* | ++ |
| | *A. sacchari*[a] | +++ |
| | *A. saprophilus* | ++ |
| | *A. swarupi*[a] | ++ |
| | *Aphelenchus avenae* | ++ |
| | *Paraphelenchus myceliophthorus* | ++ |
| | *Seinura winchesi* | ++ |

[a]Species reported from India, +++ Highly pathogenic, ++ Pathogenic

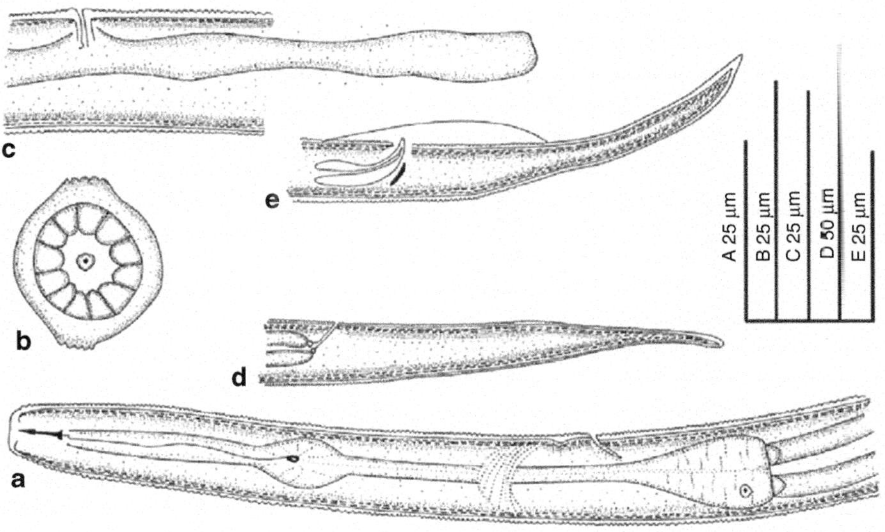


**Fig. 19.1** *Ditylenchus myceliophagus*. A—anterior end; B—cross-section of mid-body showing lateral field; C—post-vulval uterine sac; D—female posterior end; E—male posterior end

In India, it is found in Himachal Pradesh, Jammu and Kashmir, and Punjab (Thapa et al. 1981). It is one of the highly destructive species. An inoculum level of 10–1000 individuals/100 g of compost could cause 15–70% mycelial depletion within 60 days (Sharma et al. 1985). However, it is considered to be less pathogenic than *A. composticola* and *A. sacchari* (Sharma et al. 1981).

#### 19.2.1.2 *Aphelenchoides composticola*

It is one of the most widely distributed species present throughout the mushroom-growing countries of the world which is reported from Punjab and Himachal Pradesh, India. This nematode (Fig. 19.2) also has a short life cycle of 8 days at 23 °C and multiplies very fast on mushrooms (Cayrol 1967). An initial inoculum of 10 nematodes/100 g of compost completely depleted the mycelium within 40 days resulting in about 1000 times multiplication rate (Khanna and Sharma 1983). *A. composticola* was responsible for 35–60% loss in yield of mushroom (Laqman Khan 2001).

#### 19.2.1.3 *Aphelenchoides agarici*

This species was recorded from mushroom farms in Himachal Pradesh (Seth and Sharma 1986). It reproduces well on mushroom mycelium which is one of the most destructive nematode species. An initial inoculum level of 10 nematodes/100 g of compost has been shown to cause 100% mycelial depletion within 25 days at 28 °C resulting over 800 times multiplication rate under laboratory conditions (Khanna and Sharma 1988). It completes its life cycle within 8 days at 28 $\pm$ 1 °C, and the ratio between female to male was 1.5: 1(Seth 1984).

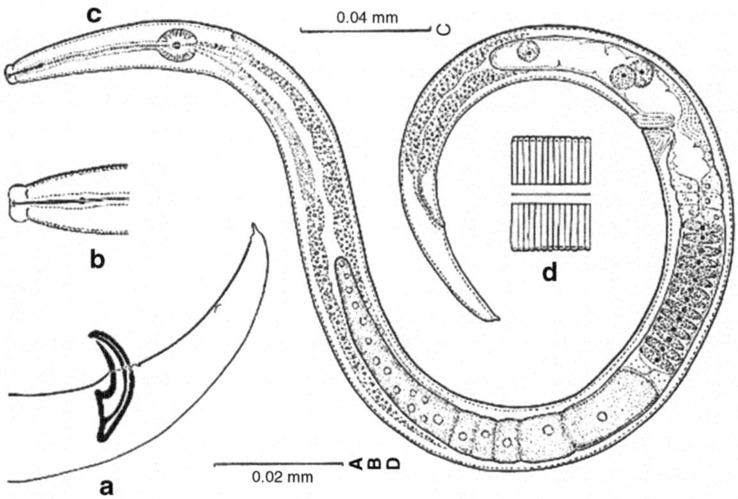


**Fig. 19.2** *Aphelenchoides composticola.* A—male tail; B—head; C—female; D—annules and lateral field

### 19.2.1.4 *Aphelenchoides myceliophagus*

It is recorded from Solan district in Himachal Pradesh. It is also one of the highly destructive nematode species infesting mushroom showing the pathogenicity status of the level of *A. composticola* (Khanna and Sharma 1988).

### 19.2.1.5 *Aphelenchoides neocomposticola*

It was reported from Shimla district of Himachal Pradesh. Khanna and Sharma (1988) found that the nematode caused much severe damage as compared to *A. agarici*, *A. myceliophagus,* and *A. composticola.*

### 19.2.1.6 *Aphelenchoides sacchari*

It was reported by Sharma et al. (1981) and found to be highly pathogenic to white button mushroom *Agaricus bisporus*. An average of 50–100% yield loss in sporophores in white button mushroom has been reported due to this nematode species (Sharma et al. 1984). It is quite prevalent on mushrooms in India (Thapa et al. 1981). Singh et al. (2003) reported that *A. sacchari* was responsible for 40.6–100% loss in yield of mushroom.

### 19.2.1.7 *Aphelenchoides swarupi*

This fungal feeding species was for the first time found in a mushroom farm at Ambala (Haryana), India, where though brown wet pinheads were observed initially, no fruiting bodies were produced. An average of about 10,000 individuals/100 g of compost was counted in the beds showing total crop failure.

#### 19.2.1.8 *Aphelenchus avenae*

This nematode is frequently present in low numbers in compost and casing soil samples taken from cropping beds.

#### 19.2.1.9 *Seinura winchesi*

In India, population of *Seinura* sp. were found from the cropping beds of a mushroom farm at Solan, Himachal Pradesh. It reduced sporophore yields even though the initial growth of mycelium was normal.

### 19.2.2 Saprophagous Nematodes

In India and other parts of the world, almost all mushroom farms were severely damaged by the saprophagous nematodes, belonging to the order Rhabditida. *Acrobeloides, Cephalobus, Diplogaster,* and *Panagrolaimus* are the major nematodes occurring normally in mushroom-growing areas. The greatest development of *Rhabditis cucumeris* (1500/g) was observed when the nematodes were introduced before spawning, and less (900/g) if the infestation and spawning were simultaneous (Dmowska et al. 1997).

### 19.2.3 Crop Losses

Yield losses to the tune of 16.3–53.5% were recorded due to *Aphelenchoides composticola* and *Ditylenchus myceliophagus* inoculated at 100 nematodes/kg compost at the time of spawning and casing in 3 kg plastic bags, respectively. Losses caused by *A. sacchari* range from 87% to 100% (Sharma et al. 1984).

### 19.2.4 Nature of Damage

The mouth parts of myceliophagous nematodes have the stylet (needle like structure) which is hollow inside and can be moved forward and backward by the contraction and relaxation of the muscles. The nematodes secrete variety of strong enzymes. These enzymes act immediately after ejection and help in penetration of stipe and to convert the cell contents to assimilable forms. The tremendous reproduction potential of is 50–100 fold/week. Rate of multiplication is faster during spawn-run period (22–28 °C) than the cropping period (14–18 °C). Beyond 30 °C, these do not reproduce. The complete destruction of 100 g mycelium was achieved only by three nematodes of *D. myceliophagous* within 70 days.

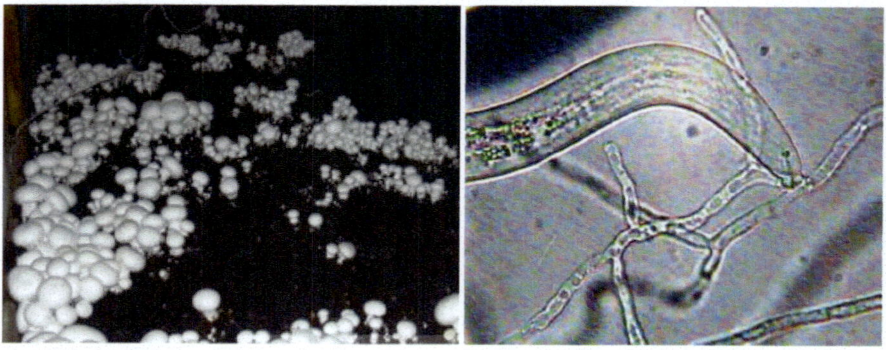

**Fig. 19.3** *Left*—severe infection of the casing layer by mushroom nematodes showing half-empty beds. *Right*—*aphelenchoides* sp. feeding on mushroom hyphae

### 19.2.5 Symptoms

Following symptoms of nematode attack appear in infected beds in succession (Fig. 19.3):

- Mycelial growth is sparse, patchy and mycelium turns stingy.
- The compost surface sinks.
- Whiteness of spawn-run slowly changes to brown.
- Mushroom flushes are poor and delayed.
- Alternate high and poor yield in successive flushes.
- Browning of pinheads.
- Decline in yield.
- Complete crop failure.
- White fungal growth over casing in patches.

### 19.2.6 Biology and Life Cycle

The life cycle of aphelenchids is very short (8–10 days), while *Ditylenchus* takes about 1 month.

### 19.2.7 Spread and Survival

The common source of nematode contamination of mushroom beds is through casing soil and compost. However, other common sources responsible for nematode contamination include damp wheat straw, manures, FYM, garden soil, spent compost, platform soil, irrigation water, sciarid flies, and contaminated implements.

When the compost is dried gradually, the nematodes survive as long as 2 years in a state of anabiosis (a restoring to life from a death-like condition), but they die rapidly if the compost is dried quickly.

### 19.2.8 Management

Prevention of the entry of nematodes into the mushroom beds rather than controlling them afterward is the key for nematode management in view the nature of the crop, limited availability of safe and potent nematicides, residue problems, and other hazards of nematicides. Integrated nematode management using available components is the only option, since curative measures are not available.

#### 19.2.8.1 Prophylactic Measures

- Purposely built mushroom houses with proper ventilation should be utilized for mushroom cropping using properly pasteurized compost made by short method.
- Throughout the cropping season, strict sanitation and hygiene measures should be followed.
- Avoid direct contact of compost with the soil by using cemented composting yard disinfected with 2% formalin, 24 h earlier to compost preparation.
- All the instruments, walls, floors, and galleries should be disinfected with 4% formalin.
- Composting ingredients should always be stored in clean area.
- Cleanliness should be maintained inside and in surroundings of mushroom farms.
- Casing mixture should be properly pasteurized.
- Manures used for composting should be thoroughly broken and allowed to decay properly.
- In long method of composting, covering compost with double PVC sheet for 24 h after third turning gives effective control of nematodes.
- Each should be provided with foot dips in the front.
- Irrigation water should be clean.
- Proper disinfection of hands and feet is a must for each worker to enter cropping room.
- All the cropping rooms should be fly proof and only recommended insecticides should be sprayed for the control of flies.
- In long method, chicken manure may be avoided. Neem cake can be added

#### 19.2.8.2 Physical Methods

Use of heat is the most successful method of nematode control in mushroom cultivation. It is recommended that to make compost nematodes free, air and bed temperature in the pasteurization room must be maintained at 60 °C for at least 2 h and cook out of mushroom house at 70 °C for 5–6 h or 80 °C for 30–60 min is necessary. Complete disinfection of appliances like handling tools and mushroom trays should be carried out by dipping them in formalin or boiling water to kill nematodes. Finally, spent substrate should be disposed-off completely. Destruction of all spent compost is of prime importance as dry infested debris is a potential source of trouble, since *D. myceliophagus* can withstand drying for up to 3 years. Lowering down the room temperature to 13 °C during cropping period retards the pathogenicity of these nematodes.

Seth (1984) tested the efficiency of boiling water against *A. agarici*; complete kill of nematodes was achieved when compost was dipped in boiling water for 15 s and above. Steam was most efficient for killing nematodes when used for 4 min or more.

#### 19.2.8.3 Cultural Methods

A large number of plants have been found to possess nematicidal properties. Enhancement of mushroom yields by 6.5–19% and reduction of the population of *A. composticola* below economic injury levels could be achieved by incorporation of dried leaves of *Azadirachta indica*, *Cannabis sativa*, *Eucalyptus tereticornis,* and *Ricinus communis* at 3 kg/100 kg of dry wheat straw, which enhanced the population of thermophilic fungi, mesophilic antibiotic producing fungi (Grewal 1989). Karanj leaves mixed in compost also reduce the nematode population. Addition of neem leaf powder at 2% on w/w basis of compost not only check nematode multiplication in mushroom-growing trays but also helps in proper mycelial growth and improved sporophore yields (Khanna and Sharma 1994). Enhancement of mushroom yields and reduction in nematode reproduction (*A. compostiocola*) was obtained by application of oil cakes like neem, Karanj (*Pongamia pinnata*), Coconut, Castor, and Groundnut at 5% on w/w basis of compost at spawning (Rao et al. 1992). Karanj cake both at 2% and 4%, coconut, niger, and neem cakes each at 4% were effective in reducing *A. composticola* population. Coconut, karanj, and neem cakes each at 2 and 4% and sesame cake at 4% were as effective as chemical treatments in reducing *Rhabditis* spp. population (Parvatha Reddy et al. 1999).

The yield of button mushroom was highest when 1% neem seed kernel extract and Achook at 400 ppm were applied in combination at the time of spawning. The *Aphelenchoides composticola* nematode population was also reduced drastically in the treated boxes compared to inoculated untreated control (Rajesh et al. 2003). Neem cake at 20 g was most efficient in reducing the population of *Aphelenchoides composticola*, and neem leaf at 16 g dose showed manurial effect on *Agaricus bisporus* in terms of yield potential (Gitanjali and Nandal 2001).

Karanj cake both at 2% and 4%, coconut, niger, and neem cakes all at 4% were effective in reducing *A. composticola* population.

Treatment of compost with neem seed kernel powder solution at 7.5 L per 100 kg was found effective. (Prepare solution by soaking 400 g of neem seed kernel powder in 10 L water overnight, strain through muslin and use.)

Dry neem leaf powder at 2% w/w basis when mixed in compost at spawning not only kept the nematode population under check but also improved button production (Khanna and Sharma 1994).

Neem cake at 5% when mixed in nematode-infested compost at spawning hindered *A. composticola* multiplication and enhanced the sporophore production significantly (Khanna and Sharma 1996).

Application of 4% (w/v) neem seed kernel extract (NSKE) at spawning at 7.5 L/ 100 kg compost can be used as prophylactic measure for the management of *Aphelenchoides* spp.

#### 19.2.8.4 Chemical Methods

Mushrooms, being indoor crop, provide little scope for pesticidal usage. Short life of the crop and deleterious effect of pesticides on mycelium and residue problems further limits the scope of pesticide application. The nematode population can be managed by using certain safe chemicals especially during long method of composting. Dichlorvos (0.04%) under polythene cover for 3–4 days, Fenamiphos EC at 20 mg/kg incorporated in compost, and Thionazin at the rate of 80 ppm have been found to be most effective for control of *A. composticola* and *Rhabditis* sp

#### 19.2.8.5 Biological Methods

Since the use of chemicals for the management of nematodes has many constraints, only alternative left is to exploit biological means. Various microorganisms have been exploited for the management of mushroom nematodes. Extensive work has been done on nematode-trapping fungi such as *Arthrobotrys oligospora* and *A. superba*, which are known to reduce the nematode population and increase the yield. Fungus *A. robusta* has been recommended against mycophagous nematodes. The nematophagous fungus *Arthrobotrys irregularis* commercially available as "Royal 300" (strain Antipolis) in France was found to be highly effective against *A. composticola* (Anon 1994).

Two nematophagous fungi *Arthrobotrys irregularis* and *Candelaretta musiformis*, isolated from spent compost, have been demonstrated as highly effective in checking nematode multiplication on mushroom mycelium in Himachal Pradesh (Khanna and Sharma 1990).

A mixed nematode population of *A. composticola* and *Rhabditis* sp. can be reduced by 42.5% in 25 days by incorporation of mass-produced *Arthrobotrys conoides* on wheat grain into compost at spawning (Grewal and Sohi 1988).

The bacterium *Bacillus thuringiensis* is found to suppress the population of *Ditylenchus myceliophagus* (Cayrol 1974).

#### 19.2.8.6 Host Resistance

*Agaricus edulis* has recorded resistance against *D. myceliophagus*. Some other edible mushrooms like *Pleurotus sajor caju* and *Stropharia rugosa annulata* have shown resistance against *A. sacchari* and *D. myceliophagus* (Thapa et al. 1987; Khanna and Sharma 1989; Chandel and Sharma 1991). *A. bitorquis* (K-30) is resistant to *A. saachari* but susceptible for *D. myceliophagus*.

## 19.3 Conclusion

The major parasitic nematodes include *Aphelenchoides composticola* and *Ditylenchus myceliophagus* which cause degeneration of the mushroom mycelium in the compost. Nematodes cause heavy losses in mushroom production. Compared to other countries, the productivity of mushrooms is significantly less in India. Oyster mushrooms are relatively resistant to nematodes as compared to the button mushrooms which are generally highly susceptible. Enhancement of yields and effective management of mushroom nematodes should be attempted by the use of plant products (leaves, nonedible oil cakes of neem, *Pongamia*, castor, etc.), prophylactic practices (sterilized production of substrates, containers, implements, gloves, etc.), and maintenance of hygienic production conditions.

## References

Anon (1994) Biennial report for AICRP on nematode pests of crops and their management. Department of Entomology, University of Horticulture & Forestry, Solan, Himachal Pradesh

Bhardwaj AK, Sharma PL, Thakur JR (1972) Mushroom nematodes. Indian Bot Rep 42:257–258

Cayrol JC (1967) E'tude de cycle e'volutif d' *Aphelenchoides composticola*. Nematologica 13:23–32

Cayrol JC (1974) Proceedings of the XII international symposium de nematologia, Society of European Nematologists, Grenada, pp 21–22

Chandel YS, Sharma NK (1991) Relative susceptibility of some commonly grown mushroom species to *Aphelenchoides sacchari*. Indian Phytopathol 44:143–144

Dmowska E, Dmoch J, Ilieva K (1997) Early interaction of the bacterivorous nematode *Rhabditis cucumeris* and the edible fungus *Agaricus bisporus* in relation to time and quantity of nematode inoculation. Int J Mush Sci 2:15–23

Gitanjali, Nandal SN (2001) Nematodes associated with white button mushroom (*Agaricus bisporus*) in Sonipat district of Haryana. Indian J Nematol 31:95–96

Grewal PS (1989) Effect of leaf-matter incorporation on *Aphelenchoides composticola* (Nematode), mycofloral composition, mushroom compost quality and yield of *Agaricus bisporus*. Ann Appl Biol 115:299–312

Grewal PS, Sohi HS (1988) A new and cheaper technique for the rapid multiplication of *Arthrobotrys conoides* and its potential as a bio-nematicide in mushroom culture. Curr Sci 57:44–46

Khanna AS, Sharma NK (1988) Pathogenicity of various *Aphelenchoides* species on *Agaricus bisporus* (Lange) singer. Indian Phytopathol 41(3):472–473

Khanna AS, Sharma NK (1989) Resistance of *Pleurotus sajor caju* against myceliophagous nematode *Aphelenchoides agarici*. Nematol Medit 17:61

Khanna AS, Sharma NK (1990) Efficacy of nematophagous fungi against myceliophagous nematodes *in vitro*. In: Proceedings of national seminar on bio-agents in nematode management. Indian Agricultural Research Institute, New Delhi, pp 83–84

Khanna AS, Sharma NK (1994) Management of *Aphelenchoides composticola* infesting white button mushroom by neem cake. Indian J Hill Fmg 7:114–115

Khanna AS, Sharma VP (1996) Management of *Aphelenchoides composticola* infesting white button mushrooms by neem cake. Indian J Hill Fmg 1(2):41–44

Laqman Khan M (2001) Dr. Y.S. Paramar University of Horticulture and Forestry, Solan, Himachal Pradesh. In: Dhawan SC et al (eds) Indian nematology-progress and prospectives. Division of Nematology, Indian Agricultural Research Institute, New Delhi, pp 153–158

Parvatha Reddy P, Rao MS, Tewari RP, Nagesh M (1999) Evaluation of oil cakes for their efficacy against mushroom nematodes in chicken manure. Pest Mangmt Hort Ecosyst 5:50–53

Rajesh V, Bajaj HK, Mann SS (2003) Integrated management of *Aphelenchoides composticola* in button mushroom with neem seed kernel water extract and Achook. In: Singh RV, Pankaj, Dhawan SC, Gaur HS (eds) Proceedings of the national symposium biodiversity and management of nematodes in cropping systems for sustainable agriculture, Jaipur, pp 228–231

Rao MS, Tewari RP, Pandey M (1992) Occurrence of mushroom nematodes in South India. Mushroom Res 1:135–136

Seth A (1984) Studies on the incidence, pathogenicity, biology and control of Aphelenchoides species associated with mushroom. Ph. D. thesis, Department of Entomology and Apiculture, Himachal Pradesh Krishi Viswa Vidyalaya, Solan

Seth A, Sharma NK (1986) Five new species of genus *Aphelenchoides* (Nematoda: Aphelenchida) infesting mushroom in North India. Indian J Nematol 16:205–215

Sharma NK, Thapa CD, Nath A (1981) Pathogenicity and identity of myceliophagus nematode infesting *Agaricus bisporus* (Lange) Sing. in Himachal Pradesh (India). Indian J Nematol 11:230–231

Sharma NK, Thapa CD, Kaur D (1984) Estimation of loss in mushroom yield due to myceliophagus nematode, *Aphelenchoides sacchari*. Indian J Nematol 14:121–124

Sharma NK, Thapa CD, Kaur D (1985) Pathogenicity and identity of *Ditylenchus myceliophagus* "Indian population" infesting *Agaricus bisporus* (Lange) Sing. in India. Indian J Nematol 15:233–234

Singh RV, Midha SK, Kumar V (2003) Major nematode pests of economically important crops in India and their management. In: Trivedi PC (ed) Advances in nematology. Scientific Publishers (India), Jodhpur, pp 233–256

Thapa CD, Sharma NK, Nath A (1981) Mushroom nematodes and their source of infestation in mushroom beds in India. Mushroom Sci 11:443–448

Thapa CD, Jandaik CL, Sharma NK (1987) Comparative resistance in some cultivated edible fungi against two mushroom nematodes occurring in India. In: Proceeding of the international conference on science and cultivation technology of edible fungi. Regional Research Laboratory, Jammu and Kashmir, pp 169–172

# Protected Cultivation of Vegetable Crops 20

**Abstract**

The production of vegetable crops like tomato, bell pepper, cucumber, and lettuce under protected cultivation worldwide is threatened by root-knot nematodes (*Meloidogyne* spp.). In some crops, the extent of crop losses attributed to nematodes is as high as 80%. The nematodes interact with fungal and bacterial pathogens and exacerbate severity of diseases leading to complete crop failure. In this chapter, crop losses, symptoms, and management using physical, cultural, chemical, biological methods, host resistance, and integrated methods for root-knot nematodes are discussed.

**Keywords**

*Meloidogyne incognita* · *M. javanica* · Crop losses · Symptoms · Management

## 20.1 Introduction

Farmers were compelled to grow some selected vegetable crops under protected cultivation (controlled conditions) in order to make available horticultural crop produce throughout the year, particularly during off-season; to export high-quality horticultural products; and to get substantial increase in yield (high productivity). Vegetable farming (tomato, capsicum, cucumbers, French bean, crucifers, gherkins, muskmelon, watermelon, etc.) inside modern greenhouses is promoted in order to improve earning potential of farmers especially during harsh winters. The nematode incidence under protected cultivation became severe and led to complete crop losses because of congenial conditions of higher temperature, humidity, and use of high agronomic inputs like fertilizers and plant growth promoters in polyhouses. The proliferation rates of nematodes in polyhouse cultivation reached up to 10–30-fold more than in the open field cultivation.


P. P. Reddy, *Nematode Diseases of Crops and their Management*,
https://doi.org/10.1007/978-981-16-3242-6_20

### 20.1.1 Major Nematode Problems

Polyhouses were observed to be severely infested with major plant parasitic nematode genera, viz. *Meloidogyne*, *Pratylenchus*, *Rotylenchulus, Helicotylenchus*, *Mesocriconema,* and *Tylenchorhynchus*. The above-ground symptoms of the most economically important root-knot nematode infection include stunting of plant growth, leaf yellowing, and wilting of plants. Root galling is the characteristic under-ground symptom caused by the root-knot nematode which interferes in absorption of water and nutrients from the soil due to disruption of conducting vessels. They feed on the roots devitalizing the plants ultimately affecting both qualitative and quantitative production. Nematode problems on all these crops under protected conditions have assumed alarming proportions leading to huge losses (up to 80%) in select crops. The nematode infestations exacerbate severity of fungal and bacterial diseases leading to complete crop failure. In view of damage potential of these tiny hidden enemies of crop plants, an urgent need is felt for developing nematode management strategies.

In most cases, problems arise from contaminated soil or soil mixture used as a component of the growing medium, monocropping, and infested planting materials. Crop rotation, fallowing, and summer plowing, which are generally employed for nematode management under field conditions, are not effective under protected cultivation. Hence, exclusion or avoidance is of prime importance for nematode management under polyhouse. It is very difficult to control nematodes once they are introduced in the polyhouse. Therefore, prevention is better than cure.

### 20.1.2 Management

The nematodes may be managed by using the following methods:

- Growing cowpea (trap crop for root-knot nematode) closely before the main vegetable crop and removing the plants out of polyhouse at 45 days after sowing.
- Raising marigold (antagonistic to nematodes) and incorporating in situ after plucking the flowers.
- If the beds are highly infested with the nematodes, bring down the population of nematodes by applying neem/*Pongamia* cake at 500 g + Carbofuran 3G at 50 g/ $m^2$.
- Treatment of seed with biopesticide *Pseudomonas fluorescens* at 20 g per kg seeds.
- Raise the seedlings in the coco-peat or any other substrates such as FYM, vermicompost, and neem/*Pongamia* cakes enriched with biopesticides (2 kg each of *Purpureocillium lilacinum*, *Trichoderma harzianum,* and *P. fluorescens* per 2000 kg of FYM/500 kg of vermicompost/200 kg of neem/*Pongamia* cakes).
- Treat the nursery beds with neem or *Pongamia* cake at 50 g, *T. harzianum* at 25 g, or *Pochonia chlamydosporia* + *P. lilacinum* at 25 g/$m^2$ at an interval of 2 months.

- Apply 2 tons of farmyard manure enriched with biopesticide *T. harzianum* per acre before planting, along with 100–200 kg of neem or *Pongamia* cake.
- Apply Carbofuran 3G at 1 kg a.i. /ha at planting and repeat after 45 days.

## 20.2 Tomato, *Solanum lycopersicum*

Productivity of tomato in greenhouse (Fig. 20.1) is increased manifold in comparison to open field. Some of the other advantages of growing vegetables in polyhouse include high economic returns, availability of fresh vegetables in winter, and improved taste, flavor, and shelf life. Unlike field-grown tomatoes which require harvesting and marketing over a short period of time (3–4 weeks), greenhouse tomatoes are harvested and marketed over a 30-week period.

### 20.2.1 Root-Knot Nematodes, *Meloidogyne incognita* and *M. javanica*

#### 20.2.1.1 Crop Losses

*M. incognita* was responsible for 30.57–46.92% loss in fruit yield of tomato, while *M. javanica* caused 77.5% loss in yield.

**Fig. 20.1** Tomato under protected cultivation

#### 20.2.1.2 Symptoms

Root-knot nematodes cause root galls on the feeder roots and sometimes affect the entire root system showing heavy galling (Fig. 20.2). This affects the uptake of nutrients and water, and the plants show wilting during warmer part of the day and results in stunting of plant growth, leaf chlorosis, and reduced yields. In tomato, dynamics of root-knot nematode showed enhanced population build-up from 1–30 juveniles per g soil within a period of 6–12 months, which is comparatively higher in contrast to the open cultivation.

#### 20.2.1.3 Management

##### 20.2.1.3.1 Physical Methods

The effectiveness of solarization, a soil disinfection technique that uses passive solar heating, to control the incidence of root-knot nematode under greenhouse condition was studied. Solarization was accomplished by the application of 0.05–0.06 mm clear polyethylene sheets to moist soil for 4 weeks during the hot season.

Solarization of moist soil with polythene mulching for 4 weeks during peak summer in closed polyhouses increase the soil temperature. Maximum soil temperatures achieved under solarization were 51 °C in the alluvial soil and 52.4 °C in the heavy clay soil type at 20 cm depth. In the non-solarized unheated greenhouse, soil temperature achieved was 39 °C (Fig. 20.3).

The effect of soil solarization treatments on the population densities of the root-knot nematode *M. incognita* was very clear. The larval density per 100 ml of soil was markedly reduced compared with the untreated control (Fig. 20.4). The soil solarization method had a significant influence to reduce the severity of root damage. Root galling index was dramatically reduced in both solarized greenhouses compared with untreated control where the highest root galling index was noticed (Fig. 20.5).

The growth of tomato plants, as indicated by visual assessment and plant height measurements, was significantly improved by the soil solarization. Tomato fruit yields in solarized greenhouses were fourfold higher than those obtained in untreated

**Fig. 20.2** Severe root galling on tomato due to the root-knot nematode infection

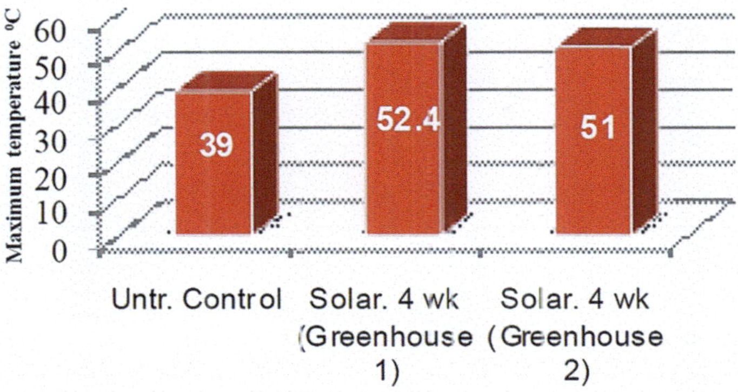


**Fig. 20.3** Maximal temperature attained during soil solarization

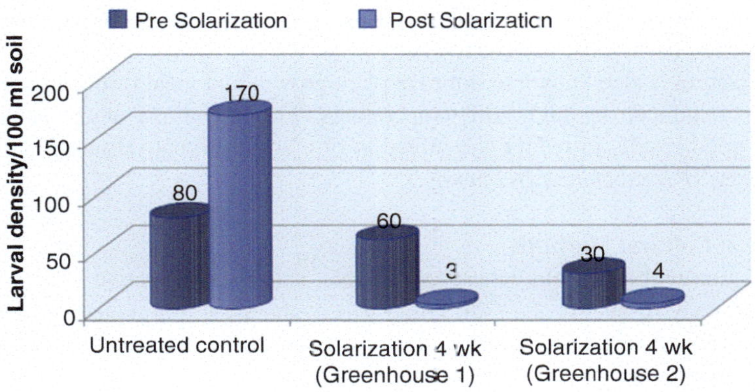


**Fig. 20.4** Effect of soil solarization on the population densities of root-knot nematodes

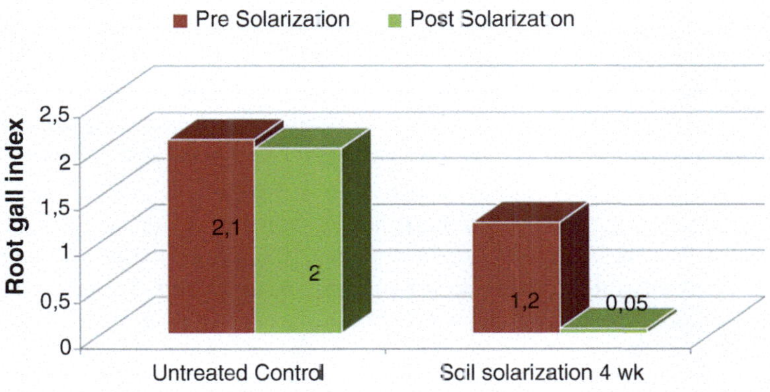


**Fig. 20.5** Effect of soil solarization on root gall index by root-knot nematode

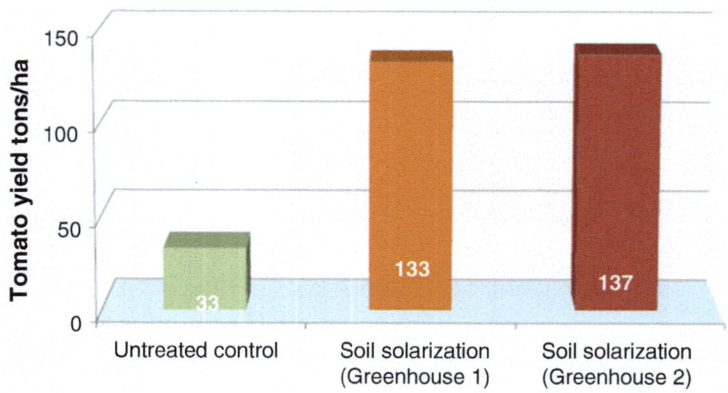


**Fig. 20.6** Effect of soil solarization on tomato fruit yield in tons/ha

control (Fig. 20.6). Such significant increases were consistent with effective control of nematode provided by solarization.

Solarization is also known to improve plant growth and yield through the release of nutrient induced by high soil temperatures. This technique at the same time showed good results to reduce the infection of a severe tomato disease like corky root disease (*Pyrenochaeta lycopersici*).

#### 20.2.1.3.2 Cultural Methods

Crop rotation with marigold, wherever possible, was found effective.

The efficacy of neem cake (320 g/m$^2$) was evaluated in polyhouse tomato. The treatment proved to be promising with benefit:cost ratio of 5.65 in tomato (Anon 2015; Chandel et al. 2014).

#### 20.2.1.3.3 Chemical Methods

Application of Carbofuran 3G at 1 kg a.i. /ha at transplanting proved beneficial.

Addition of Dazomet at 1 kg a.i. /ha to the nursery beds gave effective control of root-knot nematodes.

The efficacy of Metham sodium (40 ml/m$^2$) and Dazomet (40 g/m$^2$) was evaluated in polyhouse tomato. All the treatments proved promising. Benefit:cost ratio in Metham sodium and Dazomet was found to be 5.69 and 4.38, respectively, in tomato (Anon 2015; Chandel et al. 2014).

Metham sodium at 40 ml/m$^2$ reduced nematode reproduction, viz. no. of galls/plant, no. of egg masses/plant, no. of eggs and larvae/egg mass, and nematode population/200 ml soil, and enhanced plant growth characters. Highest yield was also recorded with Metham sodium (10.32 kg/plant) over control (5.25 kg/plant) (Giriraj et al. 2018).

#### 20.2.1.3.4 Biological Methods

Seed treatment with biopesticide *P. fluorescens* at 10 g/kg seed was found effective.

Nursery bed treatment with *T. harzianum* at 50 g/m$^2$ proved beneficial.

#### 20.2.1.3.5 Integrated Methods

Addition of enriched FYM/compost/vermicompost (with *Trichoderma harzianum*/*T. viride* + *Purpureocillium lilacinum* + *Pseudomonas fluorescens* each at 2 kg/ton of organic amendments) at 5 t/ha gave effective management of root-knot nematodes.

After 15–30 days of planting, addition of enriched neem cake (with *T. harzianum*/*T. viride* + *P. lilacinum* + *P. fluorescens* each at 2 kg/ton) at 50 g/m$^2$ gave effective management of nematodes under polyhouse conditions. Nematode management in polyhouse beds can be achieved by application of neem cake suspension enriched with bio-agents (20 kg neem cake + *T. harzianum*/*T. viride* + *P. lilacinum* + *P. fluorescens* each at 2 kg) through drip irrigation.

Application of *P. lilacinum* (2 x 10$^6$ cfu/g) as seed treatment at 10 g/kg seed followed by its soil application at 50 g/m$^2$ was found effective for the management of root-knot nematode *M. incognita*.

## 20.3 Bell Pepper, *Capsicum annuum*

Greenhouses can be a means to economically maintain a warm environment during cool season, to protect bell pepper plants from rain, wind, and high solar radiation, and to retain pollinators and beneficial insects while excluding unwanted insect pests. In greenhouses, capsicum fruits are harvested with full maturation color (Fig. 20.7), and fruit yields are greater of high quality and usually produced at a time of the year when production in the field is not possible and market prices for capsicum are highest.

**Fig. 20.7** Colored bell pepper in greenhouse

## 20.3.1 Root-Knot Nematode, *Meloidogyne incognita*

### 20.3.1.1 Symptoms

Root-knot nematodes cause root galls on the feeder roots which sometimes affect the entire root system showing heavy galling. Root galls caused by root-knot nematodes on sweet pepper are frequently small (Fig. 20.8). This affects the uptake of nutrition and water, and the plants also show wilting during warmer part of the day resulting in stunting of plant growth, leaf chlorosis, and reduced yields. In capsicum, a pathogenic bacterium *Ralstonia solanacearum* gets entry into the roots infested by root-knot nematode, and together they produce wilting disease complex that reduces yield by 60–70%.

### 20.3.1.2 Management

#### 20.3.1.2.1 Physical Methods

Solarization using plastic sheets for covering soil during peak summer in greenhouse reduces nematodes, bacteria, fungi, and weed population.

#### 20.3.1.2.2 Cultural Methods

Several cultural methods like selection of nematode-free nursery sites and destruction of infested roots after crop harvest, either alone or in combination, proved to be reasonably effective and cheap and prevent nematode reproduction on bell pepper.

#### 20.3.1.2.3 Chemical Methods

Bare-root dip treatment of capsicum seedlings in 0.1% Carbosulfan/Monocrotophos for 6 h eliminates root-knot infection.

**Fig. 20.8** *Left*—bell pepper plants infected with root-knot nematodes showing leaf yellowing and stunting. *Right*—severe root galling

#### 20.3.1.2.4 Biological Methods

Antagonistic Bacteria

Zuckermann et al. (1993) reported that significant enhancement of yield and reduction in galling in root galling due to *M. incognita* was obtained by soil incorporation of a *Bacillus thuringiensis* strain to bell pepper.

Antagonistic Fungi

Application of "Royal 350" (*Arthrobotrys irregularis* cultured on oat seed medium) at 140 g/m$^2$ a month before transplantation of capsicum resulted in good protection against root-knot nematodes. Among the egg parasites, efficacy of *Purpureocillium lilacinum* (commercially formulated as "Biocon" in the Philippines) has been found to be comparatively higher in suppressing the population of *Meloidogyne* spp. on capsicum.

Treatment of the greenhouse beds with the formulation of *Pochonia chlamydosporia* at 50 g/m$^2$ was significantly effective in reducing the galling index, number of nematodes in roots and soil, increasing the percent parasitization of eggs by bio-agent, and enhancing yields. Effective management of root-knot nematode can also be achieved by treatment of nursery with *P. chlamydosporia* and treatment of seed with *P. fluorescens* (Naik 2004).

#### 20.3.1.2.5 Host Resistance

Capsicum cvs. Mississipi-68, Santanka, Anaheim Chile, and Italian Pickling, are reported to be resistant to root-knot nematodes. Black Indica, Naharia, and Pant C1 were reported to be resistant to *M. incognita*, while All Big as moderately resistant. California Wonder and Naharia were resistant to *M. javanica*, while Early California Wonder as moderately resistant. Naharia was reported to be resistant to *M. arenaria*.

The southern root-knot nematode (*M. incognita*)-resistant "Carolina Wonder" and "Charleston Belle" are recommended for use by greenhouse growers.

#### 20.3.1.2.6 Integrated Methods

Integration of *P. fluorescens* and *P. chlamydosporia* in the greenhouse bed has proved significantly effective in reducing the root galling index (*M. incognita*), number of nematodes in the roots and soil, and increasing the yield of the capsicum crop (Naik 2004).

Integration of soil incorporation of neem-based *P. fluorescens* and *T. harzianum* followed by seed treatment with *P. fluorescens* + soil incorporation of *T. harzianum* increased the seedling stand in bell pepper (Naik 2004).

## 20.4 Cucumber, *Cucumis sativus*

Cucumber and gherkins are being extensively grown in greenhouses in many countries. The Continental or Burpless cucumber is the main type of cucumber and has long fruit (Fig. 20.9), whereas the Lebanese (mini) cucumber has shorter

**Fig. 20.9** Cucumber grown under greenhouse

fruit (parthenocarpic). Compared with field cucumbers, these two types are seedless and do not require pollination to produce fruits. Compared with field crops, protected cultivated cucumbers have a finer texture, greener in color, smoother, flesh that has less total acids (0.11% versus 0.17%) and more sugars (3% versus 2.8%). The skin is more tender, not peeled, and are easier to digest. The crops produced in greenhouses fetch higher income due to good quality produce and higher productivity per unit area as also due to being available in off-season. The cropping period gets extended, so regular supply could be maintained for a long time.

### 20.4.1 Root-Knot Nematode, *Meloidogyne incognita*

#### 20.4.1.1 Symptoms

Root-knot nematodes (*Meloidogyne* spp., especially *M. incognita*) cause stunting of plant growth, leaf yellowing, wilting of plants, and severe galls on the root system (Fig. 20.10). The disease is more common and serious on sandy soils than on clay soils; crop injury is more serious during dry seasons than seasons with adequate rainfall. Root knot is one of the most common and serious diseases of cucumbers, and all growers should practice control measures.

**Fig. 20.10** Root-knot nematode-infected cucumber seedlings (*Left*) and roots showing severe galling (*Right*)

#### 20.4.1.2 Management

##### 20.4.1.2.1 Physical Methods

Soil solarization in the month of May–July by light irrigation and covering of soil with 25 micron transparent polythene sheet for 30 days was found effective for the management of root-knot nematodes (*Meloidogyne* spp.).

##### 20.4.1.2.2 Cultural Methods

Rising of nurseries in plug trays proved beneficial. Addition of neem cake in (40 g per pit) about 15 days before the transplantation of the crop.

##### 20.4.1.2.3 Chemical Methods

Mixing of Carbofuran at 1 g per plant at the time of transplanting, if nematode infestation is high.

Sterilization of severely infested polyhouses with Dazomet (at 40 g/m$^2$) or Metham sodium (at 40 ml/m$^2$).

##### 20.4.1.2.4 Biological Methods

*Purpureocillium lilacinum* was found most effective in increasing plant growth parameters like plant height, fresh and dry shoot weight, and root growth. The reason for increased plant growth and other parameters observed could be attributed to the release of growth-promoting substances by bio-agents (Baker et al. 1986) or by production of toxic metabolites which inhibited nematodes and other deleterious microorganisms.

*P. lilacinum* caused maximum reduction in soil nematode population, root galling and egg mass production (due to parasitic activity of *P. lilacinum* on eggs and other nematode stages) followed by *P. fluorescens, P. chlamydosporia*, and marigold intercropping.

#### 20.4.1.2.5 Host Resistance

Cucumber cv. Vertina was moderately resistant to *M. incognita* (11–30 galls per root systems) at harvest.

#### 20.4.1.2.6 Integrated Methods

Soil fumigation with Metham sodium at 30 ml/$m^2$ with polythene mulch with 15 days waiting period prior to sowing/planting + neem cake 200 g/$m^2$ enriched with *P. lilacinum* at 50 g/$m^2$ ($2x10^6$ cfu/g) mixed 15 days prior to sowing/planting was effective.

Soil application of neem cake at 200 g/$m^2$ 3 weeks prior to sowing + soil application of *P. lilacinum* at 50 g/$m^2$ at sowing reduced soil nematode population and increased the yield of salad cucumber under polyhouse conditions.

Soil solarization and application of partially decomposed farmyard manure enriched with *P. lilacinum* reduced crop loss in salad cucumber under polyhouse condition. In the next crop, the population was reduced to 180 nematodes per 200 cc soil. The longevity of the crop also extended.

## 20.5 Lettuce, *Lactuca sativa*

Greenhouse-grown lettuce cvs. "Kwiek," "Emerald," and "Kloek" are grown because of their good quality, availability in off-season (November to May), and high productivity of 313 t/ha/year (Fig. 20.11).

### 20.5.1 Root-Knot Nematodes, *Meloidogyne* spp.

#### 20.5.1.1 Symptoms

Root-knot nematodes (*Meloidogyne incognita, M. hapla*, and *M. arenaria, M. javanica*) cause root galls on the feeder roots which sometimes affect the entire root system showing heavy galling (Fig. 20.12). This affects the uptake of nutrition and water, and the plants also show wilting during warmer part of the day resulting in stunting of plant growth, leaf chlorosis, and reduced yields.

#### 20.5.1.2 Management

The root-knot nematodes infecting lettuce can be managed by the use of soil solarization, nematode-free planting stock, preplanting soil treatment with Oxamyl and Dazomet, and soil incorporation of oil cakes enriched with biological control agents.

**Fig. 20.11** Rows of lettuce grown under protected cultivation

**Fig. 20.12** Root-knot nematodes *Meloidogyne incognita* (*left*) and *M. hapla* (*right*) on lettuce roots

## 20.6 Future Thrusts

The future lines of research work recommended include the following practices:

- At the time of new polyhouse establishment, ensuring nematode-free soil should be strictly adhered by testing soil for the presence nematodes or otherwise in ICAR Institutes/State Agricultural Universities.
- Region-wise dedicated referral labs should be established at State Agricultural Universities/ICAR Research Institutes for analyzing samples for the presence of nematodes.
- Sites for erecting new polyhouses should be properly treated with recommended chemicals and followed by supplementation with biological control agent enriched farm yard manure/oil cakes/vermicompost.
- There is an urgent need to grant provisional registration for new chemicals having proven efficacy against root-knot and other nematodes.
- In case of heavy population of plant parasitic nematodes species, soil fumigation with permissible chemicals should be undertaken.
- Evaluation of rootstocks of crop genotypes against target nematode species (even up to races) should be carried out, and resistant rootstocks should be employed for effective nematode management.
- Efforts on breeding vegetable varieties resistant to nematodes and suitable for polyhouse conditions should be intensified.
- Cluster approach may be followed to tackle the emergent nematode and other pest problems under polyhouse conditions.

## 20.7 Conclusion

The major plant parasitic nematodes associated with vegetable crops under polyhouse cultivation include species of *Meloidogyne*, *Rotylenchulus, Pratylenchus*, and *Helicotylenchus*. Nematodes assume new proposition in protected cultivation of vegetable crops due to moderate climate and intensive cultivation. As the accrued losses due to nematode pests are tangible, proper attention is a must for their management. Various chemicals for the management of nematodes have been used under protected cultivation, but fumigants like dazomet and Metham sodium used as preplant under tarp have filled the gap along with other methods of management. There is further need to develop viable options for nematode management, including use of biological control agents and antagonistic crop like crotalaria (a green manure crop). There is a need to assess the benefits of integrating practices such as cover crops, resistant cultivars, and environmentally compatible chemicals for effective control of root-knot and other plant parasitic nematodes in vegetable crop production systems.

## References

Anon (2015) Biennial report, AICRP on nematode pests of crops and their management, Palampur Centre, Himachal Pradesh, 47 pp

Baker R, Paulitz T, Windham MH, Elad Y (1986) Enhancement of growth of ornamentals by a biological control agent. Colorado Greenhouse Grow Assoc Res Bull 431:1

Chandel YS, Ravichandra NG, Mhase NL et al (2014) Integrated management of root-knot nematodes (*Meloidogyne spp.*) under protected cultivation. Indian J Nematol 44(1):92–96

Giriraj, Sharma MK, Nama CP (2018) Efficacy of fumigants for the management of root-knot nematode, *Meloidogyne incognita* on tomato (*Solanum esculentum* Mill.). Int J Curr Microbiol Appl Sci 7:3198–3201

Naik D (2004) Biotechnological approaches for the management of wilt disease complex in Capsicum (*Capsicum annum* L.) and eggplant (*Solanum melongena*) with special emphasis on biological control. Ph. D. thesis, Kuvempu University, Shivamogga, Karnataka, India

Zuckermann BE, Dicklow MB, Ascosta N (1993) A strain of *Bacillus thuringiensis* for the control of plant parasitic nematodes. Biocont Sci Technol 3:41–46

# Part VII

# Ornamental, Medicinal and Aromatic Crops

# Ornamental Crops

# 21

**Abstract**

The production of ornamental crops such as tuberose, gladiolus, china aster, and crossandra throughout the globe is threatened by root-knot nematodes (*Meloidogyne incognita* and *M. javanica*), lesion nematode (*Pratylenchus delattrei*), foliar nematode (*Aphelenchoides besseyi*), and needle nematode (*Longidorus africanus*). Nematodes interact with *Fusarium* wilt pathogen and cause disease complexes. In this chapter, crop losses, symptoms, biology and life cycle, spread, management using physical, cultural, chemical, biological methods, host resistance, and integrated methods for nematode pests of ornamental crops are discussed.

**Keywords**

*Meloidogyne* spp. · *Pratylenchus delattrei* · *Aphelenchoides besseyi* · Symptoms · Management

## 21.1 Tuberose, *Polianthes tuberosa*

### 21.1.1 Floral Malady, *Aphelenchoides besseyi*

In West Bengal, India, the profitable cultivation of tuberose is threatened by the severe infestation of the foliar nematode *Aphelenchoides besseyi*. Holtzmann (1968) reported for the first time this nematode on the leaves of tuberose from the Hawaii Islands. Chakraborti and Ghosh (1993) also recorded "floral malady" caused by *A. besseyi* on tuberose from the Ranaghat areas of the Nadia district of West Bengal, India. The nematode has recently got spread to Odisha presumably through bulbs. The foliar nematode has now become endemic and a major limiting factor for tuberose cultivation in Ranaghat, Haringhata, and Panskura areas of West Bengal.



P. P. Reddy, *Nematode Diseases of Crops and their Management*
https://doi.org/10.1007/978-981-16-3242-6_21

The "Calcutta single" cultivar of tuberose is more vulnerable to *A. besseyi* than the "Calcutta double" cultivar. Research results confirmed that *A. besseyi* is the primary causal agent for malformed flowers, and population of *A. besseyi* causing white tip disease in rice is the same population infecting tuberose causing floral disease (Khan 2001). The nematode species is widespread in rice in West Bengal (Das and Khan 2007). Export of the tuberose flowers to other countries is affected by the occurrence of this nematode species in the cut flower stalk.

#### 21.1.1.1 Distribution

The foliar nematode problem is restricted to tuberose in WB and Odisha.

#### 21.1.1.2 Crop Losses

Yield losses to the extent of 59% in some parts of West Bengal have been reported (Pathak and Khan 2009).

#### 21.1.1.3 Symptoms

Ectoparasitic feeding of *A. besseyi* on the scape (floral stalk) leads to the formation of prickle-like rough, irregular structure, small in size compared to healthy scapes, and becomes curled and twisted (Fig. 21.1). Endoparasitic feeding of *A. besseyi* was observed in leaves, bracts, and petals where brownish to black necrotic spots of varying shapes and sizes were developed (Fig. 21.2). The nematode forms "nematode wool" on dark brown spots.

**Fig. 21.1** *Left*—field infestation of foliar nematode on tuberose. *Right*—healthy and foliar nematode-infected tuberose spikes

**Fig. 21.2** Effects of foliar nematode infection on tuberose. (**a**) Healthy control stem with abundant white florets. (**b–d**) Infected plants showing the browned stem, leaves, and florets, as well as the distorted skin of the leaves and stem

#### 21.1.1.4 Bio-Ecology

Individuals of *A. besseyi* congregate around the dark brown spots and can form "nematode wool" (Khan and Pal 2001). The ovary part of tuberose contains large number of nematodes. Infestation of foliar nematode is particularly serious during rainy season and complete crop loss has been recorded for the cultivar "Calcutta single." However, in "Calcutta double" cultivar 30% to 40% flower stalk renders unsalable and individual flower stalk occur to the extent of 59% (Pathak and Khan 2009). Nematode population peaks (1084–2768 nematodes per flower stalk) during monsoon season (during July–November at 25–33 °C) and decreases (87 nematodes per flower stalk) during winter (January at 10–26 °C) (Khan 2004). In infested field, the nematodes concentrate in maximum numbers in the outermost scaly leaves and neck region of an infested individual bulb. More numbers of *A. besseyi* were recovered from the central bulb than obcentral and peripheral bulbs. Life cycle is completed in 10–12 days at 30 °C (Khan and Ghosh 2009).

#### 21.1.1.5 Spread and Survival

The means of foliar nematode spread is mainly through infested planting stock (infested bulbs), dry plant parts, and by irrigation and run-off water. The nematode survives in coiled anhydrobiotic condition (quiescent preadult and adult stages) in the scaly leaves outside the bulbs. The nematode can also survive in the dried scaly leaves, stems, and flowers for more than 25 months; however, they cannot survive in soil under field conditions (Khan 2004; Khan 2006; Khan and Ghosh 2011; Khan and Ghadipur 2004). Another means of survival of all stages (eggs, juveniles, and adults) of the nematode is in tuberose flower stock.

#### 21.1.1.6 Management

##### 21.1.1.6.1 Cultural Methods

Soak bulbs overnight in plain water or in 5% NSKE (home preparation from Neem). Follow clean cultivation by destroying nematode-infected plant parts.

##### 21.1.1.6.2 Chemical Methods

Dip the bulbs in Monocrotophos (36SL) at 0.05% for 6 h. After sprouting of bulbs, spray Monocrotophos (36SL) 0.05% 3–4 times at 15 days' interval. In second and third year crop, spray Monocrotophos (36SL) at 0.05% at 15 days' interval beginning April.

##### 21.1.1.6.3 Host Resistance

Grow nematode tolerant cultivars Prajwal, Phule Rajani, and Shringar. Out of the six tuberose cultivars viz. Calcutta Single, Calcutta Double, Suvasini, Swarna Rekha, Rajat Rekha, and Shringar, only Suvasini was immune to attack by *A. besseyi*.

##### 21.1.1.6.4 Integrated Methods

The treatments, viz., pre-soaking of bulbs in plain water for overnight + hot-water treatment at 50 °C for 30 min. + three sprayings with Monocrotophos 36 SL at 500 ppm at 15 days interval (repeated the same spraying in second year from April onwards) had low nematode infestation (26%) and PDI value (25%), reduction of nematode population (53%), and higher spike yield (mean 286/plot, $1.5 \times 1.8\ m^2$) as compared to untreated plots (mean 207/plot) (Pathak and Khan 2009).

With the adoption of nematode management practices, spike yield could be saved up to 38% and that could be to the extent of 59% when loose flower yield/plot was taken into consideration.

The following INM measures are recommended for the management of foliar nematode on tuberose:

- Presoaking of tuberose bulbs in plain water or in Neem seed kernel extract (NSKE 4%) for overnight followed by dipping in Monocrotophos 36SL at 500–700 ppm for 4–6 h.
- After sprouting, three to four sprayings with Monocrotophos 36SL at 500 ppm at 15 to 20 days interval is needed.
- In the second and third year crop, same spraying for 3–4 times with Monocrotophos 36SL at 500 ppm at 15 to 20 days interval should start from April to May onward to reduce the nematode infestation.
- Field sanitation is most essential for reducing infestation. Infested plant parts should be burnt or buried into the soil.
- Grow nematode tolerant tuberose cultivars, Suvasini, Phule Rajani, Prajwal, and Shringar (Khan and Ghosh 2007).
- Growing tuberose away from rice field is encouraged to avoid contamination of tuberose field.

## 21.1.2 Root-Knot Nematodes, *Meloidogyne* spp.

*M. incognita, M. javanica*, and *M. arenaria* have been reported as the major limiting factors in successful tuberose cultivation in Tamil Nadu (Sundarababu and Vadivelu 1988). Khan and Parvatha Reddy (1992) found that the major root-knot nematode pests on tuberose from Karnataka include *M. javanica* and *M. incognita*.

### 21.1.2.1 Distribution

Root-knot nematodes are distributed in Tamil Nadu, Karnataka, western UP, Odisha, and West Bengal.

### 21.1.2.2 Crop Losses

*M. incognita* was responsible for 13.25, 9.87, 14.30, 13.78, and 28.58% reduction in plant weight, number of flowers, spike length, spike weight, and number of bulblets, respectively (Khan and Parvatha Reddy 1994).

### 21.1.2.3 Symptoms

Affected plants exhibit stunting, yellowing and drying up of leaves and rotting of bulbs (Jayaraman et al. 1975). Other symptoms include less number of side shoots emerged from the nematode-infected bulbs. Sundarababu and Vadivelu (1988) reported that 65% reduction in top growth and the emergence of spike is suppressed in severely infected plants. Conspicuous galls can be observed on the root system (Fig. 21.3).

### 21.1.2.4 Biology and Life Cycle

Root-knot nematodes are soil-borne or infection is introduced through infected bulbs. After penetration, the second-stage larvae feed on conducting tissues endoparasitically and developed into adults. The sac-like adult females laid eggs in masses on root surface of galls, or inside the compound galls, while vermiform males

**Fig. 21.3** *Left*—root-knot nematode infected and healthy tuberose plants. *Right*—galls on roots of new tubers

leave the roots and come out in soil. The nematode took about 25–30 days during summer months to complete its life cycle.

#### 21.1.2.5 Spread

Root-knot nematodes are disseminated through infected planting material, irrigation water, and agricultural implements.

#### 21.1.2.6 Management

##### 21.1.2.6.1 Cultural Methods

Healthy bulbs, free from root galls should be used for planting.

Soil application of Neem cake at 1 t/ha proved beneficial.

##### 21.1.2.6.2 Chemical Methods

Dipping of cleaned bulbs in 0.05% Carbosulfan (25EC) solution for 4–6 h was found to be effective.

Field application of Carbofuran (3G)/Cartap hydrochloride at 1 kg a.i./ha proved beneficial.

Soil application of Carbofuran (3G) at 2 g/plant reduced nematode population.

##### 21.1.2.6.3 Biological Methods

Application of *P. chlamydosporia* at 100 g/4 $m^2$ significantly reduced root galling (4.9 as compared to 8.3 in control), nematode population both in soil and roots, and increased root colonization and egg parasitization by the bioagent, and flower yield (45.5 spikes/4 $m^2$ as compared to 33.4/4 $m^2$ in control) of tuberose under field conditions (Rao et al. 2004).

##### 21.1.2.6.4 Integrated Methods

Use of AMF, oil cakes of Neem, Pongamia, and Castor in combination with *Purpureocillium lilacinum* was found to be effective.

Split application (at planting and 45 DAP) of Neem cake at 1 kg/1.5 $m^2$ in combination with *P. lilacinum* (at 25 g/1.5 $m^2$ containing $18 \times 10^8$ spores/g) significantly reduced root galling (1.8 as compared to 3.2 in control). The above treatment also significantly increased number of spikes (69 as compared to 45 in control), and parasitization of egg masses by *P. lilacinum* (Nagesh et al. 1998b).

Stimulation of the plant growth and significant reduction in root-knot nematode *M. incognita* infection and multiplication was obtained by treatment of tuberose bulbs with Neem cake extract mixed with *P. lilacinum* spores. Soil application of Neem cake enriched with *T. harzianum* at 1 t/ha is also effective. Treatment of tuberose bulbs with Neem and Calotropis leaf extracts mixed with *P. lilacinum* spores significantly reduced *M. incognita* infection and multiplication besides stimulating the plant growth (Nagesh et al. 1997).

Integration of AMF (*Glomus mosseae*), a botanical (Neem cake) and a nematicide (Aldicarb), gave effective management of *M. incognita* (gall index 1.5 as compared to 4.5 in control) and increased plant weight (11.80 as compared to 3.26 in control) and yield of bulbs (2.8 as compared to 1.1 in control) (Khan and Parvatha Reddy 1994).

Combined application of *P. chlamydosporia* + *T. harzianum* gave significantly higher control of *M. incognita* on tuberose (Shylaja 2005). Soil application of *T. harzianum* ($10^9$ cfu/g) + *P. lilacinum* ($10^6$ cfu/g) gave effective control of root-knot nematodes.

### 21.1.3 Interaction of Root-Knot Nematode with *Fusarium* Wilt

#### 21.1.3.1 Interaction

Tuberose plants infected with both the soil-borne pathogens showed that the root-knot nematodes predisposed the plants to severe infection from *Fusarium* wilt and ultimately increased the death rate of tuberose plants (Rao et al. 2002). The fungal infection was observed to aggravate in the presence of *M. incognita*.

The root-knot nematode *M. incognita* reduced flower yield considerably and the tuberose plants became highly susceptible to the attack by *F. oxysporum* f. sp. *dianthi* (Rao et al. 2002).

The spike emergence was significantly delayed (65.07 and 68.33 days) due to the high population levels of nematodes at 100 and 150/100 $cm^3$ soil. The longest delay in spike emergence (81 days) was recorded when *Fusarium* was present along with the highest density of nematodes at 150/100 $cm^3$ soil. It was observed that the disease complex drastically reduced the yield of flowers in tuberose, thereby bringing down the production of flowers in this commercial crop (Shylaja 2005).

#### 21.1.3.2 Management

Tuberose plants treated with *P. chlamydosporia* + *T. harzianum* each at 40 g/4 $m^2$ gave significant reduction in root galling (4.20 as compared to 8.49 in control), nematode population both in soil and roots, disease index (1.59 as compared to 4.42 in control), and increased egg parasitization and root colonization by bioagents and spike yield (24 as compared to 18 in control) (Rao et al. 2003).

Shylaja (2005) reported that integration of *P. chlamydosporia* with *T. harzianum* gave maximum increase in plant growth (height, shoot weight, and root length), flower yield (spike length, spike weight, number of flowers/spike, flower weight, and number of spikes/plot), and least nematode population both in soil and roots, while *P. chlamydosporia* + *P. lilacinum* gave maximum root weight, least root galling, and wilt disease index.

## 21.2 Gladiolus, *Gladiolus* spp.

### 21.2.1 Root-Knot Nematode, *Meloidogyne incognita*

#### 21.2.1.1 Symptoms

The root-knot nematode causes stunting of plants and reduction in leaf count. The threshold population of 1000 juveniles of *M. incognita* per plant reduced the growth of plant markedly and there was no emergence of plants at 10,000 juveniles of *M. incognita* per plant (Chandel et al. 1997).

Severe galling on roots (Fig. 21.4) results in yellowing of leaves which subsequently leads to stunted growth. The nematode invades roots, daughter corms, and cormels which develop after flowering.

#### 21.2.1.2 Survival

The nematodes would survive in corm tissue in the soil as a source of inoculum for next season (Parvatha Reddy et al. 1979).

#### 21.2.1.3 Interaction with Other Pathogens

Incidence of gladiolus scab caused by *Pseudomonas marginata* was found to increase the population of *M. javanica* (Elgoorani et al. 1976).

#### 21.2.1.4 Management

##### 21.2.1.4.1 Physical Methods

Hot water treatment of corms at 58 °C for 30 min. Eliminates root-knot nematode infection.

##### 21.2.1.4.2 Cultural Methods

For planting, use of certified planting stock is recommended. Crop rotation or intercropping with marigold helps in reducing the nematode population.

**Fig. 21.4** Root-knot nematode infestation on gladiolus roots

Application of chicken manure at 20 t/ ha was most effective in decreasing the population of *M. incognita*.

#### 21.2.1.4.3 Chemical Methods

Soil application of Dazomet at 300 and 450 kg/ha and dip treatment of gladiolus corms in 0.03 and 0.06% Carbosulfon for 4 hr. proved effective in reducing the root-knot nematode.

#### 21.2.1.4.4 Biological Methods

*Trichoderma persicus* and *Bacillus indicus* increased the germination count, number of tillers, and number and weight of corms.

#### 21.2.1.4.5 Host Resistance

Gladiolus varieties Royal Jubilee and 77 DHM were moderately resistant.

#### 21.2.1.4.6 Integrated Methods

Soil application of Neem cake enriched with *T. harzianum* at 1 t/ha is effective.

Enhancement of the flower yield by 19–22% and significant reduction in root-knot nematode *M. incognita* population by 61–76% and reniform nematode *R. reniformis* by 65–70% in the roots of gladiolus was achieved by soil incorporation of 5 tons of FYM enriched with *P. fluorescens* (with $1 \times 10^9$ cfu/g) per ha.

Corms of gladiolus treated with Neem suspension enriched with *Pseudomonas fluorescens* strains 1 and 2 (5 g/L of water) and application of Neem cake enriched with *P. fluorescens* strains to beds at 20 g/$m^2$ was found to be effective in reducing the root population of *M. incognita* by 68% and increased the flower yield by 28% (Sowmya and Rao 2011).

Soil application of Neem cake enriched with *T. harzianum* at 1 t/ha was also found to be effective.

Both the Neem cake and the AM fungus (*Glomus fasciculatum*) increased the spike length, number of florets/spike, and the floret diameter whereas the root-knot nematode (*Meloidogyne incognita*) reduced these floral growth characters. Significantly greater reduction in root galls and enhancement of flower growth parameters were obtained by integrating neem cake with endomycorrhiza. Lower dose of cake in combination with *G. fasciculatum* is suggested as a means to protect gladiolus from root-knot damage (Hasan and Nehal Khan 2004).

### 21.2.2 Interaction of Root-Knot Nematode with *Fusarium* Wilt

#### 21.2.2.1 Interaction

The commercial production of gladiolus is limited by interaction of soil-borne pathogens like root-knot nematode (*Meloidogyne incognita*) with *Fusarium* (*Fusarium oxysporum* f. sp. *gladioli*) wilt in cultivated soils which aggravated the wilt problem causing high plant mortality in gladiolus fields.

#### 21.2.2.2 Management

Application of *P. fluorescens* effectively controlled the disease complex leading to significant improvement in plant growth and flowering (Mustafa and Khan 2004).

Management of disease complex (induced by root-knot nematode and *Fusarium* wilt fungus) until flower spikes harvest was achieved by combined soil application of gladiolus plants with Neem cake along with *P. lilacinum* + *T. harzianum/T. viride*. The corms and cormels obtained from plants treated with these combinations were free from *Fusarium* infection. Integration of *P. lilacinum* + *T. viride* gave highest bioagent colonization of galled roots (94%) followed by *P. lilacinum* + *T. harzianum* combination (69%). Similarly, combination of *P. lilacinum* + *T. viride* gave maximum parasitization of eggs (44%) followed by *P. lilacinum* + *T. harzianum* combination (38%), while egg mass parasitization was same in both the combinations (58%) (Nagesh et al. 1998a).

## 21.3 China Aster, *Callistephus chinensis*

### 21.3.1 Root-Knot Nematodes, *Meloidogyne* spp.

The root-knot nematodes are a major limiting factor both in nursery and main field (Krishnappa et al. 1980; Khan and Parvatha Reddy 1992).

#### 21.3.1.1 Symptoms

The symptoms observed on root-knot nematode-infected china aster plans include stunted plant growth, wilting, yellowing of plants, loss of luster, lose leaves, produce weak flushes of growth, and have fewer and smaller leaves than healthy plants. Infection by the root-knot nematode causes gall formation, or "knots," on roots (Fig. 21.5).

#### 21.3.1.2 Management

##### 21.3.1.2.1 Host Resistance

China aster cv. Shashank was found resistant, while Poornima was moderately resistant to root-knot nematodes. There were significantly higher total phenols and total proteins and higher activity of polyphenol oxidase in infected roots of resistant variety, compared to that in susceptible variety (Kamini) during early stages of infection. Nagesh et al. (1995) reported that the activity of phenyl alanine ammonia lyase was higher in roots of resistant variety as the infection advanced.

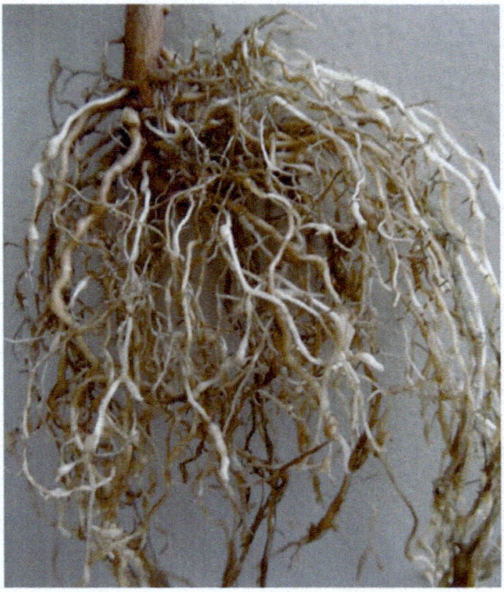

**Fig. 21.5** Galling on china aster roots

## 21.4 Crossandra, *Crossandra infundibuliformis*

*Pratylenchus delattrei*, *Meloidogyne incognita*, and *Longidorus africanus* are the major nematode pests of crossandra.

### 21.4.1 Root-Knot Nematode, *Meloidogyne incognita*

Rajendran et al. (1976) reported the root-knot nematode *M. incognita* on crossandra for the first time from India.

#### 21.4.1.1 Crop Losses

The nematode is responsible for 25.62% and 21.64% loss in number of flowers and weight of flowers, respectively (Khan and Parvatha Reddy 1994).

#### 21.4.1.2 Distribution

Root-knot nematodes are found in all crossandra growing areas like Tamil Nadu, Karnataka, and Andhra Pradesh.

#### 21.4.1.3 Symptoms

Infected plants are stunted with dried peripheral branches bearing smaller chlorotic leaves almost turning to white in later stages (Rajendran et al. 1976). Roots exhibit severe galling. Inflorescences are small and sometimes fail to produce flowers.

#### 21.4.1.4 Interaction with Other Pathogens

*M. incognita* interact synergistically with wilt fungi *Fusarium oxysporum* and *F. solani* and results in premature death of crossandra plants (Khan and Parvatha Reddy 1992). The lesion nematode *P. delattrei* interacts with *F. solani* and *F. oxysporum* in causing wilt disease complex in crossandra (Khan and Parvatha Reddy 1992; Srinivasan and Muthurkrishnan 1975). Srinivasan (1974) found that wilt symptoms appeared earlier when nematodes were added 2 weeks prior to *F. solani*.

#### 21.4.1.5 Management

##### 21.4.1.5.1 Cultural Methods

Deep plowing and fallowing in summer for about a month helps in reducing the nematode population. Planting crossandra after graminaceous crops should be avoided. Reduction in the root-knot nematode infection occurred by intercropping of crossandra with marigold or pangola grass combined with soil application of FYM.

Khan and Parvatha Reddy (1994) found that significant reduction in the root-knot and lesion nematode population both in soil and roots by intercropping of crossandra with enemy plant like marigold.

##### 21.4.1.5.2 Chemical Methods

Application of carbofuran at 2 to 3 kg a.i. per ha would be beneficial.

##### 21.4.1.5.3 Biological Methods

Khan and Parvatha Reddy (1994) reported the effective management of root-knot nematode infecting crossandra by soil incorporation of the parasitic fungus *Paecilomyces marguandi* grown on paddy seeds at 2 g per kg soil.

Promotion of better root health and plant growth along with comparatively lower nematode infection and reproduction was obtained by root colonization of crossandra with *Glomus fasciculatum* and *G. mosseae* as compared to *G. intradices* and *Aucalospora laevis*.

##### 21.4.1.5.4 Integrated Methods

Bioagents and Botanicals

Incorporation *V. lecanii* with neem cake facilitated the effective management of *M. incognita* on crossandra. Application of neem cake enriched with *T. harzianum* at 2 kg/m$^2$ (2 t/ha) in nursery beds gave effective control.

Combinations of *V. lecanii* and *Purpureocillium lilacinum* ($2 \times 10^4$ spores/ml each) with 5% neem leaf extract resulted in significantly higher plant growth parameters and crossandra flower yield (Nagesh and Parvatha Reddy 1995). Root galling and nematode multiplication factor were least, and the parasitization of eggs and egg masses was highest in *V. lecanii* + 5% neem leaf extract.

Arbuscular Mycorrhizal Fungi and Botanicals

*G. fasciculatum* and *G. mosseae* in combination with neem cake gave better control of nematodes over carbofuran treatment. The root colonization of AM fungi increased significantly in the presence of neem cake which in turn improved their efficacy in reducing *M. incognita* population in crossandra roots (Nagesh and Parvatha Reddy 1996). Integration of neem, karanj, and castor cakes with AMF, *G. mosseae* significantly enhanced plant growth parameters and flower yield of crossandra, root colonization, and sporulation of AMF. The above treatments also reduced root-knot nematode multiplication and root-galling (Nagesh and Parvatha Reddy 1996).

Bioagents, Arbuscular Mycorrhizal Fungi, and Botanicals

Integration of a bioagent (*V. lecanii*) and endomycorrhiza (*G. mosseae*) with botanicals improved the growth of crossandra and reduced the population of *M. incognita*. *G. mosseae* reduced the requirement of phosphatic fertilizer and favored the antagonistic potential of *V. lecanii* against *M. incognita*.

Two Bioagents

Combined application of *P. lilacinum* and *Pasteuria penetrans* enhanced plant growth and flower yield of crossandra besides reducing root galling due to *M. incognita* (Nagesh et al. 1995).

Bioagents and Botanicals/Chemicals

Integrated management of root-knot nematodes infecting crossandra was achieved by rational combination of a biocontrol agent (*T. harzianum* or *Pochonia chlamydosporia*) with a nematicide (aldicarb) or oil cake (neem cake) (Khan and Parvatha Reddy 1994).

### 21.4.2 Lesion Nematode, *Pratylenchus delattrei*

The lesion nematode was first recorded by Srinivasan (1974) on crossandra from Tamil Nadu.

#### 21.4.2.1 Crop Losses

Heavily infested plants do not produce tertiary spikes and thereby the flower yield is reduced by ninefold (Srinivasan and Muthurkrishnan 1975).

#### 21.4.2.2 Distribution

*P. delattrei* has been reported on crossandra from Tamil Nadu and Rajasthan.

#### 21.4.2.3 Symptoms

*P. delattrei* was frequently associated with crossandra crop exhibiting stunting, chlorosis of leaves, mottled appearance of leaves (initially brown and later pinkish), and wilting (Fig. 21.6). Later, the leaf tips dried and entire leaves turned yellow and

**Fig. 21.6** Crossandra plants infected with lesion nematode showing stunting of plants and leaf yellowing

shed. The plants were devoid of side shoots and remained completely defoliated. The roots exhibited brown to black lesions. Heavily infested plants did not produce tertiary spikes and thereby flower yield is reduced (Srinivasan and Muthurkrishnan 1975).

#### 21.4.2.4 Host Range

Besides crossandra, *P. delattrei* has been recorded on sorghum, maize, red gram, okra, sun hemp, and wheat.

#### 21.4.2.5 Life Cycle

The total life cycle of this nematode from egg to adult stage ranges between 46 to 55 days.

#### 21.4.2.6 Histopathology

Nematode moves intracellularly and remains in cortical region. Its infection results in dissolution of cell walls which leads to cavity formation in cortical region. Cell wall gets thickened and cell contents become more dense and granular.

#### 21.4.2.7 Interaction with Other Pathogens

The lesion nematode *P. delattrei* interacts with *Fusarium solani* and *F. oxysporum* in causing wilt disease complex in crossandra (Khan and Parvatha Reddy 1992). Srinivasan and Muthurkrishnan (1975) also opined that crossandra decline in Tamil Nadu occurs due to *P. delattrei* and *F. solani* complex. Srinivasan (1974) found that wilt symptoms appeared earlier when nematodes were added 2 weeks prior to *F. solani*.

#### 21.4.2.8 Survival

The nematode can survive up to 8 months in host-free moist soil, whereas in desiccated soil it could survive up to 3 months.

#### 21.4.2.9 Management

##### 21.4.2.9.1 Cultural Methods

Intercropping of crossandra with enemy plant like marigold significantly reduced the lesion nematode population both in soil and roots (Khan and Parvatha Reddy 1994). Application of farm yard manure has been found to reduce the lesion nematode population in crossandra.

##### 21.4.2.9.2 Chemical Methods

Nursery bed treatment with phorate, aldicarb, fensulfothion, and carbofuran each at 5 g per $m^2$ gave good and healthy crossandra seedlings free from *P. delattrei* (Vadivelu and Muthukrishnan 1979).

Vadivelu and Muthukrishnan (1979) reported protection of crossandra seedling from *P. delattrei* for 30 days after transplanting by root dip treatment for 8 h. in 0.05% to 0.1% solution of aldicarb, carbofuran, fensulfothion, fenamiphos, or phorate.

Srinivasan and Muthukrishnan (1976) found that carbofuran at 2.5 kg a.i. per ha was effective under field conditions. Application of phorate at 1 g per plant a week after planting checked the lesion nematodes effectively and increased the flower yield (Vadivelu and Muthukrishnan 1979).

##### 21.4.2.9.3 Biological Methods

Application of *T. viride* at 10 g/kg soil was found to be effective to maximize the flower yield and suppressing the lesion nematode population both in soil and roots (Jothi et al. 2004).

##### 21.4.2.9.4 Integrated Methods

There was a synergistic effect when the neem cake was coupled with carbofuran 3G in the management of *Pratylenchus delattrei*. The treatment resulted in better establishment of seedlings, and with increased plant biomass and flower yield (Jothi et al. 2004).

### 21.4.3 Needle Nematode, *Longidorus africanus*

In India, this nematode was reported for the first time on crossandra by Vadivelu et al. (1976) from Tamil Nadu.

#### 21.4.3.1 Symptoms

In crossandra, *L. africanus* causes stunting of plants, branching, swelling, and darkening of root tips (Muthukrishnan et al. 1977).

#### 21.4.3.2 Host Range

Sorghum, snap bean, lima bean, and sugar beets are reported as best hosts, while wheat, cotton, okra, cucumber, eggplant, and tomato as good hosts (Kolodge et al. 1987).

#### 21.4.3.3 Histopathology

*L. africanus* initiates hyperplasia in parenchymatous cells of peripheral region which subsequently progress toward the center, forking of roots and clubbing of root terminal. Due to this, meristem is pushed laterally which ultimately leads to formation of darkly stained cells (Muthukrishnan et al. 1977).

#### 21.4.3.4 Management

Pea, carrot, squash, spearmint, and onion are reported to be poor hosts, while cauliflower, cabbage, and radish as nonhosts. Either of these crops can be fitted as intercrops or in crop rotation scheme. Kolodge et al. (1987) reported reduction in the needle nematode population by rotation of crossandra with spearmint which is a poor host of *L. africanus*.

## 21.5 Conclusion

Root-knot, foliar, and lesion nematodes are the most important pests parasitizing the ornamental plant species. From many regions of the world, the most serious nematode infecting ornamental crops include the root-knot nematodes *Meloidogyne* spp. Management of nematodes using the nematicides was the common practice during the past two decades. However, because of environmental toxicity and cost of these chemicals, other control techniques need to be investigated. Strategies using biological control with microbial agents, natural plant products, soil amendments or resistant plants are considered as alternative methods for the management of nematode pests on various ornamental crops. Integrating biological control (antagonistic fungi and bacteria) with other methods could make an excellent model for nematode control and achieve the desired results.

## References

Chakraborti HS, Ghosh SC (1993) Studies on the floral malady of *Polianthes tuberosa* L. J Mycopathol Res 33:25–28

Chandel SS, Khanna AS, Khan ML (1997) Faunistic studies on nematodes associated with gladiolus L. Ann Agri Res 18(1):44–47

Das TK, Khan MR (2007) Occurrence and distribution of white tip nematode, *Aphelenchoides besseyi* in West Bengal, India. Indian J Nematol 27:94–97

Elgoorani MA, Abo-Eldahab MK, Mchair FF (1976) Interaction between root-knot nematode and *Pseudomonas marginata* on gladiolus corms. Phytopathology 64:271–272

Hasan A, Nehal Khan M (2004) Interaction of root-knot nematode, VAM fungus and neem cake in amended soils on gladiolus. Nematol Medit 32:95–99

Holtzmann OV (1968) A foliar disease of tuberose caused by *Aphelenchoides besseyi*. Pl Dis Reptr 52:56

Jayaraman V, Rajendran G, Muthukrishnan TS (1975) Occurrence of root-knot nematodes in *Polianthes tuberosa* L. in Tamil Nadu. Indian J Nematol 5:101–102

Jothi G, Sundara Babu R, Ramakrishnan S, Rajendran G (2004) Management of root lesion nematode, Pratylenchus delattrei in crossandra using oil cakes. Bioresour Technol 93 (3):257–259

Khan MR (2001) White-tip nematode, *Aphelenchoides besseyi* in rice-tuberose cropping system. In: Mukhopadhyay SK, Ghosh DC, De GC (eds) Proceedings of the national seminar on frontiers of crop management. Visva-Bharati, Sriniketan

Khan MR (2004) Observation on foliar nematode, *Aphelenchoides besseyi* in tuberose (*Polianthes tuberosa* L.) in West Bengal. Ann Plant Prot Sci 12(1):106–109

Khan MR (2006) Managing foliar nematode - a threat for tuberose cultivation. Indian Hort 51:17

Khan MR, Ghadipur MH (2004) Management of root-knot disease of some solanaceous vegetables by soil application of fly ash.In: Proceedings of National symposium on paradigms in nematological research for biodynamic farming, University of Agricultural Sciences, Bangalore, p 92

Khan MR, Ghosh S (2007) Evaluation of tuberose cultivars against foliar nematode, *Aphelenchoides besseyi*. In: National symposium on nematology in 21st. century: emerging paradigms. Nematological Society of India, Held at Assam, p 77

Khan MR, Ghosh S (2009) Control of foliar nematode, *Aphelenchoides besseyi in* tuberose (*Polianthes tuberosa* L.) from West Bengal, India. In: International symposium on nematodes in tropical ecosystems, Hanoi, Vietnam, p 22

Khan MR, Ghosh S (2011) Survival and population dynamics of foliar nematode, *Aphelenchoides besseyi* infecting tuberose in West Bengal, India. Indian J Nematol 41(1):47–51

Khan MR, Pal AK (2001) Plant parasitic nematodes associated with tuberose (*Polianthes tuberosa* L.) in West Bengal. Ann Plant Prot Sci 9(2):357–359

Khan RM, Parvatha Reddy P (1992) Nematode pests of ornamental crops. In: Bhatti DS, Walia RK (eds) Nematode pests of crops. CBS Publishers & Distributors, Delhi, pp 250–257

Khan RM, Parvatha Reddy P (1994) Nematode problems of ornamental crops and their management. In: Prakash J, Bhandary KR (eds) Floriculture-technology, trades and trends. Oxford & IBH Publishing Co. Pvt. Ltd., New Delhi, pp 468–472

Kolodge C, Radewald JD, Shibaya F (1987) Revised host range and studies on life cycle of *Longidorus africanus*. J Nematol 19:77–81

Krishnappa K, Prasad KSK, Reddy BMR (1980) A preliminary note on the plant parasitic nematodes associated with ornamental plants at Hebbal campus of University of Agricultural sciences Bangalore. Curr Res 9:101–103

Mustafa U, Khan MR (2004) Management of disease complex caused by *Fusarium oxysporum* f. sp. *gladioli* and *Meloidogyne incognita* on gladiolus. In: Proceedings of national symposium on paradigms in nematological research for biodynamic farming, University of Agricultural Sciences, Bangalore, pp 91–92

Muthukrishnan TS, Vadivelu S, Rajendran G (1977) *Longidorus africanus* and its pathogenic effect on *Crossandra undulaefolia*. Indian J Nematol 7:69–72

Nagesh M, Parvatha Reddy P (1995) Comparative efficacy of *Paecilomyces lilacinus* (Thom.) Samson and *Verticillium lecanii* (A. Zimmerman) Viegas in combination with botanicals against *Meloidogyne incognita* (Kofoid and White, 1919) Chitwood, 1949 infecting *Crossandra undulaefolia* L. J Biol Control 9:109–112

Nagesh M, Parvatha Reddy P (1996) Eco-friendly management of *Meloidogyne incognita* on *Crossandra undulaefolia* using vesicular arbuscular mycorrhiza, *Glomus mosseae* and oil cakes. Mycorrhiza News 9(1):12–14

Nagesh M, Parvatha Reddy P, Rao MS (1995) Management of *Meloidogyne incognita* on *Crossandra undulaefolia* using *Paecilomyces lilacinus* and *Pasteuria penetrans* formulations. In: National symposium on perspectives in eco-friendly approaches to plant protn. Indian Agricultural Research Institute, New Delhi, pp 47–48

Nagesh M, Parvatha Reddy P, Rao MS (1997) Integrated management of *Meloidogyne incognita* on tuberose using *Paecilomyces lilacinus* in combination with plant extracts. Nematol Medit 25:3–7

Nagesh M, Parvatha Reddy P, Ramachandran N (1998a) Integrated management of *Meloidogyne incognita* and *Fusarium oxysporum* f sp. *gladioli* in gladiolus using antagonistic fungi and neem cake. In: Mehta UK (ed) Nematology - challenges & opportunities in 21st century. Sugarcane Breeding Inst, Coimbatore, pp 263–266

Nagesh M, Parvatha Reddy P, Vijaya Kumar MV, Nagaraju BM (1998b) Management of *Meloidogyne incognita* on *Polianthes tuberosa* using parasitic fungus, *Paecilomyces lilacinus* in combination with oil cakes. In: Parvatha Reddy P, Krishna Kumar NK, Verghese A (eds) Advances in IPM for horticultural crops. Association for Advancement of Pest Management in Horticultural Ecosystems, Bangalore, pp 339–344

Parvatha Reddy P, Singh DB, Rao VR (1979) The root-knot nematode on Gladiolus from India. Curr Sci 48:82

Pathak B, Khan MR (2009) Yield loss potential of foliar nematode, *Aphelenchoides besseyi* in tuberose. Indian J Nematol 39:38–40

Rajendran G, Vadivelu S, Muthukrishnan TS (1976) Pathogenicity of *Meloidogyne incognita* on crossandra. Indian J Nematol 6:115–116

Rao MS, Naik D, Shylaja M, Parvatha Reddy P (2002). Interaction of *Meloidogyne incognita* and *Fusarium oxysporum* f. sp. *dianthi* in the manifestation of the disease complex on tuberose. In: Zonal meeting of the southern chapter of indian phytopathological society, Bangalore

Rao MS, Shylaja M, Naik D (2003) Management of nematode induced wilt disease complex in tuberose (*Polianthes tuberosa* L.) cultivar Prajwal using *Pochonia chlamydosporia* (*Verticillium chlamydosporium*) and *Trichoderma harzianum*. J Orn Hort 6:341–346

Rao MS, Shylaja M, Reddy PP (2004) Bio-management of *Meloidogyne incognita* on tuberose using a formulation of *Pochonia chlamydosporia*. Nematol Medit 32:165–167

Shylaja M (2005) Studies on the development of bio-intensive disease management strategies in tuberose (*Polyanthes tuberosa* L.) and carnations (*Dianthus caryophyllus* L.) using biological control agents. Ph. D. thesis, Kuvempu University, Shimoga, 132 pp

Sowmya DS, Rao MS (2011) Bio-management of disease complex in gladiolus by using PGPR. In: Proceedings of the second Asian PGPR conference, Beijing, pp 342–350

Srinivasan A (1974) Studies on root lesion nematode, Pratylenchus delattrei. M. Sc. (Agriculture) thesis, Tamil Nadu Agricultural University, Coimbatore

Srinivasan A, Muthukrishnan TS (1976) Control of root lesion nematode *Pratylenchus delattrei* Luc 1958 on crossandra. Indian J Nematol 6:111–114

Srinivasan A, Muthurkrishnan TS (1975) A note on nematode fungal complex in crossandra (*Crossandra undulaefolia*) in Coimbatore. Curr Sci 44:743–744

Sundarababu R, Vadivelu S (1988) Pathogenicity of *Meloidogyne* species to tuberose (*Polyanthes tuberosa* L.). Indian J Nematol 18:146–148

Vadivelu RS, Muthukrishnan TS (1979) Control of root lesion nematode, *Pratylenchus delattrei* on crossandra. Indian J Nematol 9:86

Vadivelu RS, David BV, Rajmohan N (1976) Pesticides 10:39–40

# Protected Cultivation of Flower Crops

# 22

**Abstract**

The production of flower crops like carnation, gerbera, rose, chrysanthemum, lilies, orchids, and anthuriums under protected cultivation throughout the globe is threatened by root-knot nematodes (*Meloidogyne* spp.), lesion nematodes (*Pratylenchus* spp.), spiral nematodes (*Helicotylenchus* spp.), and foliar nematodes (*Aphelenchoides* spp.). In some crops, the extent of crop losses attributed to nematodes is as high as 80%. The nematodes interact with fungal pathogens and exacerbate severity of diseases leading to complete crop failure. In this chapter, crop losses, symptoms, and management using physical, cultural, chemical, biological methods, host resistance, and integrated methods for flower crop nematodes under protected cultivation are discussed.

**Keywords**

*Meloidogyne* spp. · *Pratylenchus* spp. · *Aphelenchoides* spp. crop losses · Symptoms · Management

## 22.1 Introduction

In 1980s, the demand for high-quality, export-oriented horticultural products and need for the availability of horticultural crop produce round the year especially in off season, compelled the growers to cultivate selected flower crops in different states in India under protected cultivation. Soon the nematode incidence particularly under protected cultivation became severe and led to total crop losses because of congenial conditions of higher temperature, humidity, and use of highly intensive farming.


P. P. Reddy, *Nematode Diseases of Crops and their Management*,
https://doi.org/10.1007/978-981-16-3242-6_22

### 22.1.1 Nematode Problems

An overall average annual yield loss in major ornamental crops due to nematodes goes up to 60% under protected cultivation. Rapid spread of nematode infestation through soil, crop residues, and indiscriminate use of agrochemicals in horticultural ecosystems is a major concern for crop protection specialists and policy makers.

In severely infested plants, stunted growth, leaf yellowing, and wilting symptoms are observed. As compared to open field conditions, the nematode multiplication under poly house is as high as 10–30 times and the nematodes reach economic injury levels within a very short period.

Crops such as carnations, roses, gerbera, orchids, and anthuriums are being grown under protected cultivation (in polyhouses/greenhouses/shade nets). These crops which are grown throughout India are seriously infested with nematodes such as *Meloidogyne incognita, M. javanica* (root-knot nematodes), and *Rotylenchulus reniformis* (reniform nematode). Nematode problems on all these crops under protected conditions have assumed alarming proportions leading to huge crop losses (up to 80%) in select crops. The nematode infestations exacerbate severity of fungal diseases leading to total crop loss.

*M. incognita* infection makes the plants highly susceptible for the attack of *Fusarium oxsporum* f. sp. *dianthi*. The root-knot nematode *M. incognita* interacts with foot rot pathogen *Phytophthora parasitica* in causing root rot disease complex in gerbera resulting in yield loss of 40–60%.

### 22.1.2 Management

Management practices adopted by farmers include continual use of chemical nematicides, often at higher than recommended rates resulting in build-up of resistance. In addition, biomagnification and environment deterioration due to hazardous chemicals has rendered several cultivated ecosystems unstable and nonprofitable.

Successful management strategies for nematodes and other disease complexes have been evolved using biopesticides like *Purpureocillium lilacinum, Pochonia chlamydosporia, Trichoderma harzianum, T. viride,* and *Pseudomonas fluorescens*. Farmers who adopted these technologies reduced the use of agrochemicals to the extent of 40% to 45% and obtained 30 to 35% increased yields in gerbera and carnations.

The preparation of beds should be as follows:

- Raised beds are to be prepared and the soil should be brought to fine tilth. The bed size can vary according to the requirement and the type of crop grown.
- Add recommended doses of fertilizers, Carbofuran or Phorate at 50 g/$m^2$ + 200 g Neem/Pongamia/Mahua cake enriched by biopesticides per $m^2$.
- Further incorporate bio-pesticide enriched FYM at 2 kg/ $m^2$ or biopesticides enriched Vermicompost at 500 g/$m^2$ in top 18 cm of soil in the beds.
- Water the beds for 7–10 days for proper decomposition of these organic materials.

Enrichment of organic amendments (FYM/Neem/Pongamia/Mahua cake/ Vermicompost) should be done as follows:

One ton of farm yard manure/Neem/Pongamia/Mahua cake/Vermicompost has to be enriched by mixing with 2 kg each of *Pseudomonas fluorescens*, *Trichoderma harzianum*, and *Purpureocillium lilacinum*. It has to be covered with plastic mulch and optimum moisture of 25–30% has to be maintained by regular watering for a period of 15 days. Once in a week thoroughly mix the mixture for maximum multiplication and uniform spread of microorganisms in the entire lot of substrates.

Apply 100 g of Neem/Pongamia/Mahua cake or 250 g of Vermicompost enriched with *Pseudomonas fluorescens* + *Trichoderma harzianum* + *Purpureocillium lilacinum* to 1 $m^2$ beds in the rhizosphere region of plants.

The organic formulation containing *Pseudomonas fluorescens* and *Trichoderma harzianum* has to be sprayed on the plants at regular intervals of 20 days at a dosage of 5 g/L or 5 ml/L.

Alternately, take 20 kg of Neem/Pongamia/Mahua cake enriched in the abovementioned manner and mix it in 200 L of water, leave it for a period of 2–3 days. Filter this suspension and use it for spraying by mixing 250 ml of suspension in 1 L of water and spray at regular interval of 20 days.

The organic formulation of biological control agents has to be given through drip by drenching at 5 g/L or 5 ml/L at regular interval of 20 days.

By following the above methods, farmers can get significant increase in the yield of flower crops with the cost-benefit ratio of 1:3 and above.

## 22.2 Carnation, *Dianthus caryophyllus*

Carnation is one of the beautiful flowers native to the Mediterranean region which is commercially cultivated in greenhouse (Fig. 22.1).

### 22.2.1 Root-Knot Nematode, *Meloidogyne incognita*

#### 22.2.1.1 Crop Losses

Most of the highly fetching exotic cultivars of carnation from Europe have shown 40 to 60% mortality in polyhouse beds due to root-knot nematode infection in and around Bangalore (Nagesh and Parvatha Reddy 2005). Nagesh and Parvatha Reddy (2000) reported that *M. incognita* was responsible for 27% loss in flower yield of carnation.

#### 22.2.1.2 Symptoms

Infected plants showed poor and stunted growth coupled with slight chlorosis of the foliage and tends to wilt on warmer days (Fig. 22.2).

Upon uprooting of such plants, root system showed clear-cut galling and severely reduced root volume (Fig. 22.2). Mortality of the plants can also be observed.

**Fig. 22.1** Carnations under protected cultivation

**Fig. 22.2** Carnation roots infected with root-knot nematode showing severe galling

#### 22.2.1.3 Management

##### 22.2.1.3.1 Chemical Methods

Pre-plant treatment of beds with Chlorpyriphos and Carbofuran (each applied twice in 6 months) significantly reduced nematode populations in roots and soil up to 1 year after which the nematodes build-up.

##### 22.2.1.3.2 Biological Methods

Under polyhouse conditions, carnation plants treated with bio-agents such as *Purpureocillium lilacinum* and *Pochonia chlamydosporia* gave significantly higher reduction of root galling as compared to *T. harzianum* (Shylaja 2005).

Soil application of *P. chlamydosporia* at 50 g/m$^2$ (10$^6$ chlamydospores/g) gave significant reduction in root galling, nematode population (*M. incognita*) in soil and roots, and increased egg parasitization, root colonization of the bioagent, and flower yield (19.35% over control) of carnation under polyhouse conditions (Rao and Shylaja 2003).

Effective management of root-knot nematodes can be obtained by soil application of the organic formulation of biological control agents (*Pseudomonas fluorescens + Trichoderma harzianum + Purpureocillium lilacinum*) through drip/drenching at 5 g/L or 5 ml/L and repeated at 20 days interval.

##### 22.2.1.3.3 Host Resistance

Carnation cvs. Desco, Castelaro, Kappa, Rara, Logupink, Target, and Antalia are highly resistant to *M. incognita*.

Carnation cv. Randez was moderately tolerant (Shylaja 2005).

Carnation cv. Chabaud Giant was found highly resistant to *M. incognita* race 2 (gall index, 0.00; $J_2$ per root system, 1.00) and *M. javanica* (gall index, 0.25; $J_2$ per root system, 2.00) (Om et al. 2008).

##### 22.2.1.3.4 Integrated Methods

###### Bioagents and Botanicals

Integration of Neem cake at 0.5 kg with *T. viride* at 100 g per m$^2$ was found to be effective for the management of root-knot nematodes and increased carnation flower yields.

Three antagonistic fungi, viz., *T. harzianum*, *Verticillium lecanii*, and *P. lilacinum*, at 2 × 10$^4$ spores/ml in combination with Neem cake reduced *M. incognita* population in both soil and roots of carnation.

Enhancement of early flowering, flower diameter, stalk length, and yield, and reduction in nematode population were obtained by soil application of *P. lilacinum* at 0.5 g/kg of soil along with Neem cake at 1.0 t/ha (Johnson 2000).

###### Arbuscular Mycorrhizal Fungi (AMF) and Bioagents

Combined inoculation of AMF and *Pseudomonas fluorescens* had positive effect on root-knot nematode control on carnation (Anusuya and Vadivelu 2002).

###### Bioagents, Botanicals and Chemicals

Significant reduction in root-knot nematode, *M. incognita* population, mortality of carnation plants was obtained for 2 years by soil application of dazomet at 40 g/m$^2$) before planting followed by integration of Neem cake (1 kg/m$^2$ 15 days later) along with antagonistic fungi, *P. chlamydosporia/P. lilacinum* (at 2 × 10$^{12}$ spores/m$^2$). The antagonistic fungi established better in the beds treated with Dazomet. The above treatment also reduced root galling (0.4 and 1.6, respectively, compared to 4.2 in control), nematode multiplication rate and plant mortality, and increased spike/stem length, flower yield (68.5/$m^2$ and 63.0/$m^2$, respectively, compared to 37.5/$m^2$ in control), and root colonization with the bioagents (Nagesh and Parvatha Reddy 2005).

## 22.2.2 Interaction of Root-Knot Nematode with *Fusarium* Wilt

### 22.2.2.1 Interaction

The appearance of the wilt symptoms was accelerated when *M. incognita* was inoculated 2 weeks prior to *F. oxysporum* f. sp. *dianthi*. The rate of wilting was observed to be 4.6, while the root galling index (RGI) value was recorded to be 4.75 during 25th week of observation. It was observed that maximum plant mortality was recorded when *M. incognita* was inoculated 2 weeks prior to *F. oxysporum* f. sp. *dianthi* which was 79.2% at 25th week of observation. The plant growth parameters (plant height, plant weight) were also reduced significantly due to prior inoculation of *M. incognita* (Shylaja 2005).

### 22.2.2.2 Management

Combined application of *Pochonia chlamydosporia* + *Purpureocillium lilacinum* each at 20 g/$m^2$ gave significant increase in plant growth parameters (plant height, root length and root weight) and flower yield (stalk length, stalk weight, and flower diameter) (Shylaja 2005). The lowest root galling (1.64), wilting index (2.0), plant mortality (49.5%), and number of nematodes in both soil and roots (44.4 $J_2$ per 100 $cm^3$ soil and 10 nematodes/5 g roots) were recorded in plants treated with *P. chlamydosporia* and *P. lilacinum* (Shylaja 2005).

## 22.2.3 Spiral Nematode, *Helicotylenchus dihystera*

### 22.2.3.1 Symptoms

*Helicotylenchus dihystera* was found to be the most predominant nematode as revealed by the community analysis of carnation rhizosphere samples. An initial inoculum level of 1000 nematodes was responsible for the significant reduction in plant growth parameters as well as for delay in flowering (Fig. 22.3). Staining of the affected carnation root system showed the presence of this nematode pathogen in the cortical tissues (Fig. 22.3). The nematode was found to be an endoparasite in carnation (Jagirdar 2005).

### 22.2.3.2 Management

Bioagents like *Pseudomonas fluorescens* and *Purpureocillium lilacinum* (at 1 g/pot and 2 g/pot, respectively) effectively reduced the numbers of *H. dihystera* besides increasing the plant growth parameters (Jagirdar 2005).

**Fig. 22.3** Stunting of carnation plants due to spiral nematode infection

## 22.3 Gerbera, *Gerbera jamesonii*

Gerbera is a shade loving plant that needs about 50% shade for quality bloom production. The average yield under greenhouse is around 200 cut flowers/m$^2$/year of which 85% of I grade quality while it is 180 flowers/m$^2$/year under open condition with only 15–20% of flowers being of I grade (Fig. 22.4).

The root-knot nematode is recognized as the major limiting factor in successful production of gerbera crop.

**Fig. 22.4** Gerbera under protected cultivation

### 22.3.1 Root-Knot Nematode, *Meloidogyne incognita*

The root-knot nematode *M incognita* is one of the serious limiting factors in commercial cultivation of gerbera under polyhouse conditions.

#### 22.3.1.1 Crop Losses

Most of the highly fetching exotic cultivars of Gerbera from Europe have shown 40 to 60% mortality in polyhouse beds due to root-knot nematode infection in and around Bangalore (Nagesh and Parvatha Reddy 2000).

#### 22.3.1.2 Symptoms

Symptoms include stunting of plants, yellowing of leaves, wilting of plants, and heavy galling on roots (Fig. 22.5).

#### 22.3.1.3 Management

Soil application of *Bacillus subtilis* (BbV 57) or *Pseudomonas fluorescens* at 2.5 kg/ha at the time of planting is effective for the management of root-knot nematode.

Integration of *P. lilacinum/T. harzianum* at 0.5 L per $m^2$ (aqueous spore suspension containing $2 \times 10^4$ spores per ml) with Neem cake at 0.5 kg per $m^2$ or Fenamiphos at 2 g a.i. per $m^2$ increased plant growth parameters and flower yield of gerbera. The above treatments also increased root-knot egg parasitization by the parasitic fungi (Nagesh and Parvatha Reddy 2005).

Pre-plant treatment of beds with Dazomet (at 40 $g/m^2$) followed by the application of Neem cake (at 1 $kg/m^2$ 15 days later) along with antagonistic fungi, *P. chlamydosporia/P. lilacinum* (at $2 \times 10^{12}$ spores/$m^2$) significantly reduced root-knot nematode population (*M. incognita*), mortality of plants and suppressed the nematode infection for 2 years in gerbera. The antagonistic fungi established better

**Fig. 22.5** *Left*—gerbera bed infected with root-knot nematode showing stunting and leaf yellowing. *Right*—gerbera roots infected with root-knot nematode showing severe galling

in the beds treated with Dazomet. The above treatment also reduced root galling (0.4 and 1.4, respectively, as compared to 3.8 in control) and plant mortality and increased spike length and flower yield (58.0/m$^2$ and 55.5/m$^2$, respectively, as compared to 40.5/m$^2$ in control) (Nagesh and Parvatha Reddy 2005).

### 22.3.2 Interaction of Root-Knot Nematode with *Phytophthora* Foot Rot

#### 22.3.2.1 Interaction

Sustainable production of gerbera is seriously hampered by the disease complex caused by *M. incognita* and *Phytophthora parasitica* resulting in crop loss to the extent of 40–60%.

#### 22.3.2.2 Management

In gerbera cv. Debora, effective management of disease complex and enhancement of the flower yield by 26% was obtained by integration of soil incorporation of Neem cake enriched with either *Trichoderma harzianum* and *Pseudomonas fluorescens* [mixing 50 g of *T. harzianum* (2 × 10$^6$ cfu/g) or *P. fluorescens* (2 × 10$^8$ cfu/g) in 1 kg of Neem cake] applied at 25 g/m$^2$ (Manoj Kumar et al. 2010).

## 22.4 Rose, *Rosa* spp.

Large-scale cultivation of export quality rose cut flowers in protected condition started in early 1990s, thereby totally altering production dynamics (Fig. 22.6). Cut flower trade is worldwide dominated by Hybrid Tea roses.

Muthukrishnan et al. (1975) reported that the endoparasites like the lesion and root-knot nematodes and ectoparasites such as *Hemicycliophora labiata* and *Xiphinema basiri* were the major problems in rose cultivation.

### 22.4.1 Lesion Nematodes, *Pratylenchus* spp.

#### 22.4.1.1 Crop Losses

Sundarababu and Vadivelu (1988) observed the pathogenic potential of *P. zeae* on Edward rose. Prasad and Dasgupta (1964) found *P. pratensis* in rose plants exhibiting stunted growth.

#### 22.4.1.2 Symptoms

The affected plants showed chlorotic symptoms, poor and stunted growth with necrotic lesions on rose roots. Reduction in plant growth parameters like shoot length, shoot weight, root length, and root weight to the extent of 69.6, 36.4, 59.6, and 33.3%, respectively, was due to the lesion nematode *P. zeae*. Affected roots show patchy discoloration in cortical region, which coalesce and lead to root rotting.

**Fig. 22.6** Rose under protected cultivation

#### 22.4.1.3 Management

##### 22.4.1.3.1 Physical Methods

Hot water treatment of rose seedlings at 50 °C for 5 min or 48 °C for 10 min eliminates *P. zeae* infestation.

##### 22.4.1.3.2 Cultural Methods

The lesion nematode *P. penetrans* can be effectively managed by intercropping of roses with African marigold.

##### 22.4.1.3.3 Chemical Methods

Enhancement of flower yield by 16% and significant reduction in the lesion nematode *P. penetrans* population can be achieved by soil incorporation of aldicarb at 4 oz/1000 sq. ft. (Johnson and McClanahan 1974).

##### 22.4.1.3.4 Host Resistance

Ohkawa and Saigusa (1981) found that *Rosa indica* (cv. Major) and *R. multiflora* (cv. 60-5) were resistant to *P. penetrans* and *P. vulnus*, respectively.

### 22.4.2 Root-Knot Nematodes, *Meloidogyne* spp.

*M. incognita* was recorded on rose (*Rosa indica*) from Aligarh (Alam et al. 1973) and *M. javanica* from Delhi (Prasad and Dasgupta 1964).

#### 22.4.2.1 Symptoms

When plants are severely infected by root-knot nematodes *Meloidogyne* spp., the normal root system is reduced to a limited number of severely galled roots with a completely disorganized vascular system. Rootlets are almost completely absent.

#### 22.4.2.2 Management

##### 22.4.2.2.1 Physical Methods

Hot water treatment of rose roots at 45.5 °C for 60 min eliminates root-knot nematode infestation.

##### 22.4.2.2.2 Chemical Methods

Dipping of bare roots of rose in 0.1% solution of fenamiphos for 30 min gave considerable reduction in root-knot infection (Dale 1973).

##### 22.4.2.2.3 Host Resistance

Ohkawa and Saigusa (1981) reported that rose cv. Mavetti to be resistant against *M. hapla*. Rose cultivar Sultany can be classified as resistant host, whereas Gory cultivar as tolerant one.

##### 22.4.2.2.4 Integrated Methods

Nagesh and Janakiram (2004) reported that the enhancement of flower yield and reduction in root-knot index was obtained by soil application of dazomet (at 25 g/ $m^2$) before planting followed by integration of soil amendment with neem cake (at 1 kg/$m^2$) and *Pochonia chlamydosporia* (at $2 \times 10^{10}$ cfu/g).

## 22.5 Chrysanthemum, *Chrysanthemum indicum*

Protected cultivation of chrysanthemum is beneficial for quality production of flowers for export (Fig.22.7).

### 22.5.1 Lesion Nematodes, *Pratylenchus coffeae, P. chrysanthus*

Edward et al. (1969) reported *P. chrysanthus* associated with root rot of chrysanthemum.

#### 22.5.1.1 Symptoms

Rashid and Khan (1975) reported that *P. coffeae* is responsible for heavy root damage which subsequently led to poor growth of chrysanthemum. The above-ground symptoms include stunted growth of plants, leaf chlorosis, and wilting of plants. Flower size is also reduced.

**Fig. 22.7** Chrysanthemum under protected cultivation

#### 22.5.1.2 Histopathology

*P. coffeae* infects and remains confined to the cortical cells. Parenchymatous cells are completely destroyed due to nematode feeding.

#### 22.5.1.3 Interaction with Other Pathogens

The presence of *P. coffeae* has been found to aggravate the disease severity of *Pythium aphanidermatum* and *Rhizoctonia solani* (Hasan 1988). Edward et al. (1969) reported *P. chrysanthus* associated with root-rot of chrysanthemum.

#### 22.5.1.4 Management

Effective reduction in nematode multiplication can be achieved by soil incorporation of Neem cake enriched with *T. harzianum* at 1 t/ha.

### 22.5.2 Foliar Nematode, *Aphelenchoides ritzemabosi*

*A. ritzemabosi* is a major pest and causes considerable damage to the foliage of chrysanthemum, zinnia, salvia, aster, and dahlia (Gill and Sharma 1976).

#### 22.5.2.1 Symptoms

Leaf symptoms on infested chrysanthemum include reddish-yellow lesions on the lower leaves of young plants; in older plants these leaves are markedly chlorotic and a large area of the leaf surface becomes necrotic (Fig. 22.8). Deformed flowers and scanty foliage can be observed on foliar nematode affected plants. Leaves in the upper part of plants show slightly higher resistance than those in the lower part (Cid del Prado and Sosa Moss 1978). Cayrol and Combettes (1972) reported that the

**Fig. 22.8** Foliar nematode symptoms on chrysanthemum plant (*Left*) and leaves (*Right*)

mechanical damage due to stylet and growth hormone damage on cell growth and division are the main effects due to infection of foliar nematode.

#### 22.5.2.2 Life Cycle

In chrysanthemum leaves, a female lays about 25–30 eggs in a compact group. Eggs hatch in 3–4 days and the larvae take 9–10 days to reach maturity. The life cycle takes 10–13 days (Wallace 1960).

#### 22.5.2.3 Survival

Nematodes overwinter in dead, infested leaves on the ground or between the scales of infected buds.

#### 22.5.2.4 Management

##### 22.5.2.4.1 Physical Methods

Hot water treatment of stools at 46 °C for 5 min eliminates nematode infestation.

Sufficient chrysanthemum stools were immersed in hot water at 48 °C to bring the temperature down to 46.6 °C, no further heating being necessary, and stools were removed after 5.5 min when the temperature was down to 46.1 °C.

##### 22.5.2.4.2 Cultural Methods

Thorough cleaning up and burning of all infected plant material is essential in all control programs. Plants propagated from top cuttings selected from plants growing under dry conditions offer an opportunity to secure relatively clean stock for foundation plantings. Avoiding wetting and overlapping of leaves of adjacent plants, banding the base of stems with petroleum jelly or tree-banding grease, and destroying all plant remain, and fallen leaves are some of the practices recommended for reducing the nematode inoculum. For outdoor plantings, the ground can be kept free of nematodes by a weed-free fallow of 2–3 months during winter.

Provide sufficient space to allow good aeration between plants for immediate leaf after showers. Avoid wetting the foliage and overhead irrigation. Destroy crop residues left after harvesting.

##### 22.5.2.4.3 Chemical Methods

Dipping of chrysanthemum stools in 0.03% parathion for 20 min was most effective for eliminating *A. ritzemabosi*. Foliar application of methyl parathion, chlorpyriphos and quinalphos each at 0.05% were effective in the control of *A. ritzemabosi* infecting zinnia and chrysanthemum (Gill 1981; Gill and Walia 1980). Drenching with 0.02% thionazin at 6.8 $L/m^2$ is also effective.

##### 22.5.2.4.4 Host Resistance

Chrysanthemum cvs. Amy Shoesmith, Orange Beauty, and Orange Peach Blossom are comparatively resistant.

### 22.5.3 Root-Knot Nematodes, *Meloidogyne* spp.

Chrysanthemum production is limited by the presence of root-knot nematodes *M. arenaria* and *M. javanica* (Sen and Dasgupta 1977; Chandwani and Reddy 1967).

#### 22.5.3.1 Symptoms

Severe plant stunting, chlorosis, and extensive root galling were observed on chrysanthemum.

#### 22.5.3.2 Management

Nagesh et al. (2003) found that enhancement of flower yield by 23% to 28%, reduction in nematode reproduction (Rf, 1.4–2.12), root-knot index (1.4–2.7), increased colony-forming units (700–1070), and infectivity (32–52%) of the bioagent were obtained by soil incorporation of the bioTalc and pesta granules of *P. lilacinus* applied at 10 and 15 kg/ha to root-knot infested field.

## 22.6 Lilies, *Lilium* spp.

Lilies, especially Asiatic and Oriental types grown in greenhouse (Fig. 22.9) are most fascinating in international floriculture trade.

**Fig. 22.9** Lily under protected cultivation

The most harmful nematodes to lilies are the root lesion or meadow nematode (*Pratylenchus* spp.) and the leaf-lesion nematode (*Aphelenchoides* spp.).

## 22.6.1 Lesion Nematode, *Pratylenchus penetrans*

### 22.6.1.1 Symptoms

Foliage that prematurely yellows (Fig. 22.10) from the bottom and plants that mature early in circular areas of a field usually indicate nematode damage. Roots are greatly reduced, and most show dark lesions or dead spots (Fig. 22.10). Lesion nematode infection leads to drastic reduction in root system and roots on bulbs and significant reduction in bulb size.

### 22.6.1.2 Management

Pruning bulbs' roots eliminates 75–80% of the nematodes.

Recently harvested, infected bulb-lets are treated in a hot water bath of 43.5 °C for 1 h. After treatment, store bulbs at 28 °F for 2 months before planting again.

Rotation of lily with marigold was found to be effective for the management of lesion nematode.

Application of Phorate 10G at 80 kg/ha over the set bulbs and cover with soil at planting time was found to be effective. Nematicides applied by drip irrigation (Ethoprop, Fenamiphos, and Oxamyl) reduced nematode populations both in soil and roots. Additional reductions ($P = 0.05$) occurred as a result of three foliar applications of Oxamyl (Westerdahl et al. 1993).

**Fig. 22.10** Symptoms of lesion nematodes on lily. *Left*—leaf yellowing stunting of plant. *Middle*—necrotic lesions on roots. *Right*—lesion nematodes feeding inside roots

Application of *P. lilacinum* (at $2 \times 10^{12}$ spores/$m^2$) at 2–4 kg/ha at planting time proved beneficial.

## 22.6.2 Foliar Nematode, *Aphelenchoides fragariae*

### 22.6.2.1 Symptoms

Bunchy-top, dieback, leaf yellowing, blotchy leaves, leaf thickening and distortion, leaf withering, "blossom blasting," "blind," and retardation of blooming are some of the symptoms of foliar nematode infection on lily (Fig. 22.11).

**Fig. 22.11** *Left*—lily plant infected with leaf lesion nematode. *Right*—deformed shoot tip due to foliar nematode infection

#### 22.6.2.2 Spread and Survival

Nematode infections are transferred to other areas and from one season to the next in planting stock. The nematodes are capable of surviving for 3 months or more in crop residues or soil.

#### 22.6.2.3 Management

##### 22.6.2.3.1 Cultural Methods

Rogueing lily planting for foliar nematodes is common practice. It is difficult to detect infection at an early stage, so rouging must be repeated several times during the season.

Dipping of infected bulb-lets in hot water bath maintained at 43.5 °C for 1 h eliminates nematodes. Treatment is best on bulb-lets only, reducing the volume of treated material and avoiding plant injury, which often follows treatment of larger bulbs. After treatment, soak bulbs in fungicide to protect against rot.

The nematode can be effectively managed by avoiding overhead irrigation to prevent splashing of water between plants and extending wet conditions in the field.

Remove and destroy all plant debris at season's end.

##### 22.6.2.3.2 Chemical Methods

Apply a granular nematicide such as Fenamiphos (e.g., Nemacur) when planting bulbs.

## 22.7 Orchids, *Cymbidium, Phalaenopsis, Cattleya, Dendrobium, Vanda* spp.

Orchidaceae, commonly known as the orchid family, is a diverse and widespread family of flowering plants grown in greenhouses with blooms that are often colorful and fragrant (Fig. 22.12).

### 22.7.1 Foliar Nematodes, *Aphelenchoides besseyi* (on Vanda and *Dendrobium nobile*), *A. fragariae* (on Oncidium)

#### 22.7.1.1 Symptoms

- Damage becomes visible as tissue discoloration, formation of rots, or inhibition of growth in young buds and shoots (Fig. 22.13).
- Infected buds are slightly yellow.
- Eventually the buds dry and become deformed as they fail to develop.
- Infected flower spikes are short and barren (Fig. 22.13).
- Dark, elongated streaks develop on leaf sheaths covering the bulb (Fig. 22.13).
- Very thin, black scar tissue forms on leaf sheaths and spikes.

**Fig. 22.12** Orchid, *Phalaenopsis* under protected cultivation

**Fig. 22.13** *Left* (1)—oncidium orchid flower spikes infected with foliar nematodes (D1, D2, and D3) as compared with a healthy spike (H). *Right* (2)—oncidium orchid pseudo bulb infected with foliar nematodes showing long, dark streak on the base of the leaf sheath

- These scars represent areas of host tissue that have been damaged by nematode movement and feeding.
- The dead cells become blackened.

#### 22.7.1.2 Favorable Conditions

Foliar nematodes are common in warm, moist environments.

#### 22.7.1.3 Spread and Survival

Nematodes reside between leaf sheaths and can survive long periods with little or no moisture. They may survive in dried buds and spikes for many months.

The main means of primary nematode spread include infected planting stock, infested soil or planting substrate, while secondary means of spread include irrigation water, water splashing, and plants touching each other.

#### 22.7.1.4 Management

- Start the crop from clean seeds or tissue-cultured plantlets.
- Vanda cuttings may be dipped in hot-water maintained at 46 °C for 15 min or 49 °C for 5–10 min.
- Use nematode-free substrate.
- Crops must be grown on raised benches.
- Keep the growing environment as dry as possible.
- Air movement within the field or greenhouse should be maximized to encourage evaporation and leaf dryness.
- Remove all diseased leaves, spikes, and other infected plant parts from the field.

## 22.8 Anthurium, *Anthurium andraeanum*

Anthurium occupies important position in cut flower trade on account of its beauty. It is grown for its attractive flowers as well as foliage (Fig. 22.14). The most popular and economically important species are *Anthurium andreanum* and *A. scherzerianum* which possess attractive long-lasting inflorescence.

### 22.8.1 Burrowing Nematode, *Radopholus similis*

#### 22.8.1.1 Symptoms

Root lesions and plant decline are well-known symptoms of burrowing nematode infection. Anthurium root lesions caused by *R. similis* are brown or dark brown to black. The rots develop relatively slowly. But compared to healthy plants, the amount of functional roots in diseased plants is greatly reduced. With time, fewer new roots are produced, and gradually the entire root mass is destroyed (Fig. 22.15). The steady and progressive destruction of the root system usually causes plant decline in the second to fourth years. The leaves turn yellow and may have other

**Fig. 22.14** Anthuriums under protected cultivation

**Fig. 22.15** *Left*—burrowing nematode-infected anthurium plants showing extensive brown root rots, yellow leaves, and stunted plants. *Right*—anthurium root tip showing rot due to burrowing nematode infection

symptoms of nutrient deficiency. The plants become smaller and lose vigor, producing fewer and smaller flowers.

The development root rot occurs in the form of small, pink, orange-brown or light brown streaks or elongated rots, brownish streaks, with or without yellow borders, and then become distinctly long, blackish-brown rots. With time, these rots expand, and large sections of the root are destroyed. At times, if the nematodes feed and grow

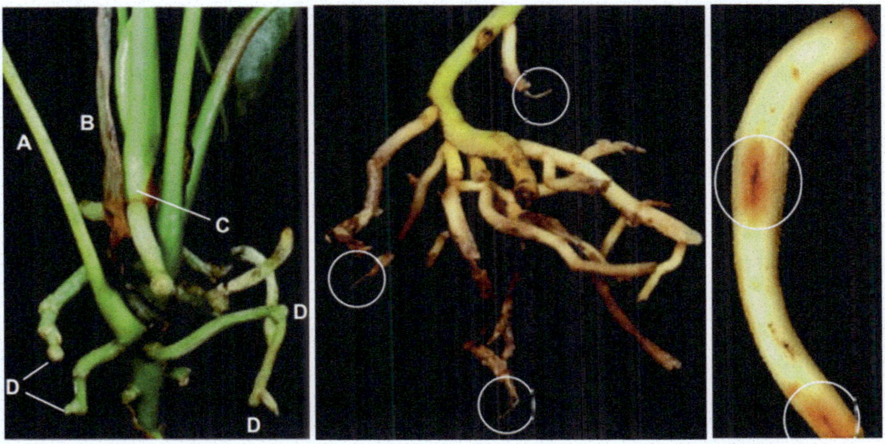


**Fig. 22.16** *Left*—typical stubby roots on an anthurium plant (A, petiole; B, sheath; C, node; D, stubby roots). *Middle*—early stages of root rot (missing root tips with exposed vascular strands). *Right*—expanding root lesions

deep into the root, only a gray streak is visible on the external surface of a white or light colored root.

Infected tips are brown (Fig.22.16) and may be missing. When the environment is dry, nematodes either migrate back into the soil or desiccate, and many rotted tips are devoid of nematodes. Nematode infection leads to formation of "stubby root" which may produce a new lateral root during favorable conditions for growth. Repetition of this cycle results in a short aerial root with several stubby tips (Fig. 22.16).

#### 22.8.1.2 Spread and Survival

The primary spread takes place through infected planting stock, while secondary spread occurs through water splashing.

#### 22.8.1.3 Biology

Burrowing nematodes emerge from eggs and mature in 18–20 days (Shurtleff and Averre 2000). A typical female may produce as many as 100 eggs in her lifetime.

#### 22.8.1.4 Management

- Clean nematode-free planting material is of paramount importance.
- Use of tissue-cultured plants.
- Treating top cuttings with 49 °C water for 12 min.
- Use nematode-free substrate (growing medium).
- A bare fallow maintained weed free for 3–6 months will eliminate the burrowing nematode.
- Anthuriums can be grown in pots or bags on benches to avoid introduction of burrowing nematodes.

- DiTera (microbial nematicide derived from the microorganism *Myrothecium verrucaria*) has given the best control when applied at 34 kg/ha once every 3 months.
- Use Fenamiphos (Nemacur) to control the burrowing nematode.
- Anthurium cv. "Kalapana" supported very low nematode reproduction. "Blushing Bride" was tolerant.

## 22.9 Future Thrusts

There is a need for emphasis on the following research aspects for nematode management in protected cultivation:

- At the time of new poly house establishment, ensuring nematode-free soil should be strictly adhered by testing soil for the presence nematodes or otherwise in ICAR Institutes/State Agricultural Universities.
- Region-wise dedicated referral labs for analyzing samples for nematodes should be established at State Agricultural Universities/ICAR Research Institutes.
- Sites for erecting new polyhouses should be properly treated with recommended chemicals followed by supplementation with biological control agent enriched farm yard manure/oil cakes/vermicompost.
- There is an urgent need to grant provisional registration for chemicals having proven efficacy against root-knot and other nematodes.
- In case of heavy population of plant parasitic nematodes species, soil fumigation with permissible chemicals should be done.
- Use of quality Neem cake, Castor cake, Coco peat, FYM, or Vermicompost enriched with bio-control agents.
- Evaluation of root-stock of ornamental crop genotypes against target nematode species (even up to races) should be carried out. The commercial scions should be grafted on resistant root-stocks.
- Efforts on breeding flower varieties suitable for polyhouse conditions should be taken up.
- Cluster approach may be followed to tackle the emergent nematode and other pest problems under polyhouse conditions.

## 22.10 Conclusion

The major plant parasitic nematodes associated with ornamental crops include species of *Meloidogyne*, *Helicotylenchus*, and *Pratylenchus*. Nematodes assume new proposition in protected cultivation of flower crops due to moderate climate and intensive cultivation. As the accrued loss due to these pests is tangible, proper attention is a must for their management. Various chemicals for the management of nematodes have been used under protected cultivation but fumigants like dazomet and metham sodium used as pre-plant under tarp have filled the gap along with other methods of management. There is further need to develop viable options for nematode management, including use of biocontrol agents and antagonistic crop like marigold (*Tagetes* spp.). Integrated management methods need to be developed using eco-friendly components like organic soil amendments (oil cakes), biological control agents, and safer biopesticides. There is a need to assess the benefits of integrating practices such as cover crops, resistant cultivars, and environmentally compatible chemicals for effective management of root-knot and other plant parasitic nematodes in ornamental production systems.

## References


Alam MM, Khan AM, Saxena SK (1973) Some new host records of root-knot nematode, *Meloidogyne incognita* (Kofoid & White, 1919) Chitwood, 1949. Indian J Nematol 3:159–160

Anusuya K, Vadivelu S (2002) Mutualistic symbiosis of VAM fungi on plant growth promoting rhizobacteria (PGPR) on micro propagated carnation. J Microbiol World 4:105–107

Cayrol JC, Combettes S (1972) Histopathological study of the chrysanthemum leaf eelworm disease in monoxenic cultures. Annales de Zoologie-Ecologie Animale 4(2):119–128

Chandwani GH, Reddy TSN (1967) The host range of root-knot nematode, *Meloidogyne javanica* in tobacco nurseries at Rajahmundry, Andhra Pradesh. Indian Phytopathol 20:383–384

Cid del Prado I, Sosa Moss C (1978) Occurrence of *Aphelenchoides ritzemabosi* on the foliage of *Chrysanthemum maximum* in Mexico. Nematropica 8:6

Dale PS (1973) Elimination of root-knot nematodes from roses by chemical bare-root dips. New Zealand J Expt Agric 1(2):121–122

Edward JC, Misra SL, Rai BB, Peter E (1969) Association of *Pratylenchus chrysanthus* nsp. with chrysanthemum root rot. Allahabad Farmer 43:175–179

Gill JS (1981) Fighting the foliar nematode of chrysanthemum. Indian Hort 25:21–22

Gill JS, Sharma NK (1976) Additional hosts of the foliar nematode, *Aphelenchoides ritzemabosi* from India. Indian J Nematol 6:169–171

Gill JS, Walia RK (1980) Control of chrysanthemum foliar nematode, *Aphelenchoides ritzemabosi* on *Zinnia elegans*. Indian J Nematol 10:86–88

Hasan A (1988) Interaction between *Pratylenchus coffeae* and *Pythium aphanidermatum* and/or *Rhizoctonia solani* on chrysanthemum. Phytopathol Z 123:227–232

Jagirdar SA (2005) Studies on plant parasitic nematodes associated with polyhouse crops. M. Sc. (Agriculture) thesis, Department of Plant Pathology , University of Agricultural Sciences, Dharwad

Johnson N (2000) Studies on nematode pests of cut flowers with special reference to Carnation, Cystera, Gladiolus and Asiatic Lily. Ph. D. thesis, Tamil Nadu Agricultural University, Coimbatore, 160 pp

Johnson PW, McClanahan RJ (1974) Nematode control and other effects of soil treatment of greenhouse rose beds with aldicarb and oxamyl. Pl Dis Reptr 58(8):730–732

Manoj Kumar R, Rao MS, Ramachandran N (2010) Biomanagement of disease complex caused by *Meloidogyne incognita* and *Phytophthora parasitica* Dastur in protected cultivation of gerbera using formulation of *Trichoderma harzianum* and *Pseudomonas fluorescens*. In: Proceedings of national conference on innovations in nematological research for agricultural sustainability–challenges and a roadmap ahead, Tamil Nadu Agricultural University, Coimbatore, pp 97–98

Muthukrishnan TS, Chandrasekaran J, Lakshmanan PL, Chinnarajan AM (1975) Occurrence of the nematode *Hemicycliophora labiata* Colbran 1960 on roses in Tamil Nadu. S Indian Hort 23:75–76

Nagesh M, Hussaini SS, Singh SP, Biswas SR (2003) Management of root-knot nematode, *Meloidogyne incognita* (Kofoid & White) Chitwood in chrysanthemum using *Paecilomyces lilacinus* (Thom.) Samson in combination with neem cake. J Biol Control 17:125–131

Nagesh M, Janakiram T (2004) Root-knot nematode problem in polyhouse roses and its management using dazomet, neem cake and *Pochonia chlamydosporia* (*Verticillium chlamydosporium*). J Orn Hort 7:147–152

Nagesh M, Parvatha Reddy P (2000) Crop loss estimation in carnation and gerbera due to the root-knot nematode, *Meloidogyne incognita* (Kofoid & White) Chitwood. Pest Managmt Hort Ecosyst 6:158–159

Nagesh M, Parvatha Reddy P (2005) Management of carnation and gerbera to control the root-knot nematode, *Meloidogyne incognita*, in commercial polyhouses. Nematol Medit 33:157–162

Ohkawa K, Saigusa T (1981) Resistance of rose rootstocks to *Meloidogyne hapla, Pratylenchus penetrans* and *P. vulnus*. Hort Sci 16:559–560

Om N, Mcsorley R, Frederick JJ (2008) Response of cut flowers and bedding plants to root-knot nematodes. Proc Fla State Hort Soc 121:370–373

Prasad SK, Dasgupta DR (1964) Plant parasitic nematodes associated with roses in Delhi. Indian J Entomol 26:12–122

Rao MS, Shylaja M (2003) Management of root-knot nematode (*Meloidogyne incognita*) in carnation using a formulation of *Pochonia chlamydosporia* (*Verticillium chlamydosporium*). J Orn Hort 6:202–206

Rashid A, Khan AM (1975) Studies on the lesion nematode, *Pratylenchus coffeae* associated with chrysanthemum. Indian Phytopathol 28:22–28

Sen K, Dasgupta MK (1977) Additional hosts of the root-knot nematode, *Meloidogyne* spp. from India. Indian J Nematol 7:74

Shurtleff M, Averre C (2000) Diagnosing plant diseases caused by nematodes. APS, St Paul

Shylaja M (2005) Studies on the development of bio-intensive disease management strategies in tuberose (*Polyanthes tuberosa* L.) and carnations (*Dianthus caryophyllus* L.) using biological control agents. Ph. D. thesis, Kuvempu University, Shimoga, 132 pp

Sundarababu R, Vadivelu S (1988) Studies on the nematodes associated with rose. Indian J Nematol 18:139

Wallace HR (1960) Observations on the behaviour of *Aphelenchoides ritzemabosi* in chrysanthemum leaves. Nematologica 5:315–321

Westerdahl BB, Giraud D, Radewald JD et al (1993) Management of *Pratylenchus penetrans* on oriental lilies with drip and foliar-applied nematicides. Suppl J Nematol 25(4S):758–767

# Medicinal Crops

# 23

**Abstract**

The production of medicinal crops such as ashwagandha, coleus, brahmi, soda apple, and henbane throughout the globe is threatened by root-knot nematodes (*Meloidogyne incognita* and *M. javanica*). Root-knot nematodes interact with *Fusarium* wilt, *Sclerotium* collar rot, and *Rhizoctonia* root rot pathogens and cause disease complexes. In this chapter, crop losses and management using physical, cultural, chemical, biological methods, host resistance, and integrated methods for root-knot nematodes on medicinal crops are discussed.

**Keywords**

*Meloidogyne* spp. · Crop losses · Symptoms · Interactions · Management

## 23.1 Introduction

Medicinal plants are being widely cultivated in view of their importance in the manufacture of drugs. Nematodes are one of the serious limiting factors responsible for significant reduction in yield and quality of medicinal crops throughout the world. The root-knot, the reniform, and lesion nematodes are the major pest species in various agro-ecosystems. The root-knot nematodes are the most damage causing nematodes of many medicinal plants. *Meloidogyne incognita* and *M. javanica* do considerable damage to *Ocimum sanctum*, *O. basilicum*, *O. gratissimum*, *O. kalamandescharicum*, etc. Likewise, *Artemisia* sp. is severely attacked by *Meloidogyne arenaria*, Geranium (*Pelargonium graveolens*), and *Mentha arvensis* are attacked by *M. hapla*. Lesion nematodes (*Pratylenchus minyus, P. pratensis,* and *P. thornei*) are some other serious nematode pests of medicinal plants. The active ingredients that are present in medicinal plants which are being used for treating


P. P. Reddy, *Nematode Diseases of Crops and their Management*,
https://doi.org/10.1007/978-981-16-3242-6_23

diseases are severely affected by infection of nematodes. Such crops loose market acceptability and market values forcing the growers to sustain economic loss.

## 23.2 Ashwagandha, *Withania somnifera*

### 23.2.1 Root-Knot Nematode, *Meloidogyne incognita*

The root-knot nematode *M. incognita* (Race 2) is an important limiting factor in the successful cultivation of ashwagandha (Pandey and Kalra 2003). The nematode is very widespread affecting more than 80% plants in India.

#### 23.2.1.1 Symptoms

The nematode infected plants typically show chlorotic, stunted, less branched with fewer and smaller leaves (Fig. 23.1), and poor response to fertilizer and irrigation. Such symptoms usually are not noticeable until severe damage to root system has been done by the nematodes. Roots of such plant were severely galled (Fig. 23.1) with reduced alkaloids. The root-knot nematodes also infect the stem that is touching the soil. The nematode-infected plants are more likely to be killed early than healthy noninfested plants.

**Fig. 23.1** Root-knot infested plant and roots of ashwagandha

#### 23.2.1.2 Management

##### 23.2.1.2.1 Cultural Methods

Neem compound, *Artemisia annua*, *Mentha,* and *Murraya koenigii* distillates were found to be highly useful to suppress the root-knot development and *M. incognita* population on *W. somnifera.*

##### 23.2.1.2.2 Biological Methods

The efficiency of *Trichoderma harzianum* was found to be comparable to that of Carbofuran (RKI = 2), followed by *Purpureocillium lilacinum*, *Arthrobotryis oligospora* and Neem compound. Besides reducing nematode infestation, the bio-control agents also enhanced the growth of the plant (Sharma and Pandey 2009).

A parasitic fungus (*Arthrobotryis oligospora*) not only attacks eggs and larvae of nematodes but also produces toxic metabolites and enhances root development.

Soil application of 100% cell extract of *Purpureocillium lilacinum* gave maximum nematode egg mortality of 83% and larval mortality of 91% (48 h after application).

Soil application of 100% cell extract of *Bacillus megaterium* was also found to be effective.

##### 23.2.1.2.3 Integrated Methods

Maximum root-knot index suppression was noticed in Vermicompost + *T. harzianum* (0.66 as compared to 4.00 in control), while maximum increase in plant yield was recorded when the soil was amended with curry leaf/*Mentha* distillates (15.0 g as compared to 5.2 g in control) (Pandey and Kalra 2003).

Root-knot nematode index was drastically impaired with combined application of three AMF (*Glomus aggregatum, G. fasciculatum,* and *G. mosseae*) simultaneously (1.3 as compared to 3.6 in control), while plant dry weight (105.9 g as compared to 77.9 g in control) and oil yield (0.48% as compared to 0.38% in control) improved when plants were inoculated with *G. mosseae* alone.

Substantial reduction in root galling was noticed in Neem oil seed cake + *T. harzianum*, Vermicompost + *T. harzianum*, cow urine + *T. harzianum* treated plots as compared to untreated control. A significant and marked improvement in plant growth and yield was also noticed in plots treated with Vermicompost + *T. harzianum*, cow urine + *T. harzianum*, Neem oil seed cake + *T. harzianum* with maximum root yield obtained from Vermicompost + *T. harzianum* treated plots (Pandey et al. 2011). It is concluded that *T. harzianum* strain Thu when inoculated with Vermicompost can be an environment friendly strategy to reduce the population of *M. incognita* on *W. somnifera* and to obtain higher root yields. The build-up of nematode population is very much reduced in the rhizosphere by the vermicompost enriched with *T. harzianum* besides providing a complete nutrition to the plant.

Ramakrishnan and Thangavelu (2010) reported increase in the biomass of economic parts of ashwagandha plants such as dry roots by 26.87% and seed 43.19% and reduced the root-knot nematode incidence by 78% over untreated control by incorporation of super *Pseudomonas* (consortia of *Pseudomonas fluorescens* strains with chitinase, ACC deaminase, and Neem mixture) at 2.5 kg/ha.

## 23.2.2 Interaction of Root-Knot Nematode with *Fusarium* Root Rot

Root-knot disease due to *Meloidogyne incognita* and the disease complex it causes in conjunction with soil-borne fungal pathogen *Fusarium solani* is an upcoming disease which appeared in severe form on ashwagandha.

Gupta et al. (2004) reported occurrence of severe form of root-rot disease complex in ashwagandha nurseries and regular crop under Lucknow (Uttar Pradesh, India) conditions. The disease was found to cause 30–50% plant mortality. Further, investigations to characterize the infecting fungus led to identification of *Fusarium solani* as the causative organism and it was a new report of root-rot disease complex in *Withania somnifera* from Uttar Pradesh.

In Karnataka, Mallesh (2008) noticed maximum disease incidence of 30.33% in Dharwad district followed by Bijapur district (21.55%). The pathogens, viz., *F. solani* and *M. incognita*, were isolated from disease complex plots.

### 23.2.2.1 Symptoms

Pots treated with *F. solani* took 25 days for expression of typical wilt/root rot symptoms and *M. incognita* took still more days for infection and for gall formation.

Plants infected with both the pathogens exhibited loss of turgidity, withering and drooping of leaves, and wilting of plants. Death and decaying of roots, brown to black discoloration of roots with pulpiness were observed (Fig. 23.2). White cottony growth of the fungus was observed near collar region.

Inoculation of root-knot nematode alone did not produce root-rot in ashwagandha. Maximum root-rot index (5.0) and more disease severity (4.0) were observed in case of ashwagandha in treatment *Meloidogyne* + *Fusarium* followed by

**Fig. 23.2** Symptoms of root-knot and root-rot complex on ashwagandha. *Left to right*—rotting, decaying, and disintegration of roots

*Fusarium* alone, that is, the expression of symptoms was early and disease severity, root-rot index was more. Plants infected with root-knot and root-rot disease complex showed rotting, decaying, and disintegration of roots.

#### 23.2.2.2 Management

Application of talc-based PGPR strains RB50 and RB31 decreased root-knot index (0.75 and 0.80 as compared to 3.06 in control) and incidence of disease complex (8.33 and 11.67 as compared to 31.67 in control) while increasing plant height, number of branches, fresh and dry biomass, and tuber yield (373.00 and 354.60 as compared to 185.60 in control) in ashwagandha (Mallesh 2008).

Biochemical analyses in the above treatments showed elevated expression of defense enzymes (peroxidase, polyphenol oxidase, and phenylalanine ammonialyase) and higher accumulation of phenolic compounds (activity being highest in respect of RB 50 treated plants) compared to respective inoculated checks. The suppression of disease complex in ashwagandha is largely due to aforementioned biocontrol mechanism as well as induction of systemic resistance by efficient PGPR strains (Lingaraju and Mallesh 2010).

## 23.3 Coleus, *Solenostemon rotundifolius*

### 23.3.1 Root-Knot Nematode, *Meloidogyne incognita*

The root-knot nematode infestation was reported on coleus from Kerala and Orissa which is responsible for damage to the crop and yield loss.

#### 23.3.1.1 Crop Losses

The dry weight of the tubers was reduced by 20% due to root-knot nematodes. The percentage of starch on fresh weight basis showed drastic reduction (16%) in the infested tubers. *M. incognita* has been reported to cause a yield reduction of up to 86% (Senthamarai et al. 2006b).

#### 23.3.1.2 Symptoms

The newly inoculated *C. forskohlii* plant exhibited stunted growth, yellow patches, and severe galling of roots due to *M. incognita*. Tuber formation was found to commensurate with the degree of galling and yield reduction was 65% under field conditions. The pathogenic study which was carried out in controlled condition revealed that the effective damage to the plant was caused by 1000 larvae/plant.

The galls on coleus roots are very big and pronounced (Fig. 23.3). The root-knot nematode damage often leads to crop failure. The infested tubers swell in size with irregular surface and cracking of the skin. Rotting sets in even before harvest, if the infestation is severe. Infested tubers rot after harvest and rarely reach market.

**Fig. 23.3** Root-knot nematode-infested plant and roots of coleus

#### 23.3.1.3 Management

##### 23.3.1.3.1 Cultural Methods

Sree Bhadra, a high yielding variety of sweet potato released by Central Tuber Crops Research Institute, Trivandrum, is identified as a resistant trap crop for the root-knot nematode. The nematodes were able to penetrate the root but giant cell formation was not induced and as a result the nematodes die inside the roots. Planting this variety in root-knot nematode-infested field helped in clearing the field free of nematodes and giving good tuber yield. Subsequently growing of susceptible crop like coleus escaped nematode damage (Mohandas 2001).

##### 23.3.1.3.2 Chemical Methods

Chemical agents such as Methyl isothiocyanate mixtures, Oxamyl, Thionazin, and Carbofuran have been found to be effective for the management of nematodes. Application of Carbofuran 3% granules at the rate of 20 kg per hectare under wet conditions near the root zone was also found to be effective.

Treatment with Fensulfothion reduced damage to 5.7% as against 21% in control. The above treatment also increased the production of healthy tubers to the extent of 47% (Pillai 1976).

##### 23.3.1.3.3 Biological Methods

Soil application of *P. fluorescens* ($10^8$ cfu/g) at 2.5 kg/ha recorded increased plant growth, tuber yield (61.57 as compared to 22.71 in control), and propagule density in soil; and reduced root galling (678/plant as compared to 1680/plant in control), egg mass production, and *M. incognita* population in soil (Senthamarai et al. 2006a).

Enhancement of crop growth and yield of *C. forskohlii* and substantial reduction of root-knot nematode population in soil and root was obtained by soil incorporation of biological control agents like *T. viride* and *P. fluorescens* (Senthamarai et al. 2008).

Increase in plant height, number of branches, fresh and dry weight of the plant as well as less root-knot index was observed due to PGPR treatments (fluorescent *Pseudomonas* sp.) as compared to control. Highest reduction in root galling (1.5 as compared to 4.0 in control) and maximum plant height, number of branches, and fresh and dry weight of the plants was achieved by soil incorporation of PGPR strains (RB50 and RB31) (Mallesh 2008).

#### 23.3.1.3.4 Integrated Methods

Integration of soil solarization in the nursery for 15 days with 150 gauge LDPE film and application of *Purpureocillium lilacinum* + Neem cake or *P. lilacinum* + *Bacillus macerans* in the main field are the best treatments in increasing plant height (64.3 and 60.3 cm compared to 40.0 cm in control), number of leaves (593.3 and 583.3 compared to 310.0 in control), weight of tubers/plant (560.0 and 546.6 g compared to 350.0 g in control), yield (11.5 and 11.3 kg/plot compared to 6.9 kg/plot in control); and in reducing root galls (0.3 and 1.0 compared to 50.6 in control), nematode population in soil (25.0 and 30.0/100 ml soil compared to 196.6/100 ml soil in control), and roots (1.0 and 1.6/5 g roots compared to 79.0/5 g roots in control) (Nisha and Sheela 2006).

Integration of dipping stem cutting in 0.1% *Pseudomonas fluorescens* + soil application of Neem cake at 400 kg/ha + growing marigold as an intercrop followed by their biomass incorporation during earthing up increased the root tuber yield by 22.7–30.0% and reduced the root-knot nematode (*M. incognita*) population by 71.2–73.8%. However, integration of *P. fluorescens* + marigold intercrop proved to be more economical (with benefit: cost ratio of 6.4–8.8) and effective management practice for the management of root-knot nematodes in medicinal coleus (Seenivasan and Devrajan 2008).

## 23.3.2 Interaction of Root-Knot Nematode with *Fusarium* Wilt

#### 23.3.2.1 Interaction

Root-knot disease due to *Meloidogyne incognita* and the disease complex it causes in conjunction with soil-borne pathogen *Fusarium chlamydosporum* is an upcoming disease which appeared in severe form in coleus (Kumar 2008).

In the interaction studies, *M. incognita* was the most aggressive pathogen compared to *F. chlamydosporum*. However, simultaneous inoculation of *M. incognita* and *F. chlamydosporum* caused greater reduction in plant growth as well as nematode multiplication (Fig. 23.4). With respect to individual pathogens, *M. incognita* caused greater reduction in plant growth compared to *F. chlamydosporum*. In case of sequential inoculation of *M. incognita* 7 days prior to *F. chlamydosporum* caused reduction in plant growth parameters.

The effect of simultaneous inoculation of *M. incognita* with *F. chlamydosporum* on coleus was additive in nature. However, when *M. incognita* was inoculated with *F. chlamydosporum* the resultant effect was almost equal to sum of individual effect. These results suggest that the nematode can predispose the coleus to infection by

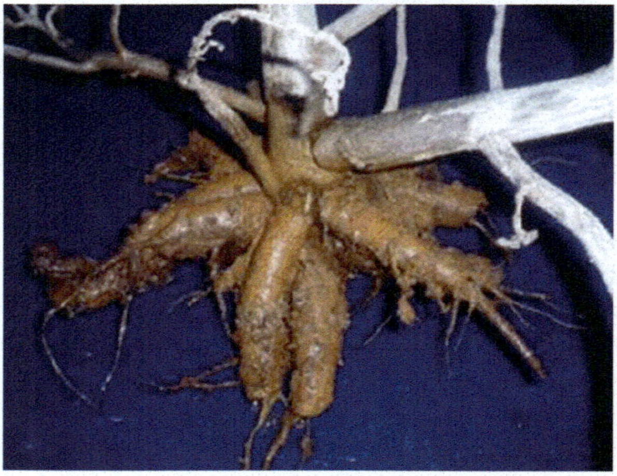

**Fig. 23.4** Symptoms of root-knot and wilt disease complex on coleus

fungus namely *F. chlamydosporum* and can aggravate the disease. Reduction in number of galls per plant and final nematode population was observed in simultaneous inoculation.

#### 23.3.2.2 Management

Talc-based bioformulations of rhizobacterial strains RB50 and RB31 decreased root-knot index (0.66 and 0.73 as compared to 2.22 in control) and the incidence of disease complex (10.00 and 13.33 as compared to 28.33 in control) while increasing plant height, number of branches, fresh and dry biomass, and tuber yield in coleus (6.74 and 5.45 kg/plot as compared to 3.00 kg/plot in control) (Mallesh 2008).

Combined application of plant products (Neem seed kernel powder at 5 g/kg of soil) with biocontrol agents (*Purpureocillium lilacinum, Trichoderma viride* + *Pseudomonas fluorescens* at 10 g/kg of soil) was found to be effective in reducing the number of galls, nematode population, number of egg masses, root-knot index, root-rot index, and improving the plant growth parameters as compared to inoculated control.

The coleus genotypes, viz., Yellow Tubers and Rabakavi Local, showed resistant reaction against *M. incognita* and *F. chlamydosporum* disease complex. Orange tubers and Nimbanur Local genotype showed moderately resistant reaction.

### 23.3.3 Interaction of Root-Knot Nematode with Collar Rot

#### 23.3.3.1 Interaction

*Rhizoctonia bataticola, Sclerotium rolfsii, F. chlamydosporum,* and *M. incognita* were found to be the most commonly associated fungi and nematode with collar rot disease complex (Ramaprasad Shresti 2005).

#### 23.3.3.2 Symptoms

Wilt symptoms were first recorded at 45 days after inoculation in treatment *M. incognita* inoculated 7 days prior to inoculation of all the fungal pathogens simultaneously (*F. chlamydosporum* + *R. bataticola* + *S. rolfsii*) (Fig. 23.5) (Ramaprasad Shresti 2005).

#### 23.3.3.3 Management

The wilt incidence, nematode population, number of galls, and colony-forming units of *F. chlamydosporum* and *R. bataticola* were significantly minimum in the plots where *T. viride* (10 ml/plant spore suspension) combined with Neem product "Neemto" (500 g/5 $m^2$) were applied compared to other treatments (Ramaprasad Shresti 2005).

The treatment *T. viride* + Neemto recorded significantly lowest percent wilt incidence (12.76) over control (35.52) followed by *T. harzianum* (18.87) and *P. fluorescens* (19.98). The number of galls/5 g of root was significantly lowest in treatment *T. viride* + Neemto (10.13), followed by carbofuran (14.93) which was on par with each other (Ramaprasad Shresti 2005).

Greenhouse studies using the plant growth promoting rhizobacterial (PGPR) fluorescent pseudomonads showed significant increase in seedling biomass besides reduction in coleus wilt complex due to combination of pathogens. Bioformulations containing RB 50 and RB 31 strains decreased root-knot index and incidence of disease complex and increased biomass and tuber yield. RB 50 and RB 31 treated plants exhibited highest activity of defense enzymes (peroxidase, polyphenol oxidase, and phenylalanine ammonialyase) and higher accumulation of phenolic compounds as compared to respective inoculated checks. Lingaraju and Mallesh (2010) opined that the induction of systemic resistance and aforementioned

**Fig. 23.5** Root-knot and collar rot disease complex on coleus. Left—infected, right—healthy

biocontrol mechanism were responsible for the suppression of disease complex in coleus by the efficient PGPR strains (RB 50 and RB 31).

### 23.3.4 Interaction of Root-Knot Nematode with Root Rot

The root-knot nematode *Meloidogyne incognita* in association with soil-borne fungus, *Macrophomina phaseolina* was responsible for root-rot disease complex in coleus at Salem, Attur, and Rasipuram in Tamil Nadu. Crop loss to the extent of 50–60% occurred from this root-rot disease complex. In order to prevent complete devastation of coleus crop from the disease complex, development of an effective management strategy on war footing is necessary.

#### 23.3.4.1 Symptoms

Simultaneous inoculation of *M. incognita* and *M. phaseolina* as well as nematode inoculation followed by fungus 15 days later caused significant reduction in tuber yield and 100% root-rot disease in medicinal coleus. The nematode multiplication was adversely affected when fungus was inoculated prior to nematode (Senthamarai et al. 2008).

#### 23.3.4.2 Management

Integrated nematode management strategy (INMS) includes dipping of stem cuttings in *Pseudomonas fluorescens* (strain Pf1) + growing marigold (*Tagetes erecta*) as intercrop and their biomass incorporation during earthing up. Biointensive disease management strategy (BDMS) include soil drenching with *P. fluorescens* (strain PfC6) at 2.5 kg/ha at planting, and repeated at 30, 60, 90, and 120 days after planting. Enhancement of yield (40.6%), reduction in nematode population in soil (73.2%), root galling (82.4%), root-knot index (1.6), egg mass production (85.9%), and fecundity (87.9%) were obtained by INMS along with BDMS treatments. The root-rot disease was also reduced by 50.4% by the above treatment. INMS treatment is highly economical with higher B:C ratio than other treatments including standard chemical check, that is, Carbofuran 3G at 1 kg a.i. /ha + soil drenching with Carbendazim 0.1%, which can be exploited commercially to manage the disease complex in growers' fields (Seenivasan 2010).

The following practices have been recommended for the management of disease complex in coleus:

- Select the coleus cuttings from disease free plants.
- Coleus cuttings should be treated with Carbendazim solution (0.1%).
- Soil drenching with Carbendazim or Propiconazole each at 0.1%.
- Soil application of FYM at 12.5 ton/ha + *Trichoderma viride* at 2.5 kg/ha before planting.
- Application of Carbofuran 3G at 1 kg a.i./ha before planting.
- Stem cuttings to be dipped in 0.1% *Pseudomonas fluorescens* at the time of planting.

- Use drip irrigation to minimize the spread of pathogens from infected plants to healthy plants.

## 23.4 Brahmi, *Bacopa mannieri*

### 23.4.1 Root-Knot Nematode, *Meloidogyne incognita*

#### 23.4.1.1 Symptoms

- Stunting in plant growth, chlorosis, gall formation in the root system, and reduction in fresh and dry weight of leaves and roots give a simple indication of the presence of root-knot nematodes (Figs. 23.6 and 23.7).
- A negative correlation exists between increasing population levels of *M. incognita* and plant growth and yield characteristics.

#### 23.4.1.2 Management

- Soil application of Neem compounds, distillation waste of Menthol mint, Curry leaf, Davana, and Vermicompost.
- Soil application of Neem cake or Sweet wormwood cake (*Artemisia annua*) at the rate of 400 kg/ha was found to be effective.
- Bioinoculants such as *Glomus fasciculatum* and *Trichoderma harzianum* alone or in combination were root-knot nematode suppressive and enhanced the growth and yield of brahmi.

**Fig. 23.6** Healthy (*Left*) and root-knot infested (*Right*) plants of brahmi

**Fig. 23.7** Root-knot infested brahmi plant

## 23.5 Soda Apple, *Solanum viarum*

### 23.5.1 Root-Knot Nematode, *Meloidogyne incognita*

#### 23.5.1.1 Losses

- Responsible for 33.6 and 11.5% reduction in plant height and root weight, respectively.

#### 23.5.1.2 Symptoms

- *M. incognita* infected *S. viarum* plants exhibited severe leaf yellowing and wilting symptoms (Shetty and Reddy 1985).
- Roots were heavily infected and deformed by prominent, bead-like galls (Fig. 23.8).

#### 23.5.1.3 Management

- Bare root-dip treatment of *S. viarum* seedlings with Fenamiphos and Aldicarb sulfone both at 1000 ppm for 2 h. was effective against *M. incognita* infection (Shetty and Reddy 1985).

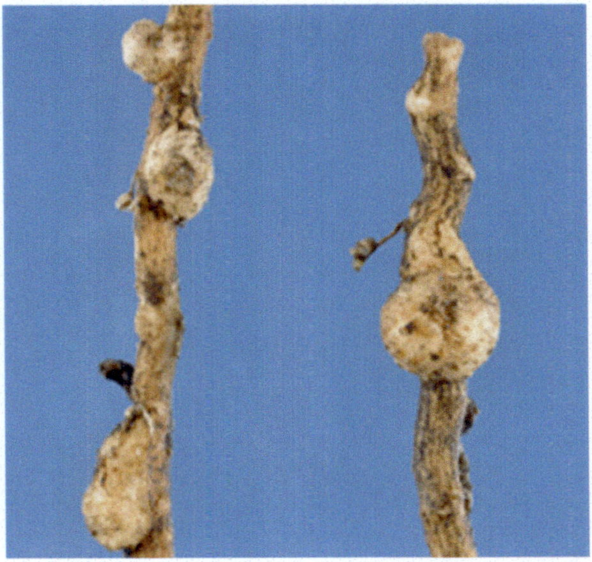

**Fig. 23.8** Root-knot nematode infestation on *Solanum viarum* roots

- Among four species of *Solanum* tested for their reaction to *M. incognita*, *S. torvum*, and *S. seaforthianum* gave resistant reaction which was reflected in the reduction in number of galls, egg masses, and fecundity of females (Shetty and Reddy 1985).

## 23.6 Henbane, *Hyoscyamus niger*

### 23.6.1 Root-Knot Nematodes, *Meloidogyne* spp.

#### 23.6.1.1 Symptoms

- Up to 60–70% plants were chlorotic and stunted showing a patchy appearance with fewer smaller leaves and flowers (Fig. 23.9).
- Varying sizes of galls were found on the root system (Fig. 23.9).

#### 23.6.1.2 Management

- Growing of nonhost crops such as *Cymbopogon flexuosus*, *C. winterianus*, and *C. martinii* can reduce the incidence of damage caused by root-knot nematode.
- Marigolds (*Tagetes minuta*) drastically suppressed root-knot nematode population with antagonistic phytochemicals in root exudates, that is, polythienyls.
- Sowing daincha (*Sesbania aculeata*) as green manuring crop decreases nematode population below the economic damage levels.
- Carbofuran (2 kg a.i. /ha) has been used to prevent root-knot nematode infestation.

**Fig. 23.9** Healthy (*Left*) and root-knot infested (*Right*) plants of henbane

- Combined application of all bio-inoculants (*Pseudomonas fluorescens, Glomus aggregatum, G. fasciculatum* and *G. mosseae*) have not only enhanced the total biomass yield of *H. niger* but also significantly decreased the multiplication of nematodes (Pandey 1997).

## 23.7 Conclusion

Since medicinal plants are being used in healthcare system, chemical methods for nematode management are becoming obsolete. There is a need for development of economical, eco-friendly, and sustainable management strategies. Some of the other alternatives for nematode management are available in medicinal and aromatic plants like resistant germplasm, useful organic materials, effective biocontrol agents, and other cultural/physical methods. Such resistant and tolerant germplasm could be exploited in future plant breeding programs for developing resistant/tolerant genotypes against major nematodes. Various organic materials are available which have been proven useful to decrease nematode incidence and enhanced medicinal plants yield and which could also be used in on-going programs for better and healthy plants. Large numbers of bioagents are available which could also be used on large scale to protect medicinal plants against various nematodes. However, much attention is needed to study and develop a new strategy to manage major nematodes in an eco-friendly way, which should be cost effective and environmentally friendly.

## References


Gupta ML, Mishra HO, Kalra A, Khanuja SPS (2004) Root-rot and wilt: a new disease of ashwagandha (*Withania somnifera*) caused by *Fusarium solani*. J Medicinal Aromatic Plants Sci 26:285–287

Kumar B (2008) Studies on root-knot and wilt complex in *Coleus forskohlii* (Wild.) Briq. caused by *Meloidogyne incognita* (Kofoid and White) chitwood and *Fusarium chlamydosporum* (Frag. and Cif.) Booth. M. Sc. (Agriculture) thesis, Department of Plant Pathology, University of Agricultural Sciences, Dharwad

Lingaraju S, Mallesh SB (2010) Emerging nematode diseases in two horticultural crops: their management through novel approaches. In: National conference on innovations in nematological research for agricultural sustainability – challenges and a roadmap ahead, Tamil Nadu Agricultural University, Coimbatore, p 51

Mallesh SB (2008) Plant growth promoting rhizobacteria, their characterization and mechanisms in the suppression of soil borne pathogens of coleus and ashwagandha. PhD thesis, University of Agricultural Sciences, Dharwad, 206 pp

Mohandas C (2001) Nematode diseases of tuber crops and their management. In: National conference on centenary of nematology in India – Appraisal & Future Plans, Indian Agricultural Research Institute, New Delhi, pp 35–36

Nisha MS, Sheela MS (2006) Bio-management of *Meloidogyne incognita* on Coleus, Solenostemon rotundifolius by integrating solarization, Paecilomyces lilacinus, Bacillus macerans and neem cake. Indian J Nematol 36:136–138

Pandey R (1997) Ecofriendly approaches for the management of root-knot nematode on black henbane. Indian J Nematol 27:175–178

Pandey R, Kalra A (2003) Root-knot disease of Ashwagandha, *Withania somnifera* and its ecofriendly and cost effective management. J Mycol Pl Pathol 33:240–245

Pandey R, Mishra AK, Tiwari S, Kalra A (2011) Nematode inhibiting organic materials and a strain of *Trichoderma harzianum* effectively manages *Meloidogyne incognita* in *Withania somnifera* fields. Biocontrol Sci Tech 21(12):1495–1499

Pillai KS (1976) Nematicidal control of root-knot nematode on *Coleus parviflorus*. J Root Crops 2:60–63

Ramakrishnan S, Thangavelu S (2010) Non-chemical management of root knot nematode in ashwagandha (*Withania somnifera* Dunal) and senna (*Cassia angustifolia* Vahl.). Indian J Nematol 39:170–174

Ramaprasad Shresti AY (2005) Studies on collar rot complex of *Coleus forskohlii* (Wild.) Briq. M. Sc. (Agriculture) thesis, University of Agricultural Sciences, Dharwad, Karnataka

Seenivasan N (2010) Biointensive management of *Meloidogyne incognita* and *Macrophomina phaseolina* disease complex in medicinal coleus. Indian J Pl Prot 38:186–192

Seenivasan N, Devrajan K (2008) Integrated approach for the management of root-knot nematode, *Meloidogyne incognita* in medicinal coleus. Indian J. Nematol 38(2):154–158

Senthamarai M, Poornima K, Subramanian S (2006a) Bio-management of root-knot nematode, *Meloidogyne incognita* on *Coleus forskohlii* Briq. Indian J Nematol 36:206–208

Senthamarai M, Poornima K, Subramanian S (2006b) Assessment of avoidable yield loss on *Coleus forskohlii* due to *Meloidogyne incognita*. Indian J Nematol 6:296–297

Senthamarai K, Poornima K, Subramanian S, Sudheer J (2008) Nematode-fungal disease complex involving *Meloidogyne incognita* and *Macrophomina phaseolina* on medicinal coleus, *Coleus forskohlii* Briq. Indian J Nematol 38:30–33

Sharma P, Pandey R (2009) Biological control of root-knot nematode; *Meloidogyne incognita* in the medicinal plant; *Withania somnifera* and the effect of biocontrol agents on plant growth. African J Agric Res 4:564–567

Shetty KD, Reddy DDR (1985) Control of root-knot nematodes on *Solanum khasianum* by chemical bare-root dips. Indian J Nematol 15:227–229

# Aromatic Crops 24

**Abstract**

The production of aromatic crops such as mints, basil, jasmine, patchouli, davana, scented geranium, and chamomile worldwide is threatened by root-knot nematodes (*Meloidogyne* spp.), lesion nematodes (*Pratylenchus* spp.), and spiral nematodes (*Helicotylenchus* spp.). Nematodes interact with *Verticillium* wilt and *Rhizoctonia* root-rot pathogens and cause disease complexes. In this chapter, crop losses, symptoms, and management using physical, cultural, chemical, biological methods, host resistance, and integrated methods for root-knot nematodes are discussed.

**Keywords**

*Meloidogyne* spp. · *Pratylenchus* spp. · *Helicotylenchus* spp. · Symptoms · Management

## 24.1 Mints, *Mentha* spp.

Among the recognized species of plant parasitic nematodes, root-knot and root-lesion nematodes have been considered most important.

### 24.1.1 Root-Knot Nematodes, *Meloidogyne incognita* and *M. javanica*

Root-knot disease of menthol/Japanese mint, spearmint, scotch spearmint, peppermint, and bergamot mint caused by *M. incognita* and *M. javanica* was observed for the first time in Lucknow and Terai region of Uttar Pradesh which reduce the



P. P. Reddy, *Nematode Diseases of Crops and their Management*,
https://doi.org/10.1007/978-981-16-3242-6_24

herbage yield and oil content (Haseeb and Pandey 1989). *M. incognita* dominates over *M. javanica* in mixed infection.

#### 24.1.1.1 Crop Losses

The root-knot nematode *M. incognita* caused 40.2% loss in herbage yield and 46.6% loss in oil yield in menthol mint (Pandey 2001). The quality of mint oil was also adversely affected due to nematode infection.

#### 24.1.1.2 Symptoms

Initial symptoms of the disease include occasional yellowing of leaves which in a month's time spread to a large portion of the foliage. Growth ceases soon after yellowing. Leaves turn yellow and thin, scorching easily and eventually turning brown. Initially symptoms appear in patches as the reduced plant growth with smaller leaf size and temporary wilting under slight stress of water during hot sun. More severe symptoms are visible in the form of stunting of plants and leaf yellowing; leaf veins remain green particularly after the initial harvest.

Enormous galls of small- and large-sized egg masses on the roots are some of under-ground symptoms. Under severe infection, initiation of lateral roots and rootlets on suckers is checked (Fig. 24.1). Haseeb and Pandey (1989) reported that the nematode-infected plants exhibit nutrient deficiency symptoms on the above-ground plant parts due to inhibition of nutrient uptake.

#### 24.1.1.3 Interaction with Other Pathogens

The severity of root-knot disease of Japanese mint was increased under field conditions in the presence of *Rhizoctonia solani*. *M. incognita*-*R. solani* disease complex caused severe reduction in growth and oil yield at lower inoculum densities, and at higher inoculum levels, young plants could not survive.

**Fig. 24.1** *Left*—Root-knot-infected sucker of menthol mint. *Right*—Root galling on root-knot nematode-infected mint

#### 24.1.1.4 Management

##### 24.1.1.4.1 Physical Methods

Complete eradication of *M. incognita* can be achieved by dipping bare roots of *M. arvensis* suckers in hot water bath maintained at 48 °C for 30 min without affecting the germination.

##### 24.1.1.4.2 Cultural Methods

###### Nematode-Free Planting Material

The nematodes which damage mints are generally endoparasitic (*M. incognita* and *P. thornei*) in nature and they can easily spread through suckers (propagating materials) from infested to noninfested planting sites. Therefore, it is essential and utmost important to use nematode-free planting material to check the spread of nematodes (Pandey et al. 2003).

###### Crop Rotation

Cultivation of some nonhost crops in rotation with Japanese mint, water logging, and rice cultivation during rainy season caused reduction in the population of root-knot nematodes. Crop rotation sequences of menthol mint with rice-potato, maize-potato, maize-mustard, red gram, rice-wheat, rice-potato, maize-mustard, and rice were effective in reducing the root-knot nematode population and in increasing the yield.

###### Organic Amendments

The severity of the nematode was reduced to a considerable extent in clayey soils. The high organic matter content in the soil suppressed the nematodes and enhanced the growth and oil yield. Soil application of neem/castor/mahua oil cakes at 1.0 g N/kg of soil was found effective against *M. incognita* and resulted in increase of oil yield and also plant growth parameters (Pandey et al. 1992). The root-knot nematode *M. hapla* can be effectively managed by soil application of neem cake 15 days before transplanting of *M. spicata*.

The application of vermicompost and different hydro-distillation wastes of *M. arvensis, M. piperita, Cymbopogon martini* (palmarosa), and *C. winterianus* (citronella) in the soil (1 month before transplantation) was found effective in reducing the nematode population under field conditions.

Maximum reduction in *M. incognita* population in menthol mint was recorded in neem cake-treated soil followed by mustard cake and *Trichoderma harzianum*. Soil incorporation of oil cakes gave maximum herbage yield (21.00 t/ha) as compared to control (14.66 t/ha). The crop produced had significantly higher oil content (2.7%) and yield (151 kg/ha) in neem cake treatment followed by mustard cake (2.7% oil and 143.6 kg/ha yield), *T. harzianum* (2.5% oil and 116.4 kg/ha yield), and *Glomus aggregatum* (2.4% oil and 112 kg/ha yield). The untreated crop gave 1.8% oil and 70.4 kg/ha yields.

##### 24.1.1.4.3 Chemical Methods

Application of Carbofuran at 3 kg a.i. /ha increased plant growth and oil yield of *M. arvensis* and reduced the root-knot nematode population.

##### 24.1.1.4.4 Biological Methods

Application of arbuscular mycorrhizal fungi such as *G. aggregatum, G. fasciculatum,* and *G. mosseae* improved plant growth, enhanced herbage and oil yield, and effectively inhibited root-knot nematode infection (Pandey et al. 1997).

##### 24.1.1.4.5 Host Resistance

The lowest root-knot (*M. incognita*) infection level was found on Kalka. Spear mint cvs. Neera and Arka were resistant to *M. incognita* (Pandey 2003). Peppermint cv. Kukrail; Spearmint cvs. Arka, Neera, and Neer-Kalka; and Bergamot mint cv. Kiran were found resistant to *M. incognita* (Pandey and Patra 2001).

##### 24.1.1.4.6 Integrated Methods

Enhancement of the growth of Japanese mint, oil yield, and reduction in population of root-knot nematodes was obtained with integration of soil incorporation of neem cake along with Carbofuran.

Pandey (2005a) reported that combined application of mustard cake along with *T. harzianum* gave significantly higher herbage and oil yields and maximum reduction in *M. incognita* population.

Integration of Carbofuran + neem cake/mustard cake + *T. harzianum*; *T. harzianum* + *G. intraradices* + *B. megaterium* + *P. fluorescens*; *G. intraradices* + *B. megaterium*; *G. intraradices* + *P. fluorescens*; *T. harzianum* + *G. intraradices*; and *T. harzianum* + *P. fluorescens* proved beneficial.

### 24.1.2 Lesion Nematodes, *Pratylenchus* spp.

Pathogenicity of *P. penetrans* on *M. piperita* has been demonstrated. *M. arvensis* favors the high population build-up of the nematode followed by *M. cardiaca, M. piperita,* and *M. citrata*.

#### 24.1.2.1 Crop Losses

*P. penetrans* has been reported to cause yield losses up to 34%.

#### 24.1.2.2 Symptoms

High population of *P. thornei* was found to be associated with different mints and caused wilting and chlorosis of the plants (Haseeb 1992). The lesion nematode causes light brown to dark-colored lesions on the suckers and roots (Fig. 24.2). In severe cases, whole underground portion becomes black and rotting of cortical portion takes place, first due to the nematode infection and later because of the invasion of secondary pathogens on necrotic cells.

**Fig. 24.2** Lesion nematode-infected mentha roots

#### 24.1.2.3 Interaction with Other Pathogens

The lesion nematodes *P. penetrans* and *P. minyus* predisposed mentha plants for higher incidence and severity of *Verticillium* wilt caused by *Verticillium albo-atrum* and *V. dahliae* f. sp. *menthae*.

Bergeson (1963) reported that when *P. penetrans*-infected *M. piperita* plants were inoculated with *V. albo-utrum,* diagnostic symptoms of wilt appeared approximately 2 weeks earlier than the plants inoculated with the fungus in the absence of nematode.

Presence of *P. miniyus* increased both incidence and severity of wilt (*V. dahliae* f. sp. *menthae*) disease, reduced the period for the development of wilt by 2–3 weeks, increased the reproduction of nematode, and reduced the dry weight of peppermint plants up to 68% when both the pathogens were present (Faulkner and Bolander 1969).

#### 24.1.2.4 Management

##### 24.1.2.4.1 Cultural Methods

Use clean planting stock to prevent introducing the nematode. Increased plant growth, oil yield of *M. piperita, M. citrata,* and *M. spicata*, and reduced multiplication of the lesion nematode (*P. thornei*) was achieved by soil incorporation of neem and mustard cakes at 1.5 t/ha (Shukla and Haseeb 1996).

##### 24.1.2.4.2 Chemical Methods

Significant reduction in the population of *P. penetrans* both in soil and root tissue of mint plants was obtained by soil application of Fenamiphos at 5.6 kg per ha, Ethoprophos (Mocap) at 6 kg/ha, and Oxamyl (Vydate) at 1 kg/ha (Rhodes 1984; Ingham et al. 1988).

#### 24.1.2.4.3 Host Resistance

Menthol mint cvs. Kosi and Kalka were resistant to *P. thornei* and had least *M. incognita* infection. Bergamot mint cv. Kiran and Spear mint cv. Neer-Kalka were having least *P. thornei* infection.

## 24.2 Basil, *Ocimum basilicum*

Root-knot nematodes are recognized as the major limiting factors in successful production of basil crop.

### 24.2.1 Root-Knot Nematode, *Meloidogyne incognita*

*Meloidogyne incognita* and *M. javanica* have been found to be the major problems in the cultivation of *Ocimum basilicum, O. canum, O. sanctum, O. gratissimum,* and *O. kilmandescharicum* (Haseeb et al. 1988; Haseeb and Pandey 1987). Balasubramanian and Rangaswami (1964) reported *O. sanctum* as a host of *M. javanica*. Krishnamurthy and Elias (1967) reported that *O. basilicum* was severely affected by *M. incognita*.

#### 24.2.1.1 Symptoms

Basil is very susceptible to root-knot nematodes *Meloidogyne* spp. They damage the roots and impede the plant's ability to take up water and nutrients from the soil resulting in plants exhibiting nutrient deficiency symptoms, wilting, and crop loss. Galling and root rot occur on plants that are heavily infected (Fig. 24.3). The

**Fig. 24.3** *Left*—Root-knot nematode-infested plant of Ocimum basilicum. *Right*—Severe root galling due to root-knot nematode infection on O. basilicum

nematode is responsible for significant reduction in foliage and root growth over a period of 10 months (Fig. 24.3).

#### 24.2.1.2 Management

Soil solarization employing a clear plastic sheet placed over tilled soil for 6 or more weeks proved beneficial. The addition of organic matter such as chicken manure (500 g per hill) may improve soil conditions and reduce the effect of nematodes and other pathogens. Rotation with nonsusceptible plants was found effective.

Haseeb et al. (1988) reported that soil application of neem oil cake as the best for reduced root galling increased plant growth and oil yield of *O. basilicum*, closely followed by mahua cake.

### 24.2.2 Foliar Nematode, *Aphelenchoides fragariae*

#### 24.2.2.1 Symptoms

Nematodes enter plant leaves through the stomata, feeding within the leaves on the spongy mesophyll cells. The nematode colonized and reproduced within the leaf, petiole, and stem tissues. The epidermis and mesophyll were invaded by the nematode which caused interveinal discoloration and necrosis and collapse of the palisade and spongy parenchyma (Fig. 24.4).

The fresh growth is likely curl, stunted, and twisted due to the feeding on the outer leaf surface. As the nematodes feed within the leaves, the affected leaf tissue turns pale green, then yellow, and later brown (Fig. 24.4). The leaf lesions are frequently restricted by the bigger veins, since nematodes cannot freely move across the veins. During mid to late summer, the prevalence of blighting of leaves may be seen.

#### 24.2.2.2 Survival

Nematodes can overwinter in buds or desiccated leaves. Adults and juveniles overwinter below ground in the soil, plant debris, and living plant tissue such as below-ground buds. For many years, the foliar nematode survives in the form of quiescent stage in crop residues.

**Fig. 24.4** Symptoms of foliar nematode infection on basil leaves

The means of nematode spread within the field is by leaf and plant contact as well as splashing of rain water. Spread can occur during vegetative propagation or when infested plant material is introduced into greenhouses.

#### 24.2.2.3 Management

- The first and most important step is to destroy infected plants.
- Do not place infected plants in a compost pile. The foliar nematodes can survive desiccation and could later be reintroduced into the production area.
- Use only nematode-free planting material and exclude nematodes from propagation areas.
- Reduce overhead watering and the duration of leaf wetness by thoughtful plant placement and spacing.
- Chlorfenapyr is reported to be effective for reducing foliar nematode populations.

## 24.3 Jasmine, *Jasminum* spp.

The root-knot and the burrowing nematodes are major problems in jasmine cultivation. Bajaj et al. (1988) recorded *Tylenchulus semipenetrans* on *J. sambac* from Haryana.

### 24.3.1 Root-Knot Nematode, *Meloidogyne incognita*

Rajendran and Rajendran (1979) reported *M. incognita* for the first time on *Jasminum. flexile* and *J. sambac* from Tamil Nadu, India.

#### 24.3.1.1 Symptoms

*M. incognita* infects *J. sambac* and *J. flexile*. It causes severe stunting of plants, drying of branches, and chlorosis of leaves (Fig. 24.5) which then drop prematurely, reducing the yield considerably. The roots exhibited very small swellings and enlarged rootlets (Fig. 24.5). The minute galls were more in *J. sambac*. Pale-colored leaves and dieback symptoms were associated (Rajendran and Rajendran 1979).

#### 24.3.1.2 Management

##### 24.3.1.2.1 Cultural Methods

The planting material should be raised in nematode-free or solarized nursery beds. Treatment of nursery beds with neem or *Pongamia* cakes at 2 kg/m$^2$ helps in reducing the nematode population.

##### 24.3.1.2.2 Chemical Methods

Sundarababu (1992) reported that the 70% reduction in root-knot nematode population and 50% enhancement of jasmine flower yield by soil incorporation of Phorate at 4 g a.i. per plant and 20 kg FYM during May and September.

**Fig. 24.5** *Left*—Root-knot nematode symptoms on jasmine plant. *Middle and right*—Jasmine roots infected with root-knot nematode

Nursery beds should be treated with Carbofuran 3G or Phorate 10G at 25 g/m$^2$ and main field with Carbofuran 3G or Phorate 10 G @ 2.5 kg a.i. /ha before planting.

#### 24.3.1.2.3 Biological Methods

*Pseudomonas fluorescens* at 10 g/plant (6 × 10$^8$ colony-forming units/g) gave maximum reduction of juveniles in soil (69.7%), number of adult females (70.1%), eggs (87.2%), gall index (2.6), and increased flower yield (36.6%) (Seenivasan and Poornima 2010).

#### 24.3.1.2.4 Integrated Methods

Cuttings or planting material should be raised in nematode-free or treated nursery beds. Application of neem cake enriched with *T. harzianum* at 1 t/ha is recommended.

The effective management of root-knot nematodes was obtained by nursery beds treatment with *Pongamia* or neem cake enriched with endomycorrhizal spores at 1 kg/m$^2$.

## 24.4 Patchouli, *Pogostemon cablin*

The important nematode problems limiting the production of patchouli crop include root-knot, lesion, and spiral nematodes.

### 24.4.1 Root-Knot Nematodes, *Meloidogyne* spp.

#### 24.4.1.1 Economic Importance

It was responsible for 47.0% and 86.7% loss in shoot weight and shade dried leaf yield, respectively (Krishna Prasad and Reddy 1984).

#### 24.4.1.2 Symptoms

Root-knot-infested plants are weak and grow slowly. Infection occurs when plants are in their early stage of development. Infected patchouli plants were stunted in their growth with yellow-colored leaves and premature drying and shedding of leaves (Krishna Prasad and Reddy 1984). When infected plants were uprooted, heavy galling of varying sizes was noticed on root system (Fig. 24.6) resulting in stunting, wilting, defoliation, and chlorosis of the plant. The compound galls formed by merging of surrounding short galls may be as large as 2–5 cm or more. The presence of egg masses outside the gall was a common phenomenon noticed in patchouli roots.

#### 24.4.1.3 Favorable Conditions

Multiplication of *M. incognita* was more in sandy soil on patchouli than clay soil.

#### 24.4.1.4 Management

##### 24.4.1.4.1 Cultural Methods

Neem cake at 4 t/ha controlled the root-knot nematodes (Sarwar et al. 1982). *Mucuna purita* as rotational crop with patchouli was found effective.

##### 24.4.1.4.2 Chemical Methods

Carbofuran at 2.0 kg a.i. /ha was effective in controlling phytonematodes infecting patchouli.

**Fig. 24.6** *Left*—Symptoms of root-knot nematode-infected patchouli plant. *Right*—Severe galling on patchouli roots

#### 24.4.1.4.3 Biological Methods

*Trichoderma harzianum* enhanced the growth, biomass, oil yield, significant reduction in the severity of root-knot disease, and population of *M. incognita* on patchouli (Pandey et al. 2009).

*Pseudomonas putida* strain BG2 (U312064.1) and *Bacillus cereus* BC1 (KX762284) promoted overall growth of patchouli, provided resistance to *Meloidogyne incognita* infection, as well as increased the essential oil yield (Borah et al. 2018).

Application of 150 kg N and 50 kg *Glomus fasciculatum*/ha enhanced fresh herbage (297.2 g/plant compared to 212.7 in control), dry herbage (92.0 g/plant compared to 64.0 in control), and oil yield (110.4 kg/ha compared to 28.0 kg/ha in control); increased mycorrhizal colonization (90.5% compared to 65.5% in control); reduced *M. incognita* population (145/200 ml soil compared to 395/200 ml soil in control); and recorded the highest cost-benefit ratio (1: 3.67) (Sumathi et al. 2006).

#### 24.4.1.4.4 Host Resistance

Planting of cultivated patchouli scion grafted on wild patchouli rootstock was effective for the management of root-knot nematodes (Krishna Prasad 1978).

#### 24.4.1.4.5 Integrated Methods

Treat nursery beds with neem/*Pongamia* cake at 1 kg/$m^2$ along with *Glomus fasciculatum* at 50 g/$m^2$.

Soil application of FYM/oil cakes enriched with *Trichoderma harzianum*.

### 24.4.2 Lesion Nematode, *Pratylenchus brachyurus*

*P. brachyurus* was a major problem on patchouli in Annamalai hills of Tamil Nadu and Hessaraghatta in Karnataka (Khan and Parvatha Reddy 1992).

#### 24.4.2.1 Symptoms

The infected plants invariably exhibited wilting symptoms.

#### 24.4.2.2 Management

The addition of compost was capable of suppressing the population of nematode *Pratylenchus* sp. as much as 58.93–67.31% and 40.74–59.19% in root and in soil, respectively. Compost positively affected the increasing amount of patchouli oil yield, and it could also reduce the loss as much as 18.34–43.36%. *Trichoderma* sp. was found to be dominant in compost in the treatment with its population of $2.1 \times 10^4$ propagules/g (Subarjah et al. 2016).

### 24.4.3 Spiral Nematode, *Helicotylenchus dihystera*

#### 24.4.3.1 Symptoms

*H. dihystera* causes heavy damage to patchouli roots and is responsible for decline.

#### 24.4.3.2 Management

Carbofuran and Phorate each at 2–3 kg a.i. per ha significantly reduced *H. dihystera* and increased leaf yield of patchouli.

## 24.5 Davana, *Artemisia pallens*

### 24.5.1 Root-Knot Nematode, *Meloidogyne incognita*

The root-knot nematode is a major problem in the cultivation of davana. *M. incognita* is responsible for more than 30% reduction in oil yield of davana (Haseeb and Pandey 1990).

#### 24.5.1.1 Symptoms

The main symptoms were chlorotic and stunted plants with less number of flower buds (which are the major source of essential oil) showing patchy appearance in the field. Their roots were severely galled by root-knot nematodes (Fig. 24.7). One larva/2 g soil has been found as economic threshold level of *M. incognita* on this crop.

**Fig. 24.7** Root-knot infested roots of davana

##### 24.5.1.2 Management

###### 24.5.1.2.1 Cultural Methods

Soil application of neem cake at 1 g N/kg soil significantly suppressed root-knot nematode population and increased oil yield (Haseeb and Butool 1991). Better efficacy of neem cake over the other organic materials, viz. mahua cake, castor cake, and nematicide like Carbofuran, has been demonstrated on *M. incognita* infecting *A. pallens* (Pandey 1994). Increase in nematicidal efficacy of neem cake and FYM for 90 to 120 days in comparison to 60 days for Fenamiphos has also been noticed (Anita and Vadivelu 1997).

A considerable enhancement of herbage biomass, flower buds, and oil yields were noticed when davana plants were treated with the distillation waste of palmarosa, citronella, curry leaf, and menthol mint and vermicompost of pyrethrum, marigold, and menthol mint. Least nematode infections were recorded in distillation waste of curry leaf, palmarosa, and lemon grass and vermicompost of pyrethrum, marigold, and menthol mint (Pandey 2005b).

###### 24.5.1.2.2 Biological Methods

A considerable enhancement of herbage biomass, flower bud, and oil yields were noticed when davana plants were treated with the bio-agent *Trichoderma harzianum* (Pandey 2005b).

## 24.6 Scented Geranium, *Pelargonium graveolens*

### 24.6.1 Root-Knot Nematode, *Meloidogyne hapla*

*Meloidogyne hapla* was found to be widely distributed on geranium in Nilgiri Hills of Tamil Nadu (Khan and Parvatha Reddy 1992).

#### 24.6.1.1 Symptoms

Root-knot nematode infection results in the formation of root galls (Fig. 24.8) which negatively affects the plant's ability to absorb water, nutrients, and other essential elements from the soil, leading to poor plant growth and yield loss.

#### 24.6.1.2 Management

Repeated application of either *P. lilacinum* at 10 g/plant or Fenamiphos at 2 kg a.i. / ha is necessary for effective management of *M. hapla* in scented geranium plantations.

**Fig. 24.8** Root-knot nematode infestation on geranium roots

## 24.7 Chamomile, *Matricaria chamomilla*

### 24.7.1 Root-Knot Nematode, *Meloidogyne incognita*

#### 24.7.1.1 Symptoms

*M. incognita* causes considerable reduction in plant growth, flower buds, and essential oil yield of chamomile (Pandey et al. 1999).

#### 24.7.1.2 Management

##### 24.7.1.2.1 Chemical Methods

Maximum reduction in root-knot severity and nematode population occurred with chemical activators such as 2-chlorosalicylic acid, O-acetylsalicylic acid, and 2-chloronicotinic acid (Pandey and Kalra 2005).

##### 24.7.1.2.2 Biological Methods

Using of *Bacillus thuringiensis kurastaki* (Dipel) seemed to be the most effective biocides in controlling *M. incognita* followed by *Bacillus subtilis*.

##### 24.7.1.2.3 Integrated Methods

Integration of *G. mosseae* and neem cake reduced the severity of root-knot disease and enhanced the growth, biomass, and flower yield by 12–37% of chamomile.

## 24.8 Conclusion

The major nematodes encountered with aromatic plants include *Pratylenchus* spp. (*P. brachyurus, P. coffeae*), *Meloidogyne* spp., *Rotylenchulus reniformis,* and *Radopholus similis*. In view of adverse effects on human, animal and environmental health, nontarget organism, and high cost, the chemical nematicides (even though effective) cannot be advocated to farmers for nematode management. The eco-friendly methods like physical methods, cultural practices, organic soil amendments, biological control agents, and resistant cultivars can be used for alternative nematode management. Resistant and tolerant germplasm could be exploited in future plant breeding programs for developing resistant/tolerant genotypes against major nematodes. Various organic materials are available which have been proven useful to decrease nematode incidence and enhanced aromatic plants yield. Large numbers of bio-agents are available which could also be used on large scale to protect aromatic plants against various nematode pathogens. However, much attention is needed to study and develop a new strategy to manage major nematodes in an eco-friendly way, which should be cost-effective and environmental friendly.

## References


Anita B, Vadivelu S (1997) Management of root-knot nematode, *Meloidogyne hapla* on scented geranium, *Pelargonium graveolans*. Indian J Nematol 27:123–125

Bajaj HK, Kanwar RS, Baghel PPS, Bhatti DS (1988) Motia (*Jasminum sambac*) - a new host record of citrus nematode, *Tylenchulus semipenetrans* Cobb, 1913. Haryana Agric Univ J Res 18(2):146–147

Balasubramanian M, Rangaswami G (1964) Studies on host range and histopathology of root-knot infections caused by *Meloidogyne javanica*. Indian Phytopath 317:126–132

Bergeson GB (1963) Influence of *Pratylenchus penetrans* alone and in combination with *Verticillium albo-atrum* on growth of peppermint. Phytopathology 53:1164–1166

Borah B, Ahmed R, Hussain M et al (2018) Suppression of root-knot disease in *Pogostemon cablin* caused by *Meloidogyne incognita* in a rhizobacteria mediated activation of phenylpropanoid pathway. Biol Control 119:43–50

Faulkner LR, Bolander WJ (1969) Interactions of *Verticillium dahliae* and *Pratylenchus minyus* in Verticillium wilt of peppermint: effect of soil temperature. Phytopathology 59:868–870

Haseeb A (1992) Nematode pests of medicinal and aromatic plants. In: Bhatti DS, Walia RK (eds) Nematode pests of crops. CBS Publishers & Distributors, Delhi, pp 239–249. 381 pp

Haseeb A, Butool F (1991) Evaluation of different pesticides and oilcake on the control of root-knot nematode (*Meloidogyne incognita*) infesting davana (*Artemisia pallens*). Curr Nematol 2:201–204

Haseeb A, Pandey R (1987) Incidence of root-knot nematodes in medicinal and aromatic plants - new host records. Nematropica 17:209–212

Haseeb A, Pandey R (1989) Observation on *Meloidogyne* spp. infesting Japanese mint - new disease records. Nematropica 19:93–97

Haseeb A, Pandey R (1990) Root-knot nematodes - a constraint to cultivation of Davana, *Artemisia pallens*. Trop Pest Mangmt 36:317–319

Haseeb A, Butool F, Pandey R (1988) Influence of different initial population densities of *Meloidogyne incognita* on the growth and oil yield of *Ocimum basilicum* cv. French. Indian J Mycol Pl Path 6:176–179

Ingham RE, Newcomb GB, Morris MA (1988) Spring nematicide application for control of root lesion and pin nematodes of peppermint, 1987. Fungic Nematic Tests 43:170

Khan RM, Parvatha Reddy P (1992) Nematode pests of ornamental crops. In: Bhatti DS, Walia RK (eds) Nematode pests of crops. CBS Publishers & Distributors, Delhi, pp 250–257. 381 pp

Krishna Prasad PR (1978) Studies on the root-knot nematodes infecting patchouli (Pogostemon cablin Bench). M. Sc. (Agriculture) thesis, University of Agricultural Sciences, Bangalore

Krishna Prasad PR, Reddy DDR (1984) Pathogenicity and analysis of crop losses on patchouli due to *Meloidogyne incognita*. Indian J Nematol 14:36–38

Krishnamurthy GVG, Elias NA (1967) Host range of *Meloidogyne incognita* causing root-knot on tobacco in Hunsur, Mysore State. Indian Phytopathol 20:374–377

Pandey R (1994) Comparative performance of oil seed cakes and pesticides in management of root-knot nematode disease of davana. Nematol Medit 22:17–19

Pandey R (2001) Influence of *Meloidogyne incognita, Pratylenchus thornei* and Tylenchorhynchus vulgaris on growth and oil yield of menthol mint. Indian J Nematol 31:111–114

Pandey R (2003) Mint nematology – current status and future needs. In: Trivedi PC (ed) Advances in nematology. Scientific Publishers (India), Jodhpur, pp 155–166

Pandey R (2005a) Field application of bio-organics in the management of *Meloidogyne incognita* in *Mentha arvensis*. Nematol Medit 33:51–54

Pandey R (2005b) Management of *Meloidogyne incognita* in *Artemisia pallens* with bioagents. Phytoparasitica 33:304–308

Pandey R, Kalra A (2005) Chemical activators: a novel and sustainable management approach for *Meloidogyne incognita* (Kofoid and White) Chitwood in *Chamomilla recutita* L. Arch Phytopathol Pl Prot 38:107–111

Pandey R, Patra NK (2001) Screening mint (*Mentha* spp.) accessions against root-knot nematode infection. J Spices Aromatic Crops 10:55–56

Pandey R, Haseeb A, Husain A (1992) Distribution, pathogenicity and management of *Meloidogyne incognita* on *Mentha arvensis* cv. MAS-1. Afro-Asian J Nematol 2:27–34

Pandey R, Singh HB, Gupta ML (1997) Antagonistic impact of vesicular arbuscular mycorrhizae (VAM) on *Meloidogyne incognita* population development in Japanese mint. Int J Trop Pl Dis 15:237–245

Pandey R, Singh HB, Kumar S (1999) Pathogenicity of *Meloidogyne incognita* (Kofoid & White) Chitwood on Matricaria chamomilla L. J Spices Aromatic Crop 8:201–203

Pandey R, Kalra A, Gupta ML, Sharma P (2003) Phytonematodes: major pests of MAPs. In: Mathur AK, Dwivedi S, Patra DD et al (eds) Proceedings of first national interactive meet on medicinal and aromatic plants. Central Institute of Medicinal and Aromatic Plants, Lucknow, pp 188–197

Pandey R, Kalra A, Gupta ML (2009) Evaluation of bio-agents and pesticide on root-knot nematode development and oil yield of patchouli. Arch Phytopathol Pl Prot 42(419):423

Rajendran B, Rajendran G (1979) Report of *Meloidogyne incognita* in *Jasminum flexile*. S Indian Hort 27:70

Rhodes HL (1984) Control of *Pratylenchus scribneri* on spearmint, *Mentha spicata* with non-fumigant nematicides. Nematropica 14:85–89

Sarwar M, Parameswaren TN, Shanamugham C, Narayana MR (1982) Appraisal of patchouli in India, root-knot problem and control. Ind Perfumer 26:117–121

Seenivasan N, Poornima K (2010) Bio-management of root-knot nematode, *Meloidogyne incognita* (Kofoid and White) Chitwood in jasmine (*Jasminum sambac* L.). *Pest Mangmt. Hort.* Ecosystems 16(1):34–40

Shukla PK, Haseeb A (1996) Effectiveness of some nematicides and oil cakes in the management of *Pratylenchus thornei* on *Mentha citrata, M. piperita* and *M. spicata*. BITE 57:307–309

Subarjah C, Himawan T, Puspitarini RD (2016) Effects of compost on nematode *Pratylenchus* sp. (Tylenchida: Pratylenchidae) population in patchouli. J Trop Life Sci 6(2):101–106

Sumathi M, Parthiban S, Samuel SD (2006) Effect of graded doses of nitrogen and VAM on herbage, oil yield and nematode population on patchouli (*Pogostemon patchouli* Pellet). Indian J Nematol 36:209–213

Sundarababu R (1992) Nematodes in jasmine. Tamil Nadu Agri Univ Newsl 21(11):2

# Part VIII

# Plantation, Spice and Tuber Crops

# Plantation Crops

# 25

**Abstract**

The most important nematode pests that are potentially damaging to plantation crops like coconut, areca nut, coffee, and tea include root-knot nematodes (*Meloidogyne* spp.), burrowing nematode (*Radopholus similis*), and lesion nematodes (*Pratylenchus* spp.). Distribution, crop losses, biology, survival and spread, and nematode management methods for plantation crops using physical, cultural, chemical, biological, host resistance, and integrated methods are discussed in this chapter.

**Keywords**

*Meloidogyne* spp. · *Radopholus similis* · *Pratylenchus* spp. · Symptoms · Management

## 25.1 Coconut, *Cocoas nucifera*

Koshy et al. (1978) reported that the single major nematode problem limiting the production of coconut crop by inciting root rot includes the burrowing nematode *Radopholus similis*.

### 25.1.1 Burrowing Nematode, *Radopholus similis*

Weischer (1967) reported the burrowing nematode on coconut palms for the first time from the Kerala State in India.



P. P. Reddy, *Nematode Diseases of Crops and their Management*,
https://doi.org/10.1007/978-981-16-3242-6_25

#### 25.1.1.1 Crop Losses

It causes 30% yield loss in coconut (Koshy and Geetha 1992). The threshold inoculum density required to cause significant reduction in various growth parameters of coconut is 100 nematodes/625 ml sandy loam soil over a period of 5 years under field conditions (Koshy 1986).

#### 25.1.1.2 Symptoms

The burrowing nematode-infested coconut palms exhibit general decline symptoms like yellowing, button shedding, reduction in number and size of leaves and leaflets, delay in flowering, and reduced yield, which are nonspecific. Symptoms on the root are more specific. *R. similis* on infestation produces isolated, elongated orange-colored lesions on tender and semi-hard roots (Fig. 25.1). Widespread root rotting was observed. The lesions coalesce and enlarge because of nematode reproduction and parasitization. Other symptoms observed on roots include surface cracks on semi-hard orange-colored roots and spongy texture of tender roots due to heavy nematode infestation. The extent of nematode multiplication may reach up to 4000 nematodes per gram of main roots. The drastic reduction in the number and mass of tertiary feeder roots on parasitization by the nematode limits plant growth (Koshy and Sosamma 1987).

#### 25.1.1.3 Host Range

Large numbers of crops like banana, black pepper, ginger, turmeric, sweet potato, groundnut, sugarcane, betel vine, etc. are known hosts of *R. similis*.

#### 25.1.1.4 Life Cycle

*R. similis* is a migratory endoparasite capable of spending its entire life cycle within the roots. Except the fourth-stage male juveniles and adult males, all other juvenile stages and adult females are found to be infective. The nematode completes its life

**Fig. 25.1** Burrowing nematode symptoms on coconut plant (*left*) and roots (*right*)

cycle (J2 to J2) within 25 days at a temperature range of 25–28 °C (Koshy and Sosamma 1977).

#### 25.1.1.5 Survival and Spread

The burrowing nematode *R. similis* is capable for surviving for 1 month in dry soil (29–39 °C) and 6 months in moist soil (27–36 °C) under field conditions. In roots of stumps of felled coconut palms, the nematode survives for up to 6 months (Sosamma 1984). The peak nematode multiplication takes place during October to November, while minimum during March to July. A mean soil temperature below 25 °C and light rain fall coupled with availability of tender fleshy roots are the factors favorable for *R. similis* multiplication (Koshy and Sosamma 1978).

Infested coconut seedlings help in the dissemination of the nematode to distant places. Apart from coconut, infested planting materials of intercrops such as areca nut, banana, pepper, ginger, and turmeric also serve as sources of inoculum.

#### 25.1.1.6 Management

##### 25.1.1.6.1 Cultural Methods

The cultural nematode management strategies include use of nematode-free planting stock of coconut and other intercrops and avoiding banana as a shade crop in coconut nurseries.

Application of 50 kg cow dung/FYM, 2 kg neem/Marotti cake and 25 kg green manure to the coconut basins in November is quite effective in bringing down the nematode population. Application of *Hydnocarpus* sp. oil cake at 8 kg/palm/year in June–July and October–November to the burrowing nematode-infested coconut palms increased the yield by 30% (Koshy 1986).

The reduction in nematode reproduction occurs when sun hemp *Crotalaria juncea* is grown as a green manure crop in the basins and interspaces of coconut and its incorporation in soil.

##### 25.1.1.6.2 Chemical Methods

In infested coconut nurseries, complete control of *R. similis* can be obtained by soil application of Fenamiphos or Phorate at 2.5 kg a.i. /ha during September, December, and May (Koshy and Sosamma 1979).

Application of Phorate at 10 g a.i. /palm in June–July and in September–October is effective in reducing the population of burrowing nematodes. Maximum increase in yield was obtained with the application of Fenamiphos at 10 g a.i. /palm.

##### 25.1.1.6.3 Biological Methods

Introduction of biocontrol agents like *Purpureocillium lilacinum* and arbuscular mycorrhizal fungus (AMF) to nursery potting mixture helps in the reduction of nematode population and at the same time imparts resistance/tolerance to nematode infestation (Koshy et al. 1993). Among the AMF, *Acaulospora bireticulata* was found to be more effective in reducing the nematode population. The damage to seedlings of coconut and reduction in reproduction rate of *R. similis* can be achieved by soil incorporation of *Cylindrocarpon effusum* (Sosamma and Koshy 1978; Koshy and Sosamma 1987).

#### 25.1.1.6.4 Host Resistance

The dwarf cvs. Kenthali and Klappawangi and the hybrids such as Java Giant x Kulasekharam Dwarf Yellow, Java x Malayan Dwarf Yellow, and San Ramon x Gangabondam recorded the least nematode multiplication and lesion indices (Sosamma et al. 1980). The cv. Philippines Ordinary was found to be least susceptible to *R. similis* with lesion index of 1.22 (Sudha 1998).

#### 25.1.1.6.5 Integrated Methods

##### Botanicals and Chemicals

Thirty percent increase in yield and 5–10% decrease in disease indices of palms affected with root wilt disease were recorded by the application of *Hydnocarpus* oil cake and Phorate granules at 10 g a.i. /palm in June–July and in October–November.

##### Cultural, Botanicals, and Chemicals

Application of farmyard manure or oil cakes, cultivation of *Crotolaria juncea* in the basin and interspaces (and used as green manure), and application of Phorate 10 g a.i./palm twice a year (during May–June and September–October) is effective against burrowing nematode.

The effective management of the burrowing nematode infecting coconut can be achieved by adopting the following recommended INM practices:

- Application of cow dung, FYM, oil cakes, and green manure to the basins.
- Growing sun hemp as a green manure crop in basins and their incorporation in soil.
- Application of Phorate at 10 g a.i./palm twice yearly.
- Avoiding banana as a shade crop in coconut nurseries.
- Use of nematode-free planting material of coconut and other intercrops.
- Use of less susceptible/tolerant cvs. Such as Kenthali and Klappawangi or their hybrids in infested areas.

### 25.1.2 Red Ring Nematode, *Bursaphelenchus cocophilus*

The "red ring" disease of coconut caused by *B. cocophilus* is very destructive disease on some islands of Caribbean which threatens the existence of this crop. The extensive losses caused by this nematode are a major catastrophe to natives of these areas. The nematode is transmitted by the palm weevil *Rhynchophorus palmarum*. The disease was first reported as occurring in Trinidad by Hart (1905). Fortunately, this nematode is not recorded so far in India.

#### 25.1.2.1 Crop Losses

At present, most Latin American and Caribbean countries show losses ranging from less than 1% to more than 20% on 1–10-year-old trees. Thirty-five percent mortality of young coconut palms has been reported in Trinidad and 80% loss in single plantation of Tobago (Esser 1969).

#### 25.1.2.2 Distribution

The disease is reported to occur in 20 countries seriously affecting their economy. At present, *B. cocophilus* has been reported from the West Indian Islands of Trinidad, Tobago, Grenada, and St. Vincent and from Latin America, Dominican Republic, Venezuela, Guyana, Surinam, French Guyana, Colombia, Ecuador, Peru, Mexico, Brazil, Panama, Nicaragua, Guatemala, Costa Rica, Honduras, Belize, and El Salvador.

#### 25.1.2.3 Symptoms

The disease is present most commonly in young coconut trees (3–10 years old) with maximum incidence in trees of 4–7 years old. Chlorosis first appears at the tips of the oldest leaves and spreads toward their bases. The brown lower leaves may break across the petiole or the lower part of the rachis, or they may become partly dislodged at the base and hang down (Fig. 25.2). Premature shedding of the nuts may occur simultaneously with the development of the leaf symptoms or slightly before. As a result of the severe damage caused by the palm weevil internally, the crown often topples over about 4–6 weeks after the appearance of the first symptoms. The trunk then remains standing in the field for several months and finally it decays (Griffith and Koshy 1989).

The most characteristic symptom of red ring disease is the internal lesions. In the beginning, about 3–5 cm beneath the stem surface, scattered reddish dots of about 1 mm diameter appear, which later coalesce to form an orange red ring of about 3 cm in width (Fig. 25.2). The ring assumes a crescent-like shape in petioles, while it

**Fig. 25.2** *Left*—Red ring nematode (*Bursaphelenchus cocophilus*)-infected coconut plant. *Right*—Cross section of trunk showing red ring symptoms

surrounds the full length of the stem and roots. Large numbers of juveniles are seen in the center of the discolored areas and adults in the periphery. The nematode infection leads to falling down of all sized nuts.

*B. cocophilus* causes little leaf disease of coconut and oil palm in Surinam and Guyana. A thermostable phytotoxin was produced due to breakdown of coconut tissue. Water uptake of infested coconut palms was less due to occlusion of xylem vessels. In coconut roots, the nematodes attack cortical tissues.

#### 25.1.2.4 Host Range

Though *B. cocophilus* is primarily found on coconut and oil palm, it has also been found to occur naturally in the date palm and cabbage palm (Hagley 1962).

#### 25.1.2.5 The Vector

The vector of the red ring nematode is the palm weevil (*Rhynchophorus palmarum*). The adult palm weevil acts as the carrier. The female weevil makes punctures in the softer tissue of the stem of the coconut palm with the help of its proboscis to lay its eggs. Normally, the punctures are made in the soft axils to a depth of 5–8 cm and the eggs are placed deep in the puncture. The nematodes are injected into the tissues of the coconut palm when the insect deposits its eggs. Larvae of the palm weevil while feeding in diseased trees get infested with the nematodes and the adult beetles which develop from them in turn become the carriers.

#### 25.1.2.6 Life Cycle

Blair (1966) carried out preliminary studies on the life history of *B. cocophilus* in immature nuts of coconut palm. When the nuts were inoculated with eggs containing active larvae, adult nematodes and eggs were recovered 9–10 days later. The nematodes completed one life cycle in 9–10 days which is probably one of the shortest life cycles reported for plant parasitic nematodes.

#### 25.1.2.7 Survival

*R. cocophilus* can survive for a maximum of 15 days in soil in the form of persistent third-stage larvae.

#### 25.1.2.8 Management

##### 25.1.2.8.1 Cultural Methods

"Cut and burn" the infested trees is one of the oldest and most effective practices used to prevent further spread of the disease.

##### 25.1.2.8.2 Chemical Methods

More recent methods involve the destruction of palms using arboricides, herbicides, or phytotoxic nematicides applied to holes made in tree-trunks. Hoyle (1968) reported that injection of cacodylic acid ("Silvisar") killed the diseased trees within 2 weeks and prevented weevil breeding, without having any effect on nematodes.

Victoria et al. (1970) used sodium and potassium arsenate to kill both diseased trees and nematodes.

Adult weevils are attracted to anaerobic fermentation products such as ethyl alcohol, n-butyl alcohol, and to certain volatile esters extracted from diseased palm tissue (Griffith 1987). This behavior can be exploited to attract weevils to poisoned baits at 25/ha.

The palm weevils entering from the nearby infested trees can be attracted and killed using the trap or guard baskets in the plantations. The guard baskets are 1 m high and 0.3 m in diameter, cylindrical in shape, and made out of 2 cm mesh wire. These baskets are filled with chunks of fresh tissue from diseased coconut trees to attract the beetle. The guard baskets are sprayed completely with about 4.5 liters of Methomyl solution and distributed on the ground in the plantations at 2.5 baskets/ha of young coconut trees. During the dry season when the weevils are most active in the cool nights, the above method is advocated. The tissue and insecticide in the guard baskets should be destroyed after 2 weeks and replaced with fresh tissue and insecticide.

#### 25.1.2.8.3 Biological Methods

Griffith (1977) reported the pathogenicity of a bacterium *Micrococcus agilis* to the palm weevil and suggested its potential use as a biocontrol agent. The rhabditid nematodes of the genera *Steinernema* and *Heterorhabditis* can heavily parasitize the weevils by maintaining a high population of nematodes in the environment (Griffith and Koshy 1989).

## 25.2 Areca Nut, *Areca catechu*

Although a number of nematodes have been reported from the rhizosphere of areca nut, the endoparasite reported as an important pest is only the burrowing nematode *Radopholus similis*.

### 25.2.1 Burrowing Nematode, *Radopholus similis*

Kumar et al. (1971) first reported the burrowing nematode *R. similis* on the roots of areca nut from Mysore, Karnataka, India. The threshold inoculum level causing significant damage to growth of areca nut was found to be 100 nematodes per seedling or one nematode in 800 g laterite soil (Koshy 1986).

#### 25.2.1.1 Crop Losses

Tenfold increase in yield was recorded by treatment with nematicides (Sundraraju and Koshy 1986).

#### 25.2.1.2 Symptoms

The burrowing nematode-infested areca palms exhibit nonspecific above-ground symptoms like general yellowing and visible reduction in growth, vigor, and yield. The characteristic symptom of *R. similis* infestation is the appearance of lesions and rotting of roots (Fig. 25.3). The nematode produces small, elongate, orange-colored lesions on the young, succulent creamy-white to light orange-colored portion of the main and lateral roots. Later these lesions coalesce and cause extensive rotting (Fig. 25.3). The thick primary roots arising from the bole of the palm exhibit large, oval, sunken dark lesions. In areca nut, the tips of lateral and tertiary roots on infestation become black.

#### 25.2.1.3 Life Cycle

At a temperature range of 21–31 °C, *R. similis* takes 25–30 days to complete one life cycle (J2 to J2) on areca nut seedlings.

#### 25.2.1.4 Ecology

The burrowing nematode *R. similis* multiplication in areca nut roots was found to be highest during October–November, while it was lowest during March–June.

**Fig. 25.3** *Left*—Burrowing nematode-infected areca nut seedlings. *Right*—Areca nut roots infected with the burrowing nematode

#### 25.2.1.5 Management

##### 25.2.1.5.1 Cultural Methods

Use of nematode-free planting materials of areca nut and other inter/mixed crops and avoiding *R. similis* susceptible inter/mixed crops like black pepper and banana in infested areas helps in reducing the nematode population

Complete control of *R. similis* was obtained on areca nut palms affected by yellow leaf disease (YLD) after 4 years of continuous application of neem cake at 1.5 kg/palm or Fensulfothion at 50 g a.i. /palm thrice a year (June, September, January). Significant increase in yield and decrease in disease indices of YLD-affected palms were recorded only in the fifth year.

##### 25.2.1.5.2 Chemical Methods

Application of Fensulfothion at 1 g a.i. /seedling thrice a year for 3 years controlled *R. similis* infestation in the seedling stage. In adult palms, reduction in nematode population and tenfold increase in yield was obtained by application of Fensulfothion at 50 g a.i. /palm in May–June, September–October, and December–January for 5 years (Sundararaju and Koshy 1986).

##### 25.2.1.5.3 Host Resistance

The areca nut hybrid VTL-11 x VTL-17 has been found to be highly resistant to *R. similis*. The cvs. Indonesia-6, Mahuva B, and Andaman-5 are tolerant to *R. similis* and hence can be recommended for infested areas (Koshy et al. 1979; Sundararaju and Koshy 1982). The yields of areca nut cvs. Indonesia-6 and Singapore (VTL-17) were 50% more as compared to the local South Kanara variety.

##### 25.2.1.5.4 Integrated Methods

- Use of nematode-free planting material of areca nut and other inter/mixed crops.
- Use of black pepper and banana (which are susceptible to *R. similis*) as inter/mixed crops in infested areas should be avoided.
- Use of tolerant/resistant varieties.
- Application of 5–10 kg of green manure/palm preferably *Glyricidia* or *Crotolaria*.
- Application of 1 kg neem oil cake/palm/year.
- Application of Phorate at 3 g a.i. /plant in the root zone of areca nut, banana, and black pepper in June–July and in September–October in areca nut-based farming systems.

Integration of neem oil cake application at 1 kg/plant with Phorate at 15 g a.i. /plant gave effective control of *R. similis* in areca nut. Sudha and Sundararaju (1998) reported that the most effective reduction of the burrowing nematode infecting areca nut was obtained by combining soil application of neem cake at 500 g/plant along with Phorate at 15 g a.i. /plant.

Soil amended with glyricidia leaves and bio-agents (*Purpureocillium lilacinum*, *Pasteuria penetrans* and AMF) was very effective in increasing plant height, leaf area, and root growth and in reducing burrowing nematode population (95%) and root lesion index.

## 25.3 Coffee, *Coffea arabica* and *Coffea canephora*

The lesion nematode, *Pratylenchus coffeae*, and root-knot nematodes are the most important pests of coffee.

### 25.3.1 Lesion Nematode, *Pratylenchus coffeae*

Zimmeman (1898) described a lesion nematode found infecting the roots of coffee plants in Java. This nematode has also been reported from coffee growing tracts of south India. *P. coffeae* destroyed over 95% of arabica coffee plantations of Java. Over 1000 hectares have been infected with this nematode in coffee-growing tracts of south India.

#### 25.3.1.1 Crop Losses

The annual loss to Indian coffee due to this nematode is estimated at $three million.

#### 25.3.1.2 Symptoms

In nurseries, *P. coffeae* causes death of coffee plants. The infested coffee plants show yellowing of leaves, loss of primary roots, and stunting of the shoot followed by gradual wilting and death (Fig. 25.4). Leaves and stems may show symptoms of mineral deficiencies. The infected plants can be easily pulled off by hand because of scanty root growth. Lesions on the roots were observed on the plants infected with *P. coffeae*.

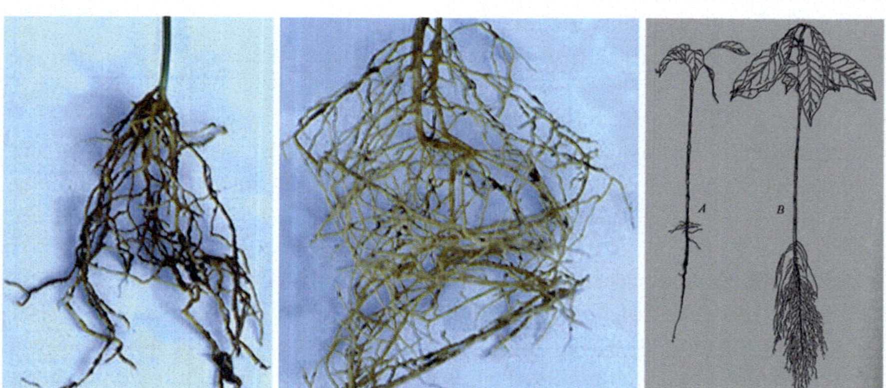


**Fig. 25.4** *Left*—Coffee seedling infested by *Pratylenchus coffeae*. *Middle*—Healthy roots. *Right*—Lesion nematode-infected (**a**) and healthy (**b**) coffee seedlings

The lesion nematode causes wounds in roots through which other pathogenic organisms, such as fungi and bacteria, enter the root tissues. The interaction of these agents results in the formation of lesions that finally destroys the root tissues.

#### 25.3.1.3 Life Cycle

*P. coffeae* is a migratory endoparasite which enters the root slightly behind the elongating zone in the piliferous region of the root cortex where it feeds, multiplies, and establishes a colony or "nest" causing brownish lesions on the root. When decay sets in due to secondary invasion by bacteria and fungi, the nematodes quit and migrate through the soil to attack fresh roots.

Zimmeman (1898) has discussed in detail the developmental stages of *P. coffeae* infesting coffee in Java. After penetration, the second-stage larvae feed endoparasitically on conducting tissues of root and develop into adults after undergoing three molts. Eggs take 15–17 days for hatching, 15–16 days for feeding and development to adults, and 15 days for egg production. The nematode takes about 45–48 days to complete its life cycle from egg to egg.

#### 25.3.1.4 Host Range

*P. coffeae* is known to attack coffee, banana, citrus, grapevine, apples, strawberry, abaca, yam, cocoa, rubber, potato, sweet potato, peas, beans, cabbage, cauliflower, tomato, chrysanthemum, caladium, dahlia, marigold, cotton, lucerne, etc.

#### 25.3.1.5 Survival and Spread

*P. coffeae* is capable of surviving for about 18 months in moist soil under laboratory conditions but failed to exist beyond 12 months under field conditions.

The nematode is disseminated through working implements and running water especially across the slope.

#### 25.3.1.6 Management

##### 25.3.1.6.1 Cultural Methods

Organic Amendments

Coffee pulp at 18 kg/ $m^2$ improved plant growth and reduced the nematode population.

Fallowing

Lordello (1968) stated that the replanting of old nematode-infested plantations be preceded by a fallow period of at least 2 years after removing the old coffee roots.

##### 25.3.1.6.2 Chemical Methods

In field trials with young trees of arabica coffee, soil application of Fenamiphos at 0.5 to 1.0 g a.i. /plant gave good control of *P. coffeae* which should be repeated at 90 days interval (Kumar 1982).

#### 25.3.1.6.3 Host Resistance

*Coffea robusta* is relatively more resistant to *P. coffeae* than *C. arabica*. Grafting of arabica scion on Robusta rootstock at 40-day-old "topee stage" plants by using wedge-cleft method can be adopted (Kumar 1988) (Fig. 25.5).

#### 25.3.1.6.4 Integrated Methods

It is advisable to change nursery site once in every 7–8 years. The nursery site should also be dug and exposed to sun. The nematode-free plants should be used for planting. Affected plants should be pulled out and burnt. The planting site should be dug and the soil exposed to the sun for at least one summer season. Susceptible weeds need to be removed periodically. The site should be planted with either robusta or arabica-robusta grafted plants.

### 25.3.2 Root-Knot Nematodes, *Meloidogyne exigua* and *M. coffeicola*

Srinivasan and D'Souza (1965) reported *M. exigua* on coffee from south India. Schieber and Sosa (1960) found that *Coffea canephora* (robusta coffee) was found highly resistant to *M. exigua*, while *C. Arabica* was susceptible.

#### 25.3.2.1 Crop Losses

In Hawaiian coffee plantations, *M. coffeicola* is responsible for 5% loss in yield. Losses due to *Meloidogyne* spp. in Brazil are estimated at 15% of that country's production.

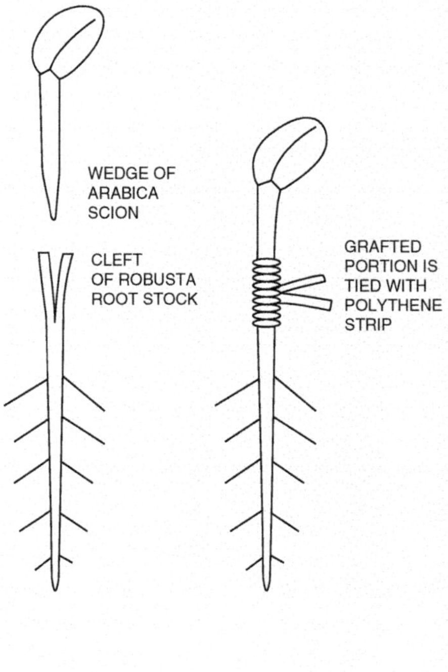


**Fig. 25.5** Grafting of arabica coffee scion on lesion nematode-resistant robusta rootstock

#### 25.3.2.2 Symptoms

Affected trees look unthrifty and have chlorotic, wilting, and dying leaves. *M. exigua* produces small elongated galls that are located primarily at the root tips (Fig. 25.6). Other symptoms observed include cracks in roots, necrosis in cortical region, and sloughing of root tissue. The effectiveness of root system is greatly diminished, with rootlets and root hairs practically eliminated. Especially, cold temperatures and drought conditions are fatal to trees. Affected plants lack vigor and are unable to withstand adverse conditions.

#### 25.3.2.3 Life Cycle

*M. exigua* takes 32–42 days at 25–30 °C to complete the life cycle. The egg masses of *M. exigua* are mostly located under the epidermis of coffee roots.

#### 25.3.2.4 Host Range

*M. exigua* is known to attack coffee, tea, and bell pepper.

#### 25.3.2.5 Survival and Spread

*M. exigua* does not survive in the soil, in the absence of the host, for more than 6 months after eradication of infested plants. The nematode spreads through the infested seedlings used for transplanting in the main field.

#### 25.3.2.6 Interaction with Other Pathogens

The disease complex involving *Rhizoctonia solani* and *M. exigua* caused more root necrosis and defoliation when both pathogens were inoculated. Extensive fungal colonization within the coffee root systems was observed.

**Fig. 25.6** Symptoms caused by root-knot nematode damage on *Coffea arabica roots*

### 25.3.2.7 Management

#### 25.3.2.7.1 Cultural Methods

At the crop renovation time, the infested coffee area may be turned to pasture. Most graminicolous plants are nonhosts to *M. exigua* and *M. coffeicola*. The decision of growing forage crops and turning back again the area to coffee will certainly reduce drastically the population of *Meloidogyne* species in 6–12 months.

Coffee can also be rotated with groundnut, castor (both immune), *Styzolobium deeringiatum* and *Crotalaria spectabilis* (both resistant), cotton, soybean, and maize (non-hosts) for 1 year to manage *M. exigua*.

#### 25.3.2.7.2 Chemical Methods

Application of Carbofuran and Fenamiphos each at 1.6 to 6.0 g a.i. /plant, in one or two applications during the year, gives effective control of root-knot nematodes. The first application should be in the beginning of the rainy season followed by the second 3 months later.

#### 25.3.2.7.3 Host Resistance

Robusta coffee (*Coffea canephora* line 2258 from Costa Rica) was found resistant to *M. exigua* and *M. incognita*. In Brazil, grafting of arabica coffee (*Coffea arabica* var. Mundo Novo or Catuai Vermelho) on line 2258 rootstock showed good growth and production under field conditions.

## 25.4 Tea, *Camellia sinensis*

### 25.4.1 Root-Knot Nematodes, *Meloidogyne* spp.

The first report of root-knot nematode in young tea was from south India, where large numbers of the seedlings were found infected (Barber 1901). *Meloidogyne javanica, M. incognita,* and *M hapla* have been found associated with young tea, while *M. brevicauda* cause severe galling of the roots of mature tea bushes in India.

### 25.4.1.1 Symptoms

#### 25.4.1.1.1 On Young Tea

Young nursery plants, both seedlings and vegetatively propagated clonal plants, are severely damaged by root-knot nematodes. Seedling plants in which both the tap root and lateral roots are severely attacked (Fig. 25.7) suffer greater damage than do clonal tea plants of equivalent age, probably because seedling tea plants possess less than half the root bulk of clonal plants of similar age. A marked increase in resistance is observed between 8 and 15 months.

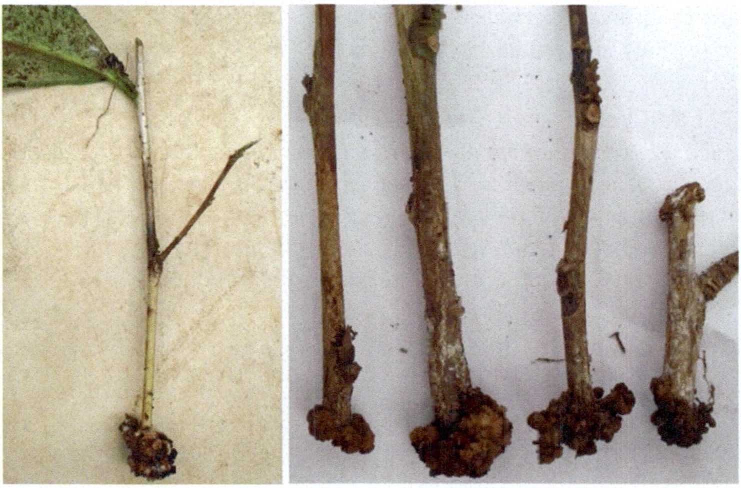

**Fig. 25.7** Heavy root galling due to root-knot nematode infection on tea seedlings

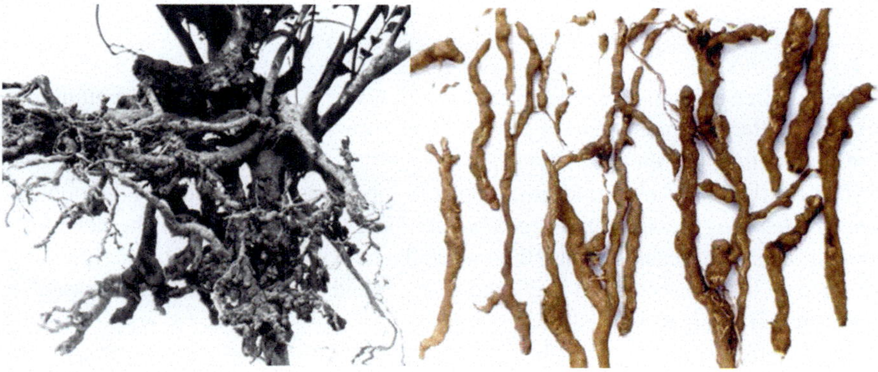

**Fig. 25.8** Mature tea bush roots infected with *Meloidogyne brevicauda*

#### 25.4.1.1.2 On Mature Tea

Root-knot nematodes (*M. brevicauda*) cause severe galling of the roots of mature tea bushes with resultant decline. Tea plants of all ages are susceptible to infection by this nematode which has been recorded in only Ceylon and south India.

The above-ground symptoms of root-knot-infected plants include very slow growth of bushes; smaller, dull, and yellow appearance of the leaves. The roots show the characteristic swelling and pitting that is typical of *M. brevicauda* (Fig. 25.8).

The effect of infection on the growth of tea bushes is most evident during the period of recovery from pruning. Healthy bushes are usually harvested about 16 weeks after pruning, but the harvesting of severely infected bushes is necessarily delayed up to a further 6 months. The production of new branches by infected bushes is very restricted, and as a consequence, the cropping capacity of the bush is reduced. The most severely affected bushes fail to recover from pruning.

#### 25.4.1.2 Life Cycle

The average size of mature female of *M. brevicauda* is about 5 or 6 times that of a mature female of *M. incognita*. Despite this massive size, the females are often observed to be empty, with only a few eggs. The mean hatch per egg mass is around 10, while in the other common species, it is 200–600 juveniles/egg mass. The rest of the life cycle is very similar to the other species of *Meloidogyne*.

#### 25.4.1.3 Management

##### 25.4.1.3.1 In Nurseries

Physical Methods

Soil solarization in nurseries is adequate to manage root-knot and other nematodes as the plants develop complete resistance by about 9–15 months (Gnanapragasam et al. 1989).

Cultural Methods

Nematode-free soil must be used to raise seedlings in the nurseries. Incorporation of organic matter into nematode-infested soil is known to suppress the incidence of root-knot nematodes on tea.

The cultivation of marigold as a preplant crop considerably reduces the population of *Meloidogyne* spp. and *Pratylenchus loosi*. The reduction in the nematode population is significantly greater than when the land is left fallow. The interplanting of marigold in tea has resulted in 7% increase in yield compared to 10% increase that followed soil fumigation.

The rehabilitation crops such as Mana grass (*Cymbopogon confertiflorus*) or Guatemala grass (*Tripsacum laxum*) (which help to improve the soil) are resistant to *Meloidogyne* spp. The soil population of parasitic nematodes decline rapidly when these grasses are grown for one or more years before replanting tea.

The severity of knotting was reduced in treatments with application of potash fertilizer by 44% by augmenting host-plant resistance (Kamunya et al. 2008).

Chemical Methods

In field trials, Carbofuran 3G at 1 g/tea seedling were found effective for the control of *M. incognita* and also caused significant increase in plant height (Basu and Gope 1985). The severity of knotting was reduced in treatments with Carbofuran by 28.7%.

Integrated Methods

Reduction in root galling, egg mass production, and fecundity of root-knot nematodes and enhancement of seedling growth parameters were obtained by integration of neem cake at 5 g/seedling + Carbofuran 3G at 1.5 g/ (Kalita and Bora 2006).

##### 25.4.1.3.2 In Field

###### Cultural Methods

The first step in the management of nematodes in replanted areas is by ensuring a very thorough uprooting of all old tea roots (up to pencil thickness). Planting of deep-rooted grass species including Mana grass and Pangola grass which are nonhosts to parasitic nematodes and which bring down the residual nematode population are grown in Sri Lanka before reverting to tea plantation after uprooting (Kerr and Vythilingam 1966).

Effective reduction of lesion nematode population can be obtained by growing trees and leguminous green manure crops (and their incorporation in soil) that help to provide shade to the tea plants, help in N fixation, and provide regular supplies of mulch in the form of leaf litter and loppings.

Plants such as marigolds (*Tagetes erecta* and *T. patula*), *Vetiveria zizanoides,* and *Eragrostis curvula* have been found to help reduce nematode populations. Intercropping marigold in tea plantations reduced the lesion nematode *Pratylenchus loosi* population and increased leaf yield by 7%.

Large inputs of organic matter including cow dung and well-decomposed plant residues have been reported to suppress nematode populations. In India, neem cake has been recommended to manage root-knot nematodes in the tea field (Muraleedharan 1991). In Sri Lanka, castor oil cake has been proved to be the most effective one for suppressing nematodes, followed by mahua, neem, and karanj cakes (Gnanapragasam 1991).

###### Chemical Methods

The recommended nematicides for the management of root-knot nematodes in mature tea include Fenamiphos (Nemacur 5% G) and Carbofuran (Furadan 3% G) at 10 g/plant applied after pruning (Gnanapragasam 1987).

## 25.5 Conclusion

The most damaging and destructive nematodes such as *Radopholus similis*, *Meloidogyne* spp., and *Pratylenchus* spp. were among those present. The common methods of control that have been recommended include: use of nematode-free planting material of plantation crops and other intercrops—avoiding crops that are susceptible to *R. similis* (e.g., banana, black pepper) as intercrops, minimum use of nematicides, use of any available resistant/tolerant cultivars, treatment with neem oil cake, either alone or in combination with chemicals, and inoculation with *Purpureocillium lilacinum.*

Management of soil-borne nematode pathogens in plantation systems requires an integrated approach, which will, in most cases, be based on cultural, physical, and biological methods. Sustainable nematode management strategies recommended include: (1) practicing regulatory methods like plant quarantine to avoid the introduction of new pests; (2) following cultural methods such as crop rotations to prevent or the build-up of pests to damaging levels; (3) using organic soil amendments to reduce pathogen populations by increasing populations of agents antagonistic to the pathogens; and (4) adopting the use of multiple cropping systems and multiple cultivars to increase biodiversity.

Developing the best management practices is fundamental to coexist with them and maintain abundant and healthy agricultural production of plantation crops. Adoption of integrated nematode management schedules is important in these crops.

## References


Barber CA (1901) Department land records and agriculture. Madras Agriculture Branch 2, Bulletin No 45, pp 227–234

Basu SD, Gope B (1985) Control of root-knot nematode *Meloidogyne incognita* (Kofoid and White) Chitwood in tea seedlings with some systemic nematicides. Two Leaves and a Bud 32:31–35

Blair GP (1966) The use of immature nuts of *Cocos nucifera* for studies on *Rhadinaphelenchus cocophilus*. Nematologica 11:590–592

Esser RP (1969) *Rhadinaphelenchus cocophilus* a potential threat to florida palms. Nematology Circular No. 9, Division of Plant Industry, Florida Department of Agriculture, Gainesville, Florida, USA

Gnanapragasam NC (1987) Integrated strategies to manage nematodes in tea. In: Proceedings of the international conference on tea quality and human health, Hangzhou, pp 117–183

Gnanapragasam NC (1991) Influence of soil amendments in reducing pathogenicity to tea by the root lesion nematode, *Pratylenchus loosi*. Proceedings of the international symposium on tea science, Shizuoka, pp 684–688

Gnanapragasam NC, Dharmasena WAM, Liyanage DD (1989) Development of age resistance in young tea to different species of *Meloidogyne*. Int Nematol Netw Newsl 6:4

Griffith R (1977) A species of bacterium pathogenic to the palm weevil, *Rhynchophorus palmarum*. 1976 Annual report red ring research division, Ministry of Agriculture, Land and Fisheries, Trinidad and Tobago, pp 31–35

Griffith R (1987) Red ring disease of coconut palm. Plant Dis 71:193–196

Griffith R, Koshy PK (1989) Nematode parasites of coconut and other palms. In: Luc M, Sikora RA, Bridge J (eds) Plant parasitic nematodes in subtropical and tropical agriculture. CAB International, Wallingford, Oxon, pp 363–386

Hagley EAC (1962) The palm weevil *Rhynchophorus palmarum* L. a probable vector of red ring disease of coconuts. Nature 193:499

Hart JH (1905) Coconut disease. Bulletin of miscellaneous information. Royal Botanic Garden, Trinidad 6:241–243

Hoyle JC (1968) A simple method for poisoning coconut trees. J Agri Sci Trinidad 68:459–465

Kalita S, Bora BC (2006) Effect of nematicides and organic amendments in the management of *Meloidogyne incognita* in nursery tea. Indian J Nematol 36:148–149

Kamunya SM, Wachira FN, Lang'at J et al (2008) Integrated management of root knot nematode (*Meloidogyne* spp.) in tea (*Camellia sinensis*) in Kenya. Int J Pest Mangmt 54:129–136

Kerr A, Vythilingam MK (1966) Replanting eelworm infested areas. Tea Quart 37:67–72

Koshy PK (1986) The burrowing nematode, *Radopholus similis* (Cobb, 1893) Thorne, 1949. In: Swarup G, Dasgupta DR (eds) Plant parasitic nematodes of India-problems and progress. Indian Agricultural Research Institute, New Delhi, pp 223–248

Koshy PK, Geetha SM (1992) Nematode pests of palms and cocoa. In: Bhatti DS, Walia RK (eds) Nematode pests of crops. CBS Publishers & Distributors, Delhi, pp 214–227

Koshy PK, Sosamma VK (1977) Identification of the physiological races of *Radopholus similis* populations infesting coconut, areca nut and banana in Kerala, South India. Indian Phytopath. 30:415–416

Koshy PK, Sosamma VK (1978) Studies on the population fluctuations of *Radopholus similis* in coconut and areca nut roots. Indian Phytopath 31:180–185

Koshy PK, Sosamma VK (1979) Control of the burrowing nematode, *Radopholus similis* (cobb, 1893) Thorne, 1949 on coconut seedlings with DBCP. Indian J Nematol 9:32–33

Koshy PK, Sosamma VK (1987) Pathogenicity of *Radopholus similis* on coconut (*Cocos nucifera* L.) seedlings under greenhouse and field conditions. Indian J Nematol 17:108–118

Koshy PK, Sundararaju P, Sosamma VK (1978) Occurrence and distribution of *Radopholus similis* (Cobb, 1893) Thorne, 1949 in South India. Indian J Nematol 8:49–58

Koshy PK, Sosamma VK, Sundararaju P (1979) Reactions of thirty one areca germplasm collections to the burrowing nematode, *Radopholus similis*. Pl Dis Reptr 63:433–435

Koshy PK, Sosamma VK, Samuel R (1993) Interactive effect of vesicular arbascular mycorrhizae (VAM) with *Radopholus similis* on coconut seedlings. In: Mehta UK (ed) Nematology – challenges and opportunities in 21st century. Sugarcane Breeding Institute, Coimbatore, pp 106–110

Kumar AC (1982) Evaluation of Nemacur 5G against coffee root-lesion nematode, *Pratylenchus coffeae*. J Coffee Res 12:1–7

Kumar AC (1988) Nematode problem of coffee and its management. Indian Coffee 52:12–19

Kumar AC, Viswanathan PRK, D'Souza GI (1971) A study of plant parasitic nematodes of certain commercial crops in coffee tracts of South India. Indian Coffee 36:1–3

Lordello LGE (1968) Nematodes das Plantas Cultivadas. Livaria Nobel SA, Sao Paulo

Muraleedharan N (1991) Non-arthropod pests – nematodes and rodents. In: Director (ed) Pest management in tea. United Planter's Association of Southern India, Tea Research Institute, Takolai, pp 77–81

Schieber E, Sosa ON (1960) Nematodes on coffee in Guatemala. Pl Dis Reptr 44:722–723

Sosamma VK (1984) Studies on the burrowing nematode of coconut. Ph. D. thesis, Kerala University, Trivandrum, 166 pp

Sosamma VK, Koshy PK (1978) A note on the association of *Cylindrocarpon* spp. with *Radopholus similis* in coconut. Indian Phytopath 31:381–382

Sosamma VK, Koshy PK, Bhaskar Rao EVV (1980) Susceptibility of coconut cultivars and hybrids to *Radopholus similis* in the field. Indian J Nematol 10:250–252

Srinivasan CS, D'souza GI (1965) A note on the occurrence of Brazilian pyroid coffee nematode, *Meloidogyne exigua* Goeldi in South India. Indian Coffee 29(6):13

Sudha S (1998) Screening of coconut cultivars against the burrowing nematode, *Radopholus similis*. In: Parvatha Reddy P, Krishna Kumar NK, Verghese A (eds) Advances in IPM of horticultural crops. Association for Advancement of Pest Management in Horticultural Ecosystems, Bangalore, pp 332–333

Sudha S, Sundararaju P (1998) Effect of neem oil cake and nematicide for the control of burrowing nematode, *Radopholus similis* in the areca nut based cropping system. In: Mehta UK (ed) Nematology - challenges & opportunities in 21st century. Sugarcane Breeding Institute, Coimbatore, pp 251–257

Sundararaju P, Koshy PK (1982) Response of Areca collections against *Radopholus similis*. Indian J Nematol 12:404–405

Sundararaju P, Koshy PK (1986) Control of *Radopholus similis* on areca nut seedlings with aldicarb, aldicarb sulfone, carbofuran and fensulfothion. Indian J Nematol 16:4–7

Victoria KJ, Sanchez PA, Barriga OR (1970) Eradication de palms de cocotero afectadas por el anillo rojo (*Rhadinaphelenchus cocophilus* (Cobb, 1919) Goodey, 1960, Nematoda: Aphelenchoididae) mediante la utilization de sustancias quimicas. Revista Inst Colomb Agropec 5:185–197

Weischer B (1967) Report to the Government of India on plant parasitic nematodes. PL 24/52, Report No TA 2332, FAO, Rome, 15 pp

Zimmeman A (1898) De nematodes der Koffieworiels. Meded Pl Tuin Batalva 27:16–41

# Spice Crops

# 26

**Abstract**

The most important nematode pests that are potentially damaging to spice crops like black pepper, cardamom, ginger, and turmeric include root-knot nematodes (*Meloidogyne* spp.) and the burrowing nematode (*Radopholus similis*). Nematodes interact with *Fusarium* wilt, *Phytophthora* foot rot, and *Rhizoctonia* rhizome rot/ root-rot pathogens in causing disease complexes. Distribution, crop losses, biology, survival and spread, and management methods for tropical fruit crops such as physical, cultural, chemical, biological, host resistance, and integrated methods on spice crops are discussed in this chapter.

**Keywords**

*Meloidogyne* spp. · *Radopholus similis* · Symptoms · Biology · Management

## 26.1 Black Pepper, *Piper nigrum*

The burrowing nematode *Radopholus similis* and root-knot nematode *Meloidogyne incognita* are the major nematodes infecting black pepper.

### 26.1.1 Burrowing Nematode, *Radopholus similis*

In Karnataka and Kerala States of India, the "slow-wilt" disease of black pepper is caused by the burrowing nematode *R. similis* (Ramana et al. 1987; Mohandas and Ramana 1987).



P. P. Reddy, *Nematode Diseases of Crops and their Management*,
https://doi.org/10.1007/978-981-16-3242-6_26

#### 26.1.1.1 Distribution

The burrowing nematode is distributed mainly in Kerala, Karnataka, and Tamil Nadu.

#### 26.1.1.2 Crop Losses

Mohandas and Ramana (1991) reported that crop loss to the extent of 59% is caused by *R. similis* in black pepper.

The burrowing nematode is responsible for annual yield loss of 24% in black pepper amounting to ₹871.84 million in India (Walia and Chakraborty 2018).

#### 26.1.1.3 Symptoms

*R. similis* on black pepper is associated with pepper yellows (slow-wilt) disease, which appears as pale yellow or whitish-yellow drooping leaves on the vines. The number of such leaves increases gradually until large numbers of leaves, or even the entire foliage, become yellow (Fig. 26.1). Other symptoms induced by *R. similis* include stunting of vines, leaf yellowing, dropping of leaves, and dieback. During low moisture regimes, the symptom expression is easily noticeable. Death of the vine takes place due to shedding of all the leaves within 3–5 years of initiation of leaf yellowing.

The tender, thin, white feeding roots show typical orange- to purple-colored lesions (Fig. 26.1). Lesions are not clearly seen on older roots, being brown in color. The root system exhibits extensive rotting, and this results in a lack of fine feeder roots from the main roots. Extensive necrosis of larger lateral roots develops subsequently.

**Fig. 26.1** Burrowing nematode-infected black pepper vine showing leaf yellowing. Necrotic root lesions formed by *R. similis* destroying roots

#### 26.1.1.4 Life Cycle

The nematode completes its life cycle within 25–30 days at temperature range of 21–31 °C. Koshy (1986) described the method to culture the black pepper isolate of *R. similis* axenically on carrot discs.

#### 26.1.1.5 Host Range

Crops like coconut and areca nut which are commonly used as live standards are good hosts of *R. similis*. Other hosts such as banana, cardamom, ginger, and turmeric, which are used as intercrops, are also susceptible to the burrowing nematode infestation.

#### 26.1.1.6 Interaction with Other Pathogens

Yellows disease in Indonesia is caused by a nematode-fungus complex involving *R. similis* and *Fusarium* spp. and possibly other fungi. *R. similis* predisposed black pepper seedlings to attack by a weakly pathogenic isolate of *Fusarium solani* causing severe root damage.

Slow decline disease of black pepper is due to a feeder root damage caused by *R. similis* and *Phytophthora capsici* in combination.

#### 26.1.1.7 Spread

The burrowing nematode is disseminated through infected planting material.

#### 26.1.1.8 Management

##### 26.1.1.8.1 Physical Methods

Soil solarization of nursery sites or potting mixture and random rotation of nursery sites are reported to be effective in black pepper.

##### 26.1.1.8.2 Cultural Methods

In order to get maximum yield, better survival, and good establishment, use of certified (nematode-free) planting stock is recommended. Mulching (with Guatemala grass) has an ameliorating effect on the symptoms of slow wilt/pepper yellows (*R. similis*). Planting nematode-free rooted cuttings, uprooting of affected vines and replanting after a period of 9–12 months, use of nonliving standards/supports, avoiding *R. similis* susceptible intercrops like banana, ginger, and turmeric, and exclusion of susceptible trees as standards for trailing the vines reduce the nematode population. Application of organic amendments like neem cake (200 g), green foliage (3–5 kg), or FYM (1 kg per vine) and earthing up in September-October are recommended. Growing marigold as a trap crop and uprooting the trap crop at flowering stage and burning reduce the incidence of nematodes. *R. similis* was not encountered when black pepper was intercropped with coffee (Venkitesan 1976).

#### 26.1.1.8.3 Chemical Methods

Nursery

Treating the planting material with Carbofuran (3G) @ 0.1 g a.i. /plant once in 2 months was found effective.

Drenching with Carbosulfan (25 EC) 0.1% at 50 ml/polybag containing potting mixture proved beneficial.

Plantations

Soil application of Carbofuran/Phorate each at 3 g a.i. /vine in May-June and again in September-October resulted in reduction of the nematode population and remission of foliar yellowing. Among the above nematicides, Phorate proved to be the best (Ramana 1986; Mohandas and Ramana 1987).

#### 26.1.1.8.4 Biological Methods

*Pasteuria penetrans, Pupureocillium lilacinum,* and *Pochonia chlamydosporia* are effective in suppressing the root-knot and burrowing nematodes. Five isolates of rhizobacteria (IISR-552, 528, 658, 853, and 859) were good growth promoters (26.7–55.6%) having dual nematicidal action (suppressing both *M. incognita* and *R. similis*).

Soil application with *P. chlamydosporia* at 20 g/vine ($2 \times 10^6$ spores/g) was found effective.

#### 26.1.1.8.5 Host Resistance

A local variety at Peringamala, Kerala, India was found not to be invaded by *R. similis*.

#### 26.1.1.8.6 Integrated Methods

Application of fertilizers at 400 kg N, 180 kg P, 480 kg K, 425 kg Ca, and 112 kg Mg in combination with mulch gave effective control of yellows disease.

Integrated methods of the burrowing nematode management that can be suggested include:

- Planting of nematode-free rooted cuttings.
- Uprooting of affected vines and replanting after a period of 9–12 months.
- Use of nonliving supports or standards.
- Exclusion of *R. similis*-susceptible trees as standards for trailing black pepper vines.
- Exclusion of susceptible intercrops such as banana, ginger, and turmeric.
- Application of organic amendments, such as neem oil cake, green foliage, or farm-yard manure.
- Earthing-up after application of nematicides, NPK fertilizers, and organic amendments.

### 26.1.2 Root-Knot Nematodes, *Meloidogyne incognita* and *M. javanica*

In Wayanad region of Kerala State in India, Butler (1906) for the first time reported root-knot nematodes on black pepper. *M. incognita* and *M. javanica* have been reported from India.

#### 26.1.2.1 Crop Losses

*M. incognita* was responsible for 46% loss in yield of black pepper (Mohandas and Ramana 1991).

#### 26.1.2.2 Symptoms

Prominent symptoms of root-knot infestation on black pepper are unthrifty growth and yellowing of leaves. Inter-veinal yellowing of the foliage is also noticeable. The leaves exhibit dense yellowish discoloration of the inter-veinal areas making the leaf veins prominent with deep green in color. Root-knot nematodes (RKN) feed on plant roots resulting in large galls or "knots" that form throughout the root system of the pepper plants (Fig. 26.2). Galls are smooth and bigger in size in few cultivars, but small galls in many cultivars and egg masses with females enclosed deep within the roots.

**Fig. 26.2** *Left*—Black pepper vine infected with root-knot nematodes showing leaf yellowing. *Right*—Black pepper roots infected with root-knot nematodes

#### 26.1.2.3 Host Range

Among commercially used standards, *Oroxylon indicum, Erythrina lithosperma, Ceiba pentandra, and Bombax malabaricum* are highly susceptible to root-knot nematodes (Koshy et al. 1977).

#### 26.1.2.4 Life Cycle

The life cycle under optimum conditions takes 3–4 weeks.

#### 26.1.2.5 Interaction with Other Pathogens

The root-knot nematode *M. incognita* in association with soil-borne fungus *Fusarium solani* was responsible for root-rot disease complex in black pepper vines. Both the organisms together were found to do more harm than either of them alone. Increased susceptibility of *M. incognita* and *M. javanica* infested pepper to *Phytophthora* infection has also been reported.

#### 26.1.2.6 Management

##### 26.1.2.6.1 Cultural Methods

Live standards for black pepper such as *Artocarpus heterophyllus, A. hirsutus, Ailanthes malabarica, Mesopsis emini, Peltophorum pterocarpum, Swietenia macrophylla, Tamarindus indica, Garuga pinnata,* and *Macaranga peltata* are resistant to *M. incognita.*

Growing of nonhost cover crops like Siratro (*Macroptilium atropurpureus*) in the interspaces and mulching with Guatemala grass (*Imperata cylindrica*) are recommended to reduce the root-knot nematode (*M. incognita*) population on black pepper in the Amazonian region (Ichniohe 1980).

Application of neem cake reduces nematode populations followed by increase in yield.

##### 26.1.2.6.2 Chemical Methods

Green berry yields can be doubled by four applications of Carbofuran incorporated into mound soil at 114 g/vine per application in black pepper fields infested with *M. incognita* and *M. javanica*. Application of Carbofuran at 10 g a.i. /vine twice a year, including at planting around cuttings, can reduce populations of *M. incognita* on black pepper. Fenamiphos at 1 g a.i. /vine followed by Carbofuran and Ethoprophos was found effective in controlling nematodes.

##### 26.1.2.6.3 Biological Methods

###### Antagonistic Bacteria

Root-knot nematodes in pepper can be managed by the application of *Bacillus macerans* at $1.2 \times 10^8$ cells/vine just before the monsoon period.

Antagonistic Fungi

Reduction in the rate of infection of root-knot nematodes to the extent of 15 and 12% was achieved by soil incorporation of *Purpureocillium lilacinum* and *Pochonia chlamydosporia*, respectively. Soil application of *P. chlamydosporia* reduced nematode population and increased the yield (5.14 kg/vine) with benefit:cost ratio of 7.12.

Biocontrol agents like *P. chlamydosporia* ($10^8$ cfu/g) or *Trichoderma harzianum* ($10^8$ cfu/g) can be applied at 50 g/vine twice a year (during April–May and September–October) to manage root-knot nematodes (Pervez 2018).

Arbuscular Mycorrhizal Fungi

Enhancement of the growth of black pepper and reduction in root-knot nematodes was obtained by soil application of AMF such as *Glomus fasciculatum*, *G. etunicatum*, and *Acaulospora laevis*. Preinoculation of *A. laevis* in black pepper cv. Panniyur-1 reduced the root-knot index (0.75) followed by *G. mosseae*/*Gigaspora margarita* (1.25) compared to control (3.5) (Anandraj et al. 1991).

#### 26.1.2.6.4 Host Resistance

The root-knot nematode infestation was less in cvs. Veliakaniakadan, Karimunda, Kalluvalli, Balankotta, Narayakodi, and Padapan.

Black pepper cv. "Pournami" released by the Indian Institute of Spices Research, Calicut, India, showed resistant reaction to *M. incognita* with a potential yield of 17,280 kg green berries/ha and has high oleoresin content (13.8%).

#### 26.1.2.6.5 Integrated Methods

Two Bioagents

Black pepper vines combinedly inoculated with *P. lilacinum* and *Pastueria penetrans* had put out 23–112% increase in plant growth over control and were very effective in the management of root-knot nematodes (Sosamma and Koshy 1995).

Devapriyanga et al. (2012) reported significant enhancement of crop yield, and reduction in nematode infestation both in soil and roots was recorded by treatment with the consortial formulation of *Bacillus subtilis* Bs 214 and *Pseudomonas fluorescens* Pf 123.

Cultural and Chemical Methods

Integration of neem cake application at 1 kg/vine along with Phorate/Carbofuran at 3 g a.i./vine during May–June and again during September-October gave effective control of nematodes infesting black pepper.

Venkitesan and Jacob (1985) reported that the reduction in *M. incognita* population by 33–38% and foliar yellowing by 83% was obtained by soil incorporation of aldicarb at 1 g a.i. /vine twice a year (May/June and October/November) in combination with fertilizers (N-100 g, P-40 g, K-140 g/vine) in two equal splits, earthing up to 50 cm radius at the base of the vines, and mulching the base of the vines with leaves.

### 26.1.3 Interaction of Root-Knot and Burrowing Nematodes with Foot Rot

#### 26.1.3.1 Interaction

Increased susceptibility of *R. similis/ M. incognita/M. javanica* infested cultivars of black pepper to *Phytophthora* infestation had been reported.

#### 26.1.3.2 Symptoms

There is a gradual reduction in the vigor and productivity of the vine which leads to death over a period of few years and hence called slow decline. The nematodes produce small, elongate lesions on the young tender roots, and later these lesions coalesce and cause extensive root rotting. The primary symptoms are pale yellow, whitish, discoloration of leaves, typical orange to purple colure lesion on young roots, root system exhibits extensive rotting, main roots are devoid of fine feeder root that rot quickly, extensive necrosis of longer lateral roots develops, and yellow patches that later turn as barren standard that have lost their vines or standard supporting dead vines without any leaves. These symptoms are well pronounced when soil moisture is depleted. In general, foliar yellowing and defoliation were low during July and high during April–May.

#### 26.1.3.3 Management

##### 26.1.3.3.1 Cultural Methods

Severely affected vines should be removed from the plantation and destroyed, as it is impossible to recover them whenever high population of nematode are noticed. Nematode-free rooted cuttings raised in fumigated or steam-sterilized nursery mixture should be used for planting in the field (Thomas and Rajeev 2015).

##### 26.1.3.3.2 Chemical Methods

The pits for planting should be treated with Phorate 10 G at15 g or Carbofuran 3 G at 50 g at the time of planting. Phorate 10 G at 30 g or Carbofuran 3 G at 100 g/vine should be applied during May/June (with the onset of south west monsoon) and September/October. Along with nematicides, the basins should be drenched with either copper oxychloride (0.2%) or potassium phosphonate (0.3%) or metalaxyl-mancozeb (0.125%) (Thomas and Rajeev 2015).

##### 26.1.3.3.3 Integrated Methods

Integrated management of foot rot (*P. capsici*) and nematodes (*M. incognita* and *R. similis*) on black pepper was achieved by:

- Mixing AMF and *T. harzianum* in solarized nursery mixture to raise healthy and robust seedlings.
- Application of *T. harzianum* and FYM in planting pit.
- Field application of Neem cake at 1 kg/vine mixed with 50 g of *T. harzianum* during August.

## 26.2 Cardamom, *Elettaria cardamomum*

Root knot nematodes *Meloidogyne* spp., the lesion nematode *Pratylenchus coffeae*, and the burrowing nematode *Radopholus similis* are the major nematode problems on cardamom (Sundararaju et al. 1979).

### 26.2.1 Root-Knot Nematodes, *Meloidogyne incognita* and *M. javanica*

#### 26.2.1.1 Crop Losses

Yield loss of 32–47% due to root-knot infestation has been reported (Ali 1984, 1986). Monetary loss due to root-knot infection is ₹4 million annually.

#### 26.2.1.2 Symptoms

In cardamom, root-knot nematodes are very serious in nurseries than in plantations. In primary nurseries, more than 50% of the germinating seeds do not emerge due to infection of the radicle and plemule by the second-stage juveniles of the root-knot nematode. The infested cardamom seedlings at the two-leaf stage show marginal yellowing and drying of leaves and severe galling of roots (Fig. 26.3). On transplantation to a secondary nursery, they exhibit curling of the unopened leaves. These leaves mostly emerge after the breaking of the pseudostem. The symptoms expressed by severely infected mature plants in a plantation include stunting of plant growth, chlorosis, narrowing of leaf blades, premature drying of leaf tips and margins, reduced tillering, excessive root branching ("Witches broom"), delay in flowering, immature fruit drop, and reduction in yield.

**Fig. 26.3** *Left*—Cardamom plant infected with root-knot nematode showing leaf yellowing. *Right*—Root-knot galls on cardamom roots

#### 26.2.1.3 Host Range

Many annual weeds and common shade trees like *Erythrina indica* and *E. lithosperma* present in cardamom plantations are susceptible to root-knot nematodes.

#### 26.2.1.4 Survival and Spread

The nematodes are spread through infested seedlings and rhizomes used for propagation.

#### 26.2.1.5 Interaction with Other Pathogens

In nurseries, presence of *M. incognita* increases incidence of rhizome rot and damping off diseases caused by *Rhizoctonia solani* (Ali 1986; Eapen 1987).

In cardamom, *M. incognita* is the predisposing factor for *R. solani* infection, damping off, and rhizome rot, prevalent in nurseries (Ali and Venugopal 1992, 1993).

#### 26.2.1.6 Management

##### 26.2.1.6.1 Physical Methods

Soil solarization for 40–45 days using 400 gauze transparent LDPE was found ideal presowing treatment for the management of nematodes and other soil-borne diseases in cardamom nurseries. Mean soil temperature during the solarization period increased by 8.7 °C in solarized beds and the germination was enhanced by 25.5%, while weed growth was suppressed by 82%. *Pythium vexans* was totally eliminated, while populations of *Rhizoctonia solani, Phyllosticta elettaria,* and nematodes were suppressed to varying levels. Solarization was also found to stimulate the growth of cardamom seedlings (Eapen 1995).

##### 26.2.1.6.2 Cultural Methods

Proper weeding schedule should be followed, since many of weeds are good hosts of nematodes.

Regular application of organic manures such as neem cake twice a year (May/-June and September) at 250–1000 g depending on the clump size reduces root-knot nematode infestation.

Application of *Tithonia diversifolia* (wild sunflower) leaves as mulch at 5 kg/plant around the vines reduced nematode population and increased the yield of cardamom.

##### 26.2.1.6.3 Chemical Methods

Carbosulfan 6G at 1 kg/ha was found to be superior treatment in reducing the population of nematodes in soil and roots followed by Cartap hydrochloride 4G at 1 kg/ha.

Application of Phorate 10 G at 30 g/clump twice, during May-June and again in September, was found effective.

#### 26.2.1.6.4 Biological Methods

Soil application of *B. macerans* ($1 \times 10^7$ cfu/g) at 30 g/plant or *P. fluorescens* ($1 \times 10^8$ cfu) 30 g/plant before monsoon proved beneficial for controlling root-knot nematodes.

*P. lilacinum* and *T. harzianum* suppressed *Meloidogyne* spp. population by 58.3–86.9% and 58.3–68.9%, respectively, in cardamom nursery. The number of quality seedlings for transplanting was greater in beds treated with this biocontrol agent (Eapen and Venugopal 1995).

#### 26.2.1.6.5 Integrated Methods

Eapen and Venugopal (1995) found that suppression of root-knot nematode population and better seedling growth of cardamom was achieved by solarization of nursery beds soil and application of bio-agents like *Trichoderma* spp. and *P. lilacinum*.

Besides soil solarization of nursery beds, subsequent soil application of *T. harzianum* reduced root-rot infection and also infection caused by the root-knot nematode *M. incognita* (Eapen and Ramana 1996).

Application of *T. harzianum* multiplied on decomposed coffee husks (7 day old) at the time of sowing at 2.5 kg/bed (4.5 m × 1.0 m) and repeated after 3 months is recommended for the control of root-knot nematodes and damping off in nurseries (Eapen and Ramana 1996).

Application of Cartap hydrochloride 4G @ 1 kg/ha and use of neem cake at 2 kg/plant before monsoon was found effective.

In order to prevent the residue problem in the cardamom capsules, soil incorporation of nematicides like Carbofuran/Phorate at 15–50 g/clump during premonsoon period followed by Neem cake application (300–500 g/clump) in mid-monsoon period is the safest approach.

Integration of the following practices can help in the successful management of root-knot nematodes:

- Changing nursery sites frequently
- Disinfecting nursery beds
- Introduction of biocontrol agents at nursery level
- Control of susceptible weeds
- Exclusion of susceptible shade trees
- Destruction of infested crop residues
- Mulching with dead leaves
- Application of Neem cake

### 26.2.2 Interaction of Root-Knot Nematode with Rhizome Rot

#### 26.2.2.1 Interaction

*M. incognita* was found to predispose cardamom seedlings to *Rhizoctonia solani* infection, which causes damping off and rhizome rot in the primary nursery (Ali and Venugopal 1992, 1993).

#### 26.2.2.2 Management

When both fungicide (Metalaxyl) and nematicide (Carbofuran) were applied for the control of rhizome rot disease and nematodes, the mortality of seedlings was least.

Eapen and Venugopal (1995) found that the most effective approach to manage *Meloidogyne* spp. and rhizome rot disease (*R. solani*) complex was by soil application of Purpureocillium *lilacinum along* with *Trichoderma* spp. in solarized cardamom nursery beds.

Soil solarization alone enhanced the germination by 25.5% and suppressed weed growth by 82.0%. Solarization also enhanced the growth and vigor of cardamom seedlings. The disease complex was suppressed by incorporation of *P. lilacinum/T. harzianum* and Phorate into the solarized nursery beds. This approach is being adopted on a large scale for the production of nematode-free cardamom seedlings.

## 26.3 Ginger, *Zingiber officinale*

Root-knot (*Meloidogyne* spp.), burrowing (*Radopholus similis*), and lesion (*Pratylenchus* spp.) nematodes are important pests of ginger.

### 26.3.1 Root-Knot Nematode, *Meloidogyne incognita*

#### 26.3.1.1 Crop Losses

Kaur (1987) estimated 41–59% yield loss in ginger when the crop was raised using apparently healthy rhizomes in nematode-infested fields.

The root-knot nematode is responsible for annual yield loss of 29–33% in ginger amounting to ₹1894.28 million in India (Walia and Chakraborty 2018).

#### 26.3.1.2 Symptoms

Heavily infested plants exhibit stunting and chlorotic leaves with marginal necrosis. The root-knot nematodes cause galling and rotting of roots and underground rhizomes (Fig. 26.4).

Infested rhizomes show brown, water-soaked lesions in the outer tissues, particularly in the angles between shoots. The $J_2$ of *M. incognita* invade the rhizome through the axils of leaf sheaths in the shoot apex. The juveniles can invade the entire length of root, but prefer penetration in the area of differentiation in fleshy fibrous roots. In the rhizomes and fleshy roots, extensive internal lesions develop.

#### 26.3.1.3 Biology and Life Cycle

Infected rhizomes and soil are the source of infection. After penetration, the juveniles feed on conducting vessels and develop endoparasitically into adults after undergoing three molts. Sac-like adult females lay eggs in masses on root surface on galls, or in advanced stages inside the compound galls, while the vermiform males leave the roots and come out in soil. The nematodes complete its life cycle in about 25–30 days during summer months; while it may be prolonged to

**Fig. 26.4** *Left*—Ginger field infested with root-knot nematodes. *Right*—Infection on daughter ginger rhizomes with root-knot nematodes

60–90 days in winter under north Indian conditions. Infection spreads to daughter rhizomes leading to severe deterioration in quality and quantity of rhizomes.

#### 26.3.1.4 Spread

Spread of the nematode over short distance occurs in water between soil particles and in surface water. Infected rhizomes used for planting, soil attached to the agricultural implements, and the labors' feet or foot ware are the means of nematode spread over long distances.

#### 26.3.1.5 Management

##### 26.3.1.5.1 Physical Methods

Hot Water Treatment

Ginger rhizomes are invariably found infested with the root-knot nematode due to monocropping of ginger in infested field. Dipping of such rhizomes in hot water at 45 °C for 3 h or at 51 °C for 10 min was found to be nematode-free. This treatment resulted in 27% increase in yield.

Solar Energy Treatment

Since the hot water treatment using water bath equipment is not feasible for farmers, a technique has been developed that is practically feasible for farmers by using 10-cm-deep galvanized sheet trays coated with black paint. The trays are filled with water up to 8 cm level and kept under sunlight from 15th May or when atmospheric temperature maximum reaches to above 42 °C for 2 weeks. The trays filled with water are exposed to sun from 8 to 11 am, the ginger rhizomes are dipped in it for further 3 h, i.e., up to 2 pm. By doing so, the temperature of water in the trays is attained to 45 °C.

#### 26.3.1.5.2 Cultural Methods

Crop Rotation

Planting ginger after taro and cassava crops proved beneficial.

Intercropping

Intercropping bell-pepper *(Capsicum annum* L) with ginger gave effective control of *Pratylenchus penetrans* and *Meloidogyne incognita* and increased ginger rhizome yield. Bell-pepper was a nonhost to *P. penetrans* and nonpreferred host to *M. incognita.* This helped to improve ginger yields by making the rhizosphere unfavorable for the development and multiplication of the major ginger nematode pests (Sharma and Bajaj 2008).

Organic Amendments

Application of neem, castor, and mustard cakes at 1.25 t/ha reduced *Rotylenchulus reniformis, Meloidogyne incognita, Hoplolaimus indicus,* and *Helicotylenchus* spp. in ginger up to 6 weeks, although there was no difference in gall index at harvest stage however, higher yield was recorded in neem cake.

Tissue Cultured Plants

In vitro culture of shoot tips from emerging buds on stored rhizomes, which were heavily infested with root-knot nematodes, resulted in plants completely free from nematodes.

#### 26.3.1.5.3 Host Resistance

IISR Mahima is a selection developed by the Indian Institute of Spices, Calicut, resistant to root-knot nematode on ginger.

Ginger cv. Rio de Jenairo was found resistant to root-knot nematodes (Gall index—1.33 to 2.00, reproduction factor—0.17 to 0.93) (Okorocha et al. 2014).

#### 26.3.1.5.4 Integrated Methods

Application of Carbofuran at 1 kg a.i. /ha at the time of planting of ginger and neem cake at 250 kg/ha after 45 days of planting gave 65% reduction in root-knot and burrowing nematode population and 48% increase in yield.

Combination of hot water treatment of ginger with either solarization or summer plowing gave better result as compared to individual treatment in terms of reduction in root-knot nematode population.

The root-knot nematode on ginger can be controlled by alternating ginger with a green manure crop (*Panicum maximum*) and applying at least 150 $m^3$/ha/annum of poultry manure, with or without additional sawdust (Stirling and Nikulin 1998).

## 26.3.2 Burrowing Nematode, *Radopholus similis*

### 26.3.2.1 Crop Losses

In Fiji, yield reductions of 40% have been found when 50% of the plants are attacked. The amount of rejected rhizomes left in the field after harvest is an indication of the seriousness of this disease. Further, the loss due to the burrowing nematode also occurs during storage of rhizomes in godowns.

### 26.3.2.2 Symptoms

Burrowing nematode attacks the roots and burrows into the rhizome. The first sign of the disease is small, water-soaked shallow areas that turn brown, join together, and develop into rots that destroy the rhizomes.

Stunted growth of plants, leaf yellowing, and reduction in number of shoots are some of the above-ground symptoms. The plants grow slowly. The top most leaves dry and die. Affected plants appear to mature earlier than normal, but this is because they die early due to nematode infection.

### 26.3.2.3 Host Range

The nematode has a large host range, including banana, betel nut, black pepper, coconut, coffee, giant swamp taro, and tea.

### 26.3.2.4 Life Cycle

Eggs are laid in the roots and rhizomes. They hatch in about 10 days into juveniles which molt several times as they grow to become adult males and females. Sac-like adult females lay eggs in a gelatinous matrix on the roots, while males move to the soil and do not feed. The time from egg to adult is about 3 weeks at 25–30 °C.

### 26.3.2.5 Spread

Spread of the nematode over short distance occurs in water between soil particles and in surface water. Infected rhizomes used for planting, soil attached to the agricultural implements, and the labors' feet or foot ware are the means of nematode spread over long distances.

### 26.3.2.6 Management

#### 26.3.2.6.1 Physical Methods

After selection of what looks like healthy "seed," place it in hot water at 51 °C for 10 minutes.

#### 26.3.2.6.2 Cultural Methods

Use certified (nematode-free) planting material. Planting material selected from healthy fields where the nematodes are not present or hot water-treated rhizomes can also be used for planting.

Application of FYM at 20 t/ha 6–8 weeks before planting was found effective.

Rotation of ginger with taro and cassava proved beneficial.

Burrowing nematode susceptible weeds should not be allowed to grow in the field.

## 26.4 Turmeric, *Curcuma longa*

Of the many plant parasitic nematodes reported in association with turmeric, the root-knot nematode *Meloidogyne* spp., the burrowing nematode *Radopholus similis*, and the lesion nematode *Pratylenchus coffeae* are important.

### 26.4.1 Root-Knot Nematodes, *Meloidogyne incognita* and *M. javanica*

*M. incognita* and *M. javanica* have been reported on turmeric of which *M. incognita* is more important.

#### 26.4.1.1 Crop Losses

In pot experiments, inoculation of 100,000 nematodes/plant resulted in 76.6% reduction in the rhizome weight after 6 months (Sukumaran and Sundararaju 1986).

The avoidable yield losses to the extent of 33.61% have been reported in turmeric due to *Meloidogyne incognita* (Ray et al. 1995).

*M. incognita* is responsible for annual yield loss of 33% in turmeric amounting to ₹1627.56 million in India (Walia and Chakraborty 2018).

#### 26.4.1.2 Symptoms

The second-stage juveniles of *M. incognita* invade the rhizome through the axils of leaf sheaths in the shoot apex (Eapen et al. 2005). The above-ground symptoms of turmeric plants infected with *M. incognita* exhibit stunted plant growth, leaf yellowing, reduction in tillering, and marginal drying of leaf tips. High populations of *M. incognita* in field cause stunting, yellowing, and withering of plants in large patches. Galling and rotting of roots can also be noticed (Fig. 26.5). Infested rhizomes have brown, water-soaked areas in the outer tissues and lose their bright yellow color (Mani et al. 1987). The affected turmeric plants mature, dry faster, and die sooner than healthy ones resulting in a poor crop stand.

#### 26.4.1.3 Survival and Spread

Infested rhizomes serve as a source of infection and means of dissemination.

#### 26.4.1.4 Management

##### 26.4.1.4.1 Physical Methods

Use of nematode-free rhizomes for fresh planting, washing seed material to free of soil, and drying in shade before planting help to reduce the inoculum. Hot water treatment of rhizomes at 50–55 °C for 10 min or 45 °C for 30 min eliminates

**Fig. 26.5** Root-knot nematode infestation on turmeric roots

nematode infection. Soil solarization of beds for 40 days during summer reduces nematode population in soil.

Rabbing was most effective for the management of root-knot nematodes on turmeric which gave maximum plant height and rhizome yield with lowest root-knot index (Patel et al. 2001).

Summer plowing at an interval of 15 days and use of hot water-treated turmeric rhizomes as seed proved promising in reducing the nematode population and in increasing the yield up to 65%.

#### 26.4.1.4.2 Cultural Methods

Use only healthy, nematode-free planting material.

Crop rotation with cereals and millets should be practiced, preferably with rice at least once in 3 years. Deep summer plowing with furrow turner plows should be practiced. Application of well-decomposed FYM/compost at 25-30 t/ha or neem cake at 2 t/ha and mulching with green leaves at 10-12 t/ha at planting and repeating the mulching during the growth period helps in reducing nematode multiplication (Kaur 1987).

#### 26.4.1.4.3 Chemical Methods

Application of Fenamiphos at 2.5 kg a.i. /ha 1 day before planting gave yield increase of 59–187% (Patel et al. 1982). Carbofuran at 1 kg a.i. /ha increased yield by 68% over control with a benefit:cost ratio of 2.0 (Gunasekaran et al. 1987). Mani et al. (1987) reported reduction in root-knot nematode population by 81.6% by in-row application of Carbofuran at 4 kg a.i. /ha to a four-month-old turmeric crop. Carbofuran at 3 kg a.i. /ha is effective in reducing the gall index (Pervez 2018).

#### 26.4.1.4.4 Biological Methods

Complete suppression of nematodes could be achieved by the application of *Fusarium oxysporum* (isolate IISR-11) at 50 g/bed of 3 $m^2$ at the time of planting.

*Purpureocillium lilacinum, Pochonia chlamydosporia,* and *Aspergillus nidulans* suppressed root-knot nematode population.

Seenivasan et al. (2001) reported enhancement of number of tillers plant growth parameters (58% increase in pseudostem height and 31% increase in number of tillers), reduction in nematode population (40%), and gall indices (25%) by rhizome treatment with *P. fluorescens* at 10 g/kg.

*Pochonia chlamydosporia* can be applied to the beds at the time of sowing at 20 g/bed ($10^6$ cfu/g) for management of nematode problems (Pervez 2018).

#### 26.4.1.4.5 Host Resistance

IISR Pragati is a selection developed by the Indian Institute of Spices, Calicut, tolerant to root-knot nematode on turmeric.

Turmeric cvs. Kodur, Chayapasupu, Duggirala, Guntur-1, Guntur-9, Rajampet, Sugandham, and Appalapadu are resistant to *M. incognita* (Gunasekaran et al. 1987).

#### 26.4.1.4.6 Integrated Methods

Soil application of *T. harzianum* + neem cake at 1 t/ha is recommended.

## 26.4.2 Burrowing Nematode, *Radopholus similis*

### 26.4.2.1 Crop Losses

Koshy and Sosamma (1975) reported turmeric as a host of *R. similis.* An initial inoculum level of 10 nematodes/plant resulted in 35% reduction in rhizome weight after 4 months. A reduction of 65% and 76% in rhizome weight was observed after 4 and 8 months, respectively, with an inoculum level of 100,000 nematodes (Sosamma et al. 1979).

### 26.4.2.2 Symptoms

Drying of leaf tips and margins is conspicuous. Infested rhizomes have shallow, water-soaked brownish areas on the surface. Nematode infestation changes the golden yellow color of healthy rhizomes to yolk-yellow color (Fig. 26.6). Infested plants age and dry faster than healthy plants. Rotting of roots takes place and the decayed roots remain devoid of cortex and stellar portions (Fig. 26.6) (Sosamma et al. 1979).

### 26.4.2.3 Survival and Spread

The nematodes are spread through infested planting material. Populations of *R. similis* from coconut are known to infest turmeric, and the use of turmeric as an intercrop in *R. similis* infested coconut and areca nut-based farming systems should be avoided.

### 26.4.2.4 Management

Avoiding use of turmeric as an intercrop in *R. similis* infested coconut and areca nut-based farming systems and use of only clean, nematode-free rhizomes for fresh

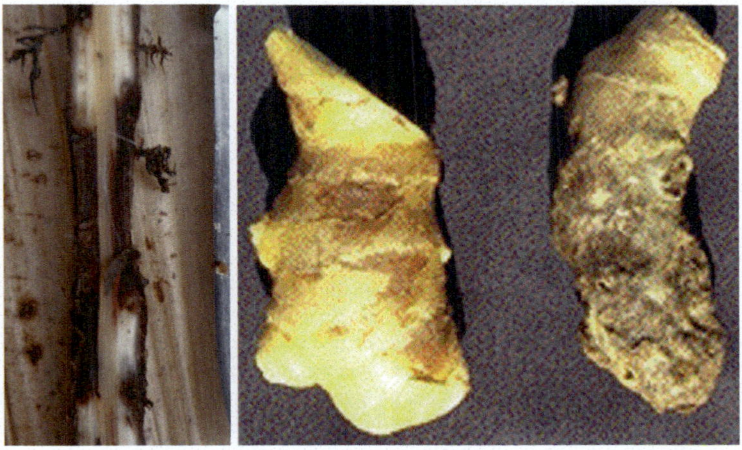

**Fig. 26.6** *Left*—Burrowing nematode-infested root showing necrotic lesions. *Right*—Healthy (*left*) and burrowing nematode-infested (*right*) rhizomes

planting helps in reducing the nematode population. Integration of neem cake + *Purpureocillium lilacinum* was found effective.

## 26.5 Conclusion

Root-knot nematodes (*Meloidogne* spp.) are among the most important group of plant parasitic nematodes, causing damage and yield losses on most of the spices. The burrowing nematode (*Radopholus similis*) and lesion nematodes (*Pratylenchus* spp.) are two other important and destructive root parasites of spice crops. Cultural practices, use of resistant cultivars, and application of chemical nematicides are primary strategies for nematode management, but yield losses persist with numerous crops. Application of microorganisms, antagonistic to *Meloidogne* spp., or of compounds produced by these microbes could provide additional opportunities for managing nematode problems in spices. The use pesticide of plant origin like clove, tobacco, and betel vine extract have been proven effective in reducing the penetration rate of the nematodes and suppressing the nematode population in roots and the percentage of rhizome-knot symptom. The active ingredients which are toxic to nematodes in organic amendments should be identified to develop botanical pesticides. There is a need to develop strategies to reduce nematode population and its reproduction rates, while at the same time to improve soil quality and hence better plant growth by the use of organic amendments as soil mulch. These parasitic nematodes could be controlled by integrating several approaches such as biocontrol agents, botanical nematicides, and enhancing cultural practices through the use of organic mulch and resistant varieties.

## References

Ali SS (1984) Preliminary observations on the effect of some systemic nematicides and neem oil cakes in a cardamom field infected with root-knot nematodes (Abstract). In: Proceedings of the sixth symposium on plantation crops, Kottayam, Kerala, pp 215–223

Ali SS (1986) Root-knot nematode problem in cardamom and its management. In: Proceedings of the second group discussion on the nematol. problems of plantation crops, Central Coffee Res. Inst., Balehonnur, pp 10–12

Ali SS, Venugopal MN (1992) Interaction between meloidogyne incognita and rhizoctonia solani in damping off or rhizome rot disease of cardamom seedlings. Nematol Medit 20:65–66

Ali SS, Venugopal MN (1993) Prevalence of damping off and rhizome rot disease in nematode infested cardamom nurseries in Karnataka. Curr Nematol 4:19–24

Anandraj M, Ramana KV, Sarma YR (1991) Interaction between vesicular arbuscular mycorrhizal fungi and *Meloidogyne incognita* in black pepper. In: Bagyaraj DJ, Manjunath A (eds) Proceedings of second national conference on mycorrhizal symbiosis and plant growth. University of Agricultural Sciences, Bangalore, pp 110–112

Butler EJ (1906) The wilt disease of pigeon-pea and pepper. Agric J India 1:25–36

Devapriyanga R, Jonathan EI, Sankari Meena K, Kavita PG (2012) Bioefficacy of *Pseudomonas* and *Bacillus* isolates against root-knot nematode in black pepper cv. Panniyur-1. Indian J Nematol 42(1):57–65

Eapen SJ (1987) Plant parasitic nematode problem in small cardamom. In: Proceedings of the third group discussion on the nematology problems of plantation crops, Sugarcane Breeding Institute, Coimbatore, pp 43–44

Eapen SJ (1995) Investigations on plant parasitic nematodes associated with cardamom. Final report of the project Nema I (813), Indian Institute of Spices, Calicut, 39 pp

Eapen SJ, Ramana KV (1996) Biological control of plant parasitic nematodes of spices. In: Anandaraj M, Peter KV (eds) Biological control in spices. Indian Institute of Spices Research, Calicut, pp 20–32

Eapen SJ, Venugopal MN (1995) Field evaluation of *Paecilomyces lilacinus* and *Trichoderma* spp. in cardamom nurseries for the control of root-knot nematodes and rhizome rot disease. In: National symptoms on nematode problems of India – an appraisal of the nematode management with eco-friendly approaches and biocomponents, Indian Agricultural Research Institute, New Delhi

Eapen SJ, Beena B, Ramana KV (2005) Tropical soil microflora of spice based cropping systems as potential antagonists of root-knot nematodes. J Invert Pathol 88(3):218–225

Gunasekaran CR, Vadivelu S, Jayaraj S (1987) Experiments on nematodes of turmeric–a review. In: Proceedings of the third group discussions on the nematol. problems of plantation crops. Sugarcane Breeding Institute, Coimbatore, pp 45–46

Ichniohe M (1980) Studies on the root-knot nematode of black pepper plantation in Amazon. Ann RepSoc Plant Protect, North Japan 31:1–8

Kaur DJ (1987) Studies on nematodes associated with ginger (*Zingiber officinale* Rosc.) in Himachal Pradesh. Ph. D. thesis, Himachal Pradesh University, Shimla, 198 pp

Koshy PK (1986) The burrowing nematode, *Radopholus similis* (Cobb, 1893) Thorne, 1949. In: Swarup G, Dasgupta DR (eds) Plant parasitic nematodes of India-problems and progress. Indian Agricultural Research Institute, New Delhi, pp 223–248

Koshy PK, Sosamma VK (1975) Host-range of the burrowing nematode *Radopholus similis* (Cobb, 1893) Thorne, 1949. Indian J Nematol 5:255–257

Koshy PK, Sosamma VK, Sundararaju P (1977) Screening of plants used as pepper standards against root-knot nematode. Indian Phytopathol 30:128–129

Mani A, Naidu PH, Madhavachari S (1987) Occurrence and control of *Meloidogyne incognita* on turmeric in Andhra Pradesh, India. Int Nematol Netw Newsl 4:13–18

Mohandas C, Ramana KV (1987) Third group discussion on the nematology problems of plantation crops. Sugarcane Breeding Institute, Coimbatore, pp 43–44

Mohandas C, Ramana K (1991) Pathogenicity of *Meloidogyne incognita* and Radopholus similis on black pepper (Piper nigrum L.). J Plant Crops 19:41–43

Okorocha EOA, Ogbuji RO, Onyenobi FI et al (2014) Screening of ginger varieties for resistance to root - knot nematodes *Meloidogyne* spp. J Adv Agric Sci Technol 2(4):59–62

Patel DJ, Makadia BM, Shah HM (1982) Occurrence of root-knot on turmeric (*Curcuma longa* L.) and its chemical control. Indian J Nematol 12:168–171

Patel BA, Patel RG, Patel HR et al (2001) Management of root-knot nematodes in turmeric. In: National congress on "centenary of nematology in India: appraisal and future plans, Indian Agricultural Research Institute, New Delhi, p 165

Pervez R (2018) Current status of plant parasitic nematodes and their management of major spice crops. Trends Hort 1(2). https://doi.org/10.24294/th.v1i2.964

Ramana KV (1986) Slow wilt disease of black pepper and the role of plant parasitic nematodes in its etiology (Abstract). In: Proceedings of the second group discussion on nematology problems of plantation crops, Central Coffee Research Station, Balehonnur, pp 17–21

Ramana KV, Mohandas C, Balakrishnan R (1987) Role of plant parasitic nematodes in the slow wilt disease complex of black pepper (*Piper nigrum* L.) in Kerala. Indian J Nematol 17:225–230

Ray S, Mohanty KC, Mohapatra SN et al (1995) Yield losses in ginger (*Zingiber officinale* Rose.) and turmeric (*Curcuma longa* L.) due to root knot nematode (*Meloidogyne incognita*). J Spices Aromatic Crops 4(1):67–69

Seenivasan N, Parameswaran S, Sridar P, Gopalakrishnan C, Gnanamurthy P (2001) Application of bioagents and neem cake as soil application for the management of root-knot nematode in turmeric. In: National Congress on "centenary of nematology in India: appraisal and future plans". Indian Agricultural Research Institute, New Delhi, p 164

Sharma GC, Bajaj BK (2008) Effect of inter-cropping bell-pepper with ginger on plant parasitic nematode populations and crop yields. Ann Appl Biol 133:199–205

Sosamma VK, Koshy PK (1995) Effect of *Pasteuria penetrans* and *Paecilomyces lilacinus* on population build up of root-knot nematode, *Meloidogyne incognita* on black pepper. In: National symposium on nematode problems of India – an appraisal of the nematode mangmt with eco-friendly approaches and biocomponents. Indian Agricultural Research Institute, New Delhi, p 47

Sosamma VK, Sundararaju P, Koshy PK (1979) Effect of *Radopholus similis* on turmeric. Indian J Nematol 9:27–31

Stirling GR, Nikulin A (1998) Crop rotation, organic amendments and nematicides for control of root-knot nematodes (*Meloidogyne incognita*) on ginger. Australasian Plant Pathol 27 (4):234–243

Sukumaran S, Sundararaju P (1986) Effect of root-knot nematode, *Meloidogyne incognita* on the growth of turmeric (Abstract). Seventh symposium of plantation crops, Coonoor, Nilgiris, Tamil Nadu, p 33

Sundararaju P, Koshy PK, Sosamma VK (1979) Plant parasitic nematodes associated with spices. J Plant Crops 7:15–26

Thomas L, Rajeev P (2015) Black pepper. ICAR- Indian Institute of spices Research, Kozhikode, Kerala. 27 pp

Venkitesan TS (1976) Studies on the burrowing nematode, *Radopholus similis* (Cobb, 1893) Thorne, 1949 on pepper (*Piper nigrum* L.) and its role in slow wilt disease. Ph. D. thesis, University of Agricultural Sciences, Bangalore

Venkitesan TS, Jacob A (1985) Integrated control of root-knot nematode, infecting pepper vines in Kerala (Abstract). Indian J Nematol 15:261–262

Walia RK, Chakraborty PK (eds) (2018) Nematodes problems of crops in India. ICAR-All India Coordinated Research Project on Nematodes in Agriculture, New Delhi. 402 pp

# Tuber Crops

# 27

**Abstract**

The most important nematode pests that are potentially damaging to tuber crops like cassava, taro, sweet potato, yams, yam bean, winged bean, and elephant foot yam include root-knot nematodes (*Meloidogyne* spp.), lesion nematodes (*Pratylenchus* spp.), and Dry rot nematode (*Scutellonema bradys*). Nematodes interact with Fusarium/bacterial wilt, Phytophthora foot rot, Rhizoctonia rhizome rot/root rot pathogens in causing disease complexes. Distribution, crop losses, biology, survival and spread, and management methods such as physical, cultural, chemical, biological, host resistance, and integrated nematode management on tuber crops are discussed in this chapter.

**Keywords**

*Meloidogyne* spp. · *Radopholus similis* · Symptoms · Biology · Management

## 27.1 Introduction

Nematodes can cause serious yield and quality reductions in most root and tuber crops. Numerous reports from the tropics and subtropics have detailed the incidence of plant-parasitic nematodes on root and tuber crops or associated crop damage. Plant-parasitic nematodes regarded as problematic on root and tuber crops include root-knot nematodes (*Meloidogyne* spp.); *Scutellonema bradys*, which causes dry rot of yam in West Africa; lesion nematodes (*Pratylenchus* spp.); and *Hirschmanniella miticausa*, causal agent of "miti miti" disease of taro in the south Pacific. Nematodes can therefore pose a significant threat to root and tuber crop production.


P. P. Reddy, *Nematode Diseases of Crops and their Management*,
https://doi.org/10.1007/978-981-16-3242-6_27

## 27.2 Cassava, *Manihot esculenta*

Important nematodes parasitizing cassava are root-knot and root lesion nematodes.

### 27.2.1 Root-Knot Nematodes, *Meloidogyne* spp.

*Meloidogyne incognita* and *M. javanica* are the most widely reported and economically important nematodes on cassava (Sorley et al. 1983).

#### 27.2.1.1 Distribution

*Meloidogyne incognita*, *M. javanica*, and *M. arenaria* have been reported from cassava from Brazil, Colombia, Dominican Republic, Hawaii, India, Ivory Coast, Nigeria, Rhodesia, Tanzania, Togo, Trinidad, and the United States.

The root-knot nematodes (RKN) occur in Latin America, the United States, West Africa, East Africa, and the Pacific (Hillocks and Wydra 2002).

Nirula (1963) reported *M. javanica* for the first time from India.

#### 27.2.1.2 Crop Losses

Jatala (1988) reported that *M. javanica* and *M. incognita* race 2 were responsible for considerable yield reduction (49% in fresh storage root weight and 22% in dry top weight) in cassava.

In Nigeria, yields were decreased by 24–38% in two susceptible cassava cvs. Bukalasa 11 and TMS 30337 (Coyne 1994).

Percentage yield loss of between 41.8% and 88.4% was recorded in *M. incognita*-infected cassava plants (Akinsanya et al. 2019).

#### 27.2.1.3 Symptoms

RKN is causing galls exceeding 1 cm diameter on susceptible cultivars. The symptoms expressed by severely infected cassava plants include stunting of plant growth, reduction in stem diameter, drastic reduction in feeder root system, and severe root galling resulting in 17–50% root yield losses and reduction in quantity and quality of the planting material (Théberge 1985; Caveness 1982). The greatest effect of the nematode may be on the storability of the harvested roots which may result in 87% post-harvest loss and drastic deterioration of roots in storage (Caveness 1982).

*Meloidogyne* spp. feeding produces physically damaging galls with eggs inside them (Fig. 27.1) which may prevent absorption of water and nutrients from the soil. It is therefore possible that extensive galling can be observed even at low densities following infection. *M. incognita* race 2 and *M. javanica* significantly reduced stalk length, stalk weight, and storage root weight of cassava after 15.5 months growing period (Caveness 1981).

**Fig. 27.1** *Left*—cassava plant infected with root-knot nematodes showing stunting of plants and leaf yellowing. *Right*—cassava storage roots deformed by root-knot nematode infection

#### 27.2.1.4 Management

##### 27.2.1.4.1 Cultural Methods

The summer fallowing which is already being practiced by farmers in many parts of cassava-growing areas in Kerala due to nonavailability of water is an excellent means of controlling nematode population. It is also essential to avoid planting susceptible crops either in the previous season or as intercrops. The root-knot nematodes *M. incognita* and *M. javanica* on cassava can also be effectively managed by soil application of Cassareep (a by-product of the cassava industry).

##### 27.2.1.4.2 Host Resistance

Cassava cvs. Sree Sahya and Narayaniyakappa were found to be resistant to root-knot nematodes (Sreeja et al. 1998). Freitas and Moura (1986) reported cassava cv. Mandiocol to be resistant to *M. incognita* and *M. javanica*.

##### 27.2.1.4.3 Integrated Methods

Soil application of Neem cake at 250 g/plant + Carbofuran at 1 g a.i./plant.

### 27.2.2 Lesion Nematodes, *Pratylenchus brachyurus*, *P. sefaensis*

*Pratylenchus brachyurus* and *P. sefaensis* were found attacking cassava. *P. brachyurus* is responsible for predisposing plants to infection by secondary soil-borne pathogens as well as disrupting tissues in cortical region of roots.

#### 27.2.2.1 Crop Losses

*P. sefaensis* reduced fresh storage root weight by 45% and dry top weight by 21%.

#### 27.2.2.2 Distribution

The lesion nematode occurs on the crop in many parts of the world including the United States, East Africa, Latin America (Brazil), and India.

#### 27.2.2.3 Symptoms

The lesion nematode produces root lesions and subsequent rotting of roots. Affected plants are stunted with yellowing of leaves and dieback of twigs and produce small or no tubers. Leaves also show nutrient deficiency symptoms (Mohandas et al. 1990).

#### 27.2.2.4 Management

A number of cassava varieties (Atitogen, Ba Pou II, Agba Tiega, Agba Boquia, and Sodjievi) show some resistance to the lesion nematode.

Rosa et al. (2014) reported that two cassava cvs. Caipora and Colonial were immune to *P. zeae*, while they are resistant to *P. brachyurus* with reproduction factor = 0.5 and 0.9, respectively.

## 27.3 Taro, *Colocasia esculenta*

### 27.3.1 Root-Knot Nematodes, *Meloidogyne incognita, M. javanica*

Root-knot nematodes are important limiting factors for successful production of *Colocasia* crop. *M. incognita* and *M. javanica* are widely reported on *Colocasia*. Nirula (1959) was the first to report root-knot nematode on *Colocasia* from India.

#### 27.3.1.1 Crop Losses

The root-knot nematode *M. incognita* caused *Colocasia* tuber yield loss of 24%, while *M. javanica* causes severe losses in India.

#### 27.3.1.2 Symptoms

The foliage of nematode infested plants at first becomes yellow and then turns brown and ultimately dies back. No corms were found in plants attacked in the earlier stages, and late-stage attack resulted in deformed and galled corms (with blister-like swellings varying in size from 2 to 15 mm) (Fig. 27.2) of little market value. Rotting of corms associated with the nematode occurs during storage. It causes reduction in yield and irregular-shaped smaller branched tubers.

#### 27.3.1.3 Survival and Spread

*Meloidogyne* spp. can be carried over from one *Colocasia* crop to next in the wide range of other host crops and weeds. As the nematodes feed and reproduce in corm

**Fig. 27.2** Root-knot nematode symptoms on taro corms and roots

tissues, they can be spread in corms and cormels if infested material is used for propagation.

#### 27.3.1.4 Management

##### 27.3.1.4.1 Physical Methods

Root-knot can be controlled in taro by dipping the corms in hot water at 51 °C for 10 min (Martin and Hancock 1985).

##### 27.3.1.4.2 Cultural Methods

The use of nematode-free planting material will prevent spread into the field. Seed corms or cormels should be free of root-knot damage. It is best if the "tops" are selected from fields without any history of infection. There is no specific effort to use nematode-free planting material in American Samoa, though farmers inadvertently do so by planting corms with no roots and less than a centimeter of corm tissue.

Rotation with crops bred for resistance to *M. incognita* or nematode-resistant cover crops such as marigolds (produce chemicals that are toxic to nematodes) planted during periods of fallow can lower nematode populations and decrease root and corm damage in subsequent taro plantings. Fields must be kept free of weeds and other plants that may act as hosts for this polyphagous parasite. Crop rotation with *Panicum maximum* var. *trichoglume* or a mixture of *P. maximum* and *Macroptilium atropurpureum* is recommended. Growing of *Colocasia* under flood or highly wet conditions reduced the root-knot nematode population.

##### 27.3.1.4.3 Chemical Methods

Nematicides (e.g., Dazomet, Oxamyl, and Fenamiphos) are commonly applied as pre-planting treatments to reduce *M. javanica* in India (Srivastava et al. 1971).

##### 27.3.1.4.4 Biological Methods

The bacterium *Pasteuria penetrans* and the fungus *Purpureocillium lilacinum* have been used as biological control agents. *P. penetrans* can be cultured easily by planting root-knot-infected tomatoes where the bacterium is found. The roots can then be harvested, chopped and dried, and incorporated into the soil where there is a nematode problem.

#### 27.3.1.4.5 Host Resistance

Mohandas and Palaniswami (1990b) reported resistance in *Colocasia* cv. "Sree Reshmi" (released high yielding resistant variety) and very high resistance in C-9 (Tamnarakannan, a popular variety among farmers) grown extensively in Central Travancore. In cv. C-9, no nematode could be detected either from root or tubers. Cv. "Dodare" was found to be completely resistant to both *M. incognita* and *M. javanica*. Enhancement of yields (roots and corms) and reduction in root-knot nematode multiplication can be achieved by growing *Colocasia* cultivars like "Mana Ulaulu" and "Piko Ulaulu." These cvs. Can be utilized as parent material in developing root-knot nematode-resistant varieties.

## 27.4 Sweet Potato, *Ipomoea batatas*

### 27.4.1 Root-Knot Nematodes, *Meloidogyne* spp.

*M. incognita* and *M. javanica* are the major nematode species damaging sweet potato in India (Ray et al. 1990).

#### 27.4.1.1 Distribution

The root-knot nematode *Meloidogyne incognita* is worldwide in distribution. It is widespread in Asia, particularly Southeast Asia, and usually occurs in warmer areas. In some countries, *M. javanica* is more dominant.

#### 27.4.1.2 Crop Losses

In sweet potato an estimated annual yield loss of 10.2% was reported. In susceptible varieties pathogenicity of *Meloidogyne incognita* showed a 50% storage root reduction at a population density of 20,000/$cm^3$. Roots in storage become unmarketable due to cracking. Gaspin and Valdez (1979) reported tuber reduction to the extent of 47.7 and 50.6% with an initial inoculum level of 20,000 eggs of *M. incognita* and *M. javanica*, respectively.

#### 27.4.1.3 Symptoms

Above-ground symptoms exhibited by sweet potato plants due to root-knot nematode include poor shoot growth, leaf chlorosis, stunting, excessive wilting during dry and hot conditions, reduced yield and quality, and sometimes premature death. Reduction in vine growth, yellowing, or abnormally abundant production of flowers was observed. Cause swelling (knotting) on the entire primary roots and heavy infection can inhibit apical growth (Fig. 27.3). The most obvious symptoms of damage on enlarged roots are longitudinal cracking and general rough appearance of the skin or formation of small bumps or blisters on enlarged roots (Fig. 27.3). The

**Fig. 27.3** Root-knot galls on sweet potato roots (*Left*), tubers (*Middle*), and cracks in sweet potato tuber due to root-knot nematode infection (*Right*)

nematode drastically reduces both above-ground and under-ground plant parts. In fibrous roots, the nematode infestation produces small swellings or knots, smaller feeder roots with shorter and fewer secondary roots and root hairs (Fig. 27.3).

Abnormally abundant production of flowers could also be due to infestation by the nematode.

#### 27.4.1.4 Biology

The nematode may complete several generations during the cultivation of this crop. Sweet potatoes are equally susceptible to all *M. incognita* races which can infect both storage roots and fibrous roots with same intensity.

#### 27.4.1.5 Interaction with Other Pathogens

*M. incognita* interacts with *Fusarium* spp. and *Ralstonia solanacearum* causing severe wilting and premature death. Although there are several *Fusarium*-resistant cultivars, their resistance is broken in the presence of *M. incognita*.

#### 27.4.1.6 Ecology

*Meloidogyne* species seem to do well in light, friable, sandy loam soil which happens to predominate and constitute the major portion of the world's sweet potato-growing areas. The root-knot nematode *M. incognita* can complete four to five generations during hot climate. Therefore, it is capable of increasing its population to a level of economic threshold in a short period.

#### 27.4.1.7 Survival and Spread

*Meloidogyne* juveniles and/or eggs survive in storage roots and can be disseminated in root, but not stem, and propagative material. Nematodes can also survive on many alternate weed hosts. The means of nematode spread include farm equipment, labors' feet, and irrigation water.

#### 27.4.1.8 Management

##### 27.4.1.8.1 Physical Methods

Hot water treatment of tubers at 45 °C for 30 h, 46.7 °C for 65 min, or 50 °C for 3–5 min eliminates root-knot nematode infection without seriously impairing the viability of roots. Hot air treatment of roots, 5–8 cm in diameter, at 50 °C for 4–8 h eliminates most root-knot nematodes without significantly affecting their keeping qualities.

##### 27.4.1.8.2 Cultural Methods

The clean cuttings of seed roots from the apex are less prone to root-knot damage when planted out than cuttings from the basal end.

Rotation of sweet potato with Groundnut or maize can significantly reduce the nematode population.

Planting of trap crops and antagonistic crops such as *Tagetes erecta* and *Crotalaria spectabilis* in nematode-infested soil is effective against the root-knot nematode.

Soil application of chicken manure is very effective reducing nematode egg masses by 56%.

##### 27.4.1.8.3 Chemical Methods

Enhanced sweet potato yields and quality of the produce and reduction in the root-knot nematode population was obtained by soil incorporation of Fensulfothion/ Ethoprophos each at 3.4 kg a.i./ha in a 30 cm band 2 weeks earlier to planting or Carbofuran at 6.7 kg a.i./ha 1–4 days prior to planting.

Soil application of Carbofuran and Ethoprophos each at 1 g/30 $cm^2$ in nursery beds enhanced total and marketable yield. The early protection against root-knot nematodes by way of establishment of the sweet potato crop can be achieved by bare root-dip treatment of the propagating stock in a solution of Oxamyl or side dressing with nematicides at the time of planting.

##### 27.4.1.8.4 Biological Methods

Galano et al. (1996) reported that application of different *Purpureocillium lilacinum* isolates grown on water hyacinth substrate, 50 days after inoculation of *M. incognita* eggs, controlled root-knot nematodes, and increased fresh vine and fibrous root weights by 32–26% and 30–35%, respectively, over untreated control in sweet potato.

##### 27.4.1.8.5 Host Resistance

Mohandas and Palaniswami (1990a) reported the availability of high degree of resistance in high-yielding released varieties of sweet potato, viz., "Sree Vardhini" and "Sree Nandini." Sweet potato varieties that were found to be resistant to *M. incognita* include "Kanjangad," "Sree Bhadra," "Sree Nandini," and "Sree Vardhini" (Mohandas and Ramakrishnan 2002).

The root-knot nematode resistance in cv. Sree Bhadra is due to hypersensitive reaction. The juvenile nematodes penetrate the root but they cannot survive inside the root. Nematodes die inside due to denial of food. The nematodes cannot be detected up to 95 days during the crop season. The crops following the cv. Sree Bhadra do not suffer from the damage of root-knot nematodes.

#### 27.4.1.8.6 Integrated Methods

Control of root-knot damage to sweet potatoes involves integration of at least three methods.

## 27.4.2 Reniform Nematode, *Rotylenchulus reniformis*

*R. reniformis* is another important nematode species reported from sweet potato from a large number of countries.

### 27.4.2.1 Crop Losses

Gaspin and Valdez (1979) reported 60.6% yield reduction with an initial inoculum level of 5000 larvae of *R. reniformis*.

### 27.4.2.2 Symptoms

The reniform nematode is associated with leaf yellowing, transitory wilting, and reduction in yield and size of swollen tubers (Fig. 27.4). Under severe infestation, especially at planting time, the plants suffer from stunted of growth and drastic reduction in feeder roots. Infestation by *R. reniformis* can manifest as deeply cracked roots that enlarge as the roots grow, reduction in number of feeder roots and/or in a general discoloration of storage roots (Fig. 27.4). Reniform nematode infestations can result in poor sizing (bulking) of crop and in a reduced number of marketable grade sweet potatoes. In mature roots, deep suberized cracks are the most noticeable symptom (Fig. 27.4).

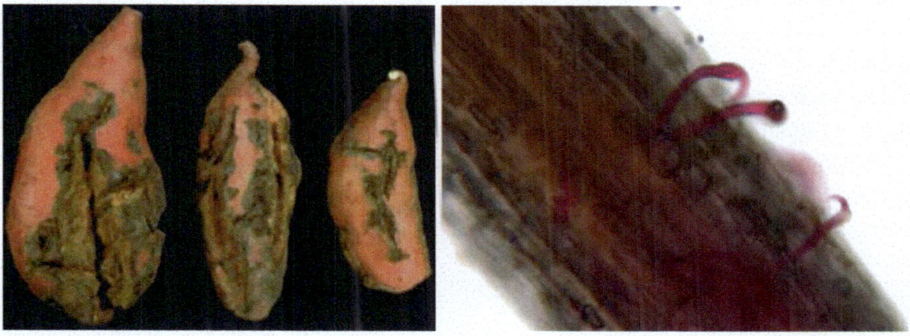

**Fig. 27.4** *Left*—reniform nematode infected sweet potato tubers. *Right*—reniform female nematodes attached to sweet potato root

#### 27.4.2.3 Biology

Juveniles of the reniform nematode are differentiated within the egg and undergo one molt before the second-stage juvenile's hatch. The second-stage larvae feed and develop endoparasitically and develop into adults after undergoing another three molts. In susceptible cultivars, the kidney-like adult female laid eggs 16 days after inoculation. The life cycle from egg to egg is from 22 to 29 days in susceptible cultivars such as "V20–436."

#### 27.4.2.4 Ecology

Male and female nematodes can survive in air dried soil kept at 20–25 °C for 7 months. Local dissemination is through infested soil. Distribution is limited by low winter temperatures, and nematode and disease development are both greater at 29.5 °C than at 15, 21.5, or 36 °C.

#### 27.4.2.5 Host Range

It has very wide host range including weeds. Major host are soybean, cowpea, cotton, pineapple, sweet potato, cassava, and other vegetable crops.

#### 27.4.2.6 Interaction with Other Pathogens

*R. reniformis* interact with *Fusarium* spp. in the development of disease complex.

#### 27.4.2.7 Survival and Spread

- Male and female nematodes can survive in air dried soil kept at 20–25 °C for 7 months.
- Local dissemination is through infested soil.

#### 27.4.2.8 Management

##### 27.4.2.8.1 Physical Methods

Hot water treatment of sweet potatoes at 50 °C for 3–5 min eliminated the reniform nematode (Martin 1970).

##### 27.4.2.8.2 Cultural Methods

Rotation of sweet potato with nonhost crops like sorghum, beet, broccoli, corn, okra, pepper, radish, spinach, turnip, and watermelon helps to manage reniform nematode.

Planting of trap and antagonistic crops like *Tagetes erecta* and *Crotalaria spectabilis* in nematode-infested soil has been found to be effective against the nematode.

##### 27.4.2.8.3 Chemical Methods

Good control of *R. reniformis* and increase in sweet potato grade have been achieved by application of Fensulfothion granules at 5.6 kg a.i./ha or Ethoprophos at 3.4–6.7 kg a.i. /ha in 30 cm band in row 2 to 3 weeks prior to planting.

#### 27.4.2.8.4 Biological Methods

*Purpureocillium lilacinum*, a fungal egg parasite, was found to be effective against the reniform nematode.

#### 27.4.2.8.5 Host Resistance

The sweet potato Selection P-104 is reported to be resistant to cracking, while variety "Goldrush" is moderately susceptible to *R. reniformis*.

### 27.4.3 Dry Rot Nematode, *Scutellonema bradys*

#### 27.4.3.1 Symptoms

- Causes significant reduction in root mass in infected plants (Fig. 27.5).
- Nematodes feed within tuber causing tissue breakdown and producing cavities.
- Initially small yellowish lesions develop beneath periderm and as infection spreads these turn into areas of dark-brown or black colored rot.
- External symptoms are slight to deep cracks on tuber skin and malformation of tuber (Fig. 27.5).

#### 27.4.3.2 Management

- Leave soil fallow for several months.
- Use certified nematode-free seed tubers.
- Dip all propagating material in hot water maintained at 50 °C for 30 min.
- Crop rotation with maize, ground nut, bell pepper, and amaranth which are immune.

**Fig. 27.5** Sweet potato tuber infected with dry rot pathogen

## 27.4.4 Lesion Nematodes, *Pratylenchus coffeae, P. brachyurus*

### 27.4.4.1 Symptoms

- Small, brown to black, necrotic lesions are also produced on storage roots, which make the roots unmarketable (Fig. 27.6).
- On fibrous roots, lesion nematodes produce small, brown necrotic lesions that are often invaded by saprophytic fungi and bacteria.
- Stunting of plant growth and substantial decrease in the quality of fleshy storage roots occurs due to necrosis of feeder roots.

### 27.4.4.2 Favorable Conditions

- Warm temperatures and sandy soils are highly congenial to the nematode damage.

### 27.4.4.3 Survival and Spread

- Survives a shorter time in wet than in dry soil.
- Preferred soil types include coarse-textured sandy loam.
- Disseminated by transportation of plant root parts or soil and by surface or irrigation water.

### 27.4.4.4 Management

- Rotation of sweet potato groundnut and rice or enhanced potassium fertilization reduced severity of root-lesion nematode infection.
- Addition of organic amendment such as chicken manure is very effective in reducing the nematode.

**Fig. 27.6** Sunken lesions induced by root-lesion nematode on sweet potato tubers

- *Purpureocillium lilacinum*, a fungal egg parasite, was found to be effective against *Pratylenchus* sp.
- Use of soil fumigant like Metham sodium is effective.

## 27.5 Yams, *Dioscorea* spp.

### 27.5.1 Yam Nematode, *Scutellonema bradys*

The dry rot of yam tubers is caused by *Scutellonema bradys* which is recognized as an economically major nematode pest resulting in high quantitative losses, as well as loss in the marketable value of the affected tubers from Kerala (Nadakal and Thomas 1967).

*Dioscorea alata*, *D. cayenensis*, *D. esculenta*, and *D. rotundata* are good hosts of the yam nematode. Sesame and cowpea support high root populations and melon can increase soil populations.

#### 27.5.1.1 Distribution

*Scutellonema bradys* is widely distributed in the tropics, especially in yam-growing areas. It has been reported from the West African countries of Nigeria, Ivory Coast, Senegal, Gambia, Ghana, Benin, and Togo and from the central African country of Cameroon. It also occurs in Cuba, Jamaica, Guatemala, Puerto Rico, Guadeloupe, Haiti, and Martinique in the Caribbean and from Venezuela, Brazil, and India (Bridge et al. 2005).

#### 27.5.1.2 Crop Losses

Weight differences between healthy and diseased tubers harvested from the field have been estimated to be 20–30% in the Ivory Coast (Bridge 1982) and 0–29% in Nigeria (Wood et al. 1980). Crop losses to the extent of 80–100% occurred when dry rot is followed by wet rot in stored yams. Bridge (1982) reported that in late harvested tubers left in dry soil, moisture loss leads to weight reduction.

#### 27.5.1.3 Symptoms

A tuber disease commonly associated with *S. bradys* is referred to as "dry rot," where the nematodes feed within the tuber causing tissue breakdown and producing cavities (Fig. 27.7). Initially small yellowish lesions develop beneath the periderm and as the infection spreads these turn into areas of dark-brown- or black-colored necrotic rot along with cracks on the tuber skin and malformation of the tuber (Fig. 27.7). The nematode cause small cracks on the surface of tubers, with the proximal parts (head) affected most. The species *D. rotundata* is generally affected most. Nematode damage to the tubers results in considerable reduction in the edible portion and marketable value of the tubers. Moisture loss is also significantly greater from diseased tubers during storage.

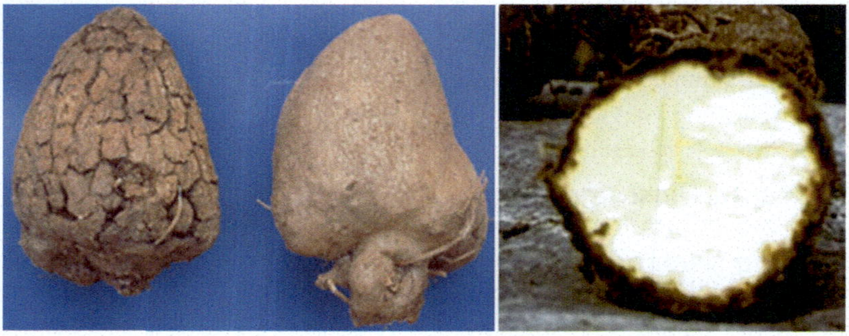

**Fig. 27.7** *Left*—yam nematode-infested and healthy tubers. *Right*—dry rot nematode infection on cut tuber

#### 27.5.1.4 Host Range

All the *Dioscorea* species grown as food crops are susceptible to *S. bradys*. A wide range of other crops and some weeds have been reported to support low root populations of *S. bradys* including yam bean (*Pachyrhizus erosus*), pigeon pea (*Cajanus cajan*), okra (*Hibiscus esculentus*), tomato (*Solanum esculentum*), sorghum (*Sorghum bicolor*), Loofah (*Luffa cylindrica*), and roselle (*Hibiscus sabdariffa*) (Adesiyan 1976). It occurs also in cassava (Missah and Peters 2001), taro (*Xanthosoma* sp., *Colocasia esculenta*), and sweet potato (*Ipomea batatas*) (Kermarrec et al. 1987). These alternative hosts permit the yam nematodes to survive in the soil even in the absence of yams. Bridge (1982) reported that the populations of *S. bradys* increased in the presence of sesame, watermelon, and cowpea.

#### 27.5.1.5 Biology

Eggs are laid in soil or plant tissues (roots and tubers) where they hatch and the juveniles develop into adults by subsequent molting. All stages seem to be infective. Dense populations can build up in the tubers with a maximum of 62,000 nematodes/10 g of tuber recorded in Nigeria but 100,000 nematodes were also reported to be found in the infested tubers in Nigeria (Bridge 1982). The *S. bradys* populations are affected by storage conditions and increase at twice the rate in tubers stored at 22–32 °C and relative humidity 40–85%, when compared to tubers stored at 16–18 °C (Adesiyan 1977).

#### 27.5.1.6 Survival and Spread

Sizeable populations of the nematode are maintained in the absence of yams probably on other host plants. Yams are propagated from whole tubers or pieces of tuber which are the principal means of spread of *S. bradys*.

#### 27.5.1.7 Interaction with Other Pathogens

The more extensive internal decay of tubers known as "wet rot," "soft rot," or "watery rot" is associated with fungal (*Botryodiplodia theobromae* and *Fusarium* spp.) and bacterial (*Erwinia* sp.) pathogens. This general decay of tubers, which is a serious problem in stored yams, is increased when tubers are wounded or damaged. The damage caused by nematodes can predispose the tubers to invasion by decay organisms resulting in complete rotting of the tubers.

#### 27.5.1.8 Management

Management of *S. bradys* can be achieved by one or more of the following measures: (1) controlling nematodes in field soil by cultural or chemical means, (2) use of planting material that is naturally free of nematodes or treatment of seed material (tubers and sets of yams) prior to planting to reduce or eliminate nematodes from propagative material, and (3) storage losses are prevented by treatment of yam tubers after harvesting (Jatala and Bridge 1990).

##### 27.5.1.8.1 Physical Methods

The dry rot nematode *S. bradys* can be managed by dipping the tubers in hot water bath maintained at 50–55 °C for 45 min without damaging *D. rotundata* tubers (IITA 2005).

##### 27.5.1.8.2 Cultural Methods

###### Phytosanitation and Clean Planting Materials

Phytosanitation is the first method used for nematode control on yam. It is necessary to separate infested tubers before storage and planting in order to prevent establishment of nematode infection. Meristem tissue culture method can also be used to produce nematode-free planting stock (Speijer et al. 2000). In yam, use of nematode-free planting material is a practical and economic means of preventing damage by *S. bradys* and their dissemination. Aerial tubers of the yam *D. bulbifera* and some forms of *D. alata*, which are used for propagation, should, however, be completely free of nematodes. A number of yam species, such as *D. alata*, *D. rotundata*, and *D. dumetorum*, can also be produced from vine cuttings (Coursey 1967). Although these methods of propagation are not practical for producing yam tubers, they can be used to produce nematode free seed tubers (IITA 2005). The use of "microsets" or "minisets," cut from mature tubers (IITA 2007), can be used to provide clean planting material if the mother seed yam tubers selected is free of nematodes.

Even true seed can be used for to produce nematode-free seed tubers of *D. rotundata* (Sadik and Okereke 1975; Jatala and Bridge 1990).

Some traditional practices in Africa, such as the use of wood ash on yam tubers or mixing cow dung in yam mounds before planting, are reported to decrease nematode numbers (Adesiyan and Adeniji 1976).

###### Agronomic Practices in the Field

Agronomic practices such as fallow (Decker et al. 1967), crop rotation with nonhost or cover crops are efficient for nematode control. A fallow of 8–12 months as well as crop rotation with nonhosts can reduce the nematode population in the soil (Adesiyan 1976). Rotation of yam with nonhosts or poor hosts such as groundnut (*Arachis hypogaea*), tobacco (*Nicotiana tabacum*), or cotton will limit damage by

*S. bradys*. Soil populations of *S. bradys* will be reduced if a nonhost or poor host crops, such as groundnut, chili pepper, tobacco, Indian spinach (*Beta vulgaris* var. *bengalensis*), cotton, maize, or sorghum, are grown prior to yams (Adesiyan 1976).

Using *Mucuna* sp. and *Crotalaria* sp. as cover crops has been reported to reduce soil nematode populations, particularly *Meloidogyne* spp. and *S. bradys* (Claudius-Cole et al. 2004). Other cover crops such as *Tagetes* spp., *Stylosanthes* spp., *Centrosema* spp., and *Aspilia* spp. have been recommended to lower nematode populations and restore fertility for yam production in Nigeria (Atu et al. 1983).

Organic and Mineral Fertilizers

It was reported that the application of mulch or organic matter might result in increased plant vigor and probably to some level, tolerance to nematodes. The use of cow dung, mixed in yam mound before planting at a rate of 1.5 kg per mound (1886.3 kg/ha), was reported to increase yields of tubers and significantly decreased nematodes numbers (Adesiyan and Adeniji 1976). The use of *Azadirachta indica* powder at a rate of 2.5 t/ha has also been reported to decrease nematode populations as well as to increase soil fertility and tuber yields (Onalo et al. 2001).

Superphosphate fertilizers were reported to reduce *S. bradys* populations in tubers of *D. alata* to a low level but did not increase yam yield (Baimey et al. 2006). The traditional practice of using wood ash on yam tubers before planting is reported to decrease nematode numbers (Adesiyan and Adeniji 1976). NPK fertilizers can reduce *S. bradys* populations in tubers of *D. alata* to a very low level.

#### 27.5.1.8.3 Chemical Methods

Bare root-dip treatment of diseased yams in Fensulfothion (Dasanit) and Thionazin (Nemafos) at 1250 ppm for 15 min and 625 ppm for 30–60 min was found to be effective in eliminating the nematode infection (Ayala and Acosta 1971). Significant increases in yield have been obtained by soaking tuber pieces of *D. alata* infected with *S. bradys* for 30 min in 1000 ppm aqueous solutions of Carbofuran and Oxamyl.

Enhancement of yield and reduction in nematode population can be obtained by soil incorporation of Oxamyl, Carbofuran, Miral, or Isazophos each at 2 kg a.i./ha 2 weeks after planting.

#### 27.5.1.8.4 Biological Methods

Antagonistic Bacteria

The yam nematode *S. bradys* can be managed by soil application of *Pasteuria penetrans* (Varghese and Mohandas 2004). Eight to ten spores were found attached in different parts of the body (4–5 spores in the head region and 3–4 spores in the esophageal region).

Plant Growth-Promoting Rhizobacteria (PGPR)
Rhizobacteria such as *Rhizobium* spp. and *Pseudomonas* spp. not only have a positive effect on plants by promoting their growth, but in addition they show a repellent effect toward nematodes (Hallmann et al. 2001).

Arbuscular Mycorrhizal Fungi (AMF)
Significantly enhanced yields were obtained by soil application of arbuscular mycorrhizal fungi *Glomus mosseae* and *G. dussii*.

#### 27.5.1.8.5 Host Resistance

Yam cv. "Florido" of *Dioscorea alata* did not seem to be susceptible to nematode attack (Ayala and Acosta 1971). One cultivar of yellow yam, *D. esculenta*, and one of *D. dumetorum* has shown some resistance to *S. bradys* (Bridge 1982; Kwoseh 2000).

#### 27.5.1.8.6 Integrated Methods

A combination of hot water treatment and phytosanitary measures as well as crop rotations was experimentally used in IITA-Ibadan, West Africa, for successful yam nematode control (IITA 2005).

The effective management of yam nematode can be obtained by combining use of nematode-free planting stock, rotation of yam with yam nematode-resistant crops, fallowing, and removal of crop residues.

## 27.5.2 Root-Knot Nematodes, *Meloidogyne* spp.

The root-knot nematode is another important nematode occurring on yams and is present worldwide. Yams have been found to be infested by *M. incognita*, *M. javanica*, *M. arenaria*, and *M. hapla*.

### 27.5.2.1 Crop Losses

*M. javanica* populations of 30,000 nematodes/plant can reduce yields of *D. opposita* by over 50%. It is estimated that there is a reduction of 39–52% in the price of galled tubers compared to healthy ones (Nwauzor and Fawole 1981). The proportion of galled tubers collected from yam barns and markets in Nigeria can be as high as 90% for *D. alata* and 70% for *D. rotundata* (Adesiyan and Odihirin 1978).

Nwauzor and Fawole (1981) recorded losses of 25–75% due to *Meloidogyne* spp. infection on yam within a storage period of 16 weeks in Nigeria.

### 27.5.2.2 Symptoms

Yam seedlings infested with root-knot nematodes can be severely stunted and chlorotic; very young seedlings may be killed by severe infestation. Foliar symptoms, such as early yellowing, leaf fall, and inhibited vine growth, have been observed on yam infested with *Meloidogyne* spp. (Nwauzor and Fawole 1981). The roots swell in size at the site of infection producing the characteristic and typical

**Fig. 27.8** Root-knot nematode-infected yam feeder roots and tubers

knots on the roots (Fig. 27.8). There is sometimes a proliferation of tuberous roots on these galls (hirsutism). *D. alata* is more sensitive to this type of nematodes than *D. rotundata*. Other symptoms of root-knot nematode infection include non-development of tuber forming roots, drastic reduction in tuber formation, and smaller size of infected tubers. Rotting of tubers has been reported.

The root-knot nematode infests the roots and tubers causing reduction in size of tubers, irregular wart-like projections on the surface, and branching of tubers with irregular shape. This nematode also continues to inflict post-harvest damage in storage.

#### 27.5.2.3 Biology

The root-knot nematode *M. incognita* took 35 days at 28 °C to complete the life cycle on *D. alata* or *D. rotundata* tubers (Nwauzor and Fawole 1981). The depth of nematode concentration in soil is between 4 and 6 mm with few at 14 mm, while it is 2 mm with none beyond the 8 mm depth.

#### 27.5.2.4 Host Range

Susceptible yam hosts of *M. incognita* are *D. alata*, *D. cayenensis*, *D. esculenta*, *D. bulbifera*, *D. composita*, *D. floribunda*, *D. trifida*, *D. prehensilis*, *D. spiculiflora*, and *D. rotundata*. Hosts of *M. javanica* are *D. alata* and *D. rotundata*.

#### 27.5.2.5 Survival and Spread

Where *Meloidogyne* juveniles and/or eggs survive in stored tubers, they will be spread in propagative material. However, *Meloidogyne* spp. have extremely wide host ranges and damaging populations will come from field soil having survived on other weed hosts or be introduced into yam fields on infested seedlings of other crops.

#### 27.5.2.6 Management

##### 27.5.2.6.1 Physical Methods

Hot water treatment of corms at 45–51 °C for 30 min eliminates root-knot nematode infection.

##### 27.5.2.6.2 Cultural Methods

The root-knot nematode can be controlled by growing sweet potato cv. "Shree Bhadra" as a trap crop.

Intercropping yam miniset (40,000/ha) with groundnut SAMNUT 10 at (160,000/ha or 120,000/ha), nematode damage of seed yams will be reduced by about 60% (Nwauzor 1996).

Use of nematode-free tubers for propagation will effectively reduce the incidence and damage.

##### 27.5.2.6.3 Chemical Methods

Application of Carbofuran at 3 kg a.i. /ha gave effective control of root-knot nematodes. *M. javanica* on *D. rotundata* can be managed by soil application of Oxamyl at 3 kg a.i./ha applied at planting and repeated at 3- and 4-week intervals. Soil incorporation of Oxamyl at 3 kg a.i./ha applied at planting followed by subsequent applications of ammonium sulfate or calcium nitrate each at 60 kg N and repeated at 3–4 week intervals was found to be effective against both *M. javanica* and *Pratylenchus brachyurus* and enhanced tuber yields by 40%. These treatments also reduce the incidence of rot in stored yams associated with nematodes.

##### 27.5.2.6.4 Biological Methods

Application of *Trichoderma viride* suspensions at a concentration of $1.7 \times 10^7$ spores/ml 1 month after planting resulted in significant reduction in *Meloidogyne* populations.

At harvest tuber galling was significantly lower on plantlets inoculated with arbuscular mycorrhizal fungi (AMF) species. AMF inoculation suppressed the densities of *Meloidogyne* spp. both in yam roots and in soil. For individual yam cultivars, only *Glomus dussii* significantly suppressed *Meloidogyne* spp. soil density on cv. "TDa98-01183." However, application of either AMF species (*Glomus mosseae* or *G. dussii*) improved tuber quality of plants. When challenged with *Meloidogyne* spp., cv. "TDa98-01183" plantlets pre-inoculated with either AMF species (*Glomus mosseae* or *G. dussii*) yielded heavier tubers than the non-mycorrhizal control plantlets.

##### 27.5.2.6.5 Host Resistance

The root-knot nematode-resistant yam varieties include "Sree Kirthi" and "Sree Latha" developed by Central Tuber Crops Research Institute, Thiruvanthapuram. The only yam species consistently found to be resistant to attack by *M. incognita* is the cluster yam, *D. dumetorum*. *D. alata* cv. "Obunenyi" is reported to be resistant to *M. incognita*; and *D. cayenensis* can be resistant to *M. incognita* and *M. javanica*.

### 27.5.3 Lesion Nematode, *Pratylenchus coffeae*

#### 27.5.3.1 Symptoms

- Young and adult nematodes feed on roots.
- Entry into the developing tubers causes a shallow brown, dry rot that continues in storage (Fig. 27.9).
- When the cells die, the nematodes migrate through the root or tuber in search of healthier parts, or they return to the soil and search for another root.
- The plants are stunted and/or die early due to root damage.

#### 27.5.3.2 Survival and Spread

- They remain alive in stored yam tubers.
- The means of short distance nematode spread is through soil water, while long distance spread is through infecting planting stock.

#### 27.5.3.3 Management

- Use nematode-free healthy planting material.
- Hot water treatment of yam tubers at 51 °C for 10 min.
- Lesion nematode can be managed by not planting yam for at least 3 years. If possible, allow the land to fallow naturally or plant marigolds or cover crops, for example, green panic (*Panicum maximum*), siratro (*Macroptilium atropurpureum*) or velvet bean, and *Mucuna* sp.

## 27.6 Yam Bean, *Pachyrhizus erosus*

### 27.6.1 Root-Knot Nematode, *Meloidogyne arenaria*

Farmers cite root-knot nematodes as one of the most important problems affecting tuber quality in the jíquima cultivar group (*P. tuberosus*).

**Fig. 27.9** Lesion nematode infestation on yam tubers

#### 27.6.1.1 Symptoms

The symptoms, which they describe as "warts" (locally known as "verruga" or "peste") (Fig. 27.10), give the affected tubers a bitter taste. Yam bean accession belonging to the jíquima cultivar group (*P. tuberosus*) did not produce any marketable tubers, owing to nematode (*Meloidogyne*) damage under field conditions. Duke (1981) mentions the species *Meloidogyne arenaria* as an important cause of tuber damage. Of the farms monitored by INIAP, 57% reported the presence of the nematode *Meloidogyne* spp. in their soils.

The root-knot nematode *Meloidogyne* sp. completely destroyed all tubers in the test plants. The most severe tuber damage observed was rotting due to lack of irrigation management and/or nematodes. *Meloidogyne arenaria* cause dramatic local yield reduction, especially on sandy soils (Duke 1981).

#### 27.6.1.2 Management

Although nematodes may cause tuber damage, only crop rotation and the use of noninfested fields are used as control measures.

## 27.7 Winged Bean, *Psophocarpus tetragonolobus*

### 27.7.1 Root-Knot Nematodes, *Meloidogyne* spp.

*Meloidogyne incognita*, *M. javanica*, and *M. arenaria* have been identified to be responsible for nematode disorders in the winged bean.

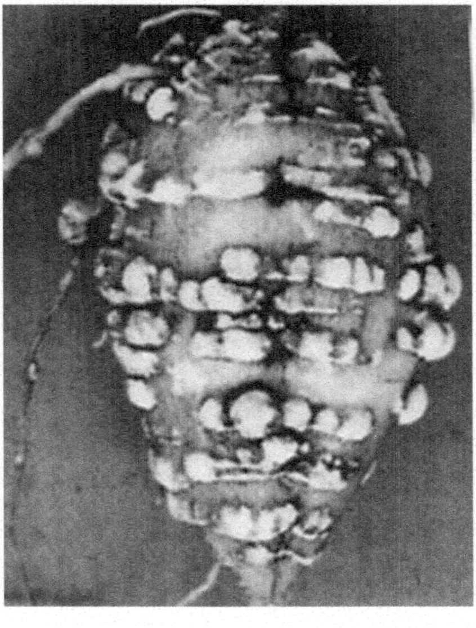

**Fig. 27.10** Root-knot nematode infection on yam bean tuber

#### 27.7.1.1 Crop Losses

Root-knot nematodes have been reported to cause up to 70% losses in tuberous roots. The damage caused by *M. incognita* is higher than *M. javanica*.

#### 27.7.1.2 Symptoms

The root-knot nematodes (*M. incognita and M. javanica*) can cause stunting of plants, yellowing of leaves, and severe galling of roots; this not only damages the roots but also reduces tuber production (Fig. 27.11) and may affect pod and seed yield. The nematodes are cosmopolitan and have a wide host range. Damage to winged bean has been reported in Papua New Guinea, the Philippines, Ivory Coast, Indonesia, and Mauritius, but the problem arises wherever the plant is grown and is especially troublesome in sandy soils. No assessment of reduction in yield of pods, seeds, or leaves was made.

A survey in January 1978 showed that the field was very heavily infested. Many plants had been killed and tubers were reduced in number (an estimated 50% of the tubers of the Papua cultivar and 60–70% of the Ghana cultivar failed to develop). At harvest time, in November, heady galled roots were found in the field plot. On some roots, the attack was severe enough to render the tubers unfit for consumption.

Root-knot nematode damage on tubers was evident, even in a field that had a very low level of infestation at sowing time resulting in decreased yield of beans, pods, and other edible parts.

#### 27.7.1.3 Management

It appeared imperative to control the nematodes either by chemicals, or by using resistant cultivars, or by rotation with nonhost plants. The root-knot nematodes infestation on winged bean tubers and roots can be effectively managed by soil application of Carbofuran 3G at 5 g/plant during planting time. However, it is not known whether this was due to the nematicide or to a low level of infestation in the soil. In the Philippines, the nematode can be controlled by flooding for 30–40 days.

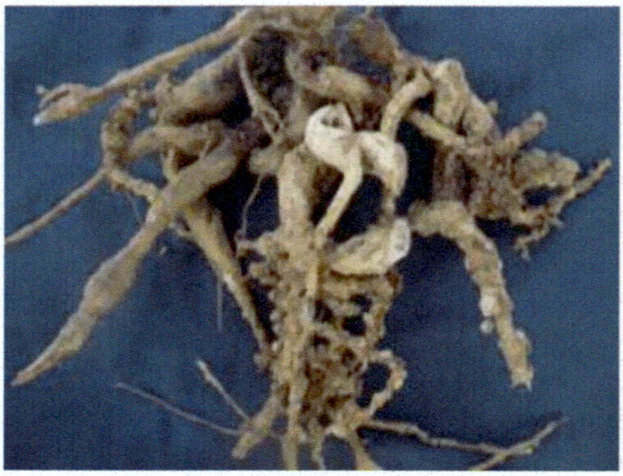

**Fig. 27.11** Root-knot nematode infestation on winged bean tubers and roots

## 27.8 Elephant Foot Yam, *Amorphophallus paeoniifolius*

### 27.8.1 Root-Knot Nematode, *Meloidogyne incognita*

*Amorphophallus* spp. are highly susceptible to *M. incognita*. The nematode infestation is very high in Kerala, Andhra Pradesh, and Tamil Nadu. The root-knot nematodes have been recorded in high numbers from Kerala, Tamil Nadu, and Andhra Pradesh leading to crop failure in many farms.

#### 27.8.1.1 Symptoms

In *Amorphophallus* the dry rot of tubers is caused by the root-knot nematode. Infestation sets in early attacking the roots and then infests the tuber tissue. The nematode multiplies inside the tuber tissue and cause dry rot. In store the dry rot continues to be serious resulting in damage of the whole infested tuber. The roots produce typical root-knot symptoms. In corms, the galls appear as irregular projections which harbor adult females and eggs. The area of infestation in tuber was discolored when the infestation was severe, and the infested area dried up resembling dry rot. Infested tubers were deformed and smaller. The degree of infestation was very high in cormels compared with that in corms.

#### 27.8.1.2 Management

Rotation of *Amorphophallus* with banana, cereals, rice, and sugarcane was found to be effective in East and West Godavari District of Andhra Pradesh, while rotation of *Amorphophallus* with resistant *Colocasia* was found to be effective in Tanjore District of Tamil Nadu. *Amorphophallus* crop is grown only once in 3 years in the same field to manage root-knot nematodes in Kerala.

Rotation of *Amorphophallus* with sweet potato and cassava or soil incorporation of cassava leaves and its dry leaf powder gave effective management of root-knot nematodes. Incorporation of the powder was found to be effective in reducing the root-knot nematode problem in *Amorphophallus*.

Seed material treatment with *Bacillus macerans* at 3 g/kg or *Purpureocellium lilacinum* at 3 g/kg proved beneficial for root-knot nematode management.

## 27.9 Conclusion

Although root and tuber crops constitute the basic staple diet of much of the worlds' population, nematological information regarding these crops is sorely lacking. With the intensification of root and tuber crops and the monocropping of high yielding cultivars, such as cassava for industrial purposes, nematode problems are likely to become increasingly important production constraints. The major nematode problems on root and tuber crops include root-knot nematodes *Meloidogyne* spp., lesion nematodes *Pratylenchus* spp., reniform nematode *Rotylenchulus reniformis*, and yam nematode *Scutellonema bradys*.

Tropical root and tuber crops are largely propagated vegetatively and that a major carryover of nematode pests between cropping seasons occurs mainly through such planting material. There is a need to ensure a focused attention to address these nematode problems. The development of sustainable healthy seed production systems is essential to overcome nematode and other pest and disease problems of these crops. There is a need for supplying the certified nematode-free planting stock to the farmers in order to increase crop production. The assessment of nematode problems on minor root and tuber crops and their economic importance in production systems requires greater attention.

For better progress, it is clear that suitable, adoptable, and relevant management options need to be made available. Simple techniques, such as the disinfection of planting material by dipping them in boiling water, have been developed and are being adopted by farmers. The use of host resistance and biological control is critical in reducing the impact of nematode pests, and is attracting increasing attention. Development of reliable diagnostic tools, nematode-resistant cultivars, and quality biological control agents are the need of the hour. There is surprising interest from both national and international companies to bring new biologically based products onto the market. There are no shortages of realized and potential biotechnological approaches to interfere with nematode parasitism using modern crop improvement tools for cost-effective and reliable options.

## References

Adesiyan SO (1976) Host range studies of the yam nematode *Scutellonema bradys*. Nematologica 22:289–297

Adesiyan SO (1977) Penetration and multiplication of *Scutellonema bradys* in yams (*Dioscorea* spp.). Nematol Medit 5:313–317

Adesiyan SO, Adeniji MO (1976) Studies on some aspects of yam nematode (*Scutellonema bradys* ). *Ghana*. J Agric Sci 9:131–136

Adesiyan SO, Odihirin RA (1978) Root-knot nematodes and pest of yams (*Dioscorea* spp.) in southern Nigeria. Nematologica 24:132–134

Akinsanya AK, Afolami SO, Kulakow P, Coyne D (2019) The root-knot nematode, *Meloidogyne incognita*, profoundly affects the production of popular biofortified cassava cultivars. Nematology 6:667–676

Atu UG, Odurukwe SO, Ogbuji RO (1983) Root-knot nematode damage to *Dioscorea rotundata*. Plant Dis 67:814–815

Ayala A, Acosta N (1971) Observation on yam (*Dioscorea alata*) nematodes. Nematropica 1:39–40

Baimey H, Coyne DL, Labuschagne N (2006) Effect of fertiliser application on yam nematode (*Scutellonema bradys*) multiplication and consequent damage to yam (*Dioscorea* spp.) under field and storage conditions in Benin. Int J Pest Mangmt 52(1):63–70

Bridge J (1982) Nematodes of yam. In: Miege J, Lyonga SN (eds) Yams. Clarendon Press, Oxford, Ignamea, pp 253–264

Bridge J, Coyne D, Kwoseh CK (2005) Nematode parasites of tropical root and tuber crops (excluding potatoes). In: Luc M, Sikora RA, Bridge J (eds) Plant parasitic nematodes in subtropical and tropical agriculture, 2nd edn. CAB International Publishing, Cambridge, MA, pp 221–258

Caveness FE (1981) Root-knot nematodes on cassava. Annual report IITA. IITA, Ibadan, Nigeria, pp 64–65

Caveness FE (1982) Root-knot nematodes as parasites of cassava. IITA Rese Briefs 3:2–3

Claudius-Cole AO, Asiedu R, Fawole B (2004) Cultural control of Scutellonema bradys on Dioscorea spp. In: Akporoda MO, Ekanayake IJ (eds) Proceedings of the international symposium for root and tuber crops, Ibadan, 12–16 November 2001, pp 522–526

Coursey DG (1967) Yams: an account of the nature, origins, cultivation, and utilization of the useful members of dioscoreaceae. Longmans, Green and Co Ltd, London. 230 pp

Coyne DL (1994) Nematode pests of cassava. African Crop Sci J 2 355–359

Decker H, Casamayor GR, Bosch D (1967) Observaciones sobre la presencia del nematode *Scutellonema bradys* en el tuberculo die name en la provincea de Oriente (Cuba). Centro, Boletin de Ciencias y Tecnologia, Universidad Central de Las Villas 2:67–75

Duke JA (1981) Handbook of legumes of world economic importance. Plenum Press, New York/London

Freitas OMBL, Moura RM (1986) Response of cultivars of cassava (*Manihot esculenta* Crantz) to *Meloidogyne incognita* and *M. javanica* (Nematoda: Heteroderidae) and comparisons with hydrogen cyanide content. Nematol Bras 10:109–131

Galano CD, Gapasin RM, Lim JL (1996) Efficacy of *Paecilomyces lilacinus* isolates for the control of root-knot nematode [*Meloidogyne incognita* (Kofoid and White) Chitwood] in sweet potato. Ann Trop Res 18:12–23

Gaspin RM, Valdez RB (1979) Pathogenicity of *Meloidogyne* spp. and *Rotylenchulus reniformis* on sweet potato. Ann Trop Res 1:20–25

Hallmann J, Quadt-Hallmann A, Miller WG et al (2001) Endophytic colonization of plants by the biocontrol agent *Rhizobium etli* G12 in relation to *Meloidogyne incognita* infection. Phytopathology 91:415–422

Hillocks RJ, Wydra K (2002) Bacterial, fungal and nematode diseases In: Hillocks RJ, Thresh JM, Bellotti AC (eds) Cassava: biology, production and utilization. CAB International

IITA (2005) Annual report. Plant Health Management, Ibadan

IITA (2007) Annual report. Plant Health Management, Ibadan

Jatala P (1988) Nematodes in tuber and root crops and their management – improvement of sweet potato (*Ipomea batatas*) in East Africa with some references of other tuber and root crops. Report of the workshop on sweet potato improvement in Africa. ILRAD, UNDP Project CIAT – CIP – IITA, Nairobi, Kenya, pp 91–100

Jatala P, Bridge J (1990) Nematode parasites of roots and tuber crops. In: Luc M, Sikora RA, Bridge J (eds) Plant parasitic nematodes in tropical and subtropical agriculture. CAB International, Wallingford, pp 137–180

Kermarrec A, Castagnone-Sereno P, Degras L et al (1987) Nouvelle distribution of *Scutellonema bradys* (Tylenchida: Hoplolaiminae) dans le Caraibe. Le cas des Antilles Francaises Medelingen van de Faculteit Landbouwwetenschappen Rijksuniversiteit Gent 52:617–624

Kwoseh CK (2000) Identification of resistance to major nematode pests of yams (*Dioscorea* spp.) in West Africa. Ph. D. thesis, The University of Reading, Reading, 196 pp

Martin WJ (1970) Elimination of root-knot and reniform nematodes and scurf infection from rootlets of sweet potato plants by hot water treatment. Pl Dis Reptr 54:1056–1058

Martin FN, Hancock JG (1985) Chemical factors in soils suppressive to *Pythium ultimum*. In: Parker CA et al (eds) Ecology and management of soil-borne plant pathogens. American Phytopathology. Society, St Paul, pp 113–116

Missah A, Peters JC (2001) Plant parasitic nematodes associated with soils, intercrops and tubers of yam in Ghana. In: Proceedings of the 7th triennial symposium. of the International Society for Tropical Root Crops-Africa Branch, Cotonou, Benin, 11–17 October 1998, pp 639–644

Mohandas C, Palaniswami MS (1990a) Resistance in sweet potato (*Ipomea batatas* L) to *Meloidogyne incognita* (Kofoid and white) Chitwood in India. J Root Crops 16:148–149

Mohandas C, Palaniswami MS (1990b) Effect of *Meloidogyne incognita* on three popular cultivars of Colocasia esculenta (Abstract). National symposium on recent advances in the production and utilization of tropical tuber crops, ISRC, Central Tuber Crops Research Institute, Thiruvananthapuram

Mohandas C, Ramakrishnan S (2002) Grow sweet potato and manage root-knot nematode problems of vegetables (Abstract). In: International conference on vegetables, PNASF, Bangalore, p 256
Mohandas C, Palaniswami MS, Potty VP (1990) Survey, identification and pathogenicity of nematodes in tuber crops. Annual report of Central Tuber Crops Research Institute (1989–90), Thiruvananthapuram, pp 72–73
Nadakal AM, Thomas AN (1967) Observations on nematodes associated with dry rot of *Dioscorea alata* L. Sci Cult 33:142–143
Nirula KK (1959) Root-knot nematode on *Colocasia*. Curr Sci 28:125–126
Nirula KK (1963) Collateral host plants of root-knot nematodes. Curr Sci 32(25):221–222
Nwauzor EC (1996) Use of groundnut (*Arachis hypogaea* L) as an intercrop for nematode control in seed yam production by minisett techniques. Final report, National Agricultural Research Project (NARP). Priority Research and REFILLS Projects, pp 22–26
Nwauzor EC, Fawole B (1981) Root knot nematodes on yams (Dioscorea spp.) in eastern Nigeria. In: Proceedings of the 3rd resource planning on root-knot nematodes, *Meloidogyne* spp., Region IV and V, Ibadan, pp 161–167
Onalo JI, Asiedu R, Adisiyan SO (2001) Control of the yam nematode (*Scutellonema bradys*) with neem fruit powder. In Proceedings of the 7th triennial symposium of the international society for tropical root crops-Africa branch, Cotonou, Benin, 11–17 October 1998, pp 634–637
Ray S, Sahoo NK, Mohanty K (1990) Plant parasitic nematodes associated with tuber crops in Orissa (Abstract). In: National symposium on recent advances in the production and utilization of tropical tuber crops, ISRC, Central Tuber Crops Research Institute, Thiruvanthapuram
Rosa JM, de Oliveira SA, Jordão AL et al (2014) Plant parasitic nematodes on cassava cultivated in the Brazilian Amazon. Acta Amaz 44(2):271–275
Sadik S, Okereke OU (1975) Flowering, pollen grain germination, fruiting, seed germination and seedling development of white yam, *Dioscorea rotundata* Poir. Ann Bot 39:597–604
Sorley R, Ohair SK, Parrado JL (1983) Nematodes of cassava, *Manihot esculenta* Crantz. Nematropica 13:261–287
Speijer PP, Ssango F, Vuylsteke D (2000) Evaluation of host plant response to nematodes in *Musa* germplasm in Uganda. Acta Hort 540:225–232
Sreeja P, Nageswari S, Mohandas C (1998) Reaction of cassava cultivars to root-knot nematode, *Meloidogyne incognita* (Kofoid and white, 1919) Chitwood, 1949. In: Mehta UK (ed) Nematology: challenges and opportunities in 21st century. Sugarcane Breeding Institute, Coimbatore, pp 137–139
Srivastava AS, Verma RS, Pandey RC (1971) Comparative efficacy of different nematicides tested against root-knot nematode infesting *Colocasia antiquorum. Labdev.* J Sci Technol 9B:142–144
Théberge RL (1985) Cassava nematodes. In: Théberge RL (ed) Common african pests and diseases of cassava, yam, ,sweet potato and cocoyam. IITA Ibadan, Nigeria, pp 26–29
Varghese R, Mohandas C (2004) *Scutellonema bradys,* a new host of *Pasteuria penetrans. Indian.* J Nematol 34(2):220–221
Wood TG, Smith RW, Johnson RA, Komolafe PO (1980) Termites damage and crop loss studies in Nigeria- pre-harvest losses to yam due to termites and other soil pests. Trop Pest Mangmt 26:355–370

# Part IX

# Future Thrusts and Conclusion

# The Way Ahead

# 28

**Abstract**

Plant parasitic nematodes constitute one of the major limiting factors for cultivation of crop plants. The changes in agricultural situations have tremendous effects on the emergence of new nematode problems in India. Nematode pathogens could be tackled with the intelligent planning of basic research, cultural practices, chemical methods, biological control agents, host plant resistance, and integrated methods. Future lines of research work have been outlined. There is a need for development of strategies for mass awareness (transfer of technology) on nematode problems to the target beneficiaries which is of paramount importance for protecting crops from these unseen enemies. Extending recent milestone advancements to effect solutions to sustainable nematode management will provide a foundation for the future.

**Keywords**

Sustainable nematode management · Research thrusts · Biological methods · Integrated methods · Transfer of technology

## 28.1 Introduction

Plant parasitic nematodes (PPNs) constitute one of the major limiting factors for cultivation of crops. The changes in agricultural situations have tremendous effects on the emergence of new nematode problems in India. The recent outbreaks of *M. graminicola* on rice in Karnataka, West Bengal, Orissa and Assam; *floral malady* (*Aphelenchoides besseyi*) on tuberose in West Bengal and Odisha; *Kalahasty malady* (*Tylenchorhynchus brevilineatus*) on groundnut in Andhra Pradesh; *Meloidogyne indica* on kagzi lime and Bt cotton in Gujarat; *M. enterolobii* on guava in Tamil Nadu besides 10 other states; *M. incognita* in pomegranate in Maharashtra,



P. P. Reddy, *Nematode Diseases of Crops and their Management*, https://doi.org/10.1007/978-981-16-3242-6_28

Karnataka, Gujarat and Andhra Pradesh; *M. arenaria* on groundnut in Gujarat; cyst nematodes (*Globodera rostochiensis* and *G. pallida*) on potato in Jammu and Kashmir and Uttarakhand besides Tamil Nadu and Kerala; and *Pratylenchus thornei* in wheat, maize, chickpea, and soybean are few examples of current threats and serious concerns. However, PPNs are neglected pests of crops and considered as of low priority for crop production and protection in India.

The scientists, plant protection specialists, and administrators have not recognized and very frequently underestimated the economic significance of nematodes in agriculture and their damage potential. As an important component of integrated pest management, nematode pathogens cannot be ignored and they could be tackled with the intelligent planning of nematode suppressive crop sequences, summer plowing, organic manuring, clean cultivation, adjusting sowing time, water and irrigation management and sensible use of nematicides. The increasing concerns on the ill effects of chemical pesticides on the environment have driven recent research interests on the use of several alternative strategies like botanicals, biopesticides and cropping system research for management of nematodes. Some of the successes have been obtained for managing plant parasitic nematodes with Neem-based formulations, fungal formulation of *Purpureocillium lilacinum* and *Trichoderma* spp. and bacterial formulations of *Pasteuria penetrans* and *Pseudomonas fluorescens* strains. However, wide adoption of bioagents in field scale is still a fantasy rather than reality because of their inconsistent efficacy in different agro-ecological situations. Integration of more than one option can be explored on the basis of their compatibility, economic viability and availability to the farmers. In order to provide the greater benefits to the farmers, the major challenge in future is the development of holistic strategies not only to manage nematodes but also the disease complexes they induce along with other soil-borne pathogens such as fungi, bacteria and viruses.

## 28.2 Perspectives in Sustainable Nematode Management

### 28.2.1 Basic Research

There is great need for basic work on the relations between nematodes and plants in every branch of the subject in order to provide the foundation from which to offer solutions to the practical problems of nematode diseases in agriculture.

Through surveys and identifications, determine more accurately the numbers and kinds of nematodes which actually exist and pinpoint their distribution. Success of crop rotations and resistant cultivars depend upon knowing what nematodes occur in which areas and fields.

With regard to viruses transmitted by soil-borne nematodes, it is advisable to explore the possibility of finding dorylaimid nematodes from the rhizosphere and their role in transmitting soil-borne viruses mainly in crops like fruits, vegetables and ornamentals.

Other areas of further research include the characterization of nematode diagnostics using modern techniques like molecular characterization has to be carried out for formulating effective management strategies.

### 28.2.2 Cultural Methods

Currently management of plant nematode problems is primarily dependent on cultural practices (crop rotation, clean cultivation, fallowing, summer plowing, tolerant varieties etc.), organic amendments, botanicals (neem, karanj, biocontrol agents, soil antagonists) or bionematicides. There is a need for giving more emphasis for development of economical, eco-friendly and sustainable nematode management strategies like encouraging use of organic soil amendments enriched with biological control agents, bionematicides, cropping systems suitable for nematode suppression, precision crop protection, and nanotechnology.

Among the non-aggressive approaches of nematode management approaches, low input agricultural practices like crop rotation with non-host, poor host and antagonistic crops, resistant varieties, identification of nematode suppressive crop sequences, organic farming practices, summer plowing, clean cultivation, irrigation management etc. should be promoted among the growers. Amending soil with various organic sources like oil and de-oiled cakes, farmyard manure, farm compost, green manures etc. may reduce soil nematode populations. The effects of biofumigation using crucifers have not yet been understood on soil nematodes. Some new molecules are being investigated for their field efficacy against phytonematodes.

There is need for development of management approaches for multiple nematode species occurring in the same field, sequential susceptible host crops, and sequential susceptible host crops. Cash or cover crops with resistance to *M. incognita* and crops with biofumigation potential can be used to reduce population densities of root-knot nematodes. Mycorrhizal fungal inoculations of potting mixes during production of seedlings may improve early crop establishment. Other approaches to nematode management utilize soil suppressiveness. Soil-borne nematode and other pathogens occurring in some soils can be managed by following reduced tillage combined with crop rotation. Thus, sustainable nematode management strategies require holistic approaches that consider entire production systems rather than focus on a single crop in its year of production.

Our future research programs need to be more based towards development of sustainable and non-chemical methods for the management of nematode pests. There is a need for development of cropping systems based nematode management technologies. Detailed evaluation of the entire cropping system, though more cumbersome than single crop-approaches, is essential to integrate novel tools and strategies successfully into these systems. Integrated nematode management using eco-friendly components can be utilized in place of "Silver-bullet" approaches. In order to realize the sustainable production goal with minimal environmental impacts, there is a need to exploit the advantages of innovative management strategies and cultural methods in improving sustainable agricultural production.

In some intensive cropping systems the nematode management costs can be reduced by adopting the site specific nematode management approach (precision nematode management) using geographical information systems (GIS) and global positioning systems (GPS) or nanotechnology approach.

In order to achieve sustainable food and fiber production, there is a need for adopting ecology-based crop nematode and disease management strategies by acquiring increased knowledge and understanding. Cover crops and carefully developed rotations that include non-hosts and durable host resistance result in increased soil organic matter; this, in turn, contributes very significantly to the increase of antagonists of plant pathogens. Benefits of the greater organic matter on soil quality includes a reduction of soil-borne diseases including nematodes, greater pools of P and K, elevated microbial biomass and activity, and increased soil water holding capacity.

Crop production is an intensive production system with narrow profit margins. Hence, there is a need for utilization of crops with more economic value in rotation rather than traditional maize and soybean rotation with low economic value for nematode management.

Further research has to be undertaken to evaluate the efficacy of some promising botanicals for the management of nematodes like root-knot under field conditions under different agro-ecologies and improve their application technology. Other areas of further research include the use of such promising botanicals in integrated pest management strategy and evaluation of their effects on other soil-borne plant diseases.

### 28.2.3 Chemical Methods

We should continue research on nematode management to include efforts to obtain new generation nematicides (which may not kill nematode but may delay the development and reproduction) which are less costly, more effective, can be used at lower dosage, and cause less pollution in the environment. Efficient placement of chemicals is of paramount importance, since the nematodes in a field not uniformly distributed, but have a clustered distribution.

### 28.2.4 Biological Methods

Advancement of biological control to its fullest potential has to be explored, utilizing bioagents plus organic amendments, sewage, sludge, etc. This would include cultural methods for biocontrol agents, understand their nature, make them practical, and genetically modify the biocontrol agents to improve their effectiveness. Many fungal and bacterial bioagents have been found effective under in vitro conditions. Efforts should be made to test them under field conditions. Further, our efforts should be directed towards developing mass production technologies for effective bioagents and easy field application techniques.

Integration of two or more biological control agents has been found highly effective in nematode management and yield enhancement. Thus the combination of highly toxic fungus, *Aspergillus niger* (killed most of the infective second stage juveniles) and an egg parasite, *Cladosporium oxysporum* (invaded and killed the eggs in egg sac) both at half the doses reduced significantly more *M. incognita* population and exhibited better plant growth than when either of the fungal bioagents in brinjal (Goswami and Sharma 2001).

Some promising biocontrol agent like *Purpureocillium lilacinum* and *Pasteuria penetrans* are the most effective agent against root-knot (*Meloidogyne* spp.), pigeon pea cyst (*Heterodera cajani*), reniform (*Rotylenchulus reniformis*) and citrus (*Tylenchulus semipenetrans*) nematodes. Therefore, these biocontrol agents could be solely used for management of the field problem of nematodes infecting crops.

Therefore, role of antagonistic soil fungi (like *Trichoderma* spp.) and plant growth promoting rhizospheric bacteria (like *Pseudomonas fluorescens*) in complex disease management need intensive research and some possible approaches could be validated under field conditions.

### 28.2.5 Host Resistance

There is much hope that the development and deployment of high-quality crop cultivars with effective levels of genetic resistance will find widespread acceptance among growers and reduce the impact of nematodes on crop yields. Further, it is generally acknowledged that greater effort on the part of private sector breeders for nematode resistance will be dependent on development of efficient, high-throughput marker-assisted selection protocols. However, in the final analysis it will likely be the yield potential and quality of these resistant cultivars, rather than the level of resistance itself, that will determine growers' acceptance and whether host resistance plays a major more important role in future nematode management systems in crops than it does today.

Although valuable improvement has been obtained in development of nematode resistant horticultural crop cultivars, there is a need for breeding varieties with combined resistance to different nematode species and to other organisms such as soil-borne fungi and bacteria which cause wilt and root rot. In the integrated nematode management strategy, the use of resistant cultivars as one of the component is highly economical to farmers. Hence, priority needs to be given for developing host plant resistance using either conventional breeding or biotechnological approaches. The development of transgenic plants with durable nematode resistance keeping pace with the evolution of resistance breaking populations should be considered.

Phytoalexins effectively induce plant resistance mechanisms to nematodes, particularly to sedentary ones. However, the mechanism by which phytoalexins are accumulated in plants is not known and need to be worked out on priority basis.

Development of a better understanding of the biological aspects of all our more important plant nematodes is the need of the hour. This would include the molecular genetics of closely related species and races; host-parasite relationships and histochemical details which may provide clues to the nature of plant resistance to nematodes; ecological factors, such as temperature, moisture and soil types which may influence nematode development and survival which in turn affect control measures; and make full use of current biotechnologies as appropriate in this and other nematology research.

There is a need for developing additional sources of resistance to several nematodes. Recent advances in engineered resistance to *M. incognita* in Arabidopsis based on RNA interference (Huang et al. 2006) and the recent announcement by a company that it was working to develop nematode resistance in maize by RNA interference technology needs further encouragement.

### 28.2.6 Integrated Methods

The implementation of the INM technologies that are already developed would require scientists/trained technicians and efficient extension media through ICAR Institutes, State Agriculture Universities, and NGOs. Since integrated approach may involve more than one discipline, it is therefore, essential that professional scientists must collaborate in its planning and execution.

Development of INM strategies need additional components like genetically modified crops with genes for nematode resistance, quality biological control agents, and new generation nematicides. Improved data management and decision-support systems should facilitate the integration of new and traditional IPM strategies and tactics.

Research opportunities remain in the area of integration of compatible strategies/tactics. Eco-friendly components include soil solarization, cropping systems (crop rotation, intercropping, etc.), cultural practices (fallowing, destruction of crop residues, organic soil amendments), safer chemicals, bioagents, and resistant varieties. Carefully developed research-extension priorities, and the use of some type of data/decision-support system, should facilitate the development and deployment of effective IPM programs. In future, there is a need to focus on aspects like interdisciplinary approach for development of integrated pest (insect pests, disease pathogens like fungi, bacteria, viruses, and nematodes) management, sustainable crop-production systems with critical input use efficiency, and development and transfer of technologies.

## 28.3 Evolving Research Thrusts

In looking to the future, what are some of the things that we need to do from a realistic and practical stand point of view?

- Web-based state-wise distribution maps of agriculturally important nematode fauna in the country.
- Impact of economically important plant parasitic nematode populations on crop health.
- Population dynamics of economically important nematode pests under different cropping systems and role of cropping systems in nematode management.
- Identification and development of nematode resistant cultivars.
- Front line demonstrations at farmers' fields to demonstrate and validate the most effective, low cost, and eco-friendly nematode management technologies generated in the project.
- Designing nematode management technologies for organic farming systems.
- Pest risk analysis to address phytosanitary needs against plant parasitic nematodes under WTO regime of international trade in agricultural commodities.
- Carry out researches on use of indigenous microorganisms for their antagonistic role against plant parasitic nematodes.
- Nematode problems and their management in protected horticultural crops.
- Development of a national nematode diagnostic facility based on morphological and molecular characterization of nematodes.
- New nematode control measures which are economical, eco-friendly and sustainable are required due to the likely removal of effective nematicides from the market because of increasing concerns of possible effects on human health and the environment.

## 28.4 Transfer of Technology

Generating awareness among farmers is crucial for preventing crop loss by plant parasitic nematodes (PPNs). Many important PPNs have a wide host range and can attack a number of crops, like cereals, pulses, oilseeds, fruits, fibers, fodder, plantations, spices, medicinal and aromatics, etc. The nematodes are hidden enemies, since they are present in soil. Hence, the farmers, administrators, and policymakers are unaware of the damage they cause to crop plants. They should be made aware of the nematode damage symptoms, biosecurity threats, and their management methods through transfer of technology by extension agencies.

Application of Information and Communication technology (IC&T) is essential for effective nematode awareness programs. In many counties abroad, IC&T have been very successfully applied by the agriculture extension agencies. Field days and crop clinics should be organized and newsletters, newspaper articles, website news to be published for effective dissemination of information about nematodes among farmers and stakeholders. Nematode problems could be easily disseminated through IC&T approach like TV and radio programs (e.g. Krishi Darshan), newspaper articles, extension publications such as leaf-lets, book-lets, extension bulletins in regional languages and organizing Krishi Mela, seminars, exhibitions, field demonstrations, e-mail, SMS, social media (Facebook, Twitter, Blogs), etc.

PPNs have local or regional distribution on certain crops and their nature of damage, survival, disseminations and management practices are known. These should be made aware to the farmers through ICAR Institutes, SAUs, State Agriculture and Horticulture Departments, KVKs, and NGO's.

### 28.4.1 What Needs to Be Done?

- Mass awareness on phytonematodes and it capabilities to pose threat for cultivation of crops.
- Imparting training on nematode as an important pathogen to extension officers dealing with crop production and protection, equip them with a low cost publication for circulation among the farmers at least before *Rabi* and *Kharif* seasons every year.
- Adoption of village to demonstrate the effectiveness of recommendations for building confidence among the growers on nematode pest management practices.
- Nematode management recommendations should be refined for their feasibility, utility and compatibility in IPM systems.
- MNCs should recognize nematode as a potential pathogen and invest on development of new molecules effective to control nematodes and that should be available everywhere.

## 28.5 Conclusion

Plant-parasitic nematodes are increasingly recognized as economically important crop pests. Since 1990, nematode management strategies have been developed by public and private partners through research and development efforts. Because the economics of crop production rather than ability to suppress nematode population densities largely govern the strategies used by most growers, many effective approaches to nematode management are under-utilized.

Improved methods to identify and quantify nematodes will also result in faster development of the tactical options available for control. Nematode management technologies developed should be refined under local field conditions before adoption in fields by the farmers. Screening for resistance to nematodes requires large-scale estimation of population development on germplasm lines. Similarly, screening of new chemicals to kill or disrupt nematode life cycles is often restricted to a few species such as root-knot nematodes that cause visible symptoms which are easily evaluated. Rapid, reliable species and race identification is critical for selection of appropriate rotation crops or resistant cultivars and for regulatory management of plant shipments (Dunn et al. 1989).

We have made much progress in nematology; but we still have a long way to go in the battle against these destructive and insidious plant nematodes. Through our continued dedication and strong research efforts, greater knowledge and better understanding of nematodes will be achieved. This will result in further reductions in losses caused by nematodes and a significant increase of food and fiber in the world.

Extending recent milestone advancements to effect solutions to sustainable nematode-pest-crop management— including durable host resistance for root-knot and cyst nematodes, and other taxa—will provide a foundation for the future.

## References

Dunn RA, Eisenback JD, Radewald JD, Westerdahl BB (1989) Extension nematology: taxonomy and the dilemma of diagnosis (Abstr.). J Nematol 21(4):558–559

Goswami BK, Sharma SB (2001) Application of *Aspergillus terreus* and *Paecilomyces lilacinus* for the management of *Meloidogyne incognita* on tomato. Int J Nematol 4:42–46

Huang G, Allen R, Davis EL et al (2006) Engineering broad root-knot resistance in transgenic plants by RNAi silencing of a conserved and essential root-knot nematode parasitism gene. Proc Nat Acad Sci 103:14302–14306

# Index




P. P. Reddy, *Nematode Diseases of Crops and their Management*,
https://doi.org/10.1007/978-981-16-3242-6

The manufacturer's authorised representative in the EU is Springer
Nature Customer Service Centre GmbH, Europaplatz 3, 69115 Heidelberg,
Germany. If you have any concerns regarding our products, please
contact ProductSafety@springernature.com

Printed and bound by CPI Group (UK) Ltd, Croydon, CR0 4YY
30/09/2026
02225241-0016